Operator Theory: Advances and Applications
Vol. 166

Symplectic Geometry and Quantum Mechanics

Maurice de Gosson

Advances in Partial Differential Equations

Birkhäuser Verlag
Basel · Boston · Berlin

Maurice de Gosson
Institut für Mathematik
Universität Potsdam
Postfach 60 15 53
14415 Potsdam
Germany
e-mail: maurice.degosson@gmail.com

2000 Mathematics Subject Classification 34D20, 47N50, 53Dxx, 81Sxx

A CIP catalogue record for this book is available from the
Library of Congress, Washington D.C., USA

Bibliographic information published by Die Deutsche Bibliothek
Die Deutsche Bibliothek lists this publication in the Deutsche Nationalbibliografie; detailed bibliographic data is available in the Internet at <http://dnb.ddb.de>

ISBN 3-7643-7574-4 Birkhäuser Verlag, Basel – Boston – Berlin

Member of the BertelsmannSpringer Publishing Group
Printed on acid-free paper produced from chlorine-free pulp. TCF ∞
Cover design: Heinz Hiltbrunner, Basel
Printed in Germany
ISBN-10: 3-7643-7574-4
ISBN-13: 978-3-7643-7574-4
9 8 7 6 5 4 3 2 1

www.birkhauser.ch

I dedicate this work,
with all my love
and gratitude,

to Charlyne,

mathematician
'honoris causa'

Contents

Preface

Introduction

We have been experiencing since the 1970s a process of "symplectization" of Science especially since it has been realized that symplectic geometry is the natural language of both classical mechanics in its Hamiltonian formulation, and of its refinement, *quantum mechanics*. The purpose of this book is to provide core material in the symplectic treatment of quantum mechanics, in both its semi-classical and in its "full-blown" operator-theoretical formulation, with a special emphasis on so-called phase-space techniques. It is also intended to be a work of reference for the reading of more advanced texts in the rapidly expanding areas of symplectic geometry and topology, where the prerequisites are too often assumed to be "well-known" by the reader. This book will therefore be useful for both pure mathematicians and mathematical physicists. My dearest wish is that the somewhat novel presentation of some well-established topics (for example the uncertainty principle and Schrödinger's equation) will perhaps shed some new light on the fascinating subject of quantization and may open new perspectives for future interdisciplinary research.

I have tried to present a balanced account of topics playing a central role in the "symplectization of quantum mechanics" but of course this book in great part represents my own tastes. Some important topics are lacking (or are only alluded to): for instance Kirillov theory, coadjoint orbits, or spectral theory. We will moreover almost exclusively be working in flat symplectic space: the slight loss in generality is, from my point of view, compensated by the fact that simple things are not hidden behind complicated "intrinsic" notation.

The reader will find the style in which this book has been written very traditional: I have been following the classical pattern "Definition–Lemma–Theorem–Corollary". Some readers will inevitably find this way of writing medieval practice; it is still, in my opinion, the best way to make a mathematical text easily accessible. Since this book is intended to be used in graduate courses as well as for reference, we have included in the text carefully chosen exercises to enhance the understanding of the concepts that are introduced. Some of these exercises should be viewed as useful complements: the reader is encouraged to spend some time on them (solutions of selected exercises are given at the end of the book).

Organization

This book consists of three parts which can to a large extent be read independently of each other:

- The first part (partly based on my monograph [61] *Maslov Classes, Metaplectic Representation and Lagrangian Quantization*) is joint work with Serge de Gosson. It is intended to be a rigorous presentation of the basics of symplectic geometry (Chapters I and II) and of its multiply-oriented extension "q-symplectic geometry" (Chapter III); complete proofs are given, and some new results are presented. The basic tool for the understanding and study of q-symplectic geometry is the Arnold–Leray–Maslov (For short: *ALM*) index and its topological and combinatorial properties. In Chapter IV we study and extend to the degenerate case diverse Lagrangian and symplectic intersection indices with a special emphasis on the Conley–Zehnder index; the latter not only plays an important role in the modern study of periodic Hamiltonian orbits, but is also essential in the theory of the metaplectic group and its applications to the study of quantum systems with chaotic classical counterpart. A remarkable fact is that all these intersection indices are easily reduced to one mathematical object, the ALM index.
- In the second part we begin by studying thoroughly the notion of phase of a Lagrangian manifold (Chapter V). That notion, together with the properties of the ALM index defined in Chapter III, allows us to view quantized Lagrangian manifolds as those on which one can define a generalized notion of wave function. Another attractive feature of the phase of a Lagrangian manifold is that it allows a geometric definition of the Heisenberg–Weyl operators, and hence of the Heisenberg group and algebra; these are studied in detail in Chapter VI, together with the related notions of Weyl operator and Wigner–Moyal transform, which are the keys to quantum mechanics in phase space. In Chapter VII we study the metaplectic group and the associated Maslov indices, which are, surprisingly enough, related to the ALM index in a crucial way.
- In the third and last part we begin by giving a rigorous geometrical treatment of the uncertainty principle of quantum mechanics Chapter VIII). We show that this principle can be expressed in terms of the notion of symplectic capacity, which is closely related to Williamson's diagonalization theorem in the linear case, and to Gromov's non-squeezing theorem in the general case. We thereafter (Chapter IX) expose in detail the machinery of Hilbert–Schmidt and trace-class operators, which allows us a rigorous mathematical treatment of the fundamental notion of density matrix. Finally, in Chapter X (and this is definitely one of the novelties compared to traditional texts) we extend the Weyl pseudo-differential calculus to phase space, using Stone and von Neumann's theorem on the irreducible representations of the Heisenberg group. This allows us to derive by a rigorous method a "Schrödinger equation

in phase space" whose solutions are related to those of the usual Schrödinger equation by a "wave-packet transform" generalizing the physicist's Bargmann transform.

For the reader's convenience I have reviewed some classical topics in a series of Appendices at the end of the book (Classical Lie Groups, Covering Spaces, Pseudo-Differential Operators, Elementary Probability Theory). I hope that this arrangement will help the beginner concentrate on the main text with a minimum of distraction and without being sidetracked by technicalities.

Prerequisites

The mathematical prerequisites for reading with profit most of this book are relatively modest: solid undergraduate courses in linear algebra and advanced calculus, as well as the most basic notions of topology and functional analysis (Hilbert spaces, distribution theory) in principle suffice. Since we will be dealing with problems having their origin in some parts of modern physics, some familiarity with the basics of classical and quantum mechanics is of course helpful.

Bibliography

A few words about the bibliography: I have done my very best to give an accurate and comprehensive list of references. Inevitably, there are omissions; I apologize in advance for these. Some of these omissions are due to sheer ignorance; on the other hand this book exposes techniques and results from diverse fields of mathematics (and mathematical physics); to give a *complete* account of *all* contributions is an impossible task!

Enough said. The book – and the work! – is now yours.

Acknowledgements

While the main part of this book was written during my exile in Sweden, in the lovely little city of Karlskrona, it has benefitted from many visits to various institutions in Europe, Japan, USA, and Brazil. It is my duty – and great pleasure – to thank Professor B.-W. Schulze (Potsdam) for extremely valuable comments and constructive criticisms, and for his kind hospitality: the first part of this book originates in research having been done in the magnificent environment of the *Neues Palais* in the *Sans-Souci park* where the Department of Mathematics of the University of Potsdam is located.

I would like to express very special thanks to Ernst Binz (Mannheim) and Kenro Furutani (Tokyo) for having read and commented upon a preliminary version of this book. Their advice and encouragements have been instrumental. (I am, needless to say, soley responsible for remaining errors or misconceptions!)

I also wish to acknowledge debt and to express my warmest thanks for useful discussions, criticisms, and encouragements to (in alphabetical order): Bernhelm Booss-Bavnbek (Roskilde), Matthias Brack (Regensburg), Serge de Gosson (Växjö), Jürgen Eichhorn (Greifswald), Karl Gustafson (Boulder), Katarina Habermann (Göttingen), Basil Hiley (London), Ronnie Lee (Yale), Robert Littlejohn (Berkeley), James Meiss (Boulder), Walter Schempp (Siegen), Gijs Tuynman (Lille).

The third part of this book (the one devoted to quantum mechanics in phase space) has been deeply influenced by the work of Robert Littlejohn on semiclassical mechanics: may he find here my gratitude for having opened my eyes on the use of Weyl calculus and symplectic methods in quantum mechanics!

I had the opportunity to expose parts of this book during both a graduate course I gave as an Ulam Visiting Professor at the University of Colorado at Boulder during the fall term 2001, and during a course as "First Faculty in Residence" at the same University during the summer session 2003 (I take the opportunity to thank J. Meiss for his kind hospitality and for having provided me with extremely congenial environment).

This work has been partially supported by the State of São Paulo Research Foundation (Fapesp) grant 2005/51766–7; a great thanks to P. Piccione (University of São Paulo) for his hospitality and fascinating conversations about Morse theory and the Maslov index.

About the Author

Maurice A. de Gosson is scientifically affiliated with Professor B.-W. Schulze's *Partielle Differentialgleichungen und Komplexe Analysis* research group of the University of Potsdam. He has held longer visiting positions at several prestigious Universities, among them Yale University, the University of Colorado at Boulder, at the Science University of Tokyo, at the Université Paul Sabatier of Toulouse, at Potsdam University, at Utrecht University, and at the University of São Paulo. He regularly visits his *alma mater*, the University of Paris 6. De Gosson has done extensive research in the area of symplectic geometry, the combinatorial theory of the Maslov index, the theory of the metaplectic group, partial differential equations, and mathematical physics. He obtained his Ph.D. in 1978 on the subject of microlocal analysis at the University of Nice under the supervision of J. Chazarain, and his Habilitation in 1992 at the University Pierre et Marie Curie in Paris, under the mentorship of the late Jean Leray (Collège de France).

Notation

Our notation is as standard (and simple) as conflicting usages in the mathematical and physical literature allow.

Number sets

$\mathbb{R}$ (resp. $\mathbb{C}$) is the set of all real (resp. complex) numbers. We denote respectively by

$$\mathbb{N} = \{1, 2, \dots\} \quad \text{and} \quad \mathbb{Z} = \{\dots, -2, -1, 0, +1, +2, \dots\}$$

the set of positive integers and the set of all integers.

Classical matrix groups

Let $\mathbb{K} = \mathbb{R}$ or $\mathbb{C}$.

$M(m, \mathbb{K})$ is the algebra of all $m \times m$ matrices with entries in $\mathbb{K}$.

$\mathrm{GL}(m, \mathbb{K})$ is the general linear group. It consists of all invertible matrices in $M(m, \mathbb{K})$.

$\mathrm{SL}(m, \mathbb{K})$ is the special linear group: it is the subgroup of $\mathrm{GL}(m, \mathbb{K})$ consisting of all the matrices with determinant equal to 1.

$\mathrm{Sym}(m, \mathbb{K})$ is the vector space of all symmetric matrices in $M(m, \mathbb{K})$; it has dimension $\frac{1}{2}m(m+1)$; $\mathrm{Sym}_+(2n, \mathbb{R})$ is the subset of $\mathrm{Sym}(m, \mathbb{K})$ consisting of the positive definite symmetric matrices.

$\mathrm{Sp}(n) = \mathrm{Sp}(n, \mathbb{R})$ is the standard (real) symplectic group; it is the subgroup of $\mathrm{SL}(2n, \mathbb{R})$ consisting of all matrices S such that $S^T JS = J$ where J is the "standard symplectic matrix" defined by

$$J = \begin{bmatrix} 0 & I \\ -I & 0 \end{bmatrix}.$$

$\mathrm{U}(n, \mathbb{C})$ is the unitary group; it consists of all $U \in M(n, \mathbb{C})$ such that $UU^* = U^*U = I$ ($U^* = \bar{U}^T$ is the adjoint of U).

$\mathrm{U}(n)$ is the image in $\mathrm{Sp}(n)$ of $\mathrm{U}(n,\mathbb{C})$ by the monomorphism

$$A + iB \longmapsto \begin{bmatrix} A & -B \\ B & A \end{bmatrix}.$$

Vector calculus

The elements of $\mathbb{R}^m$ should be viewed as column vectors

$$x = \begin{bmatrix} x_1 \\ \vdots \\ x_n \end{bmatrix}$$

when displayed; for typographic simplicity we will usually write $x = (x_1, \ldots, x_n)$ in the text. The Euclidean scalar product $\langle \cdot, \cdot \rangle$ and norm $|\cdot|$ on $\mathbb{R}^m$ are defined by

$$\langle x, y \rangle = x^T y = \sum_{j=1}^{m} x_j y_j \ , \ |x| = \sqrt{\langle x, x \rangle}.$$

The gradient operator in the variables $x_1, \ldots, x_n$ will be denoted by

$$\partial_x \quad \text{or} \quad \begin{bmatrix} \frac{\partial}{\partial x_1} \\ \vdots \\ \frac{\partial}{\partial x_m} \end{bmatrix}.$$

Let f and g be differentiable functions $\mathbb{R}^m \longrightarrow \mathbb{R}^m$; in matrix form the chain rule is

$$\partial(g \circ f)(x) = (Df(x))^T \partial f(x) \tag{1}$$

where $Df(x)$ is the Jacobian matrix of f: if $f = (f_1, \ldots, f_m)$ is a differentiable mapping $\mathbb{R}^m \longrightarrow \mathbb{R}^m$, then

$$Df = \begin{bmatrix} \frac{\partial f_1}{\partial x_1} & \frac{\partial f_1}{\partial x_2} & \cdots & \frac{\partial f_1}{\partial x_m} \\ \frac{\partial f_2}{\partial x_1} & \frac{\partial f_2}{\partial x_2} & \cdots & \frac{\partial f_2}{\partial x_m} \\ \vdots & \vdots & \ddots & \vdots \\ \frac{\partial f_m}{\partial x_1} & \frac{\partial f_n}{\partial x_2} & \cdots & \frac{\partial f_m}{\partial x_m} \end{bmatrix}. \tag{2}$$

Let $y = f(x)$; we will indifferently use the notation

$$Df(x) \ , \ \frac{\partial y}{\partial x} \ , \ \frac{\partial(y_1, \ldots, y_m)}{\partial(x_1, \ldots, x_m)}$$

for the Jacobian matrix. If f is invertible, the inverse function theorem says that

$$D(f^{-1})(y) = [Df(x)]^{-1}. \tag{3}$$

If $f : \mathbb{R}^m \longrightarrow \mathbb{R}$ is a twice continuously differentiable function, its Hessian calculated at a point x is the symmetric matrix of second derivatives

$$D^2 f(x) = \begin{bmatrix} \frac{\partial^2 f}{\partial x_1^2} & \frac{\partial^2 f}{\partial x_1 \partial x_2} & \cdots & \frac{\partial^2 f}{\partial x_1 \partial x_n} \\ \frac{\partial^2 f}{\partial x_2 \partial x_1} & \frac{\partial^2 f}{\partial x_2^2} & \cdots & \frac{\partial^2 f}{\partial x_2 \partial x_n} \\ \vdots & \vdots & \ddots & \vdots \\ \frac{\partial^2 f}{\partial x_n \partial x_1} & \frac{\partial^2 f}{\partial x_n \partial x_2} & \cdots & \frac{\partial^2 f}{\partial x_n^2} \end{bmatrix}. \tag{4}$$

Notice that the Jacobian and Hessian matrices are related by the formula

$$D(\partial f)(x) = D^2 f(x). \tag{5}$$

Also note the following useful formulae:

$$\langle A\partial_x, \partial_x \rangle \, e^{-\frac{1}{2}\langle Mx,x\rangle} = [\langle MAMx, x\rangle - \operatorname{Tr}(AM)] \, e^{-\frac{1}{2}\langle Mx,x\rangle}, \tag{6}$$

$$\langle Bx, \partial_x \rangle \, e^{-\frac{1}{2}\langle Mx,x\rangle} = \langle MBx, x\rangle \, e^{-\frac{1}{2}\langle Mx,x\rangle} \tag{7}$$

where A, B, and M are symmetric. matrices.

Function spaces and multi-index notation

We will use "multi-index" notation: for $\alpha = (\alpha_1, \ldots, \alpha_n)$ in $\mathbb{N}^m$ we set $|\alpha| = \alpha_1 + \cdots + \alpha_m$ (it is the "length" of the "multi-index" α) and

$$x^\alpha = x_1^{\alpha_1} \cdots x_n^{\alpha_n} \quad , \quad \partial^\alpha = \partial_{x_1}^{\alpha_1} \cdots \partial_{x_n}^{\alpha_n};$$

we will set $D_x^\alpha = i^{|\alpha|}\partial_x^\alpha$.

We denote by $C^k(\mathbb{R}^m)$ the vector space of k times continuously differentiable functions $\mathbb{R}^m \longrightarrow \mathbb{C}$; k is an integer ≥ 1 or ∞. The subspace of $C^k(\mathbb{R}^m)$ consisting of the compactly supported functions is denoted by $C_0^k(\mathbb{R}^m)$.

$\mathcal{S}(\mathbb{R}_x^n)$ is the Schwartz space of rapidly decreasing functions: $\Psi \in \mathcal{S}(\mathbb{R}_x^n)$ if and only if for every pair (α, β) of multi-indices there exist $K_{\alpha\beta} > 0$ such that

$$|x^\alpha \partial_x^\beta \Psi(x)| \leq K_{\alpha\beta} \quad \text{for all } x \in \mathbb{R}^n. \tag{8}$$

In particular, every C^∞ function on $\mathbb{R}^n$ vanishing outside a bounded set is in $\mathcal{S}(\mathbb{R}_x^n)$: $C_0^\infty(\mathbb{R}_x^n) \subset \mathcal{S}(\mathbb{R}_x^n)$. Taking the best constants $K_{\alpha\beta}$ in (8) we obtain a family of semi-norms on $\mathcal{S}(\mathbb{R}_x^n)$ and one shows that $\mathcal{S}(\mathbb{R}_x^n)$ is then a Fréchet space, and we have continuous inclusions

$$C_0^\infty(\mathbb{R}_x^n) \subset \mathcal{S}(\mathbb{R}_x^n) \subset L^2(\mathbb{R}_x^n).$$

The dual of $\mathcal{S}(\mathbb{R}^m)$ (*i.e.*, the space of tempered distributions) is denoted by $\mathcal{S}'(\mathbb{R}^m)$.

For $f \in \mathcal{S}(\mathbb{R}^m)$ and $\alpha > 0$ we define the "α-Fourier transform" Ff of f by

$$Ff(y) = \left(\tfrac{1}{2\pi\alpha}\right)^{m/2} \int e^{-\frac{i}{\alpha}\langle y,x\rangle} f(x) d^m x;$$

its inverse is

$$f(x) = \left(\tfrac{1}{2\pi\alpha}\right)^{m/2} \int e^{\frac{i}{\alpha}\langle y,x\rangle} Ff(y) d^m y.$$

Also recall that if M is a real symmetric $m \times m$ matrix then

$$\int e^{-\langle Mu,u\rangle} d^m u = \pi^{m/2} (\det M)^{-1/2}.$$

Combinatorial notation

Let X be a set, k an integer ≥ 0, $(G,+)$ an Abelian group. By definition, a $(G$-valued$)$ k-cochain on X (or just *cochain* when the context is clear) is a mapping

$$c : X^{k+1} \longrightarrow G.$$

To every k-cochain one associates its *coboundary*: it is the $(k+1)$-cochain ∂c defined by

$$\partial c(x_0, \dots, x_{k+1}) = \sum_{j=0}^{k+1} (-1)^j c(x_0, \dots, \hat{x}_j, \dots, x_{k+1}), \tag{9}$$

where the cap ^ suppresses the term it covers. The operator

$$\partial_k : \{k\text{-cochains}\} \longrightarrow \{(k+1)\text{-cochains}\}$$

defined by (9) is called the *coboundary operator*; we will use the collective notation ∂ whenever its range is obvious. The coboundary operator satisfies the important (but easy to prove) equality $\partial^2 c = 0$ for every cochain c. A cochain c is called *coboundary* if there exists a cochain m such that $c = \partial m$; a cochain c is called a *cocycle* if $\partial c = 0$; obviously every coboundary is a cocycle.

Part I

Symplectic Geometry

Chapter 1

Symplectic Spaces and Lagrangian Planes

The main thrust of this chapter is to familiarize the reader with the notions of symplectic geometry that will be used in the rest of the book. There are many good texts on symplectic geometry, especially since the topic has become so fashionable. To cite a few (in alphabetical order): Abraham and Marsden [1], Cannas da Silva [21], Libermann and Marle [110], the first Chapter of McDuff and Salamon [114], Vaisman [168]; the latter contains an interesting study of characteristic classes intervening in symplectic geometry. Bryant's lecture notes in [46] contains many interesting modern topics and extensions and can be read with profit even by the beginner. A nicely written review of symplectic geometry and of its applications is Gotay and Isenberg's paper [76] on the "symplectization of science". Other interesting reviews of symplectic geometry can be found in Weinstein [177, 178].

1.1 Symplectic Vector Spaces

We will deal exclusively with finite-dimensional real symplectic spaces. We begin by discussing the notion of symplectic form on a vector space. Symplectic forms allow the definition of symplectic bases, which are the analogues of orthonormal bases in Euclidean geometry.

1.1.1 Generalities

Let E be a real vector space; its generic vector will be denoted by z. A symplectic form (or: *skew-product*) on E is a mapping $\omega : E \times E \longrightarrow \mathbb{R}$ which is

- linear in each of its components:

$$\omega(\alpha_1 z_1 + \alpha_2 z_2, z') = \alpha_1 \omega(z_1, z') + \alpha_2 \omega(z_2, z'),$$
$$\omega(z, \alpha_1 z_1' + \alpha_2 z_2', z') = \alpha_1 \omega(z, z_1') + \alpha_2 \omega(z, z_2')$$

for all $z, z', z_1, z_1', z_2, z_2'$ in E and $\alpha_1, \alpha_1', \alpha_2, \alpha_2'$ in $\mathbb{R}$;
- antisymmetric (one also says *skew-symmetric*):

$$\omega(z, z') = -\omega(z', z) \text{ for all } z, z' \in E$$

(equivalently, in view of the bilinearity of ω: $\omega(z, z) = 0$ for all $z \in E$):
- non-degenerate:

$$\omega(z, z') = 0 \text{ for all } z \in E \text{ if and only if } z' = 0.$$

Definition 1.1. A real symplectic space is a pair (E, ω) where E is a real vector space on $\mathbb{R}$ and ω a symplectic form. The dimension of (E, ω) is, by definition, the dimension of E.

The most basic – and important – example of a finite-dimensional symplectic space is the *standard symplectic space* $(\mathbb{R}_z^{2n}, \sigma)$ where σ (the *standard symplectic form*) is defined by

$$\sigma(z, z') = \sum_{j=1}^{n} p_j x_j' - p_j' x_j \tag{1.1}$$

when $z = (x_1, \ldots, x_n; p_1, \ldots, p_n)$ and $z' = (x_1', \ldots, x_n'; p_1', \ldots, p_n')$. In particular, when $n = 1$,

$$\sigma(z, z') = -\det(z, z').$$

In the general case $\sigma(z, z')$ is (up to the sign) the sum of the areas of the parallelograms spanned by the projections of z and z' on the coordinate planes x_j, p_j.

Here is a coordinate-free variant of the standard symplectic space: set $X = \mathbb{R}^n$ and define a mapping $\xi : X \oplus X^* \longrightarrow \mathbb{R}$ by

$$\xi(z, z') = \langle p, x' \rangle - \langle p', x \rangle \tag{1.2}$$

if $z = (x, p), z' = (x', p')$. That mapping is then a symplectic form on $X \oplus X^*$. Expressing z and z' in the canonical bases of X and X^* then identifies $(\mathbb{R}_z^{2n}, \sigma)$ with $(X \oplus X^*, \xi)$. While we will only deal with finite-dimensional symplectic spaces, it is easy to check that formula (1.2) easily generalizes to the infinite-dimensional case. Let in fact X be a real Hilbert space and X^* its dual. Define an antisymmetric bilinear form ξ on $X \oplus X^*$ by the formula (1.2) where $\langle \cdot, \cdot \rangle$ is again the duality bracket. Then ξ is a symplectic form on $X \oplus X^*$.

Remark 1.2. Let Φ be the mapping $E \longrightarrow E^*$ which to every $z \in E$ associates the linear form Φ_z defined by

$$\Phi_z(z') = \omega(z, z'). \tag{1.3}$$

The non-degeneracy of the symplectic form can be restated as follows:

$$\omega \text{ is non-degenerate} \Longleftrightarrow \Phi \text{ is a monomorphism } E \longrightarrow E^*.$$

We will say that two symplectic spaces (E,ω) and (E',ω') are isomorphic if there exists a vector space isomorphism $s : E \longrightarrow E'$ such that

$$\omega'(s(z),s(z')) = \omega'(z,z')$$

for all z, z' in E; two isomorphic symplectic spaces thus have the same dimension. We will see below that, conversely, two finite-dimensional symplectic spaces are always isomorphic in the sense above if they have same dimension; the proof of this property requires the notion of symplectic basis, studied in the next subsection.

Let (E_1,ω_1) and (E_2,ω_2) be two arbitrary symplectic spaces. The mapping

$$\omega = \omega_1 \oplus \omega_2 : E_1 \oplus E_2 \longrightarrow \mathbb{R}$$

defined by

$$\omega(z_1 \oplus z_2; z_1' \oplus z_2') = \omega_1(z_1,z_1') + \omega_2(z_2,z_2') \tag{1.4}$$

for $z_1 \oplus z_2, z_1' \oplus z_2' \in E_1 \oplus E_2$ is obviously antisymmetric and bilinear. It is also non-degenerate: assume that

$$\omega(z_1 \oplus z_2; z_1' \oplus z_2') = 0 \text{ for all } z_1' \oplus z_2' \in E_1 \oplus E_2;$$

then, in particular, $\omega_1(z_1,z_1') = \omega_2(z_2,z_2') = 0$ for all (z_1',z_2') and hence $z_1 = z_2 = 0$. The pair

$$(E,\omega) = (E_1 \oplus E_2, \omega_1 \oplus \omega_2)$$

is thus a symplectic space; it is called the *direct sum* of (E_1,ω_1) and (E_2,ω_2).

Example 1.3. Let $(\mathbb{R}_z^{2n},\sigma)$ be the standard symplectic space. Then we can define on $\mathbb{R}_z^{2n} \oplus \mathbb{R}_z^{2n}$ two symplectic forms $\sigma^{\oplus}$ and $\sigma^{\ominus}$ by

$$\begin{aligned}
\sigma^{\oplus}(z_1,z_2;z_1',z_2') &= \sigma(z_1,z_1') + \sigma(z_2,z_2'),\\
\sigma^{\ominus}(z_1,z_2;z_1',z_2') &= \sigma(z_1,z_1') - \sigma(z_2,z_2').
\end{aligned}$$

The corresponding symplectic spaces are denoted $(\mathbb{R}_z^{2n} \oplus \mathbb{R}_z^{2n},\sigma^{\oplus})$ and $(\mathbb{R}_z^{2n} \oplus \mathbb{R}_z^{2n},\sigma^{\ominus})$.

Let us briefly discuss the notion of complex structure on a vector space; we refer to the literature, for instance Hofer–Zehnder [91] or McDuff–Salamon [114], where this notion is emphasized and studied in detail.

We begin by noting that the standard symplectic form σ on $\mathbb{R}_z^{2n}$ can be expressed in matrix form as

$$\sigma(z,z') = (z')^T Jz \quad , \quad J = \begin{bmatrix} 0 & I \\ -I & 0 \end{bmatrix}, \tag{1.5}$$

where 0 and i stand for the $n\times n$ zero and identity matrices. The matrix J is called the standard symplectic matrix. Alternatively, we can view $\mathbb{R}^{2n}_z$ as the complex vector space $\mathbb{C}^n$ by identifying (x,p) with $x+ip$. The standard symplectic form can with this convention be written as

$$\sigma(z,z') = \operatorname{Im}\langle z,z'\rangle_{\mathbb{C}^n} \tag{1.6}$$

where $\langle\cdot,\cdot\rangle_{\mathbb{C}^n}$ is the usual (Hermitian) scalar product on $\mathbb{C}^n$. Notice that multiplication of $x+ip$ by i then corresponds to multiplication of (x,p) by J. These considerations lead to the following definition:

Definition 1.4. A "complex structure" on a vector space E is any linear isomorphism $j : E \longrightarrow E$ such that $j^2 = -I$.

Since $\det(j^2) = (-1)^{\dim E} > 0$ we must have $\dim E = 2n$ so that only even-dimensional vector spaces can have a complex structure. It turns out that the existence of a complex structure on E identifies it with the standard symplectic space as the following exercises show:

Exercise 1.5. Let j be a complex structure on a vector space E. Show that one can define on E a structure of complex vector space $E^{\mathbb{C}}$ by setting

$$(\alpha + i\beta)z = \alpha + \beta jz. \tag{1.7}$$

[Hint: the condition $j^2 = -I$ is necessary to ensure that $u(u'z) = (uu')z$ for all $u, u' \in \mathbb{C}$].

Exercise 1.6. Let f be a linear mapping $E^{\mathbb{C}} \longrightarrow E^{\mathbb{C}}$ such that $f\circ j = j\circ f$.

(i) Assume that the matrix of f in a basis $\mathcal{B}^{\mathbb{C}} = \{e_1,\dots,e_n\}$ of $E^{\mathbb{C}}$ is $U = A+iB$ (A, B real $n\times n$ matrices). Viewing $f : E^{\mathbb{C}} \longrightarrow E^{\mathbb{C}}$ as a real endomorphism $f^{\mathbb{R}}$ shows that the matrix of $f^{\mathbb{R}}$ in the basis $\mathcal{B} = \{j(e_1),\dots,j(e_n); e_1,\dots,e_n;\}$ is then

$$U = \begin{bmatrix} A & -B \\ B & A \end{bmatrix}.$$

(ii) Show that $\det f^{\mathbb{R}} = |\det f|^2$.

Here is an example of a nonstandard symplectic structure. Let B be an antisymmetric (real) $n\times n$ matrix: $B^T = -B$ and set

$$J_B = \begin{bmatrix} -B & I \\ -I & 0 \end{bmatrix}.$$

We have

$$J_B^2 = \begin{bmatrix} B^2 - I & -B \\ B & -I \end{bmatrix}$$

hence $J_B^2 \neq -I$ if $B \neq 0$. We can however associate to J_B the symplectic form σ_B defined by

$$\sigma_B(z,z') = \sigma(z,z') - \langle Bx, x'\rangle\,; \tag{1.8}$$

this symplectic form intervenes in the study of electromagnetism (more generally in the study of any Galilean invariant Hamiltonian system). The scalar product $-\langle Bx, x'\rangle$ is therefore sometimes called the "magnetic term"; see Guillemin and Sternberg [85] for a thorough discussion of σ_B.

1.1.2 Symplectic bases

We begin by observing that the dimension of a finite-dimensional symplectic vector is always even: choosing a scalar product $\langle\cdot,\cdot\rangle_E$ on E, there exists an endomorphism j of E such that $\omega(z,z') = \langle j(z), z'\rangle_E$ and the antisymmetry of ω is then equivalent to $j^T = -j$ where ^{T} denotes here transposition with respect to $\langle\cdot,\cdot\rangle_E$; hence

$$\det j = (-1)^{\dim E} \det j^T = (-1)^{\dim E} \det j.$$

The non-degeneracy of ω implies that $\det j \neq 0$ so that $(-1)^{\dim E} = 1$, hence $\dim E = 2n$ for some integer n, as claimed.

Definition 1.7. A set $\mathcal{B}$ of vectors

$$\mathcal{B} = \{e_1, \dots, e_n\} \cup \{f_1, \dots, f_n\}$$

of E is called a "*symplectic basis*" *of* (E,ω) if the conditions

$$\omega(e_i, e_j) = \omega(f_i, f_j) = 0 \quad , \quad \omega(f_i, e_j) = \delta_{ij} \text{ for } 1 \leq i,j \leq n \tag{1.9}$$

hold (δ_{ij} is the Kronecker index: $\delta_{ij} = 1$ if $i = j$ and $\delta_{ij} = 0$ if $i \neq j$).

We leave it to the reader to check that the conditions (1.9) automatically ensure the linear independence of the vectors e_i, f_j for $1 \leq i,j \leq n$ (hence a symplectic basis is a basis in the usual sense).

Here is a basic (and obvious) example of a symplectic basis: define vectors $e_1, \dots, e_n$ and $f_1, \dots, f_n$ in $\mathbb{R}^{2n}_z$ by

$$e_i = (c_i, 0) \quad , \quad e_i = (0, c_i)$$

where (c_i) is the canonical basis of $\mathbb{R}^n$. (For instance, if $n = 1$, $e_1 = (1,0)$ and $f_1 = (0,1)$.) These vectors form the canonical basis

$$\mathcal{B} = \{e_1, \dots, e_n\} \cup \{f_1, \dots, f_n\}$$

of the standard symplectic space $(\mathbb{R}^{2n}_z, \sigma)$; one immediately checks that they satisfy the conditions $\sigma(e_i, e_j) = 0$, $\sigma(f_i, f_j) = 0$, and $\sigma(f_i, e_j) = \delta_{ij}$ for $1 \leq i,j \leq n$. This basis is called the canonical symplectic basis.

Remark 1.8. It is not immediately obvious that each symplectic space has a symplectic basis; that this is however true will be established in Section 1.2, where we will in addition prove the symplectic equivalent of the Gram–Schmidt orthonormalization process.

Taking for granted the existence of symplectic bases we can prove that all symplectic vector spaces of the same finite dimension $2n$ are isomorphic: let (E,ω) and (E',ω') have symplectic bases $\{e_i, f_j; 1 \le i,j \le n\}$ and $\{e'_i, f'_j; 1 \le i,j \le n\}$ and consider the linear isomorphism $s : E \longrightarrow E'$ defined by the conditions $s(e_i) = e'_i$ and $s(f_i) = f'_i$ for $1 \le i \le n$. That s is symplectic is clear since we have

$$\begin{aligned}
\omega'(s(e_i), s(e_j)) &= \omega'(e'_i, e'_j) = 0,\\
\omega'(s(f_i), s(f_j)) &= \omega'(f'_i, f'_j) = 0,\\
\omega'(s(f_j), s(e_i)) &= \omega'(f'_j, e'_i) = \delta_{ij}
\end{aligned}$$

for $1 \le i,j \le n$.

The set of all symplectic automorphisms $(E,\omega) \longrightarrow (E,\omega)$ form a group $\mathrm{Sp}(E,\omega)$ – the symplectic group of (E,ω) – for the composition law. Indeed, the identity is obviously symplectic, and so is the composition of two symplectic transformations. If $\omega(s(z), s(z')) = \omega(z,z')$ then, replacing z and z' by $s^{-1}(z)$ and $s^{-1}(z')$, we have $\omega(z,z') = \omega(s^{-1}(z), s^{-1}(z'))$ so that s^{-1} is symplectic as well.

It turns out that all symplectic groups corresponding to symplectic spaces of the same dimension are isomorphic:

Proposition 1.9. *Let (E,ω) and (E',ω') be two symplectic spaces of the same dimension $2n$. The symplectic groups $\mathrm{Sp}(E,\omega)$ and $\mathrm{Sp}(E',\omega')$ are isomorphic.*

Proof. Let Φ be a symplectic isomorphism $(E,\omega) \longrightarrow (E',\omega')$ and define a mapping $f_\Phi : \mathrm{Sp}(E,\omega) \longrightarrow \mathrm{Sp}(E',\omega')$ by $f_\Phi(s) = f \circ s \circ f^{-1}$. Clearly $f_\Phi(ss') = f_\Phi(s)\Phi(s')$ hence f_Φ is a group monomorphism. The condition $f_\Phi(S) = I$ (the identity in $\mathrm{Sp}(E',\omega')$) is equivalent to $f \circ s = f$ and hence to $s = I$ (the identity in $\mathrm{Sp}(E,\omega)$); f_Φ is thus injective. It is also surjective because $s = f^{-1} \circ s' \circ f$ is a solution of the equation $f \circ s \circ f^{-1} = s'$. □

These results show that it is no restriction to study finite-dimensional symplectic geometry by singling out one particular symplectic space, for instance the standard symplectic space, or its variants. This will be done in the next section.

Note that if $\mathcal{B}_1 = \{e_{1i}, f_{1j}; 1 \le i,j \le n_1\}$ and $\mathcal{B}_2 = \{e_{2k}, f_{2\ell}; 1 \le k,\ell \le n_2\}$ are symplectic bases of (E_1,ω_1) and (E_2,ω_2), then

$$\mathcal{B} = \{e_{1i} \oplus e_{2k}, f_{1j} \oplus f_{2\ell} : 1 \le i,j \le n_1 + n_2\}$$

is a symplectic basis of $(E_1 \oplus E_2, \omega_1 \oplus \omega_2)$.

Exercise 1.10. Construct explicitly an isomorphism

$$(\mathbb{R}^{2n}_z \oplus \mathbb{R}^{2n}_z, \sigma^{\oplus}) \longrightarrow (\mathbb{R}^{2n}_z \oplus \mathbb{R}^{2n}_z, \sigma^{\ominus})$$

where the symplectic forms $\sigma^{\oplus}$ and $\sigma^{\ominus}$ are defined as in Example 1.3.

Exercise 1.11. Denote by $\mathrm{Sp}^{\oplus}(2n)$ and $\mathrm{Sp}^{\ominus}(2n)$ the symplectic groups of $(\mathbb{R}_z^{2n} \oplus \mathbb{R}_z^{2n}, \sigma^{\oplus})$ and $(\mathbb{R}_z^{2n} \oplus \mathbb{R}_z^{2n}, \sigma^{\ominus})$, respectively. Show that

$$S^{\ominus} \in \mathrm{Sp}^{\ominus}(4n, \mathbb{R}) \Longleftrightarrow I^{\ominus} S^{\ominus} \in \mathrm{Sp}^{\oplus}(4n, \mathbb{R})$$

where $I^{\ominus} = \begin{bmatrix} I & 0 \\ 0 & C \end{bmatrix}$ and $C(x,p) = (x, -p)$.

1.1.3 Differential interpretation of σ

A differential two-form on a vector space $\mathbb{R}^m$ is the assignment to every $x \in \mathbb{R}^m$ of a linear combination

$$\beta_x = \sum_{i<j\leq m} b_{ij}(x) dx_i \wedge dx_j$$

where the b_{ij} are (usually) chosen to be C^∞ functions, and the wedge product $dx_i \wedge dx_j$ is defined by

$$dx_i \wedge dx_j = dx_i \otimes dx_j - dx_j \otimes dx_i$$

where $dx_i : \mathbb{R}^m \longrightarrow \mathbb{R}$ is the projection on the ith coordinate. Returning to $\mathbb{R}_z^{2n}$, we have

$$dp_j \wedge dx_j(z, z') = p_j x'_j - p'_j x_j$$

hence we can identify the standard symplectic form σ with the differential 2-form

$$dp \wedge dx = \sum_{j=1}^{n} dp_j \wedge dx_j = d(\sum_{j=1}^{n} p_j dx_j);$$

the differential one-form

$$pdx = \sum_{j=1}^{n} p_j dx_j$$

plays a fundamental role in both classical and quantum mechanics; it is sometimes called the (reduced) *action form* in physics and the *Liouville form* in mathematics[1].

Since we are in the business of differential form, let us make the following remark: the exterior derivative of $dp_j \wedge dx_j$ is

$$d(dp_j \wedge dx_j) = d(dp_j) \wedge dx_j + dp_j \wedge d(dx_j) = 0$$

so that we have

$$d\sigma = d(dp \wedge dx) = 0.$$

[1]Some authors call it the *tautological one-form*.

The standard symplectic form is thus a closed non-degenerate 2-form on $\mathbb{R}_z^{2n}$. This remark is the starting point of the generalization of the notion of symplectic form to a class of manifolds: a symplectic manifold is a pair (M, ω) where M is a differential manifold M and ω a non-degenerate closed 2-form on M. This means that every tangent plane T_zM carries a symplectic form ω_z varying smoothly with $z \in M$. As a consequence, a symplectic manifold always has even dimension (we will not discuss the infinite-dimensional case).

One basic example of a symplectic manifold is the cotangent bundle $T^*\mathbb{V}^n$ of a manifold $\mathbb{V}^n$; the symplectic form is here the "canonical 2-form" on $T^*\mathbb{V}^n$, defined as follows: let $\pi : T^*\mathbb{V}^n \longrightarrow \mathbb{V}^n$ be the projection to the base and define a 1-form λ on $T^*\mathbb{V}^n$ by $\lambda_z(X) = p(\pi_*(X))$ for a tangent vector $\mathbb{V}^n$ to $T^*\mathbb{V}^n$ at $z = (z, p)$. The form λ is called the "canonical 1-form" on $T^*\mathbb{V}^n$; its exterior derivative $\omega = d\lambda$ is called the "canonical 2-form" on $T^*\mathbb{V}^n$ and one easily checks that it indeed is a symplectic form (in local coordinates $\lambda = pdx$ and $\sigma = dp \wedge dx$). The symplectic manifold $(T^*\mathbb{V}^n, \omega)$ is in a sense the most straightforward non-linear version of the standard symplectic space (to which it reduces when $\mathbb{V}^n = \mathbb{R}_x^n$ since $T^*\mathbb{R}_x^n$ is just $\mathbb{R}_x^n \times (\mathbb{R}_x^n)^* \equiv \mathbb{R}_z^{2n}$). Observe that $T^*\mathbb{V}^n$ never is a compact manifold.

A symplectic manifold is always orientable: the non-degeneracy of ω namely implies that the $2n$-form

$$\omega^{\wedge n} = \underbrace{\omega \wedge \cdots \wedge \omega}_{n \text{ factors}}$$

never vanishes on M and is thus a volume form on M. We will call the exterior power $\omega^{\wedge n}$ the *symplectic volume form.* When M is the standard symplectic space, then the usual volume form on $\mathbb{R}_z^{2n}$,

$$\mathrm{Vol}_{2n} = (dp_1 \wedge \cdots \wedge dp_n) \wedge (dx_1 \wedge \cdots \wedge dx_n),$$

is related to the symplectic volume form by

$$\mathrm{Vol}_{2n} = (-1)^{n(n-1)/2} \frac{1}{n!} \sigma^{\wedge n}. \tag{1.10}$$

Notice that, as a consequence, every cotangent bundle $T^*\mathbb{V}^n$ is an oriented manifold!

The following exercise proposes a simple example of a symplectic manifold which is not a vector space; this example shows at the same time that an arbitrary even-dimensional manifold need not carry a symplectic structure:

Exercise 1.12.

(i) Show that the sphere S^2 equipped with the standard area form $\sigma_u(z, z') = \det(u, z, z')$ is a symplectic manifold.

(ii) Show that the spheres S^{2n} are never symplectic manifolds for $n > 1$. [Hint: $H^k(S^{2n}) = 0$ for $k \neq 0$ and $k \neq 2n$.]

1.2 Skew-Orthogonality

All vectors in a symplectic space (E, ω) are skew-orthogonal (one also says "isotropic") in view of the antisymmetry of a symplectic form: $\sigma(z, z') = 0$ for all $z \in E$. The notion of length therefore does not make sense in symplectic geometry (whereas the notion of area does). The notion "skew orthogonality" is extremely interesting in the sense that it allows the definition of subspaces of a symplectic space having special properties. We begin by defining the notion of a symplectic basis, which is the equivalent of an orthonormal basis in Euclidean geometry.

1.2.1 Isotropic and Lagrangian subspaces

Let M be an arbitrary subset of a symplectic space (E, ω). The *skew-orthogonal* set to M (one also says *annihilator*) is by definition the set

$$M^{\omega} = \{z \in E : \omega(z, z') = 0, \ \forall z' \in M\}.$$

Notice that we always have

$$M \subset N \Longrightarrow N^{\omega} \subset M^{\omega} \text{ and } (M^{\omega})^{\omega} \subset M.$$

It is traditional to classify subsets M of a symplectic space (E, ω) as follows:

$M \subset E$ is said to be:

- *Isotropic* if $M^{\omega} \supset M : \omega(z, z') = 0$ for all $z, z' \in M$;
- *Coisotropic* (or: *involutive*) if $M^{\omega} \subset M$;
- *Lagrangian* if M is both isotropic and co-isotropic: $M^{\omega} = M$;
- *Symplectic if* $M \cap M^{\omega} = 0$.

Notice that the non-degeneracy of a symplectic form is equivalent to saying that, in a symplectic space, the only vector that is skew-orthogonal to all other vectors is 0.

The following proposition describes some straightforward but useful properties of the skew-orthogonal set of a linear subspace of a symplectic space:

Proposition 1.13.

(i) *If M is a linear subspace of E, then so is M^{ω} and*

$$\dim M + \dim M^{\omega} = \dim E \textit{ and } (M^{\omega})^{\omega} = M. \tag{1.11}$$

(ii) *If M_1, M_2 are linear subspaces of a symplectic space (E, ω), then*

$$(M_1 + M_2)^{\omega} = M_1^{\omega} \cap M_2^{\omega} \ , \ (M_1 \cap M_2)^{\omega} = M_1^{\omega} + M_2^{\omega}. \tag{1.12}$$

Proof. (i) That M^ω is a linear subspace of E is clear. Let $\Phi : E \longrightarrow E^*$ be the linear mapping (1.3); since the dimension of E is finite the non-degeneracy of ω implies that Φ is an isomorphism. Let $\{e_1, \dots, e_k\}$ be a basis of M; we have

$$M^\omega = \bigcap_{j=1}^{k} \ker(\Phi(e_j))$$

so that M^ω is defined by k independent linear equations, hence

$$\dim M^\omega = \dim E - k = \dim E - \dim M$$

which proves the first formula (1.11). Applying that formula to the subspace $(M^\omega)^\omega$ we get

$$\dim(M^\omega)^\omega = \dim E - \dim M^\omega = \dim M$$

and hence $M = (M^\omega)^\omega$ since $(M^\omega)^\omega \subset M$ whether M is linear or not.

(ii) It is sufficient to prove the first equality (1.12) since the second follows by duality, replacing M_1 by M_1^ω and M_2 by M_2^ω and using the first formula (1.11). Assume that $z \in (M_1 + M_2)^\omega$; then $\omega(z, z_1 + z_2) = 0$ for all $z_1 \in M_1, z_2 \in M_2$. In particular $\omega(z, z_1) = \omega(z, z_2) = 0$ so that we have both $z \in M_1^\omega$ and $z \in M_2^\omega$, proving that $(M_1 + M_2)^\omega \subset M_1^\omega \cap M_2^\omega$. If conversely $z \in M_1^\omega \cap M_2^\omega$, then $\omega(z, z_1) = \omega(z, z_2) = 0$ for all $z_1 \in M_1, z_2 \in M_2$ and hence $\omega(z, z') = 0$ for all $z' \in M_1 + M_2$. Thus $z \in (M_1 + M_2)^\omega$ and $M_1^\omega \cap M_2^\omega \subset (M_1 + M_2)^\omega$. □

Let M be a linear subspace of (E, ω) such that $M \cap M^\omega = \{0\}$; in the terminology introduced above M is a "symplectic subset of E".

Exercise 1.14. If $M \cap M^\omega = \{0\}$, then $(M, \omega_{|M})$ and $(M^\omega, \omega_{|M^\omega})$ are complementary symplectic spaces of (E, ω):

$$(E, \omega) = (M \oplus M^\omega, \omega_{|M} \oplus \omega_{|M^\omega}). \tag{1.13}$$

[Hint: M^ω is a linear subspace of E so it suffices to check that the restriction $\omega_{|M}$ is non-degenerate.]

1.2.2 The symplectic Gram–Schmidt theorem

The following result is a symplectic version of the Gram–Schmidt orthonormalization process of Euclidean geometry. Because of its importance and its many applications we give it the status of a theorem:

Theorem 1.15. *Let A and B be two (possibly empty) subsets of $\{1, \dots, n\}$. For any two subsets $\mathcal{E} = \{e_i : i \in A\}$, $\mathcal{F} = \{f_j : j \in B\}$ of the symplectic space (E, ω) ($\dim E = 2n$), such that the elements of $\mathcal{E}$ and $\mathcal{F}$ satisfy the relations*

$$\omega(e_i, e_j) = \omega(f_i, f_j) = 0 \ , \ \omega(f_i, e_j) = \delta_{ij} \text{ for } (i, j) \in A \times B, \tag{1.14}$$

there exists a symplectic basis $\mathcal{B}$ of (E, ω) containing $\mathcal{E} \cup \mathcal{F}$.

Proof. We will distinguish three cases.

(i) *The case* $A = B = \emptyset$. Choose a vector $e_1 \neq 0$ in E and let f_1 be another vector with $\omega(f_1, e_1) \neq 0$ (the existence of f_1 follows from the non-degeneracy of ω). These vectors are linearly independent, which proves the theorem in the considered case when $n = 1$. Suppose $n > 1$ and let M be the subspace of E spanned by $\{e_1, f_1\}$ and set $E_1 = M^\omega$; in view of the first formula (1.11) we have $\dim M + \dim E_1 = 2n$. Since $\omega(f_1, e_1) \neq 0$ we have $E_1 \cap M = 0$, hence $E = E_1 \oplus M$, and the restriction ω_1 of ω to E_1 is non-degenerate (because if $z_1 \in E_1$ is such that $\omega_1(z_1, z) = 0$ for all $z \in E_1$, then $z_1 \in E_1^\omega = M$ and hence $z_1 = 0$); (E_1, ω_1) is thus a symplectic space of dimension $2(n-1)$. Repeating the construction above $n-1$ times we obtain a strictly decreasing sequence

$$(E, \omega) \supset (E_1, \omega_1) \supset \cdots \supset (E_{n-1}, \omega_{n-1})$$

of symplectic spaces with $\dim E_k = 2(n-k)$ and also an increasing sequence

$$\{e_1, f_1\} \subset \{e_1, e_2; f_1; f_2\} \subset \cdots \subset \{e_1, \dots, e_n; f_1, \dots, f_n\}$$

of sets of linearly independent vectors in E, each set satisfying the relations (1.14).

(ii) *The case* $A = B \neq \emptyset$. We may assume without restricting the argument that $A = B = \{1, 2, \dots, k\}$. Let M be the subspace spanned by $\{e_1, \dots, e_k; f_1, \dots, f_k\}$. As in the first case we find that $E = M \oplus M^\omega$ and that the restrictions ω_M and ω_{M^ω} of ω to M and M^ω, respectively, are symplectic forms.

Let $\{e_{k+1}, \dots, e_n; f_{k+1}, \dots, f_n\}$ be a symplectic basis of M^ω; then

$$\mathcal{B} = \{e_1, \dots, e_n; f_1, \dots, f_n\}$$

is a symplectic basis of E.

(iii) *The case* $B \backslash A \neq \emptyset$ (*or* $B \backslash A \neq \emptyset$). Suppose for instance $k \in B \backslash A$ and choose $e_k \in E$ such that $\omega(e_i, e_k) = 0$ for $i \in A$ and $\omega(f_j, e_k) = \delta_{jk}$ for $j \in B$. Then $\mathcal{E} \cup \mathcal{F} \cup \{e_k\}$ is a system of linearly independent vectors: the equality

$$\lambda_k e_k + \sum_{i \in A} \lambda_i e_i + \sum_{j \in B} \mu_j e_j = 0$$

implies that we have

$$\lambda_k \omega(f_k, e_k) + \sum_{i \in A} \lambda_i \omega(f_k, e_i) + \sum_{j \in B} \mu_j \omega(f_k, e_j) = \lambda_k = 0$$

and hence also $\lambda_i = \mu_j = 0$. Repeating this procedure as many times as necessary, we are led back to the case $A = B \neq \emptyset$. □

Remark 1.16. The proof above shows that we can construct symplectic subspaces of (E, ω) having any given even dimension $2m < \dim E$ containing any pair of vectors e, f such that $\omega(f, e) = 1$. In fact, $M = \operatorname{Span}\{e, f\}$ is a two-dimensional symplectic subspace ("symplectic plane") of (E, ω). In the standard symplectic space $(\mathbb{R}_z^{2n}, \sigma)$ every plane x_j, p_j of "conjugate coordinates" is a symplectic plane.

The following exercise shows that the symplectic form is essentially the standard symplectic form in any symplectic basis:

Exercise 1.17. Show that if x_j, p_j, x'_j, p'_j are the coordinates of z, z' in any given symplectic basis, then $\omega(z, z')$ takes the standard form

$$\omega(z, z') = \sum_{j=1}^{n} p_j x'_j - p'_j x_j. \tag{1.15}$$

(Thus showing again the "naturality" of the standard symplectic space.)

It follows from the theorem above that if (E, ω) and (E', ω') are two symplectic spaces with the same dimension $2n$, there always exists a symplectic isomorphism $\Phi : (E, \omega) \longrightarrow (E', \omega')$. Let in fact

$$\mathcal{B} = \{e_1, \dots, e_n\} \cup \{f_1, \dots, f_n\} \ , \ \mathcal{B}' = \{e'_1, \dots, e'_n\} \cup \{f'_1, \dots, f'_n\}$$

be symplectic bases of (E, ω) and (E', ω'), respectively. The linear mapping $\Phi : E \longrightarrow E'$ defined by $\Phi(e_j) = e'_j$ and $\Phi(f_j) = f'_j$ $(1 \leq j \leq n)$ is a symplectic isomorphism.

This result, together with the fact that any skew-product takes the standard form in a symplectic basis shows why it is no restriction to develop symplectic geometry from the standard symplectic space: all symplectic spaces of a given dimension are just isomorphic copies of $(\mathbb{R}^{2n}_z, \sigma)$ (this is actually already apparent from Exercise 1.17).

We end this subsection by briefly discussing the restrictions of symplectic transformations to subspaces:

Proposition 1.18. *Let $(F, \omega_{|F})$ and $(F', \omega_{|F'})$ be two symplectic subspaces of (E, ω). If $\dim F = \dim F'$, there exists a symplectic automorphism of (E, ω) whose restriction $\varphi_{|F}$ is a symplectic isomorphism $\varphi_{|F} : (F, \omega_{|F}) \longrightarrow (F', \omega_{|F'})$.*

Proof. Assume that the common dimension of F and F' is $2k$ and let

$$\begin{aligned} \mathcal{B}_{(k)} &= \{e_1, \dots, e_k\} \cup \{f_1, \dots, f_k\}, \\ \mathcal{B}'_{(k)} &= \{e'_1, \dots, e'_k\} \cup \{f'_1, \dots, f'_k\} \end{aligned}$$

be symplectic bases of F and F', respectively. In view of Theorem 1.15 we may complete $\mathcal{B}_{(k)}$ and $\mathcal{B}_{(k')}$ into full symplectic bases $\mathcal{B}$ and $\mathcal{B}'$ of (E, ω). Define a symplectic automorphism Φ of E by requiring that $\Phi(e_i) = e'_i$ and $\Phi(f_j) = f'_j$. The restriction $\varphi = \Phi_{|F}$ is a symplectic isomorphism $F \longrightarrow F'$. □

Let us now work in the standard symplectic space $(\mathbb{R}^{2n}_z, \sigma)$; everything can however be generalized to vector spaces with a symplectic form associated to a complex structure. We leave this generalization to the reader as an exercise.

Definition 1.19. A basis of $(\mathbb{R}^{2n}_z, \sigma)$ which is both symplectic and orthogonal (for the scalar product $\langle z, z' \rangle = \sigma(Jz, z')$) is called an orthosymplectic basis.

The canonical basis is trivially an orthosymplectic basis. It is easy to construct orthosymplectic bases starting from an arbitrary set of vectors $\{e'_1, \ldots, e'_n\}$ satisfying the conditions $\sigma(e'_i, e'_j) = 0$: let ℓ be the vector space (Lagrangian plane) spanned by these vectors; using the classical Gram–Schmidt orthonormalization process we can construct an orthonormal basis $\{e_1, \ldots, e_n\}$ of ℓ. Define now $f_1 = -Je_1, \ldots, f_n = -Je_n$. The vectors f_i are orthogonal to the vectors e_j and are mutually orthogonal because J is a rotation; in addition

$$\sigma(f_i, f_j) = \sigma(e_i, e_j) = 0 \quad , \quad \sigma(f_i, e_j) = \langle e_i, e_j \rangle = \delta_{ij},$$

hence the basis

$$\mathcal{B} = \{e_1, \ldots, e_n\} \cup \{f_1, \ldots, f_n\}$$

is both orthogonal and symplectic.

We leave it to the reader as an exercise to generalize this construction to any set

$$\{e_1, \ldots, e_k\} \cup \{f_1, \ldots, f_m\}$$

of normed pairwise orthogonal vectors satisfying in addition the symplectic conditions $\sigma(f_i, f_j) = \sigma(e_i, e_j) = 0$ and $\sigma(f_i, e_j) = \delta_{ij}$.

1.3 The Lagrangian Grassmannian

Recall that a subset of (E, ω) is isotropic if ω vanishes identically on it. An isotropic subspace ℓ of (E, ω) having dimension $n = \frac{1}{2} \dim E$ is called a Lagrangian plane. Equivalently, a Lagrangian plane in (E, ω) is a linear subspace of E which is both isotropic and co-isotropic.

1.3.1 Lagrangian planes

It follows from Theorem 1.15 that there always exists a Lagrangian plane containing a given isotropic subspace: let $\{e_1, \ldots, e_k\}$ be a basis of such a subspace; complete that basis into a full symplectic basis

$$\mathcal{B} = \{e_1, \ldots, e_n\} \cup \{f_1, \ldots, f_n\}$$

of (E, ω); the space spanned by $\{e_1, \ldots, e_n\}$ is then a Lagrangian plane. Notice that we have actually constructed in this way a pair (ℓ, ℓ') of Lagrangian planes such that $\ell \cap \ell' = 0$, namely

$$\ell = \operatorname{Span}\{e_1, \ldots, e_n\} \quad , \quad \ell' = \operatorname{Span}\{f_1, \ldots, f_n\}.$$

Since Lagrangian planes will play a recurring role in the rest of this book it is perhaps appropriate to summarize some terminology and notation:

Definition 1.20. The set of all Lagrangian planes in a symplectic space (E,ω) is denoted by $\mathrm{Lag}(E,\omega)$ and is called the "Lagrangian Grassmannian of (E,ω)". When (E,ω) is the standard symplectic space $(\mathbb{R}_z^{2n},\sigma)$ the Lagrangian Grassmannian is denoted by $\mathrm{Lag}(n)$, and we will use the notation

$$\ell_X = \mathbb{R}_x^n \times 0 \quad \text{and} \quad \ell_P = 0 \times \mathbb{R}_p^n.$$

ℓ_X and ℓ_P are called the "horizontal" and "vertical" Lagrangian planes in $(\mathbb{R}_z^{2n},\sigma)$.

Other common notation for the Lagrangian Grassmannian is $\Lambda(E,\omega)$ or $\Lambda(n,\mathbb{R})$.

Example 1.21. Suppose $n=1$; $\mathrm{Lag}(1)$ consists of all straight lines passing through the origin in the symplectic plane $(\mathbb{R}_z^{2n},-\det)$.

When $n>1$ the Lagrangian Grassmannian is a proper subset of the set of all n-dimensional planes of $(\mathbb{R}_z^{2n},\sigma)$.

Let us study the equation of a Lagrangian plane in the standard symplectic space.

In what follows we work in an arbitrary symplectic basis

$$\mathcal{B} = \{e_1,\dots,e_n\} \cup \{f_1,\dots,f_n\}$$

of the standard symplectic space; the corresponding coordinates are denoted by x and p.

Proposition 1.22. *Let ℓ be an n-dimensional linear subspace ℓ of the standard symplectic space $(\mathbb{R}_z^{2n},\sigma)$.*

(i) *ℓ is a Lagrangian plane if and only if it can be represented by an equation*

$$Xx + Pp = 0 \quad \textit{with } \mathrm{rank}(X,P) = n \textit{ and } XP^T = PX^T. \tag{1.16}$$

(ii) *Let $\mathcal{B} = \{e_1,\dots,e_n\} \cup \{f_1,\dots,f_n\}$ be a symplectic basis and assume that $\ell = \mathrm{Span}\{f_1,\dots,f_n\}$; then there exists a symmetric matrix $M \in M(n,\mathbb{R})$ such that the Lagrangian plane ℓ is represented by the equation $p = Mx$ in the coordinates defined by $\mathcal{B}$.*

Proof. (i) We first remark that $Xx+Pp=0$ represents an n-dimensional space if and only if

$$\mathrm{rank}(X,P) = \mathrm{rank}(X^T,P^T)^T = n. \tag{1.17}$$

Assume that in addition $X^TP = P^TX$ and parametrize ℓ by setting $x = P^Tu$, $p=-X^Tu$. It follows that if z,z' are two vectors of ℓ, then

$$\sigma(z,z') = \langle -X^Tu, P^Tu'\rangle - \langle -X^Tu', P^Tu\rangle = 0$$

so that (1.16) indeed is the equation of a Lagrangian plane. Reversing the argument shows that if $Xx+Pp=0$ represents an n-dimensional space, then the condition

$\sigma(z, z') = 0$ for all vectors z, z' of that space implies that we must have $XP^T = PX^T$.

(ii) It is clear from (i) that $p = Mx$ represents a Lagrangian plane ℓ; it is also clear that this plane ℓ is transversal to $\operatorname{Span}\{f_1, \ldots, f_n\}$. The converse follows from the observation that if $\ell : Xx + Pp = 0$ is transversal to $\operatorname{Span}\{f_1, \ldots, f_n\}$, then P must invertible; the property follows taking $M = -P^{-1}X$ which is symmetric since $XP^T = PX^T$. □

Two Lagrangian planes are said to be *transversal* if $\ell \cap \ell' = 0$; since $\dim \ell = \dim \ell' = \frac{1}{2} \dim E$ this is equivalent to saying that $E = \ell \oplus \ell'$. For instance the horizontal and vertical Lagrangian planes $\ell_X = \mathbb{R}^n_x \times 0$ and $\ell_P = 0 \times \mathbb{R}^n_p$ are obviously transversal in $(\mathbb{R}^{2n}_z, \sigma)$.

Part (ii) of Proposition 1.22 above implies:

Corollary 1.23.

(i) *An n-plane ℓ in $(\mathbb{R}^{2n}_z, \sigma)$ is a Lagrangian plane transversal to ℓ_P if and only if there exists a symmetric matrix $M \in M(n, \mathbb{R})$ such that $\ell : p = Mx$.*

(ii) *For any n-plane $\ell : Xx + Pp = 0$ in $\mathbb{R}^{2n}_z$ we have*

$$\dim(\ell \cap \ell_P) = n - \operatorname{rank}(P). \tag{1.18}$$

(iii) *For any symplectic matrix*

$$s = \begin{bmatrix} A & B \\ C & D \end{bmatrix}$$

the rank of B is given by the formula

$$\operatorname{rank}(B) = n - \dim(S\ell_P \cap \ell_P). \tag{1.19}$$

Proof. (i) The condition is necessary, taking for $\mathcal{B}$ the canonical symplectic bases. If conversely ℓ is the graph of a symmetric matrix M, then it is immediate to check that $\sigma(z; z') = 0$ for all $z \in \ell$.

(ii) The intersection $\ell \cap \ell_P$ consists of all (x, p) which satisfy both conditions $Xx + Pp = 0$ and $x = 0$. It follows that

$$(x, p) \in \ell \cap \ell_P \Longleftrightarrow Pp = 0$$

and hence (1.18).

(iii) Formula (1.19) follows from the trivial equivalence

$$(x, p) \in S\ell_P \cap \ell_P \Longleftrightarrow Bp = 0.$$

□

Theorem 1.15 allows us to construct at will pairs of transverse Lagrangian planes: choose any pair of vectors $\{e_1, f_1\}$ such that $\omega(f_1, e_1) = 1$; in view of Theorem 1.15 we can find a symplectic basis $\mathcal{B} = \{e_1, \ldots, e_n\} \cup \{f_1, \ldots, f_n\}$ of (E, ω) and the spaces $\ell = \operatorname{Span}\{e_1, \ldots, e_n\}$ and $\ell' = \operatorname{Span}\{f_1, \ldots, f_n\}$ are then transversal Lagrangian planes in E. Conversely:

Proposition 1.24. *Suppose that ℓ_1 and ℓ_2 are two transversal Lagrangian planes in (E, ω). If $\{e_1, \ldots, e_n\}$ is a basis of ℓ_1, then there exists a basis $\{f_1, \ldots, f_n\}$ of ℓ_2 such that $\{e_1, \ldots, e_n; f_1, \ldots, f_n\}$ is a symplectic basis of (E, ω).*

Proof. It suffices to proceed as in the first case of the proof of Theorem 1.15 and to construct an increasing sequence of sets

$$\{e_1, f_1\} \subset \{e_1, e_2; f_1; f_2\} \subset \cdots \subset \{e_1, \ldots, e_n; f_1, \ldots, f_n\}$$

such that $\operatorname{Span}\{f_1, \ldots, f_n\} = \ell_2$ and $\omega(f_i, e_j) = \delta_{ij}$ for $1 \leq i, j \leq n$. □

Let us end this long subsection by stating a result on canonical coordinates for a Lagrangian plane

Proposition 1.25. *Let $\ell \in \operatorname{Lag}(E, \omega)$ and $\mathcal{B} = \{e_1, \ldots, e_n; f_1, \ldots, f_n\}$ a symplectic basis of (E, ω). There exists $I \subset \{1, 2, \ldots, n\}$ such that the restriction to ℓ of the orthogonal projection $P_I : E \longrightarrow \ell_I$ is an isomorphism $\ell \longrightarrow \ell_I$; ℓ_I is the Lagrangian plane generated by the vectors e_i and f_j for $i \in I$ and $j \notin J$. Denoting by x_i, p_j the coordinates in the basis $\mathcal{B}$, the Lagrangian plane ℓ can thus be represented by equations $x_i = 0$, $p_j = 0$ with $i \in I$ and $j \notin J$ ("canonical coordinates").*

We omit the proof of this result here and refer to Maslov [119] or Mischenko *et al.* [124]. Alternatively it can be derived from Corollary 1.23 by reducing the proof to the case where (E, ω) is the standard symplectic space.

1.3.2 The action of $\operatorname{Sp}(n)$ on $\operatorname{Lag}(n)$

Let us prove the following important result on the action of $\operatorname{Sp}(n)$ and its subgroup $\operatorname{U}(n)$ on the Lagrangian Grassmannian $\operatorname{Lag}(n)$.

Theorem 1.26. *The action of $\operatorname{U}(n)$ and $\operatorname{Sp}(n)$ on $\operatorname{Lag}(n)$ has the following properties:*

(i) *$\operatorname{U}(n)$ (and hence $\operatorname{Sp}(n)$) acts transitively on $\operatorname{Lag}(n)$: for every pair (ℓ, ℓ') of Lagrangian planes there exists $U \in \operatorname{U}(n)$ such that $\ell' = U\ell$.*

(ii) *The group $\operatorname{Sp}(n)$ acts transitively on the set of all pairs of transverse Lagrangian planes: if (ℓ_1, ℓ_1') and (ℓ_2, ℓ_2') are such that $\ell_1 \cap \ell_1' = \ell_1 \cap \ell_1' = 0$, then there exits $S \in \operatorname{Sp}(n)$ such that $(\ell_2, \ell_2') = (S\ell_1, S\ell_1')$.*

Proof. (i) Let $\mathcal{O} = \{e_1, \ldots, e_n\}$ and $\mathcal{O}' = \{e_1', \ldots, e_n'\}$ be orthonormal bases of ℓ and ℓ', respectively. Then $\mathcal{B} = \mathcal{O} \cup J\mathcal{O}$ and $\mathcal{B}' = \mathcal{O}' \cup J\mathcal{O}'$ are orthosymplectic bases of $(\mathbb{R}_z^{2n}, \sigma)$. There exists $U \in O(2n)$ such that $U(e_i) = e'_i$ and $U(f_i) = f'_i$ where $f_i = Je_i$, $f'_i = Je'_i$. We have $U \in \mathrm{Sp}(n)$ hence

$$U \in \mathrm{O}(2n) \cap \mathrm{Sp}(n) = \mathrm{U}(n)$$

((2.11) in Proposition 2.12).

(ii) Choose a basis $\{e_{11}, \dots, e_{1n}\}$ of ℓ_1 and a basis $\{f_{11}, \dots, f_{1n}\}$ of ℓ'_1 such that $\{e_{1i}, f_{1j}\}_{1\leq i,j\leq n}$ is a symplectic basis of $(\mathbb{R}_z^{2n}, \sigma)$. Similarly choose bases of ℓ_2 and ℓ'_2 whose union $\{e_{2i}, f_{2j}\}_{1\leq i,j\leq n}$ is also a symplectic basis. Define a linear mapping $S : \mathbb{R}_z^{2n} \longrightarrow \mathbb{R}_z^{2n}$ by $S(e_{1i}) = e_{2i}$ and $S(f_{1i}) = f_{2i}$ for $1 \leq i \leq n$. We have $S \in \mathrm{Sp}(n)$ and $(\ell_2, \ell'_2) = (S\ell_1, S\ell'_1)$. □

We will see in the next section that the existence of an integer-valued function measuring the relative position of triples of Lagrangian planes implies that $\mathrm{Sp}(n)$ cannot act transitively on triples (and *a fortiori*,on k-uples, $k \geq 3$) of Lagrangian planes.

For two integers $n_1, n_2 > 0$ consider the direct sum

$$\mathrm{Sp}(n_1) \oplus \mathrm{Sp}(n_2) = \{(S_1, S_2) : S_1 \in \mathrm{Sp}(n_1), S_1 \in \mathrm{Sp}(n_1)\}$$

equipped with the composition law

$$(S_1 \oplus S_2)(S'_1 \oplus S'_2) = S_1S'_1 \oplus S_2S'_2.$$

Setting $n = n_1 + n_2$, then $\mathrm{Sp}(n_1) \oplus \mathrm{Sp}(n_2)$ acts on the Lagrangian Grassmannian $\mathrm{Lag}(n)$. We have in particular a natural action

$$\mathrm{Sp}(n_1) \oplus \mathrm{Sp}(n_2) : \mathrm{Lag}(n_1) \oplus \mathrm{Lag}(n_2) \longrightarrow \mathrm{Lag}(n_1) \oplus \mathrm{Lag}(n_2)$$

where $\mathrm{Lag}(n_1) \oplus \mathrm{Lag}(n_2)$ is the set of all direct sums $\ell_1 \oplus \ell_2$ with $\ell_1 \in \mathrm{Lag}(n_1)$, $\ell_2 \in \mathrm{Lag}(n_2)$; this action is defined by the obvious formula

$$(S_1 \oplus S_2)(\ell_1 \oplus \ell_2) = S_1\ell_1 \oplus S_2\ell_2.$$

Observe that $\mathrm{Lag}(n_1) \oplus \mathrm{Lag}(n_2)$ is a subset of $\mathrm{Lag}(n)$:

$$\mathrm{Lag}(n_1) \oplus \mathrm{Lag}(n_2) \subset \mathrm{Lag}(n)$$

since $\sigma_1 \oplus \sigma_2$ vanishes on each $\ell_1 \oplus \ell_2$.

1.4 The Signature of a Triple of Lagrangian Planes

In this section we introduce a very useful index which measures the relative position of a triple of Lagrangian planes, due to Wall [174] and redefined by Kashiwara (see Lion–Vergne [111]). Related notions are defined in Dazord [28] and Demazure [29]. This index is a refinement of the notion of index of inertia of Leray [107] in the sense that is defined for arbitrary triples, while Leray's definition only works when some transversality condition is imposed (the same restriction applies to the index used in Guillemin–Sternberg [84]).

1.4.1 First properties

Let us introduce the following terminology and notation: let Q be a quadratic form on some real Euclidean space. The associated symmetric matrix $M = D^2Q$ (the Hessian matrix of Q) has μ^+ positive eigenvalues and μ^- negative eigenvalues. We will call the difference $\mu^+ - \mu^-$ the signature of the quadratic form Q and denote it by $\operatorname{sign} Q$:

$$\operatorname{sign} Q = \mu^+ - \mu^-.$$

Definition 1.27. Let (ℓ, ℓ', ℓ'') be an arbitrary triple of Lagrangian planes in a symplectic space (E, ω). The "Wall–Kashiwara index" (or: signature) of the triple (ℓ, ℓ', ℓ'') is the signature of the quadratic form

$$Q(z, z', z'') = \omega(z, z') + \omega(z', z'') + \omega(z'', z)$$

on $\ell \oplus \ell' \oplus \ell''$. This signature is denoted by $\tau(\ell, \ell', \ell'')$.

Let us illustrate this definition in the case $n = 1$, with $\omega = -\det$. The quadratic form Q is here

$$Q(z, z', z'') = -\det(z, z') - \det(z', z'') - \det(z'', z).$$

Choosing $\ell = \ell_X = \mathbb{R}_x \times 0$, $\ell'' = \ell_P = 0 \times \mathbb{R}_p$, and $\ell' = \ell_a : p = ax$, we have

$$Q = -axx' - p''x' + p''x.$$

After diagonalization this quadratic form becomes

$$Q = Z^2 - (X^2 + (\operatorname{sign} a)Y^2)$$

and hence, by a straightforward calculation:

$$\tau(\ell_X, \ell_a, \ell_P) = \begin{cases} -1 \text{ if } a > 0, \\ 0 \text{ if } a = 0, \\ +1 \text{ if } a < 0 \end{cases} \tag{1.20}$$

(this formula will be generalized to $(\mathbb{R}_z^{2n}, \sigma)$ in Corollary 1.31). The signature $\tau(\ell, \ell', \ell'')$ of three lines is thus 0 if any two of them coincide, -1 if the line ℓ' lies "between" ℓ and ℓ'' (the plane being oriented in the usual way), and $+1$ if it lies outside. An essential observation is that we would get the same values for an arbitrary triple ℓ, ℓ', ℓ'' of lines having the same relative positions as ℓ_X, ℓ_a, ℓ_P because one can always reduce the general case to that of the triple (ℓ_X, ℓ_a, ℓ_P), by using a matrix with determinant 1. Thus the signature is here what one sometimes calls the "cyclic order" of three lines. We leave it to the reader to verify that the following formula holds:

$$\tau(\ell, \ell', \ell'') = 2\left[\frac{\theta - \theta'}{2\pi}\right]_{\text{anti}} - 2\left[\frac{\theta - \theta''}{2\pi}\right]_{\text{anti}} + 2\left[\frac{\theta' - \theta''}{2\pi}\right]_{\text{anti}}. \tag{1.21}$$

The line $\ell : x\cos\alpha + p\sin\alpha = 0$ is here identified with $\theta = 2\alpha$ and $[\cdot]_{\rm anti}$ is the symmetrized integer part function, that is

$$[s]_{\rm anti} = \tfrac{1}{2}([s] - [-s]) \ , \ \ [s] = k \ \text{ if } \ k \leq s < k+1$$

(for k an integer).

Let us now return to the general case. The following properties of the signature are immediate:

- τ is $\mathrm{Sp}(E,\omega)$-invariant: for every $S \in \mathrm{Sp}(E,\omega)$ and $\ell, \ell', \ell'' \in (\mathrm{Lag}(E,\omega))^3$ we have
$$\tau(S\ell, S\ell', S\ell'') = \tau(\ell, \ell', \ell'') \tag{1.22}$$
(because $\sigma(Sz, Sz') = \sigma(z, z')$, and so on);
- τ is totally antisymmetric: for any permutation $\mathfrak{p}$ of the set $\{1,2,3\}$ we have
$$\tau(\ell_{\mathfrak{p}(1)}, \ell_{\mathfrak{p}(2)}, \ell_{\mathfrak{p}(3)}) = (-1)^{\mathrm{sgn}(\mathfrak{p})}\tau(\ell_1, \ell_2, \ell_3) \tag{1.23}$$
where $\mathrm{sgn}(\mathfrak{p}) = 0$ if $\mathfrak{p}$ is even, 1 if $\mathfrak{p}$ is odd (this immediately follows from the antisymmetry of σ).
- Let τ' and τ'' be the signature in $\mathrm{Lag}(E',\omega')$ and $\mathrm{Lag}(E'',\omega'')$ respectively, and τ the signature in $\mathrm{Lag}(E,\omega)$ with $(E,\omega) = (E',\omega') \oplus (E'',\omega'')$. Then
$$\tau(\ell_1' \oplus \ell_1'', \ell_2' \oplus \ell_2'', \ell_3' \oplus \ell_3'') = \tau'(\ell_1', \ell_2', \ell_3') + \tau''(\ell_1'', \ell_2'', \ell_3'')$$
for $(\ell_1', \ell_2', \ell_3') \in (\mathrm{Lag}(E',\omega'))^3$ and $(\ell_1'', \ell_2'', \ell_3'') \in (\mathrm{Lag}(E'',\omega''))^3$.

Remark 1.28. Cappell, Lee, and Miller [22] have shown the following truly remarkable property: if $(\chi_n)_{n\geq 1}$ is a family of functions $\chi_n : (\mathrm{Lag}(n))^3 \longrightarrow \mathbb{Z}$ satisfying the properties above, then each χ_n is proportional to the Wall–Kashiwara signature $\tau_n = \tau$ on $\mathrm{Lag}(n)$. Adding an appropriate normalization condition χ_n is then identified with τ.

Here are two results which sometimes simplify calculations of the signature:

Proposition 1.29. *Assume that $\ell \cap \ell'' = 0$, then $\tau(\ell, \ell', \ell'')$ is the signature of the quadratic form*

$$Q'(z') = \omega(P(\ell,\ell'')z', z') = \omega(z', P(\ell'',\ell)z')$$

on ℓ', where $P(\ell,\ell'')$ is the projection onto ℓ along ℓ'' and $P(\ell'',\ell) = I - P(\ell,\ell'')$ is the projection on ℓ'' along ℓ.

Proof. We have

$$\begin{aligned} Q(z,z',z'') &= \omega(z,z') + \omega(z',z'') + \omega(z'',z) \\ &= \omega(z, P(\ell'',\ell)z') + \omega(P(\ell,\ell'')z', z'') + \omega(z'',z) \\ &= \omega(P(\ell,\ell'')z', P(\ell'',\ell)z') - \omega(z - P(\ell,\ell'')z', z'' - P(\ell'',\ell)z'). \end{aligned}$$

Let $u = z - P(\ell, \ell'')z'$, $u' = z'$, $u'' = z'' - P(\ell'', \ell)z'$; the signature of Q is then the signature of the quadratic form

$$(u, u', u'') \longmapsto \omega(P(\ell, \ell'')u', P(\ell'', \ell)u') - \omega(u, u''),$$

hence the result since the signature of the quadratic form $(u, u'') \longmapsto \omega(u, u'')$ is obviously equal to zero. □

Proposition 1.30. *Let (ℓ, ℓ', ℓ'') be a triple of Lagrangian planes such that an $\ell = \ell \cap \ell' + \ell \cap \ell''$. Then $\tau(\ell, \ell', \ell'') = 0$.*

Proof. Let $E' \subset \ell \cap \ell'$ and $E'' \subset \ell \cap \ell''$ be subspaces such that $\ell = E' \oplus E''$. Let

$$(z, z', z'') \in \ell \times \ell' \times \ell''$$

and write $z = u' + u''$, $(u, u') \in E' \times E''$. We have

$$\sigma(z, z') = \sigma(u' + u'', z') = \sigma(u'', z'),$$
$$\sigma(z'', z) = \sigma(z'', u' + u'') = \sigma(u'', z)$$

and hence

$$Q(z, z', z'') = \sigma(u'', z') + \sigma(z', z'') + \sigma(z'', u').$$

Since $\sigma(u', u'') = 0$ this is

$$Q(z, z', z'') = \sigma(z' - u', z'' - u'')$$

so that $\tau(\ell, \ell', \ell'')$ is the signature of the quadratic form $(y', y'') \longmapsto \sigma(y', y'')$ on $\ell' \times \ell''$; this signature is equal to zero, hence the result. □

The following consequence of Proposition 1.29 generalizes formula (1.20):

Corollary 1.31. *Let (E, ω) be the standard symplectic space $(\mathbb{R}^{2n}_z, \sigma)$. Let $\ell_X = \mathbb{R}^n \times 0$, $\ell_P = 0 \times \mathbb{R}^n$, and $\ell_A = \{(x, Ax),\ x \in \mathbb{R}^n\}$, A being a symmetric linear mapping $\mathbb{R}^n \longmapsto \mathbb{R}^n$. Then*

$$\tau(\ell_P, \ell_A, \ell_X) = \operatorname{sign}(A) \quad , \quad \tau(\ell_X, \ell_A, \ell_P) = -\operatorname{sign}(A). \tag{1.24}$$

Proof. Formulae (1.24) are equivalent in view of the antisymmetry of τ. In view of the proposition above $\tau(\ell_P, \ell_A, \ell_X)$ is the signature of the quadratic form Q' on ℓ_A given by

$$Q'(z) = \sigma(P(\ell_P, \ell_A, \ell_X)z, P(\ell_X, \ell_P)z)$$

hence $Q'(z) = \langle x, Ax \rangle$ and the corollary follows. □

1.4.2 The cocycle property of τ

Less obvious – but of paramount importance for the general theory of the Arnol'd–Leray–Maslov index that we will develop in Chapter 3 – is the following "cocycle property" of the Wall–Kashiwara signature, the first proof of which apparently appeared in Lion–Vergne [111]; we are following the latter with a few simplifications. A precursor to τ is Leray's index of inertia of a triple of pairwise transverse Lagrangian planes (see de Gosson [54, 57, 61] for a detailed study and comparison with the signature). Some authors call the Wall–Kashiwara signature "Maslov triple index"; it is not a very good terminology because it is misleading since it amounts, at the end of the day, to identifying cocycles and coboundaries as we have explained in [54, 57].

Theorem 1.32. *For $\ell_1, \ell_2, \ell_3, \ell_4$ in $\operatorname{Lag}(E,\omega)$ we have*

$$\tau(\ell_1,\ell_2,\ell_3)-\tau(\ell_2,\ell_3,\ell_4)+\tau(\ell_1,\ell_3,\ell_4)-\tau(\ell_1,\ell_2,\ell_4)=0. \tag{1.25}$$

Proof. We begin by rewriting the quadratic form $\mathcal{Q}$ defining τ in a more tractable form. Let ℓ,ℓ',ℓ'' be three arbitrary Lagrangian planes and choose a symplectic basis

$$\mathcal{B}=\{e_1,\dots,e_n\}\cup\{f_1,\dots,f_n\}$$

of (E,ω) such that

$$\ell\cap\ell_0=\ell'\cap\ell_0=\ell''\cap\ell_0=0$$

where $\ell_0=\operatorname{Span}\{f_1,\dots,f_n\}$. Let $\dim E=2n$ and write a vector z in the basis $\mathcal{B}$ as

$$z=\sum_{i=1}^{n}x_ie_i+\sum_{j=1}^{n}p_{ji}f_j;$$

there exist symmetric matrices M,M',M'' such that

$$\ell:p=Mx \quad , \quad \ell':p=M'x \quad , \quad \ell'':p=M''x$$

(Proposition 1.22, (ii)). The integer $\tau(\ell,\ell',\ell'')$ being a symplectic invariant, it is the signature of the quadratic form

$$\mathcal{R}(z,z',z'')=\sigma(x,Mx;x',Mx')+ \\ \sigma(x',M'x';x'',M''x'')+\sigma(x'',M''x'';x,Mx)$$

which we can rewrite after a straightforward calculation as

$$\mathcal{R}(z,z',z'')=\tfrac{1}{2}X^TRX \quad , \quad X=(x,x',x'')$$

where R is the symmetric matrix

$$R=\begin{bmatrix}0 & M-M' & M''-M\\ M-M' & 0 & M'-M''\\ M''-M & M'-M'' & 0\end{bmatrix}.$$

The quadratic form $\mathcal{R}$ has the same signature $\tau(\ell,\ell',\ell'')$ as $\mathcal{R}\circ V$ for any invertible matrix V; choosing

$$V = \begin{bmatrix} 0 & I & I \\ I & 0 & I \\ I & I & 0 \end{bmatrix}$$

the matrix of the quadratic form $\mathcal{R}\circ V$ is

$$\tfrac{1}{2}V^T RV = \begin{bmatrix} M'-M'' & 0 & 0 \\ 0 & M''-M & 0 \\ 0 & 0 & M-M' \end{bmatrix}$$

and hence

$$\tau(\ell,\ell',\ell'') = \operatorname{sign}(M-M') + \operatorname{sign}(M'-M'') + \operatorname{sign}(M''-M). \tag{1.26}$$

The theorem now easily follows: writing, with obvious notation

$$\begin{aligned} \tau(\ell_1,\ell_2,\ell_3) &= \operatorname{sign}(M_1-M_2) + \operatorname{sign}(M_2-M_3) + \operatorname{sign}(M_3-M_1), \\ \tau(\ell_2,\ell_3,\ell_4) &= \operatorname{sign}(M_2-M_3) + \operatorname{sign}(M_3-M_4) + \operatorname{sign}(M_4-M_2), \\ \tau(\ell_1,\ell_3,\ell_4) &= \operatorname{sign}(M_1-M_3) + \operatorname{sign}(M_3-M_4) + \operatorname{sign}(M_4-M_1), \end{aligned}$$

we get, since $\operatorname{sign}(M_i-M_j) = -\operatorname{sign}(M_j-M_i)$:

$$\begin{aligned} &\tau(\ell_1,\ell_2,\ell_3) - \tau(\ell_2,\ell_3,\ell_4) + \tau(\ell_1,\ell_3,\ell_4) \\ &\qquad = \operatorname{sign}(M_1-M_2) + \operatorname{sign}(M_2-M_4) + \operatorname{sign}(M_4-M_1) \end{aligned}$$

that is

$$\tau(\ell_1,\ell_2,\ell_3) - \tau(\ell_2,\ell_3,\ell_4) + \tau(\ell_1,\ell_3,\ell_4) = \tau(\ell_1,\ell_2,\ell_4). \qquad \square$$

Formula (1.25) has the following combinatorial interpretation. Let us view the Wall–Kashiwara index as a 2-cochain on $\operatorname{Lag}(E,\omega)$ and denote by ∂ the "coboundary operator" (see (9) in the Notation section in the Preface). Then, by definition of ∂,

$$\partial\tau(\ell_1,\ell_2,\ell_3) = \tau(\ell_1,\ell_2,\ell_3) - \tau(\ell_2,\ell_3,\ell_4) + \tau(\ell_1,\ell_3,\ell_4) - \tau(\ell_1,\ell_2,\ell_4)$$

so that Theorem 1.32 can be restated in concise form as:

$$\partial\tau = 0, \textit{ that is: } \tau \textit{ is 2-cocycle on } \operatorname{Lag}(n).$$

1.4.3 Topological properties of τ

Consider three lines ℓ,ℓ',ℓ'' through the origin in the symplectic plane. As discussed in the beginning of the section the signature $\tau(\ell,\ell',\ell'')$ determines the relative positions of these lines. If we now move these three lines continuously, in such a way that their intersections do not change, the signature will remain unaltered. The same property remains true in higher dimensions. To prove this, we need the following elementary lemma which describes the kernel of the quadratic form defining τ; in order to avoid a deluge of multiple "primes" in the proof we slightly change notation and write (ℓ_1, ℓ_2, ℓ_3) instead of (ℓ, ℓ', ℓ'') so that the defining quadratic form becomes

$$Q(z_1, z_2, z_3) = \sigma(z_1, z_2) + \sigma(z_2, z_3) + \sigma(z_3, z_1)$$

with $(z_1, z_2, z_3) \in \ell_1 \times \ell_2 \times \ell_3$.

Recall that the kernel of a quadratic form is the kernel of the matrix of the associated bilinear form.

Lemma 1.33. *Let* $\operatorname{Ker} Q$ *be the kernel of the quadratic form* Q*. There exists an isomorphism*

$$\operatorname{Ker} Q \cong (\ell_1 \cap \ell_2) \times (\ell_2 \cap \ell_3) \times (\ell_3 \cap \ell_1). \tag{1.27}$$

Proof. Let A be the matrix of Q. The condition $u \in \operatorname{Ker} Q$ is equivalent to

$$v^T A u = 0 \ \text{ for all } v \in \ell_1 \times \ell_2 \times \ell_3. \tag{1.28}$$

In view of the obvious identity

$$(u+v)A(u+v)^T = vAv^T$$

valid for every u in $\operatorname{Ker} Q$, formula (1.28) is equivalent to the condition:

$$(u+v)A(u+v)^T - vAv^T = 0 \text{ for all } v \in \ell_1 \times \ell_2 \times \ell_3 \tag{1.29}$$

that is, to

$$\begin{aligned} &Q(z_1 + z_1', z_2 + z_2', z_3 + z_3') - Q(z_1' + z_2', z_3') \\ &= \omega(z_1, z_2') + \omega(z_2, z_3') + \omega(z_1' + z_2) + \omega(z_2', z_3) + \omega(z_3', z_1) \\ &= \omega(z_1 - z_3, z_2') + \omega(z_2 - z_1, z_3') + \omega(z_3 - z_2, z_1') \\ &= 0. \end{aligned} \tag{1.30}$$

Taking successively $z_1' = z_3' = 0$, $z_1' = z_2' = 0$, and $z_2' = z_3' = 0$ the equality (1.30) then implies

$$\begin{aligned} \omega(z_1 - z_3, z_2') &= 0 \ \text{ for all } \ z_2' \in \ell_2, \\ \omega(z_2 - z_1, z_3') &= 0 \ \text{ for all } \ z_3' \in \ell_3, \\ \omega(z_3 - z_2, z_3') &= 0 \ \text{ for all } \ z_1' \in \ell_1, \end{aligned}$$

hence, since ℓ_1, ℓ_2, ℓ_3 are Lagrangian planes:

$$(z_3 - z_2, z_1 - z_3, z_2 - z_1) \in \ell_1 \times \ell_2 \times \ell_3.$$

It follows that

$$\begin{aligned}(z_1+z_2-z_3&=z_1+(z_2-z_3)=(z_1-z_3)+z_2\in\ell_1\cap\ell_2,\\(z_2+z_3-z_1&=z_2+(z_3-z_1)=(z_2-z_1)+z_3\in\ell_2\cap\ell_3,\\(z_3+z_1-z_2&=z_3+(z_1-z_2)=(z_3-z_2)+z_1\in\ell_3\cap\ell_1.\end{aligned}$$

The restriction to $\operatorname{Ker} Q$ of the automorphism of z_3 defined by $(z_1, z_2, z_3) \longmapsto (z_1', z_2', z_3')$ with

$$z_1'=z_1+z_2-z_3,\quad z_2'=z_2+z_3-z_1,\quad z_3'=z_3+z_1-z_2$$

is thus an isomorphism of $\operatorname{Ker} Q$ onto $(\ell_1\cap\ell_2)\times(\ell_2\cap\ell_3)\times(\ell_3\cap\ell_1)$. □

We are now in position to prove the main topological property of the signature. Let us introduce the following notation: if k, k', k'' are three integers such that $0 \leq k, k', k'' \leq n$, we define a subset $\operatorname{Lag}^3_{k,k',k''}(n)$ of $\operatorname{Lag}^3(n)$ by

$$\operatorname{Lag}^3_{k,k',k''}(E,\omega)=\{(\ell,\ell',\ell''):\dim(\ell\cap\ell')=k,\dim(\ell'\cap\ell'')=k',\dim(\ell''\cap\ell)=k''\}.$$

Proposition 1.34. *The Wall–Kashiwara signature has the following properties:*

(i) *It is locally constant on each set* $\operatorname{Lag}^3_{k,k',k''}(E,\omega)$*;*

(ii) *If the triple (ℓ,ℓ',ℓ'') move continuously in $\operatorname{Lag}^3(E,\omega)$ in such a way that $\dim\ell\cap\ell'$, $\dim\ell'\cap\ell''=k'$, and $\dim\ell''\cap\ell$ do not change, then $\tau(\ell,\ell',\ell'')$ remains constant;*

(iii) *We have*

$$\tau(\ell,\ell',\ell'')=n+\dim\ell\cap\ell'+\dim\ell'\cap\ell''+\dim\ell''\cap\ell \mod 2. \tag{1.31}$$

Proof. Properties (i) and (ii) are equivalent since $\operatorname{Lag}(E,\omega)$ (and hence also $\operatorname{Lag}^3(E,\omega)$) is connected. Property (iii) implies (i), and hence (ii). It is therefore sufficient to prove the congruence (1.31). Let A be the matrix in the proof of Lemma 1.33. In view of the isomorphism statement (1.27) we have

$$\operatorname{rank}(A)=3n-(\dim\ell\cap\ell'+\dim\ell'\cap\ell''+\dim\ell''\cap\ell).$$

Let (τ^+,τ^-) be the signature of A, so that (by definition) $\tau(\ell,\ell',\ell'')=\tau^+-\tau^-$; since $\operatorname{rank}(A)=\tau^++\tau^-$ we thus have

$$\tau(\ell,\ell',\ell'')\equiv\operatorname{rank}(A)\mod 2$$

hence (1.31). □

Remark 1.35. Define a 1-cochain dim on $\operatorname{Lag}(n)$ by $\dim(\ell,\ell')=\dim(\ell\cap\ell')$. In view of the obvious relation

$$\dim\ell\cap\ell'+\dim\ell'\cap\ell''+\dim\ell''\cap\ell\equiv\dim\ell\cap\ell'-\dim\ell'\cap\ell''+\dim\ell''\cap\ell \mod 2,$$

we can rewrite formula (1.31) as

$$\tau(\ell,\ell',\ell'')\equiv\partial\dim(\ell,\ell',\ell'')\ ,\ \mod 2. \tag{1.32}$$

For short: $\tau=\partial\dim$, mod 2.

Chapter 2

The Symplectic Group

In this second chapter we study in some detail the symplectic group of a symplectic space (E, ω), with a special emphasis on the standard symplectic group $\mathrm{Sp}(n)$, corresponding to the case $(E, \omega) = (\mathbb{R}_z^{2n}, \sigma)$.

There exists an immense literature devoted to the symplectic group. A few classical references from my own bookshelf are Libermann and Marle [110], Guillemin and Sternberg [84, 85] and Abraham and Marsden [1]; also see the first chapter in Long [113] where the reader will find an interesting study of various normal forms. The reader who likes explicit calculations with symplectic block matrices will love Kauderer's book [101], which deals with applications of symplectic matrices to various aspects of mathematical physics, including special relativity. Those interested in applications to the rapidly expanding field of quantum optics could consult with profit the very nicely written pamphlet by Arvind *et al.* [5].

2.1 The Standard Symplectic Group

Let us begin by working in the standard symplectic space $(\mathbb{R}_z^{2n}, \sigma)$.

Definition 2.1. The group of all automorphisms s of $(\mathbb{R}_z^{2n}, \sigma)$ such that

$$\sigma(sz, sz') = \sigma(z, z')$$

for all $z, z' \in \mathbb{R}_z^{2n}$ is denoted by $\mathrm{Sp}(n)$ and called the "standard symplectic group" (one also frequently finds the notation $\mathrm{Sp}(2n)$ or $\mathrm{Sp}(2n, \mathbb{R})$ in the literature).

It follows from Proposition 1.9 that $\mathrm{Sp}(n)$ is isomorphic to the symplectic group $\mathrm{Sp}(E, \omega)$ of any $2n$-dimensional symplectic space.

The notion of linear symplectic transformation can be extended to diffeomorphisms:

Definition 2.2. Let (E,ω), (E',ω') be two symplectic vector spaces. A diffeomorphism $f:(E,\omega)\longrightarrow(E',\omega')$ is called a "symplectomorphism[1]" if the differential $d_z f$ is a linear symplectic mapping $E\longrightarrow E'$ for every $z\in E$. [In the physical literature one often says "canonical transformation" in place of "symplectomorphism".]

It follows from the chain rule that the composition $g\circ f$ of two symplectomorphisms $f:(E,\omega)\longrightarrow(E',\omega')$ and $g:(E',\omega')\longrightarrow(E'',\omega'')$ is a symplectomorphism $(E,\omega)\longrightarrow(E'',\omega'')$. When

$$(E,\omega)=(E',\omega')=(\mathbb{R}^{2n}_z,\sigma)$$

a diffeomorphism f of $(\mathbb{R}^{2n}_z,\sigma)$ is a symplectomorphism if and only if its Jacobian matrix (calculated in any symplectic basis) is in $\mathrm{Sp}(n)$. Summarizing:

$$f \textit{ is a symplectomorphism of } (\mathbb{R}^{2n}_z,\sigma)$$
$$\Longleftrightarrow$$
$$Df(z)\in\mathrm{Sp}(n) \textit{ for every } z\in(\mathbb{R}^{2n}_z,\sigma).$$

It follows directly from the chain rule $D(g\circ f)(z)=Dg(f(z)Df(z)$ that the symplectomorphisms of the standard symplectic space $(\mathbb{R}^{2n}_z,\sigma)$ form a group. That group is denoted by $\mathrm{Symp}(n)$.

Definition 2.3. Let (E,ω) be a symplectic space. The group of all linear symplectomorphisms of (E,ω) is denoted by $\mathrm{Sp}(E,\omega)$ and called the "symplectic group of (E,ω)".

The following exercise produces infinitely many examples of linear symplectomorphisms:

The notion of symplectomorphism extends in the obvious way to symplectic manifolds: if (M,ω) and (M',ω') are two such manifolds, then a diffeomorphism $f:M\longrightarrow M'$ is called a symplectomorphism if it preserves the symplectic structures on M and M', that is if $f^*\omega'=\omega$ where $f^*\omega'$ (the "pull-back of ω' by f) is defined by

$$f^*\omega'(z_0)(Z,Z')=\omega'(f(z_0))((d_{z_0}f)Z,(d_{z_0}f)Z')$$

for every $z_0\in M$ and $Z,Z'\in T_{z_0}M$.

If f and g are symplectomorphisms $(M,\omega)\longrightarrow(M',\omega')$ and $(M',\omega')\longrightarrow(M'',\omega'')$, then $g\circ f$ is a symplectomorphism $(M,\omega)\longrightarrow(M'',\omega'')$.

The symplectomorphisms $(M,\omega)\longrightarrow(M,\omega)$ obviously form a group, denoted by $\mathrm{Symp}(M,\omega)$, whose study is very active and far from being completed; see [91, 114, 132]. We will study in some detail its subgroup $\mathrm{Ham}(n)$ later on.

[1]The word was reputedly coined by J.-M. Souriau.

2.1.1 Symplectic matrices

For practical purposes it is often advantageous to work in coordinates and to represent the elements of $\mathrm{Sp}(n)$ by matrices.

Recall that definition (1.1) of the standard symplectic form can be rewritten in matrix form as

$$\sigma(z,z') = (z')^T Jz = \langle Jz, z'\rangle \tag{2.1}$$

where J is the standard symplectic matrix

$$J = \begin{bmatrix} 0 & I \\ -I & 0 \end{bmatrix}. \tag{2.2}$$

Notice that $J^T = -J$ and $J^2 = -I$.

Choose a symplectic basis in $(\mathbb{R}_z^{2n}, \sigma)$; we will identify a linear mapping $s: \mathbb{R}_z^{2n} \longrightarrow \mathbb{R}_z^{2n}$ with its matrix S in that basis. In view of (2.1) we have

$$S \in \mathrm{Sp}(n) \Longleftrightarrow S^T JS = J$$

where S^T is the transpose of S. Since

$$\det S^T JS = \det S^2 \det J = \det J$$

it follows that $\det S$ can, a priori, take any of the two values ± 1. It turns out, however, that

$$S \in \ \mathrm{Sp}(n) \Longrightarrow \det S = 1.$$

There are many ways of showing this; none of them is really totally trivial. Here is an algebraic proof making use of the notion of a Pfaffian (we will give an alternative proof later on). Recall that to every antisymmetric matrix A one associates a polynomial $\mathrm{Pf}(A)$ ("the Pfaffian of A") in the entries of A, what has the following properties:

$$\mathrm{Pf}(S^T AS) = (\det S)\,\mathrm{Pf}(A) \quad , \quad \mathrm{Pf}(J) = 1.$$

Choose now $A = J$ and $S \in \mathrm{Sp}(n)$. Since $S^T JS = J$ we have

$$\mathrm{Pf}(S^T JS) = \det S = 1$$

which was to be proven.

Remark 2.4. The group $\mathrm{Sp}(n)$ is stable under transposition: the condition $S \in \mathrm{Sp}(n)$ is equivalent to $S^T JS = J$; since S^{-1} also is in $\mathrm{Sp}(n)$ we have $(S^{-1})^T JS^{-1} = J$; taking the inverses of both sides of this equality we get $SJ^{-1}S^T = J^{-1}$, that is $SJS^T = J$, so that $S^T \in \mathrm{Sp}(n)$. It follows that we have the equivalences

$$S \in \ \mathrm{Sp}(n) \Longleftrightarrow S^T JS = J \Longleftrightarrow SJS^T = J. \tag{2.3}$$

A symplectic basis of $(\mathbb{R}^{2n}_z, \sigma)$ being chosen, we can always write $S \in \mathrm{Sp}(n)$ in block-matrix form

$$S = \begin{bmatrix} A & B \\ C & D \end{bmatrix} \tag{2.4}$$

where the entries A, B, C, D are $n\times n$ matrices. The conditions (2.3) are then easily seen, by a direct calculation, equivalent to the two following sets of equivalent conditions[2]:

$$A^T C,\ B^T D \ \text{ symmetric, and } A^T D - C^T B = I, \tag{2.5}$$

$$AB^T,\ CD^T \ \text{ symmetric, and } AD^T - BC^T = I. \tag{2.6}$$

It follows from the second of these sets of conditions that the inverse of S is

$$S^{-1} = \begin{bmatrix} D^T & -B^T \\ -C^T & A^T \end{bmatrix}. \tag{2.7}$$

Example 2.5. Here are three useful classes of symplectic matrices: if P and L are, respectively, a symmetric and an invertible $n \times n$ matrix, we set

$$V_P = \begin{bmatrix} I & 0 \\ -P & I \end{bmatrix},\ U_P = \begin{bmatrix} -P & I \\ -I & 0 \end{bmatrix},\ M_L = \begin{bmatrix} L^{-1} & 0 \\ 0 & L^T \end{bmatrix}. \tag{2.8}$$

The matrices V_P are sometimes called "symplectic shears".

It turns out – as we shall prove later on – that both sets

$$\mathcal{G} = \{J\} \cup \{V_P : P \in \mathrm{Sym}(n, \mathbb{R})\} \cup \{M_L : L \in \mathrm{GL}(n, \mathbb{R})\}$$

and

$$\mathcal{G}' = \{J\} \cup \{U_P : P \in \mathrm{Sym}(n, \mathbb{R})\} \cup \{M_L : L \in \mathrm{GL}(n, \mathbb{R})\}$$

generate the symplectic group $\mathrm{Sp}(n)$.

The reader is encouraged to use conditions (2.5)–(2.6) in the two exercises below. In the third exercise he is asked to prove that the matrices $AA^T + BB^T$ and $CC^T + DD^T$ are invertible if S is symplectic.

Exercise 2.6. Show that the $2n \times 2n$ matrix

$$S = \begin{bmatrix} P & I - Q \\ -(I - P) & P \end{bmatrix}$$

is in $\mathrm{Sp}(n)$ if and only if P is an orthogonal projector (*i.e.*, $P^2 = P$ and $P^T = P$).

Exercise 2.7. Let X and Y be two symmetric $n \times n$ matrices, X invertible. Show that

$$S = \begin{bmatrix} X + YX^{-1}Y & YX^{-1} \\ X^{-1}Y & X^{-1} \end{bmatrix}$$

is a symplectic matrix.

[2]These conditions are sometimes called the "Luneburg relations" in theoretical optics.

Exercise 2.8. Show that if the block-matrix (2.4) is symplectic, then

(i) $A+iB$ and $C+iD$ are invertible. [Hint: calculate $(A+iB)(B^T-iA^T)$ and assume that $A+iB$ is not invertible.]

(ii) Deduce from (i) that if (2.4) is symplectic, then AA^T+BB^T and CC^T+DD^T are invertible.

We can also form direct sums of symplectic groups. Consider for instance $(\mathbb{R}^{2n_1},\sigma_1)$ and $(\mathbb{R}^{2n_2},\sigma_2)$, the standard symplectic spaces of dimension $2n_1$ and $2n_2$; let $\mathrm{Sp}(n_1)$ and $\mathrm{Sp}(n_2)$ be the respective symplectic groups. The direct sum $\mathrm{Sp}(n_1)\oplus\mathrm{Sp}(n_2)$ is the group of automorphisms of

$$(\mathbb{R}^{2n}_z,\sigma)=(\mathbb{R}^{2n_1}\oplus\mathbb{R}^{2n_2},\sigma_1\oplus\sigma_2)$$

defined, for $z_1\in\mathbb{R}^{2n_1}$ and $z_2\in\mathbb{R}^{2n_2}$, by

$$(s_1\oplus s_2)(z_1\oplus z_2)=s_1z_1\oplus s_2z_2.$$

It is evidently a subgroup of $\mathrm{Sp}(n)$:

$$\mathrm{Sp}(n_1)\oplus\mathrm{Sp}(n_2)\subset\ \mathrm{Sp}(n)$$

which can be expressed in terms of block-matrices as follows: let

$$S_1=\begin{bmatrix}A_1 & B_1\\ C_1 & D_1\end{bmatrix}\quad\text{and}\quad S_2=\begin{bmatrix}A_2 & B_2\\ C_2 & D_2\end{bmatrix}$$

be elements of $\mathrm{Sp}(n_1)$ and $\mathrm{Sp}(n_2)$, respectively. Then

$$S_1\oplus S_2=\begin{bmatrix}A_1 & 0 & B_1 & 0\\ 0 & A_2 & 0 & B_2\\ C_1 & 0 & D_1 & 0\\ 0 & C_2 & 0 & D_2\end{bmatrix}\in\mathrm{Sp}(n_1+n_2). \tag{2.9}$$

The mapping $(S_1,S_2)\longmapsto S_1\oplus S_2$ thus defined is a group monomorphism

$$\mathrm{Sp}(n_1)\oplus\mathrm{Sp}(n_2)\longrightarrow\ \mathrm{Sp}(n).$$

The elements of $\mathrm{Sp}(n)$ are linear isomorphisms; we will sometimes also consider affine symplectic isomorphisms. Let $S\in\mathrm{Sp}(n)$ and denote by $T(z_0)$ the translation $z\longmapsto z+z_0$ in $\mathbb{R}^{2n}_z$. The composed mappings

$$T(z_0)S=ST(S^{-1}z_0)\ \text{ and }\ ST(z_0)=T(Sz_0)S$$

are both symplectomorphisms, as is easily seen by calculating their Jacobians. These transformations form a group.

Definition 2.9. The semi-direct product $\mathrm{Sp}(n)\ltimes_s T(2n)$ of the symplectic group and the group of translations in $\mathbb{R}^{2n}_z$ is called the affine (or: inhomogeneous) symplectic group, and is denoted by $\mathrm{ISp}(n)$.

For practical calculations it is often useful to identify $\mathrm{ISp}(n)$ with a matrix group:

Exercise 2.10. Show that the group of all matrices

$$[S, z_0] \equiv \begin{bmatrix} S & z_0 \\ 0_{1\times 2n} & 1 \end{bmatrix}$$

is isomorphic to $\mathrm{ISp}(n)$ (here $0_{1\times 2n}$ is the $2n$-column matrix with all entries equal to zero).

Let us now briefly discuss the eigenvalues of a symplectic matrix. It has been known for a long time that the eigenvalues of symplectic matrices play a fundamental role in the study of Hamiltonian periodic orbits; this is because the stability of these orbits depends in a crucial way on the structure of the associated linearized system. It turns out that these eigenvalues also play an essential role in the understanding of symplectic squeezing theorems, which we study later in this book.

Let us first prove the following result:

Proposition 2.11. *Let $S \in \mathrm{Sp}(n)$.*

(i) *If λ is an eigenvalue of S, then so are $\bar{\lambda}$ and $1/\lambda$ (and hence also $1/\bar{\lambda}$);*

(ii) *if the eigenvalue λ of S has multiplicity k, then so has $1/\lambda$.*

(iii) *S and S^{-1} have the same eigenvalues.*

Proof. (i) We are going to show that the characteristic polynomial $P_S(\lambda) = \det(S - \lambda I)$ of S satisfies the reflexivity relation

$$P_S(\lambda) = \lambda^{2n} P_S(1/\lambda); \tag{2.10}$$

Property (i) will follow, since for real matrices, eigenvalues appear in conjugate pairs. Since $S^T JS = J$ we have $S = -J(S^T)^{-1}J$ and hence

$$\begin{aligned} P_S(\lambda) &= \det(-J(S^T)^{-1}J - \lambda I) \\ &= \det(-(S^T)^{-1}J + \lambda I) \\ &= \det(-J + \lambda S) \\ &= \lambda^{2n} \det(S - \lambda^{-1} I) \end{aligned}$$

which is precisely (2.10).

(ii) Let $P_S^{(j)}$ be the jth derivative of the polynomial P_S. If λ_0 has multiplicity k; then $P_S^{(j)}(\lambda_0) = 0$ for $0 \leq j \leq k-1$ and $P_S^{(k)}(\lambda) \neq 0$. In view of (2.10) we also have $P_S^{(j)}(1/\lambda) = 0$ for $0 \leq j \leq k-1$ and $P_S^{(k)}(1/\lambda) \neq 0$.

Property (iii) immediately follows from (ii). □

Notice that an immediate consequence of this result is that if ± 1 is an eigenvalue of $S \in \mathrm{Sp}(n)$, then its multiplicity is necessarily even.

We will see in the next subsection (Proposition 2.13) that any positive-definite symmetric symplectic matrix can be diagonalized using an orthogonal transformation which is at the same time symplectic.

2.1.2 The unitary group $\mathrm{U}(n)$

The complex structure associated to the standard symplectic matrix J is very simple: it is defined by

$$(\alpha + i\beta)z = \alpha + \beta Jz$$

and corresponds to the trivial identification $z = (x,p) \equiv x + ip$. The unitary group $\mathrm{U}(n,\mathbb{C})$ acts in a natural way on $(\mathbb{R}^{2n}_z, \sigma)$ (*cf.* Exercises 1.5 and 1.6) and that action preserves the symplectic structure. Let us make this statement somewhat more explicit:

Proposition 2.12. *The monomorphism $\mu : M(n,\mathbb{C}) \longrightarrow M(2n,\mathbb{R})$ defined by $u = A + iB \longmapsto \mu(u)$ with*

$$\mu(u) = \begin{bmatrix} A & -B \\ B & A \end{bmatrix}$$

(A and B real) identifies the unitary group $\mathrm{U}(n,\mathbb{C})$ with the subgroup

$$\mathrm{U}(n) = \mathrm{Sp}(n) \cap O(2n,\mathbb{R}) \tag{2.11}$$

of $\mathrm{Sp}(n)$.

Proof. In view of (2.7) the inverse of $U = \mu(u)$, $u \in \mathrm{U}(n,\mathbb{C})$, is

$$U^{-1} = \begin{bmatrix} A^T & B^T \\ -B^T & A^T \end{bmatrix} = U^T,$$

hence $U \in O(2n,\mathbb{R})$ which proves the inclusion $\mathrm{U}(n) \subset \mathrm{Sp}(n) \cap O(2n,\mathbb{R})$. Suppose conversely that $U \in \mathrm{Sp}(n) \cap O(2n,\mathbb{R})$. Then

$$JU = (U^T)^{-1}J = UJ$$

which implies that $U \in \mathrm{U}(n)$ so that $\mathrm{Sp}(n) \cap O(2n,\mathbb{R}) \subset \mathrm{U}(n)$. □

We will loosely talk about $\mathrm{U}(n)$ as of the "unitary group" when there is no risk of confusion; notice that it immediately follows from conditions (2.5), (2.6) that we have the equivalences:

$$A + iB \in \mathrm{U}(n) \tag{2.12}$$

$$\Longleftrightarrow$$

$$A^T B \text{ symmetric and } A^T A + B^T B = I \tag{2.13}$$

$$\Longleftrightarrow$$

$$AB^T \text{ symmetric and } AA^T + BB^T = I; \tag{2.14}$$

of course these conditions are just the same thing as the conditions

$$(A+iB)^*(A+iB) = (A+iB)(A+iB)^* = I$$

for the matrix $A+iB$ to be unitary.

In particular, taking $B=0$ we see the matrices

$$R = \begin{bmatrix} A & 0 \\ 0 & A \end{bmatrix} \quad \text{with} \quad AA^T = A^TA = I \tag{2.15}$$

also are symplectic, and form a subgroup $\mathrm{O}(n)$ of $\mathrm{U}(n)$ which we identify with the rotation group $\mathrm{O}(n,\mathbb{R})$. We thus have the chain of inclusions

$$\mathrm{O}(n) \subset \mathrm{U}(n) \subset \mathrm{Sp}(n).$$

Let us end this subsection by mentioning that it is sometimes useful to identify elements of $\mathrm{Sp}(n)$ with complex symplectic matrices. The group $\mathrm{Sp}(n,\mathbb{C})$ is defined, in analogy with $\mathrm{Sp}(n)$, by the condition

$$\mathrm{Sp}(n,\mathbb{C}) = \{M \in M(2n,\mathbb{C}) : M^TJM = J\}.$$

Let now K be the complex matrix

$$K = \frac{1}{\sqrt{2}} \begin{bmatrix} I & iI \\ iI & I \end{bmatrix} \in \mathrm{U}(2n,\mathbb{C})$$

and consider the mapping

$$\mathrm{Sp}(n) \longrightarrow \mathrm{Sp}(n,\mathbb{C}) \quad , \quad S \longmapsto S_c = K^{-1}SK.$$

One verifies by a straightforward calculation left to the reader as an exercise that $S_c \in \mathrm{Sp}(n,\mathbb{C})$. Notice that if $U \in \mathrm{U}(n)$, then

$$U_c = \begin{bmatrix} U & 0 \\ 0 & U^* \end{bmatrix}.$$

We know from elementary linear algebra that one can diagonalize a symmetric matrix using orthogonal transformations. From the properties of the eigenvalues of a symplectic matrix follows that, when this matrix is in addition symplectic and positive definite, this diagonalization can be achieved using a symplectic rotation:

Proposition 2.13. *Let S be a positive definite and symmetric symplectic matrix. Let $\lambda_1 \leq \cdots \leq \lambda_n \leq 1$ be the n smallest eigenvalues of S and set*

$$\Lambda = \mathrm{diag}[\lambda_1,\ldots,\lambda_n; 1/\lambda_1,\ldots,1/\lambda_n]. \tag{2.16}$$

There exists $U \in \mathrm{U}(n)$ such that $S = U^T\Lambda U$.

Proof. Since $S > 0$ its eigenvalues occur in pairs $(\lambda, 1/\lambda)$ of positive numbers (Proposition 2.11); if $\lambda_1 \leq \cdots \leq \lambda_n$ are n eigenvalues then $1/\lambda_1, \ldots, 1/\lambda_n$ are the other n eigenvalues. Let now U be an orthogonal matrix such that $S = U^T \Lambda U$ with, Λ being given by (2.16). We claim that $U \in \mathrm{U}(n)$. It suffices to show that we can write U in the form

$$U = \begin{bmatrix} A & -B \\ B & A \end{bmatrix}$$

with

$$AB^T = B^T A \quad , \quad AA^T + BB^T = I. \tag{2.17}$$

Let $e_1, \ldots, e_n$ be n orthonormal eigenvectors of U corresponding to the eigenvalues $\lambda_1, \ldots, \lambda_n$. Since $SJ = JS^{-1}$ (because S is both symplectic and symmetric) we have, for $1 \leq k \leq n$,

$$SJe_k = JS^{-1}e_k = \frac{1}{\lambda_j} Je_k,$$

hence $\pm Je_1, \ldots, \pm Je_n$ are the orthonormal eigenvectors of U corresponding to the remaining n eigenvalues $1/\lambda_1, \ldots, 1/\lambda_n$. Write now the $2n \times n$ matrix $(e_1, \ldots, e_n)$ as

$$[e_1, \ldots, e_n] = \begin{bmatrix} A \\ B \end{bmatrix}$$

where A and B are $n \times n$ matrices; we have

$$[-Je_1, \ldots, -Je_n] = -J \begin{bmatrix} A \\ B \end{bmatrix} = \begin{bmatrix} -B \\ A \end{bmatrix},$$

hence U is indeed of the type

$$U = [e_1, \ldots, e_n; -Je_1, \ldots, -Je_n] = \begin{bmatrix} A & -B \\ B & A \end{bmatrix}.$$

The symplectic conditions (2.17) are automatically satisfied since $U^T U = I$. □

An immediate consequence of Proposition 2.13 is that the square root of a positive-definite symmetric symplectic matrix is also symplectic. More generally:

Corollary 2.14.

(i) *For every $\alpha \in \mathbb{R}$ there exists a unique $R \in \mathrm{Sp}(n)$, $R > 0$, $R = R^T$, such that $S = R^\alpha$.*

(ii) *Conversely, if $R \in \mathrm{Sp}(n)$ is positive definite, then $R^\alpha \in \mathrm{Sp}(n)$ for every $\alpha \in \mathbb{R}$.*

Proof. (i) Set $R = U^T \Lambda^{1/\alpha} U$; then $R^\alpha = U^T \Lambda U = S$.

(ii) It suffices to note that we have

$$R^\alpha = (U^T \Lambda U)^\alpha = U^T \Lambda^\alpha U \in \mathrm{Sp}(n).$$

□

2.1.3 The symplectic algebra

$\mathrm{Sp}(n)$ is a Lie group; we will call its Lie algebra the "symplectic algebra", and denote it by $\mathfrak{sp}(n)$. There is a one-to-one correspondence between the elements of $\mathfrak{sp}(n)$ and the one-parameter groups in $\mathrm{Sp}(n)$. This correspondence is the starting point of linear Hamiltonian mechanics.

Let

$$\Phi : \mathrm{GL}(2n, \mathbb{R}) \longrightarrow \mathbb{R}^{4n^2}$$

be the continuous mapping defined by $\Phi(M) = M^T JM - J$. Since $S \in \mathrm{Sp}(n)$ if and only if $S^T JS = J$ we have $\mathrm{Sp}(n) = \Phi^{-1}(0)$ and $\mathrm{Sp}(n)$ is thus a closed subgroup of $\mathrm{GL}(2n, \mathbb{R})$, hence a "classical Lie group". The set of all real matrices X such that the exponential $\exp(tX)$ is in $\mathrm{Sp}(n)$ is the Lie algebra of $\mathrm{Sp}(n)$; we will call it the "symplectic algebra" and denote it by $\mathfrak{sp}(n)$:

$$X \in \mathfrak{sp}(n) \Longleftrightarrow S_t = \exp(tX) \in \mathrm{Sp}(n) \text{ for all } t \in \mathbb{R}. \tag{2.18}$$

The one-parameter family (S_t) thus defined is a group: $S_t S_{t'} = S_{t+t'}$ and $S_t^{-1} = S_{-t}$.

The following result gives an explicit description of the elements of the symplectic algebra:

Proposition 2.15. *Let X be a real $2n \times 2n$ matrix.*

(i) *We have*

$$X \in \mathfrak{sp}(n) \Longleftrightarrow XJ + JX^T = 0 \Longleftrightarrow X^T J + JX = 0. \tag{2.19}$$

(ii) *Equivalently, $\mathfrak{sp}(n)$ consists of all block-matrices X such that*

$$X = \begin{bmatrix} U & V \\ W & -U^T \end{bmatrix} \text{ with } V = V^T \text{ and } W = W^T. \tag{2.20}$$

Proof. Let (S_t) be a differentiable one-parameter subgroup of $\mathrm{Sp}(n)$ and a $2n \times 2n$ real matrix X such that $S_t = \exp(tX)$. Since S_t is symplectic we have $S_t J (S_t)^T = J$, that is

$$\exp(tX) J \exp(tX^T) = J.$$

Differentiating both sides of this equality with respect to t and then setting $t = 0$ we get $XJ + JX^T = 0$, and applying the same argument to the transpose S_t^T we get $X^T J + JX = 0$ as well. Suppose conversely that X is such that $XJ + JX^T = 0$ and let us show that $X \in \mathfrak{sp}(n)$. For this it suffices to prove that $S_t = \exp(tX)$ is in $\mathrm{Sp}(n)$ for every t. The condition $X^T J + JX = 0$ is equivalent to $X^T = JXJ$, hence $S_t^T = \exp(tJXJ)$; since $J^2 = -I$ we have $(JXJ)^k = (-1)^{k+1} JX^k J$ and hence

$$\exp(tJXJ) = -\sum_{k=0}^{\infty} \frac{(-t)^k}{k!} (JXJ)^k = -Je^{-tX} J.$$

It follows that

$$S_t^T J S_t = (-Je^{-tX}J)Je^{tX} = J$$

so that $S_t \in \mathrm{Sp}(n)$ as claimed. □

Remark 2.16. The symmetric matrices of order n forming an $n(n+1)/2$-dimensional vector space (2.20) implies, by dimension count, that $\mathfrak{sp}(n)$ has dimension $n(2n+1)$. Since $\mathrm{Sp}(n)$ is connected we consequently have

$$\dim \mathrm{Sp}(n) = \dim \mathfrak{sp}(n) = n(2n+1). \tag{2.21}$$

The following exercise proposes to determine a set of generators of the Lie algebra $\mathfrak{sp}(n)$:

Exercise 2.17.

(i) Let $\Delta_{jk} = (\delta_{jk})_{1\leq j,k\leq n}$ ($\delta_{jk} = 0$ if $j \neq k$, $\delta_{jk} = 1$). Show that the matrices

$$X_{jk} = \begin{bmatrix} \Delta_{jk} & 0 \\ 0 & -\Delta_{jk} \end{bmatrix}, \; Y_{jk} = \frac{1}{2}\begin{bmatrix} 0 & \Delta_{jk} + \Delta_{kj} \\ 0 & 0 \end{bmatrix},$$

$$Z_{jk} = \frac{1}{2}\begin{bmatrix} 0 & 0 \\ \Delta_{jk} + \Delta_{kj} & 0 \end{bmatrix} \quad (1 \leq j \leq k \leq n)$$

form a basis of $\mathfrak{sp}(n)$.

(ii) Show, using (i) that every $Z \in \mathfrak{sp}(n)$ can be written in the form $[X, Y] = XY - YX$ with $X, Y \in \mathfrak{sp}(n)$.

One should be careful to note that the exponential mapping

$$\exp : \mathfrak{sp}(n) \longrightarrow \mathrm{Sp}(n)$$

is neither surjective nor injective. This is easily seen in the case $n = 1$. We claim that

$$S = \exp X \quad \text{with } X \in \mathfrak{sp}(1) \Longrightarrow \operatorname{Tr} S \geq -2. \tag{2.22}$$

(We are following Frankel's argument in [43].) In view of (2.20) we have $X \in \mathfrak{sp}(1)$ if and only $\operatorname{Tr} X = 0$, so that Hamilton–Cayley's equation for X is just $X^2 + \lambda I = 0$ where $\lambda = \det X$. Expanding $\exp X$ in power series it is easy to see that

$$\exp X = \cos\sqrt{\lambda} I + \frac{1}{\sqrt{\lambda}} \sin\sqrt{\lambda} X \quad \text{if } \lambda > 0,$$

$$\exp X = \cosh\sqrt{-\lambda} I + \frac{1}{\sqrt{-\lambda}} \sinh\sqrt{-\lambda} X \quad \text{if } \lambda < 0.$$

Since $\operatorname{Tr} X = 0$ we see that in the case $\lambda > 0$ we have

$$\operatorname{Tr}(\exp X) = 2\cos\sqrt{\lambda} \geq -2$$

and in the case $\lambda < 0$,

$$\operatorname{Tr}(\exp X) = 2\cosh\sqrt{\lambda} \geq 1.$$

However:

Proposition 2.18. *A symplectic matrix S is symmetric positive definite if and only if $S = \exp X$ with $X \in \mathfrak{sp}(n)$ and $X = X^T$. The mapping* exp *is a diffeomorphism*

$$\mathfrak{sp}(n) \cap \mathrm{Sym}(2n, \mathbb{R}) \longrightarrow \mathrm{Sp}(n) \cap \mathrm{Sym}_+(2n, \mathbb{R})$$

($\mathrm{Sym}_+(2n, \mathbb{R})$ is the set of positive definite symmetric matrices).

Proof. If $X \in \mathfrak{sp}(n)$ and $X = X^T$, then S is both symplectic and symmetric positive definite. Assume conversely that S is symplectic and symmetric positive definite. The exponential mapping is a diffeomorphism $\exp : \mathrm{Sym}(2n, \mathbb{R}) \longrightarrow \mathrm{Sym}_+(2n, \mathbb{R})$ (the positive definite symmetric matrices) hence there exists a unique $X \in \mathrm{Sym}(2n, \mathbb{R})$ such that $S = \exp X$. Let us show that $X \in \mathfrak{sp}(n)$. Since $S = S^T$ we have $SJS = J$ and hence $S = -JS^{-1}J$. Because $-J = J^{-1}$ it follows that

$$\exp X = J^{-1}(\exp(-X))J = \exp(-J^{-1}XJ)$$

and $J^{-1}XJ$ being symmetric, we conclude that $X = J^{-1}XJ$, that is $JX = -XJ$, showing that $X \in \mathfrak{sp}(n)$. □

We will refine the result above in Subsection 2.2.1, Proposition 2.22, by using the Cartan decomposition theorem. This will in particular allow us to obtain a precise formula for calculating X in terms of the logarithm of $S = \exp X$.

2.2 Factorization Results in Sp(n)

Factorization (or "decomposition") theorems for matrices are very useful since they often allow us to reduce lengthy or complicated calculations to simpler typical cases. In this section we study three particular factorization procedures for symplectic matrices.

2.2.1 Polar and Cartan decomposition in Sp(n)

Any matrix $M \in \mathrm{GL}(m, \mathbb{R})$ can be written uniquely as $M = RP$ (or PR) where R is orthogonal and P positive definite: this is the classical polar decomposition theorem from elementary linear algebra. Let us specialize this result to the symplectic case; we begin with a rather weak result:

Proposition 2.19. *For every $S \in \mathrm{Sp}(n)$ there exists a unique $U \in \mathrm{U}(n)$ and a unique $R \in \mathrm{Sp}(n)$, R symmetric positive definite, such that $S = RU$ (resp. $S = UR$).*

Proof. Set $R = S^TS$ and define U by $S = (S^TS)^{-1/2}U$; since $(S^TS)^{-1/2} \in \mathrm{Sp}(n)$ in view of Corollary 2.14, we have $U \in \mathrm{Sp}(n)$. On the other hand

$$UU^T = (S^TS)^{-1/2}SS^T(S^TS)^{-1/2} = I$$

so that we actually have

$$U \in \mathrm{Sp}(n) \cap O(2n) = \mathrm{U}(n).$$

That we can alternatively write $S = UR$ (with different choices of U and R than above) follows by applying the result above to S^T. □

We are going to make Proposition 2.19 precise. For this we need a suitable notion of logarithm for invertible matrices. Recall (Proposition A.2 in Appendix A) that if $\mathbb{K} = \mathbb{R}$ or $\mathbb{C}$ and M is an invertible $m \times m$ matrix with entries in $\mathbb{K}$, then there exists an $m \times m$ matrix L such that $M = e^L$.

Let us define

$$\operatorname{Log} M = \int_{-\infty}^{0} \left[(\lambda I - M)^{-1} - (\lambda - 1)^{-1} I\right] d\lambda; \tag{2.23}$$

it is straightforward to check that when $m = 1$ and M is a scalar $\lambda > 0$ formula (2.23) reduces to the usual logarithm $\operatorname{Log} \lambda$.

Exercise 2.20. Show that more generally for any $\mu > 0$ we have

$$\operatorname{Log}(\lambda I) = (\operatorname{Log} \lambda) I. \tag{2.24}$$

It turns out that formula (2.23) defines a *bona fide* logarithm for matrices having no eigenvalues on the negative half-axis:

Proposition 2.21. *Assume that M has no eigenvalues $\lambda \leq 0$. Then*

(i) $\operatorname{Log} M$ *defined by* (2.23) *exists;*

(ii) *We have*

$$e^{\operatorname{Log} M} = M \quad , \quad (\operatorname{Log} M)^T = \operatorname{Log} M^T$$

and also

$$\operatorname{Log} M^{-1} = -\operatorname{Log} M \quad , \quad \operatorname{Log}(AMA^{-1}) = A(\operatorname{Log} M)A^{-1} \tag{2.25}$$

for every invertible matrix A.

Proof. It is no restriction to assume that $M = \lambda I + N$ with $\lambda > 0$ (*cf.* the proof of Proposition A.2 in Appendix A). Set

$$f(M) = \int_{-\infty}^{0} \left[(\lambda - M)^{-1} - (\lambda - 1)^{-1} I\right] d\lambda.$$

We have

$$(\lambda I - M)^{-1} = ((\lambda - \mu)I - N)^{-1} = \sum_{j=0}^{k_0} (\lambda - \mu)^{-k+1} N^j$$

and hence

$$f(M) = \int_{-\infty}^{0} \left(\frac{1}{\lambda - \mu} - \frac{1}{\lambda - 1} \right) I d\lambda + \sum_{k=1}^{k_0} \left(\int_{-\infty}^{0} (\lambda - \mu)^{-k-1} d\lambda \right) N^k$$

that is, calculating explicitly the integrals,

$$\begin{aligned} f(M) &= (\operatorname{Log} \mu) I + \sum_{k=1}^{k_0} \frac{(-1)^{k+1}}{k} (\mu^{-1} N)^k \\ &= (\operatorname{Log} \mu) I + \sum_{k=1}^{k_0} \frac{(-1)^{k+1}}{k} (\mu^{-1} M - I)^k . \end{aligned}$$

Direct substitution of the sum in the right-hand side in the power series for the exponential yields the matrix $\mu^{-1}M$; hence $\exp f(M) = M$ which we set out to prove. Formulae (2.25) readily follow from definition (2.23) of the logarithm, and so does the equality $(\operatorname{Log} M)^T = \operatorname{Log} M^T$. □

The following consequence of Proposition 2.21, which refines Proposition 2.18, will be instrumental in the proof of the symplectic version of Cartan's decomposition theorem:

Proposition 2.22. *If $S \in \mathrm{Sp}(n)$ is positive definite, then $X = \operatorname{Log} S$ belongs to the symplectic Lie algebra $\mathfrak{sp}(n)$. That is, for every $S \in \mathrm{Sp}(n) \cap \mathrm{Sym}_+(2n, \mathbb{R})$ (the set of symmetric positive definite symplectic matrices) we have*

$$S = e^{\operatorname{Log} S} \quad , \quad \operatorname{Log} S \in \mathfrak{sp}(n).$$

Proof. Since S is symplectic we have $S^{-1} = JS^TJ^{-1}$; taking the logarithm of both sides of this equality, and using Proposition 2.21 together with the equality $J^{-1} = -J$ we get

$$X = -J(\operatorname{Log} S^T)J^{-1} = J(\operatorname{Log} S^T)J.$$

We claim that $XJ + JX^T = 0$; the result will follow. We have

$$XJ = -J \operatorname{Log} S^T = (J^{-1}(\operatorname{Log} S^T)J)J = -(\operatorname{Log} S^{-1})J$$

hence, using the fact that $\operatorname{Log} S^T = (\operatorname{Log} S)^T$,

$$XJ = (\operatorname{Log} S)J = -JX^T$$

proving our claim. □

Let us refine the results above by using Cartan' decomposition theorem from the theory of Lie groups (see Appendix A):

Proposition 2.23. *Every $S \in \mathrm{Sp}(n)$ can be written $S = Ue^X$ where $U \in \mathrm{U}(n)$ and $X = \frac{1}{2} \operatorname{Log}(S^T S)$ is in $\mathfrak{sp}(n) \cap \mathrm{Sym}(2n, \mathbb{R})$.*

Proof. The symplectic matrix $S^T S$ has no negative eigenvalues, hence its logarithm $\mathrm{Log}(S^T S)$ exists and is in $\mathfrak{sp}(n)$ in view of Proposition 2.22; it is moreover obviously symmetric. It follows that $X \in \mathfrak{sp}(n)$ and hence e^X and R are both in $\mathrm{Sp}(n)$. Since we also have $U \in O(2n)$ in view of Cartan's theorem, the proposition follows since we have $\mathrm{Sp}(n) \cap O(2n) = \mathrm{U}(n)$. □

A first consequence of the results above is that the symplectic group $\mathrm{Sp}(n)$ is contractible to its subgroup $\mathrm{U}(n)$ (which, by the way, gives a new proof of the fact that $\mathrm{Sp}(n)$ is connected):

Corollary 2.24.
(i) *The standard symplectic group* $\mathrm{Sp}(n)$ *can be retracted to the unitary group* $\mathrm{U}(n)$.
(ii) *The set* $\mathrm{Sp}(n) \cap \mathrm{Sym}_+(2n, \mathbb{R})$ *is contractible to a point.*

Proof. (i) Let $t \longmapsto S(t)$, $0 \leq t \leq 1$, be a loop in $\mathrm{Sp}(n)$; in view of Proposition 2.23 we can write $S(t) = U(t)e^{X(t)}$ where $U(t) \in \mathrm{U}(n)$ and $X(t) = \frac{1}{2}\mathrm{Log}(S^T(t)S(t))$. Since $t \longmapsto S(t)$ is continuous, so is $t \longmapsto X(t)$ and hence also $t \longmapsto U(t)$. Consider now the continuous mapping $h : [0,1] \times [0,1] \longrightarrow \mathrm{Sp}(n)$ defined by

$$h(t,t') = U(t)e^{(1-t')X(t)} \ , \ \ 0 \leq t \leq t' \leq 1.$$

This mapping is a homotopy between the loops $t \longmapsto h(t,0) = S(t)$ and $t \longmapsto h(t,1) = R(t)$; obviously $h(t,t') \in \mathrm{Sp}(n)$ hence (i).

Part (ii) follows, taking $R(t) = 1$ in the argument above. □

This result can actually be proven without invoking the consequences of Cartan's theorem:

Exercise 2.25. Prove that $\mathrm{U}(n)$ is a deformation retract of $\mathrm{Sp}(n)$ using symplectic diagonalization (Proposition 2.13).

It follows from Corollary 2.24 that the fundamental group $\pi_1[\mathrm{Sp}(n)]$ is isomorphic to $\pi_1[\mathrm{U}(n,\mathbb{C})]$, that is to the integer group $(\mathbb{Z},+)$. Let us make a precise construction of the isomorphism $\pi_1[\mathrm{Sp}(n)] \cong \pi_1[\mathrm{U}(n,\mathbb{C})]$.

Proposition 2.26. *The mapping* $\Delta : \mathrm{Sp}(n) \longrightarrow S^1$ *defined by* $\Delta(S) = \det u$ *where* u *is the image in* $\mathrm{U}(n,\mathbb{C})$ *of* $U = S(S^T S)^{-1/2} \in \mathrm{U}(n)$ *induces an isomorphism*

$$\Delta_* : \pi_1[\mathrm{Sp}(n)] \cong \pi_1[\mathrm{U}(n,\mathbb{C})]$$

and hence an isomorphism $\pi_1[\mathrm{Sp}(n)] \cong \pi_1[S^1] \equiv (\mathbb{Z},+)$.

Proof. In view of Corollary 2.24 above and its proof, any loop $t \longmapsto S(t) = R(t)e^{X(t)}$ in $\mathrm{Sp}(n)$ is homotopic to the loop $t \longmapsto R(t)$ in $\mathrm{U}(n)$. Now $S^T(t)S(t) = e^{2X(t)}$ (because $X(t)$ is in $\mathfrak{sp}(n) \cap \mathrm{Sym}(2n,\mathbb{R})$) and hence

$$R(t) = S(t)(S^T(t)S(t))^{-1/2}.$$

The result follows, composing Δ_* with the isomorphism $\pi_1[\mathrm{U}(n,\mathbb{C})] \cong \pi_1[S^1]$ induced by the determinant map (see Lemma 3.6 in Chapter 3, Subsection 3.1.2). □

Let us next study two useful factorizations of symplectic matrices that will be used several times in the rest of this book: the so-called "pre-Iwasawa factorization", reminiscent of the Iwasawa decomposition in Lie group theory, and factorization by free symplectic matrices. The latter will play an important role in the theory of the metaplectic group in Chapter 7.

2.2.2 The "pre-Iwasawa" factorization

We denote by $St(\ell)$ the stabilizer (or: isotropy subgroup) of $\ell \in \mathrm{Lag}(n)$ in $\mathrm{Sp}(n)$: it is the subgroup of $\mathrm{Sp}(n)$ consisting of all symplectic matrices S such that $S\ell = \ell$.

Exercise 2.27. Show that if $\ell, \ell' \in \mathrm{Lag}(n)$, then the stabilizers $St(\ell)$ and $St(\ell')$ are conjugate subgroups of $\mathrm{Sp}(n)$.

Exercise 2.28. Show that the stabilizer of $\ell_P = 0 \times \mathbb{R}^n$ in $\mathrm{Sp}(n)$ consists of all matrices $S = V_P M_L$ where V_P and M_L are defined by (2.8). educe from this that $St(\ell)$ has two connected components.

Let us now prove:

Proposition 2.29. *Every $S \in \mathrm{Sp}(n)$ can be written (uniquely) as a product $S = RU$ (resp. $S = UR$) where $R \in St(\ell)$ and $U \in \mathrm{U}(n)$.*

Proof. We begin by noting that $S \in St(\ell)$ if and only if $S^T \in St(J\ell)$. Assume in fact that $S\ell = \ell$; since $S^T JS = J$ we have $(S^T)^{-1} J\ell = JS\ell = J\ell$, hence $(S^T)^{-1} \in St(J\ell)$; since $St(J\ell)$ is a group we also have $S^T \in St(J\ell)$. This shows that if $S \in St(\ell)$ then $S^T \in St(J\ell)$. In the same way $S^T \in St(J\ell)$ implies $S \in St(\ell)$, hence the claim. Let us now prove the statement of the proposition. Since $\mathrm{U}(n)$ acts transitively on $\mathrm{Lag}(n)$ there exists $U \in \mathrm{U}(n)$ such that $S^T(J\ell) = U^T(J\ell)$ and hence $S^T = U^T S_1$ for some $S_1 \in St(J\ell)$. By transposition we have $S = RU$ where $R = S_1^T$ and hence $R \in St(\ell)$. □

Write now $S \in \mathrm{Sp}(n)$ in the usual block-form:

$$S = \begin{bmatrix} A & B \\ C & D \end{bmatrix}. \tag{2.26}$$

Taking into account Exercise 2.28 above, it follows from Proposition 2.29 that S can always be factored as

$$S = \begin{bmatrix} I & 0 \\ P & I \end{bmatrix} \begin{bmatrix} L^T & 0 \\ 0 & L^{-1} \end{bmatrix} \begin{bmatrix} X & Y \\ -Y & X \end{bmatrix}$$

where $P = P^T$ and $X + iY$ is unitary. The following result gives explicit formulae for the calculation of P, L, X and Y; it shows that L can actually be chosen *symmetric*:

Corollary 2.30. *Let S be the symplectic matrix* (2.26).

(i) *S can be written, in a unique way, as the product*

$$S = \begin{bmatrix} I & 0 \\ P & I \end{bmatrix} \begin{bmatrix} L & 0 \\ 0 & L^{-1} \end{bmatrix} \begin{bmatrix} X & Y \\ -Y & X \end{bmatrix} \tag{2.27}$$

where $P = P^T$, $L = L^T$, X and Y are given by the formulae

$$P = (CA^T + DB^T)(AA^T + BB^T)^{-1}, \tag{2.28}$$

$$L = (AA^T + BB^T)^{1/2}, \tag{2.29}$$

$$X + iY = (AA^T + BB^T)^{-1/2}(A + iB). \tag{2.30}$$

(ii) *Equivalently:*

$$S = \begin{bmatrix} L & 0 \\ Q & L^{-1} \end{bmatrix} \begin{bmatrix} X & Y \\ -Y & X \end{bmatrix} \tag{2.31}$$

with L as in (2.29) and $Q = PL$ that is:

$$Q = (CA^T + DB^T)(AA^T + BB^T)^{-1/2}. \tag{2.32}$$

Proof. Part (ii) of the corollary immediately follows from the formulae (2.27)–(2.27). Let us prove (i). Expanding the matrix product in the right-hand side of (2.27) we see that we must have $A = LX$ and $B = LY$. These conditions, together with the fact that $X + iY$ is unitary, imply that

$$AA^T + BB^T = L(XX^T + YY^T)L^T = LL^T,$$

hence $\det(AA^T + BB^T) \neq 0$ (*cf.* Exercise 2.8). Let us *choose* L and $X + iY$ as in formulas (2.29), (2.30). The matrix L is then evidently symmetric and we have $X + iY \in \mathrm{U}(n, \mathbb{C})$; to prove the corollary it thus suffices to show that

$$\begin{bmatrix} A & B \\ C & D \end{bmatrix} \begin{bmatrix} X^T & -Y^T \\ Y^T & X^T \end{bmatrix} \begin{bmatrix} L^{-1} & 0 \\ 0 & L \end{bmatrix} = \begin{bmatrix} I & 0 \\ P & I \end{bmatrix} \tag{2.33}$$

where P is given by (2.28); we notice that the matrix P is then automatically symmetric since the condition

$$\begin{bmatrix} I & 0 \\ P & I \end{bmatrix} \in \mathrm{Sp}(n)$$

is equivalent to $P = P^T$ (see the conditions (2.6) characterizing symplectic matrices). Expanding the product on the left-hand side of (2.33) this amounts to verifying the group of equalities

$$\begin{aligned} CX^TL^{-1} + DY^TL^{-1} &= (CA^T + DB^T)(AA^T + BB^T)^{-1}, \\ AX^TL^{-1} + BY^TL^{-1} &= -CY^TL + DX^TL = I, \\ -AY^TL + BX^TL &= 0. \end{aligned}$$

Now, taking formulae (2.29) and (2.30) into account,

$$CX^TL^{-1} + DY^TL^{-1} = CA^T(AA^T + BB^T)^{-1} + DB^T(AA^T + BB^T)^{-1},$$

that is

$$CX^TL^{-1} + DY^TL^{-1} = (CA^T + DB^T)(AA^T + BB^T)^{-1}$$

which verifies the first equality. Similarly,

$$AX^TL^{-1} + BY^TL^{-1} = AA^T(AA^T + BB^T)^{-1} + BB^T(AA^T + BB^T)^{-1},$$

that is

$$AX^TL^{-1} + BY^TL^{-1} = I.$$

We also have

$$-CY^TL + DX^TL = -CB^T + DA^T = I$$

(the second equality because S is symplectic in view of condition (2.6)); finally

$$-AY^TL + BX^TL = -AB^T + BA^T = 0$$

using once again condition (2.6). □

Remark 2.31. When the symplectic matrix S in addition is symmetric, it is of the type

$$S = \begin{bmatrix} A & B \\ B^T & D \end{bmatrix}, \quad A = A^T \text{ and } D = D^T$$

and the formulas (2.28), (2.29), (2.30) take the very simple form

$$P = (AB + BD)(A^2 + B^2)^{-1}, \tag{2.34}$$

$$L = (A^2 + B^2)^{1/2}, \tag{2.35}$$

$$X + iY = (A^2 + B^2)^{-1/2}(A + iB). \tag{2.36}$$

Exercise 2.32. Verify formulae (2.28)–(2.30) in the case $n = 1$, *i.e.*, when $S = \begin{bmatrix} a & b \\ c & d \end{bmatrix}$ with $ad - bc = 1$.

2.2.3 Free symplectic matrices

The notion of free symplectic matrix plays a very important role in many practical issues. For instance, it is the key to our definition of the metaplectic group. A noticeable fact is, in addition, that every symplectic matrix can be written as the product of exactly two free symplectic matrices.

Definition 2.33. Let ℓ be an arbitrary Lagrangian plane in $(\mathbb{R}^{2n}_z, \sigma)$ and $S \in \mathrm{Sp}(n)$. We say that the matrix S is "free relatively to ℓ" if $S\ell \cap \ell = 0$. When $\ell = \ell_P = 0 \times \mathbb{R}^n_p$ we simply say that S is a "free symplectic matrix".

That it suffices to consider free symplectic matrices up to conjugation follows from the next exercise:

Exercise 2.34. Show that $S \in \mathrm{Sp}(n)$ is free relatively to ℓ if and only if $S_0^{-1}SS_0$ is a free symplectic matrix for every $S_0 \in \mathrm{Sp}(n)$ such that $S_0\ell = \ell_P$.

Writing S as a block-matrix one has:

$$S = \begin{bmatrix} A & B \\ C & D \end{bmatrix} \text{ is free } \Longleftrightarrow \det B \neq 0. \tag{2.37}$$

Suppose in fact that $z \in S\ell_P \cap \ell_P$; this is equivalent to $x = 0$ and $Bp = 0$, that is to $z = 0$. It follows from condition (2.37) that

$$S \textit{ is free } \Longleftrightarrow \det\left(\frac{\partial x}{\partial p'}(z_0)\right) \neq 0. \tag{2.38}$$

This suggests the following extension of Definition 2.33:

Definition 2.35. Let f be a symplectomorphism of $(\mathbb{R}^{2n}_z, \sigma)$ defined in a neighborhood of some point z_0. We will say that f is free at the point z_0 if $\det(\partial x/\partial p'(z_0)) \neq 0$. Equivalently: the symplectic matrix $S = Df(z_0)$ is free.

The equivalence of both conditions follows from the observation that the Jacobian matrix

$$Df(z_0) = \begin{bmatrix} \frac{\partial x}{\partial x'}(z_0) & \frac{\partial x}{\partial p'}(z_0) \\ \frac{\partial p}{\partial x'}(z_0) & \frac{\partial p}{\partial p'}(z_0) \end{bmatrix}$$

is indeed free if and only if its upper right corner $\frac{\partial x}{\partial p'}(z_0)$ is invertible.

A very useful property is that every symplectic matrix is the product of two free symplectic matrices. This is a particular case of the following very useful result which will yield a precise factorization result for symplectic matrices, and is in addition the key to many of the properties of the metaplectic group we will study later on:

Proposition 2.36. *For every $(S, \ell_0) \in \mathrm{Sp}(n) \times \mathrm{Lag}(n)$ there exist two matrices S_1, S_2 such that $S = S_1S_2$ and $S_1\ell_0 \cap \ell_0 = S_2\ell_0 \cap \ell_0 = 0$. In particular, choosing $\ell_0 = \ell_P$, every symplectic matrix is the product of two free symplectic matrices.*

Proof. The second assertion follows from the first choosing $\ell_0 = 0 \times \mathbb{R}^n_p$. Recall that $\mathrm{Sp}(n)$ acts transitively on the set of all pairs (ℓ, ℓ') such that $\ell \cap \ell' = 0$. Choose ℓ' transverse to both ℓ_0 and $S\ell$. There exists $S_1 \in \mathrm{Sp}(n)$ such that $S_1(\ell_0, \ell') = (\ell', S\ell_0)$, that is $S_1\ell_0 = \ell'$ and $S\ell_0 = S_1\ell'$. Since $\mathrm{Sp}(n)$ acts transitively on $\mathrm{Lag}(n)$ we can find S_2' such that $\ell' = S_2'\ell_0$ and hence $S\ell_0 = S_1 S_2' \ell_0$. It follows that there exists $S'' \in \mathrm{Sp}(n)$ such that $S''\ell_0 = \ell_0$ and $S = S_1 S_2' S''$. Set $S_2 = S_2' S''$; then $S = S_1 S_2$ and we have

$$S_1\ell_0 \cap \ell_0 = \ell' \cap \ell_0 = 0 \quad , \quad S_2\ell_0 \cap \ell_0 = S_2'\ell_0 \cap \ell_0 = \ell' \cap \ell_0 = 0.$$

The proposition follows. □

Another interesting property of free symplectic matrices are that they can be "generated" by a function W defined on $\mathbb{R}^n_x \times \mathbb{R}^n_x$, in the sense that:

$$(x, p) = S(x', p') \Longleftrightarrow p = \partial_x W(x, x') \text{ and } p' = -\partial_{x'} W(x, x'). \tag{2.39}$$

Suppose that

$$S = \begin{bmatrix} A & B \\ C & D \end{bmatrix} \tag{2.40}$$

is a free symplectic matrix. We claim that a generating function for S is the quadratic form

$$W(x, x') = \tfrac{1}{2} \langle DB^{-1}x, x \rangle - \langle B^{-1}x, x' \rangle + \tfrac{1}{2} \langle B^{-1}Ax', x' \rangle . \tag{2.41}$$

In fact,

$$\begin{aligned} \partial_x W(x, x') &= DB^{-1}x - (B^{-1})^T x', \\ \partial_{x'} W(x, x') &= -B^{-1}x' + B^{-1}Ax' \end{aligned}$$

and hence, solving in x and p,

$$x = Ax' + Bp' \quad , \quad p = Cx' + Dp'.$$

Notice that the matrices DB^{-1} and $B^{-1}A$ are symmetric in view of (2.5)); in fact if conversely W is a quadratic form of the type

$$W(x, x') = \tfrac{1}{2} \langle Px, x \rangle - \langle Lx, x' \rangle + \tfrac{1}{2} \langle Qx', x' \rangle \tag{2.42}$$

with $P = P^T$, $Q = Q^T$, and $\det L \neq 0$, then the matrix

$$S_W = \begin{bmatrix} L^{-1}Q & L^{-1} \\ PL^{-1}Q - L^T & L^{-1}P \end{bmatrix} \tag{2.43}$$

is a free symplectic matrix whose generating function is (2.42). To see this, it suffices to remark that we have

$$(x, p) = S_W(x', p') \Longleftrightarrow p = Px - L^T x' \text{ and } p' = Lx - Qx'$$

and to solve the equations $p = Px - L^T x'$ and $p' = Lx - Qx'$ in x, p.

If S_W is a free symplectic matrix, then its inverse $(S_W)^{-1}$ is also a free symplectic matrix, in fact:

$$(S_W)^{-1} = S_{W^*} \quad , \quad W^*(x, x') = -W(x', x). \tag{2.44}$$

This follows from the observation that if

$$S = S_W = \begin{bmatrix} A & B \\ C & D \end{bmatrix}$$

is free then its inverse

$$S_W^{-1} = \begin{bmatrix} D^T & -B^T \\ -C^T & A^T \end{bmatrix}$$

(see (2.7)) is also free; it is generated by the function

$$\begin{aligned} W^*(x, x') &= -\tfrac{1}{2}\left\langle A^T(B^T)^{-1}x, x\right\rangle + \left\langle (B^T)^{-1}x, x'\right\rangle - \tfrac{1}{2}\left\langle (B^T)^{-1}D^T x', x'\right\rangle \\ &= -\tfrac{1}{2}\left\langle B^{-1}Ax, x\right\rangle + \left\langle B^{-1}x', x\right\rangle - \tfrac{1}{2}\left\langle DB^{-1}x', x'\right\rangle \\ &= -W(x', x). \end{aligned}$$

There is thus a bijective correspondence between free symplectic matrices in Sp(n) and quadratic polynomials of the type W above. Since every such polynomial is determined by a triple (P, L, Q), P and Q symmetric and $\det L \neq 0$, it follows that the subset of Sp(n) consisting of all free symplectic matrices is a submanifold of Sp(n) with dimension $(n+1)(2n-1)$. In particular, $\mathrm{Sp}_0(n)$ has codimension 1 in Sp(n).

An element of ISp(n) is free if it satisfies (2.38); let us characterize this property in terms of generating functions:

Proposition 2.37. *Let $[S, z_0]$ be an affine symplectic transformation. Then:*

(i) *$[S, z_0]$ is free if and only if S is free: $S = S_W$. A free generating function of $f = T(z_0) \circ S_W$ is the inhomogeneous quadratic polynomial*

$$W_{z_0}(x, x') = W(x - x_0, x') + \langle p_0, x\rangle \tag{2.45}$$

($z_0 = (x_0, p_0)$) where W is a free generating function for S.

(ii) *Conversely, if W is the generating function of a symplectic transformation S, then any polynomial*

$$W_{z_0}(x, x') = W(x, x') + \langle \alpha, x\rangle + \langle \alpha', x'\rangle \tag{2.46}$$

($\alpha, \alpha' \in \mathbb{R}^n_x$) is a generating function of an affine symplectic transformation, the translation vector $z_0 = (x_0, p_0)$ being

$$(x_0, p_0) = (B\alpha, Da + \beta) \tag{2.47}$$

when $S = \begin{bmatrix} A & B \\ C & D \end{bmatrix}$.

Proof. Let W_{z_0} be defined by (2.45), and set

$$(x',p') = S(x'',p''), \quad (x,p) = T(z_0)(x',p').$$

We have

$$\begin{aligned} pdx - p'dx' &= (pdx - p''dx'') + (p''dx'' - p'dx') \\ &= (pdx - (p-p_0)d(x-x_0) + dW(x'',x') \\ &= d(\langle p_0, x\rangle + W(x-x_0,x')) \end{aligned}$$

which shows that W_{z_0} is a generating function. Finally, formula (2.47) is obtained by a direct computation, expanding the quadratic form $W(x-x_0,x')$ in its variables. □

Corollary 2.38. *Let $f = [S_W, z_0]$ be a free affine symplectic transformation, and set $(x,p) = f(x',p')$. The function Φ_{z_0} defined by*

$$\Phi_{z_0}(x,x') = \tfrac{1}{2}\langle p,x\rangle - \tfrac{1}{2}\langle p',x'\rangle + \tfrac{1}{2}\sigma(z,z_0) \tag{2.48}$$

is also a free generating function for f; in fact:

$$\Phi_{z_0}(x,x') = W_{z_0}(x,x') + \tfrac{1}{2}\langle p_0,x_0\rangle . \tag{2.49}$$

Proof. Setting $(x'',p'') = S(x,p)$, the generating function W satisfies

$$W(x'',x') = \tfrac{1}{2}\langle p'',x''\rangle - \tfrac{1}{2}\langle p',x'\rangle)$$

in view of Euler's formula for homogeneous functions. Let Φ_{z_0} be defined by formula (2.48); in view of (2.45) we have

$$W_{z_0}(x,x') - \Phi_{z_0}(x,x') = \tfrac{1}{2}\langle p_0,x\rangle - \tfrac{1}{2}\langle p,x_0\rangle - \tfrac{1}{2}\langle p_0,x_0\rangle$$

which is (2.49); this proves the corollary since all generating functions of a symplectic transformation are equal up to an additive constant. □

We are going to establish a few factorization results for symplectic matrices.

Recall (Example 2.5) that if P and L are, respectively, a symmetric and an invertible $n \times n$ matrix, then

$$V_P = \begin{bmatrix} I & 0 \\ -P & I \end{bmatrix}, \; U_P = \begin{bmatrix} -P & I \\ -I & 0 \end{bmatrix}, \; M_L = \begin{bmatrix} L^{-1} & 0 \\ 0 & L^T \end{bmatrix}. \tag{2.50}$$

Proposition 2.39. *If S is a free symplectic matrix (2.37), then*

$$S = V_{-DB^{-1}} M_{B^{-1}} U_{-B^{-1}A} \tag{2.51}$$

and

$$S = V_{-DB^{-1}} M_{B^{-1}} J V_{-B^{-1}A}. \tag{2.52}$$

Proof. We begin by noting that

$$S = \begin{bmatrix} I & 0 \\ DB^{-1} & I \end{bmatrix} \begin{bmatrix} B & 0 \\ 0 & DB^{-1}A - C \end{bmatrix} \begin{bmatrix} B^{-1}A & I \\ -I & 0 \end{bmatrix} \tag{2.53}$$

for any matrix (2.37), symplectic or not. If now S is symplectic, then the middle factor in the right-hand side of (2.53) also is symplectic, since the first and the third factors obviously are. Taking the condition $AD^T - BC^T = I$ in (2.6) into account, we have $DB^{-1}A - C = (B^T)^{-1}$and hence

$$\begin{bmatrix} B & 0 \\ 0 & DB^{-1}A - C \end{bmatrix} = \begin{bmatrix} B & 0 \\ 0 & (B^T)^{-1} \end{bmatrix}$$

so that

$$S = \begin{bmatrix} I & 0 \\ DB^{-1} & I \end{bmatrix} \begin{bmatrix} B & 0 \\ 0 & (B^T)^{-1} \end{bmatrix} \begin{bmatrix} B^{-1}A & I \\ -I & 0 \end{bmatrix}. \tag{2.54}$$

The factorization (2.51) follows (both DB^{-1} and $B^{-1}A$ are symmetric, as a consequence of the relations $B^TD = D^TB$ and $B^TA = A^TB$ in (2.5)). Noting that

$$\begin{bmatrix} B^{-1}A & I \\ -I & 0 \end{bmatrix} = \begin{bmatrix} 0 & I \\ -I & 0 \end{bmatrix} \begin{bmatrix} I & 0 \\ B^{-1}A & I \end{bmatrix}$$

the factorization (2.52) follows as well. □

Conversely, if a matrix S can be written in the form $V_{-P}M_LJV_{-Q}$, then it is a free symplectic matrix; in fact:

$$S = S_W = \begin{bmatrix} L^{-1}Q & L^{-1} \\ PL^{-1}Q - L^T & L^{-1}P \end{bmatrix} \tag{2.55}$$

as is checked by a straightforward calculation.

From this result together with Proposition 2.36 follows that every element of Sp(n) is the product of symplectic matrices of the type V_P, M_L and J. More precisely:

Corollary 2.40. *Each of the sets*

$$\{V_P, M_L, J : P = P^T,\ \det L \neq 0\} \quad \textit{and} \quad \{U_P, M_L : P = P^T,\ \det L \neq 0\}$$

generates Sp(n).

Proof. Taking $\ell_0 = 0 \times \mathbb{R}^n$ in Proposition 2.36 every $S \in \mathrm{Sp}(n)$ is the product of two free symplectic matrices. It now suffices to apply Proposition 2.39. □

Let us finally mention that the notion of free generating function extends without any particular difficulty to the case of symplectomorphisms; this will be useful to us when we discuss Hamilton–Jacobi theory. Suppose in fact that f is a free symplectomorphism in some neighborhood $\mathcal{U}$ of a phase space point. We then have $dp \wedge dx = dp' \wedge dx'$ and this is equivalent, by Poincaré's lemma, to the existence of a function $G \in C^\infty(\mathbb{R}^{2n}_z)$ such that

$$pdx = p'dx' + dG(x', p').$$

Assume now that $Df(z')$ is free for $z' \in \mathcal{U}$; then the condition $\det(\partial x/\partial p') \neq 0$ implies, by the implicit function theorem, that we can locally solve the equation $x = x(x', p')$ in p', so that $p' = p'(x, x')$ and hence $G(x', p')$ is, for $(x', p') \in \mathcal{U}$, a function of x, x' only: $G(x', p') = G(x', p'(x, x'))$. Calling this function W:

$$W(x, x') = G(x', p'(x, x'))$$

we thus have

$$pdx = p'dx' + dW(x, x') = p'dx' + \partial_x W(x, x')dx + \partial_{x'} W(x, x')dx'$$

which requires $p = \partial_x W(x, x')$ and $p' = -\partial_{x'} W(x, x')$ and f is hence free in $\mathcal{U}$. The proof of the converse goes along the same lines and is therefore left to the reader. Since f is a symplectomorphism we have $dp \wedge dx = dp' \wedge dx'$ and this is equivalent, by Poincaré's lemma, to the existence of a function $G \in C^\infty(\mathbb{R}^{2n}_z)$ such that

$$pdx = p'dx' + dG(x', p').$$

Assume now that $Df(z')$ is free for $z' \in \mathcal{U}$; then the condition $\det(\partial x/\partial p') \neq 0$ implies, by the implicit function theorem, that we can locally solve the equation $x = x(x', p')$ in p', so that $p' = p'(x, x')$ and hence $G(x', p')$ is, for $(x', p') \in \mathcal{U}$, a function of x, x' only: $G(x', p') = G(x', p'(x, x'))$. Calling this function W:

$$W(x, x') = G(x', p'(x, x'))$$

we thus have

$$pdx = p'dx' + dW(x, x') = p'dx' + \partial_x W(x, x')dx + \partial_{x'} W(x, x')dx'$$

which requires $p = \partial_x W(x, x')$ and $p' = -\partial_{x'} W(x, x')$ and f is hence free in $\mathcal{U}$. The proof of the converse goes along the same lines and is therefore left to the reader.

2.3 Hamiltonian Mechanics

Physically speaking, Hamiltonian mechanics is a paraphrase (and generalization!) of Newton's second law, popularly expressed as "*force equals mass times acceleration*[3]". The symplectic formulation of Hamiltonian mechanics can be retraced (in embryonic form) to the work of Lagrange between 1808 and 1811; what we today call "Hamilton's equations" were in fact written down by Lagrange who used the

[3]This somewhat unfortunate formulation is due to Kirchhoff.

letter H to denote the "Hamiltonian" to honor Huygens[4] – not Hamilton, who was still in his early childhood at that time! It is however undoubtedly Hamilton's great merit to have recognized the importance of these equations, and to use them with great efficiency in the study of planetary motion, and of light propagation.

Those eager to learn how physicists use Hamiltonian mechanics are referred to the successive editions (1950, 1980, 2002) of Goldstein's classical treatise [53] (the last edition with co-workers). Here are a few references for Hamiltonian mechanics from the symplectic viewpoint: one of the first to cover the topic in a rather exhaustive way are the books by Abraham and Marsden [1], and Arnol'd [3]; a very complete treatment of "symplectic mechanics" is to be found in the treatise by Libermann and Marle [110] already mentioned in Chapter 1; it contains very detailed discussions of some difficult topics; the same applies to Godbillon's little book [50]. In [64] we have given a discussion of the notion of "Maxwell Hamiltonian" following previous work of Souriau and others.

2.3.1 Hamiltonian flows

We will call "Hamiltonian function" (or simply "Hamiltonian") any real function $H \in C^\infty(\mathbb{R}^{2n}_z \times \mathbb{R}_t)$ (although most of the properties we will prove remain valid under the assumption $H \in C^k(\mathbb{R}^{2n}_z \times \mathbb{R}_t)$ with $k \geq 2$: we leave to the reader as an exercise in ordinary differential equations to state minimal smoothness assumptions for the validity of our results).

The *Hamilton equations*

$$\dot{x}_j(t) = \partial_{p_j} H(x(t), p(t), t) \quad , \quad \dot{p}_j(t) = -\partial_{x_j} H(x(t), p(t), t) \tag{2.56}$$

associated with H form a (generally non-autonomous) system of $2n$ differential equations. The conditions of existence of the solutions of Hamilton's equations, as well as for which initial points they are defined, are determined by the theory of ordinary differential equations (or "dynamical systems", as it is now called). See Abraham–Marsden [1], Ch. 1, §2.1, for a general discussion of these topics, including the important notion of "flow box".

The equations (2.56) can be written economically as

$$\dot{z} = J\partial_z H(z,t) \tag{2.57}$$

where J is the standard symplectic matrix. Defining the *Hamilton vector field* by

$$X_H = J\partial_z H = (\partial_p H, -\partial_x H) \tag{2.58}$$

(the operator $J\partial_z$ is often called the *symplectic gradient*), Hamilton's equations are equivalent to

$$\sigma(X_H(z,t), \cdot) + d_z H = 0. \tag{2.59}$$

In fact, for every $z' \in \mathbb{R}^{2n}_z$,

$$\sigma(X_H(z,t), z') = -\langle \partial_x H(z,t), x' \rangle - \langle \partial_p H(z,t), p' \rangle = -\langle \partial_z H(z,t), z' \rangle$$

[4]See Lagrange's *Mécanique Analytique*, Vol. I, pp. 217–226 and 267–270.

which is the same thing as (2.59). This formula is the gate to Hamiltonian mechanics on symplectic manifolds. In fact, formula (2.59) can be rewritten concisely as

$$i_{X_H}\sigma + dH = 0 \tag{2.60}$$

where $i_{X_H(\cdot,t)}$ is the contraction operator:

$$i_{X_H(\cdot,t)}\sigma(z)(z') = \sigma(X_H(z,t), z').$$

The interest of formula (2.60) comes from the fact that it is intrinsic (*i.e.*, independent of any choice of coordinates), and allows the definition of Hamilton vector fields on symplectic manifolds: if (M,ω) is a symplectic manifold and $H \in C^\infty(M \times \mathbb{R}_t)$ then, by definition, the Hamiltonian vector field $X_H(\cdot,t)$ is the vector field defined by (2.60).

One should be careful to note that when the Hamiltonian function H is effectively time-dependent (which is usually the case) then X_H is not a "true" vector field, but rather a family of vector fields on $\mathbb{R}^{2n}_z$ depending smoothly on the parameter t. We can however define the notion of flow associated to X_H:

Definition 2.41. Let $t \longmapsto z_t$ be the solution of Hamilton's equations for H passing through a point z at time $t = 0$, and let f_t^H be the mapping $\mathbb{R}^{2n}_z \longrightarrow \mathbb{R}^{2n}_z$ defined by $f_t^H(z) = z_t$. The family $(f_t^H) = (f_t^H)_{t\in\mathbb{R}}$ is called the "flow determined by the Hamiltonian function H" or the "flow determined by the vector field X_H".

A *caveat*: the usual group property

$$f_0^H = I \ , \ f_t^H \circ f_{t'}^H = f_{t+t'}^H \ , \ (f_t^H)^{-1} = f_{-t}^H \tag{2.61}$$

of flows only holds when H is time-independent; in general $f_t^H \circ f_{t'}^H \neq f_{t+t'}^H$ and $(f_t^H)^{-1} \neq f_{-t}^H$ (but of course we still have the identity $f_0^H = I$).

For notational and expository simplicity we will implicitly assume (unless otherwise specified) that for every $z_0 \in \mathbb{R}^{2n}_z$ there exists a unique solution $t \longmapsto z_t$ of the system (2.57) passing through z_0 at time $t = 0$. The modifications to diverse statements when global existence (in time or space) does not hold are rather obvious and are therefore left to the reader.

As we noted in previous subsection the flow of a time-dependent Hamiltonian vector field is not a one-parameter group; this fact sometimes leads to technical complications when one wants to perform certain calculations. For this reason it is helpful to introduce two (related) notions, those of *suspended Hamilton flow* and *time-dependent Hamilton flow*. We begin by noting that Hamilton's equations $\dot{z} = J\partial_z H(z,t)$ can be rewritten as

$$\frac{d}{dt}(z(t),t) = \tilde{X}_H(z(t),t) \tag{2.62}$$

where

$$\tilde{X}_H = (J\partial_z H, 1) = (\partial_p H, -\partial_x H, 1). \tag{2.63}$$

Definition 2.42.

(i) The vector field $\tilde{X}_H$ on the "*extended phase space*"

$$\mathbb{R}^{2n+1}_{z,t} \equiv \mathbb{R}^{2n}_z \times \mathbb{R}_t$$

is called the "*suspended Hamilton vector field*"; its flow $(\tilde{f}^H_t)$ is called the "*suspended Hamilton flow*" *determined by* H.

(ii) The two-parameter family of mappings $\mathbb{R}^{2n}_z \longrightarrow \mathbb{R}^{2n}_z$ defined by the formula

$$(f^H_{t,t'}(z'), t) = \tilde{f}^H_{t-t'}(z', t') \tag{2.64}$$

is called the "time-dependent flow" determined by H.

Notice that by definition $\tilde{f}^H_t$ thus satisfies

$$\tfrac{d}{dt}\tilde{f}^H_t = \tilde{X}_H(\tilde{f}^H_t). \tag{2.65}$$

The point with introducing $\tilde{X}_H$ is that it is a true vector field on extended phase-space the while X_H is, as pointed out above, rather a family of vector fields parametrized by t as soon as H is time-dependent. The system (2.65) being autonomous in its own right, the mappings $\tilde{f}^H_t$ satisfy the usual group properties:

$$\tilde{f}^H_t \circ \tilde{f}^H_{t'} = \tilde{f}^H_{t+t'} \quad , \quad (\tilde{f}^H_t)^{-1} = \tilde{f}^H_{-t} \quad , \quad \tilde{f}^H_0 = I. \tag{2.66}$$

Notice that the time-dependent flow has the following immediate interpretation: $f^H_{t,t'}$ is the mapping $\mathbb{R}^{2n}_z \longrightarrow \mathbb{R}^{2n}_z$ which takes the point z' at time t' to the point z at time t, the motion occurring along the solution curve to Hamilton equations $\dot{z} = J\partial_z H(z,t)$ passing through these two points. Formula (2.64) is equivalent to

$$\tilde{f}^H_t(z', t') = (f^H_{t+t',t'}(z'), t + t'). \tag{2.67}$$

Note that it immediately follows from the group properties (2.66) of the suspended flow that we have:

$$f^H_{t,t'} = I \quad , \quad f^H_{t,t'} \circ f^H_{t',t''} = f^H_{t,t''} \quad , \quad (f^H_{t,t'})^{-1} = f^H_{t',t} \tag{2.68}$$

for all times t, t' and t''. When H does not depend on t we have $f^H_{t,t'} = f^H_{t-t'}$; in particular $f^H_{t,0} = f^H_t$.

Let H be some (possibly time-dependent) Hamiltonian function and $f^H_t = f^H_{t,0}$. We say that f^H_t is a *free symplectomorphism* at a point $z_0 \in \mathbb{R}^{2n}_z$ if $Df^H_t(z_0)$ is a free symplectic matrix. Of course f^H_t is never free at $t = 0$ since f^H_0 is the identity. In Proposition 2.44 we will give a necessary and sufficient condition for the symplectomorphisms $f^H_{t,t'}$ to be free. Let us first prove the following lemma, the proof of which makes use of the notion of generating function:

Lemma 2.43. *The symplectomorphism* $f : \mathbb{R}^{2n}_z \longrightarrow \mathbb{R}^{2n}_z$ *is free in a neighborhood* $\mathcal{U}$ *of* $z_0 \in \mathbb{R}^{2n}_z$ *if and only if* $Df(z')$ *is a free symplectic matrix for* $z' \in \mathcal{U}$*, that is, if and only if* $\det(\partial x/\partial p') \neq 0$.

Proof. Set $z = f(z')$; we have

$$Df(z') = \begin{bmatrix} \dfrac{\partial x}{\partial x'}(z') & \dfrac{\partial x}{\partial p'}(z') \\ \dfrac{\partial p}{\partial x'}(z') & \dfrac{\partial p}{\partial p'}(z') \end{bmatrix}$$

and the symplectic matrix $Df(z')$ is thus free for $z' \in \mathcal{U}$ if and only if

$$\det \frac{\partial x}{\partial p'}(z') \neq 0.$$

We next make the following crucial observation: since f is a symplectomorphism we have $dp \wedge dx = dp' \wedge dx'$ and this is equivalent, by Poincaré's lemma, to the existence of a function $G \in C^\infty(\mathbb{R}^{2n}_z)$ such that

$$pdx = p'dx' + dG(x', p').$$

Assume now that $Df(z')$ is free for $z' \in \mathcal{U}$; then the condition $\det(\partial x/\partial p') \neq 0$ implies, by the implicit function theorem, that we can locally solve the equation $x = x(x', p')$ in p', so that $p' = p'(x, x')$ and hence $G(x', p')$ is, for $(x', p') \in \mathcal{U}$, a function of x, x' only: $G(x', p') = G(x', p'(x, x'))$. Calling this function W:

$$W(x, x') = G(x', p'(x, x'))$$

we thus have

$$pdx = p'dx' + dW(x, x') = p'dx' + \partial_x W(x, x')dx + \partial_{x'} W(x, x')dx'$$

which requires $p = \partial_x W(x, x')$ and $p' = -\partial_{x'} W(x, x')$ and f is hence free in $\mathcal{U}$. The proof of the converse goes along the same lines and is therefore left to the reader. □

We will use the notation H_{pp}, H_{xp}, and H_{xx} for the matrices of second derivatives of H in the corresponding variables.

Proposition 2.44. *There exists $\varepsilon > 0$ such that f_t^H is free at $z_0 \in \mathbb{R}^{2n}_z$ for $0 < |t - t_0| \leq \varepsilon$ if and only if $\det H_{pp}(z_0, t_0) \neq 0$. In particular there exists $\varepsilon > 0$ such that $f_t^H(z_0)$ is free for $0 < |t| \leq \varepsilon$ if and only if $\det H_{pp}(z_0, 0) \neq 0$.*

Proof. Let $t \longmapsto z(t) = (x(t), p(t))$ be the solution to Hamilton's equations

$$\dot{x} = \partial_p H(z, t) \quad , \quad \dot{p} = -\partial_x H(z, t)$$

with initial condition $z(t_0) = z_0$. A second-order Taylor expansion in t yields

$$z(t) = z_0 + (t - t_0)X_H(z_0, t_0) + O((t - t_0)^2);$$

and hence

$$x(t) = x_0 + (t - t_0)\partial_p H(z_0, t_0) + O((t - t_0)^2).$$

It follows that

$$\frac{\partial x(t)}{\partial p} = (t - t_0)H_{pp}(z_0, t_0) + O((t - t_0)^2),$$

hence there exists $\varepsilon > 0$ such that $\partial x(t)/\partial p$ is invertible in $[t_0 - \varepsilon, t_0[\cap]t_0, t_0 + \varepsilon]$ if and only if $H_{pp}(z_0, t_0)$ is invertible; in view of Lemma 2.43 this is equivalent to saying that f_t^H is free at z_0. □

Example 2.45. The result above applies when the Hamiltonian H is of the "physical type"

$$H(z, t) = \sum_{j=1}^{n} \frac{1}{2m_j} p_j^2 + U(x, t)$$

since we have

$$H_{pp}(z_0, t_0) = \operatorname{diag}[\tfrac{1}{2m_1}, \ldots, \tfrac{1}{2m_n}].$$

In this case f_t^H is free for small non-zero t near each z_0 where it is defined.

2.3.2 The variational equation

An essential feature of Hamiltonian flows is that they consist of symplectomorphisms. We are going to give an elementary proof of this property; it relies on the fact that the mapping $t \longmapsto Df_{t,t'}^H(z)$ is, for fixed t', the solution of a differential equation, the *variational equation*, and which plays an important role in many aspects of Hamiltonian mechanics (in particular the study of periodic Hamiltonian orbits, see for instance Abraham and Marsden [1]).

Proposition 2.46. *For fixed z set $S_{t,t'}^H(z) = Df_{t,t'}^H(z)$.*

(i) *The function $t \longmapsto S_{t,t'}(z)$ satisfies the variational equation*

$$\frac{d}{dt} S_{t,t'}^H(z) = JD^2H(f_{t,t'}^H(z), t)S_{t,t'}^H(z) \quad , \quad S_{t,t}^H(z) = I \tag{2.69}$$

where $D^2H(f_{t,t'}^H(z))$ is the Hessian matrix of H calculated at $f_{t,t'}^H(z)$;

(ii) *We have $S_{t,t'}^H(z) \in \mathrm{Sp}(n)$ for every z and t, t' for which it is defined, hence $f_{t,t'}^H$ is a symplectomorphism.*

Proof. (i) It is sufficient to consider the case $t' = 0$. Set $f_{t,0}^H = f_t^H$ and $S_{t,t'}^H = S_t$. Taking Hamilton's equation into account the time-derivative of the Jacobian matrix $S_t(z)$ is

$$\frac{d}{dt} S_t(z) = \frac{d}{dt}(Df_t^H(z)) = D\left(\frac{d}{dt} f_t^H(z)\right),$$

that is

$$\frac{d}{dt}S_t(z) = D(X_H(f_t^H(z))).$$

Using the fact that $X_H = J\partial_z H$ together with the chain rule, we have

$$\begin{aligned} D(X_H(f_t^H(z))) &= D(J\partial_z H)(f_t^H(z), t) \\ &= JD(\partial_z H)(f_t^H(z), t) \\ &= J(D^2H)(f_t^H(z), t)Df_t^H(z), \end{aligned}$$

hence $S_t(z)$ satisfies the variational equation (2.69), proving (i).

(ii) Set $S_t = S_t(z)$ and $A_t = (S_t)^T JS_t$; using the product rule together with (2.69) we have

$$\begin{aligned} \frac{dA_t}{dt} &= \frac{d(S_t)^T}{dt} JS_t + (S_t)^T J \frac{dS_t}{dt} \\ &= (S_t)^T D^2H(z,t)S_t - (S_t)^T D^2H(z,t)S_t \\ &= 0. \end{aligned}$$

It follows that the matrix $A_t = (S_t)^T JS_t$ is constant in t, hence $A_t(z) = A_0(z) = J$ so that $(S_t)^T JS_t = J$ proving that $S_t \in \mathrm{Sp}(n)$. □

Exercise 2.47. Let $t \longmapsto X_t$ be a C^∞ mapping $\mathbb{R} \longrightarrow \mathfrak{sp}(n)$ and $t \longmapsto S_t$ a solution of the differential system

$$\frac{d}{dt}S_t = X_t S_t \quad , \quad S_0 = I.$$

Show that $S_t \in \mathrm{Sp}(n)$ for every $t \in \mathbb{R}$.

Exercise 2.48. Assume that H is time-independent. Show that the f_t^H are symplectomorphisms using formula (2.60) together with Cartan's homotopy formula $i_X d\alpha + d(i_X\alpha) = 0$, valid for all vector fields X and differential forms α. Can you extend the proof to include the case where H is time-dependent? [Hint: use the suspended Hamilton vector field.]

Hamilton's equations are covariant (*i.e.*, they retain their form) under symplectomorphisms. Let us begin by proving the following general result about vector fields which we will use several times in this chapter. If X is a vector field and f a diffeomorphism we denote by $Y = f^*X$ the vector field defined by

$$Y(u) = D(f^{-1})(f(u))X(f(u)) = [Df(u)]^{-1}X(f(u)). \tag{2.70}$$

(f^*X is called the "pull-back" of the vector field X by the diffeomorphism f.)

Lemma 2.49. *Let X be a vector field on $\mathbb{R}^m$ and (φ_t^X) its flow. Let f be a diffeomorphism $\mathbb{R}^m \longrightarrow \mathbb{R}^m$. The family (φ_t^Y) of diffeomorphisms defined by*

$$\varphi_t^Y = f^{-1} \circ \varphi_t^X \circ f \tag{2.71}$$

is the flow of the vector field $Y = (Df)^{-1}(X \circ f)$.

Proof. We obviously have $\varphi_0^Y = I$; in view of the chain rule

$$\begin{aligned}\frac{d}{dt}\varphi_t^Y(x) &= D(f^{-1})(\varphi_t^X(f(x)))X(\varphi_t^X(f(x))\\ &= (Df)^{-1}(\varphi_t^Y(x))X(f(\varphi_t^Y(x)))\end{aligned}$$

and hence $\frac{d}{dt}\varphi_t^Y(x) = Y(\varphi_t^Y(x))$ which we set out to prove. □

Specializing to the Hamiltonian case, this lemma yields:

Proposition 2.50. *Let f be a symplectomorphism.*

(i) *We have*

$$X_{H\circ f}(z) = [Df(z)]^{-1}(X_H \circ f)(z). \tag{2.72}$$

(ii) *The flows (f_t^H) and $(f_t^{H\circ f})$ are conjugate by f:*

$$f_t^{H\circ f} = f^{-1}\circ f_t^H \circ f \tag{2.73}$$

and thus $f^*X_H = X_{H\circ f}$ when f is symplectic.

Proof. Let us prove (i); the assertion (ii) will follow in view of Lemma 2.49 above. Set $K = H\circ f$. By the chain rule

$$\partial_z K(z) = [Df(z)]^T(\partial_z H)(f(z))$$

hence the vector field $X_K = J\partial_z K$ is given by

$$X_K(z) = J[Df(z)]^T\partial_z H(f(z)).$$

Since $Df(z)$ is symplectic we have $J[Df(z)]^T = [Df(z)]^{-1}J$ and thus

$$X_K(z) = [Df(z)]^{-1}J\partial_z H(f(z))$$

which is (2.72). □

Remark 2.51. Proposition 2.50 can be restated as follows: set $(x',p') = f(x,p)$ and $K = H\circ f$; if f is a symplectomorphism then we have the equivalence

$$\begin{gathered}\dot{x}' = \partial_{p'}K(x',p') \text{ and } \dot{p}' = -\partial_{x'}K(x',p')\\ \Longleftrightarrow\\ \dot{x} = \partial_p H(x,p) \text{ and } \dot{p} = -\partial_x H(x,p).\end{gathered} \tag{2.74}$$

Exercise 2.52.

(i) Show that the change of variables $(x,p)\longmapsto(I,\theta)$ defined by $x = \sqrt{2I}\cos\theta$, $p = \sqrt{2I}\sin\theta$ is symplectic.

(ii) Apply Proposition 2.50 to this change of variables to solve Hamilton's equation for $H = \frac{1}{2}\omega(x^2+p^2)$.

An interesting fact is that there is a wide class of Hamiltonian functions whose time-dependent flows (f_t^H) consist of free symplectomorphisms if t is not too large (and different from zero). This is the case for instance when H is of the "physical" type

$$H(z,t)=\sum_{j=1}^{n}\frac{1}{2m_j}(p_j-A_j(x,t))^2+U(x,t) \tag{2.75}$$

($m_j>0$, A_j and U smooth). More generally:

Lemma 2.53. *There exists $\varepsilon>0$ such that f_t^H is free at $z_0\in\mathbb{R}_z^{2n}$ for $0<|t|\leq\varepsilon$ if and only if $\det H_{pp}(z_0,0)\neq 0$, and hence, in particular, when H is of the type (2.75).*

Proof. Let $t\longmapsto z(t)$ be the solution to Hamilton's equations

$$\dot{x}=\partial_p H(z,t)\quad,\quad \dot{p}=-\partial_x H(z,t)$$

with initial condition $z(0)=z_0$. A second-order Taylor expansion at time $t=0$ yields

$$z(t)=z_0+tX_H(z_0,0)+O((t)^2)$$

where $X_H=J\partial_z H$ is the Hamiltonian vector field of H; in particular

$$x(t)=x_0+t\partial_p H(z_0,0)+O(t^2)$$

and hence

$$\frac{\partial x(t)}{\partial p}=tH_{pp}(z_0,0)+O(t^2)$$

where H_{pp} denotes the matrix of derivatives of H in the variables p_j. It follows that there exists $\varepsilon>0$ such that $\partial x(t)/\partial p$ is invertible in $[-\varepsilon,0[\cap]0,\varepsilon]$ if and only if $H_{pp}(z_0,0)$ is invertible; in view of Lemma 2.43 this is equivalent to saying that f_t^H is free at z_0. □

We will use this result in Chapter 5, Subsection 5.2.2, when we discuss the Hamilton–Jacobi equation.

2.3.3 The group Ham(n)

The group Ham(n) is the connected component of the group Symp(n) of all symplectomorphisms of $(\mathbb{R}_z^{2n},\sigma)$. Each of its points is the value of a Hamiltonian flow at some time t. The study of the various algebraic and topological properties of the group Ham(n) is a very active area of current research; see Hofer and Zehnder [91], McDuff and Salamon [114], or Polterovich [132].

We will say that a symplectomorphism f of the standard symplectic space $(\mathbb{R}_z^{2n},\sigma)$ is *Hamiltonian* if there exists a function $H\in C^\infty(\mathbb{R}_{z,t}^{2n+1},\mathbb{R})$ and a number

$a \in \mathbb{R}$ such that $f = f_a^H$. Taking $a = 0$ it is clear that the identity is a Hamiltonian symplectomorphism. The set of all Hamiltonian symplectomorphisms is denoted by Ham(n). We are going to see that it is a connected and normal subgroup of Symp(n); let us first prove a preparatory result which is interesting in its own right:

Proposition 2.54. *Let* (f_t^H) *and* (f_t^K) *be Hamiltonian flows. Then:*

$$f_t^H f_t^K = f_t^{H\#K} \quad \textit{with} \quad H\#K(z,t) = H(z,t) + K((f_t^H)^{-1}(z),t), \tag{2.76}$$

$$(f_t^H)^{-1} = f_t^{\bar{H}} \quad \textit{with} \quad \bar{H}(z,t) = -H(f_t^H(z),t). \tag{2.77}$$

Proof. Let us first prove (2.76). By the product and chain rules we have

$$\frac{d}{dt}(f_t^H f_t^K) = (\frac{d}{dt}f_t^H)f_t^K + (Df_t^H)f_t^K\frac{d}{dt}f_t^K = X_H(f_t^H f_t^K) + (Df_t^H)f_t^K \circ X_K(f_t^K)$$

and it thus suffices to show that

$$(Df_t^H)f_t^K \circ X_K(f_t^K) = X_{K\circ(f_t^H)^{-1}}(f_t^K). \tag{2.78}$$

Writing

$$(Df_t^H)f_t^K \circ X_K(f_t^K) = (Df_t^H)((f_t^H)^{-1}f_t^H f_t^K) \circ X_K((f_t^H)^{-1}f_t^H f_t^K)$$

the equality (2.78) follows from the transformation formula (2.72) in Proposition 2.50. Formula (2.77) is now an easy consequence of (2.76), noting that $(f_t^H f_t^{\bar{H}})$ is the flow determined by the Hamiltonian

$$K(z,t) = H(z,t) + \bar{H}((f_t^H)^{-1}(z),t) = 0;$$

$f_t^H f_t^{\bar{H}}$ is thus the identity, so that $(f_t^H)^{-1} = f_t^{\bar{H}}$ as claimed. □

Let us now show that Ham(n) is a group, as claimed:

Proposition 2.55. Ham(n) *is a normal and connected subgroup of the group* Symp(n) *of all symplectomorphisms of* $(\mathbb{R}_z^{2n},\sigma)$.

Proof. Let us show that if $f, g \in$ Ham(n) then $fg^{-1} \in$ Ham(n). We begin by remarking that if $f = f_a^H$ for some $a \neq 0$, then we also have $f = f_1^{H^a}$ where $H^a(z,t) = aH(z,at)$. In fact, setting $t^a = at$ we have

$$\frac{dz^a}{dt} = J\partial_z H^a(z^a,t) \Longleftrightarrow \frac{dz^a}{dt^a} = J\partial_z H(z^a,t^a)$$

and hence $f_t^{H^a} = f_{at}^H$. We may thus assume that $f = f_1^H$ and $g = f_1^K$ for some Hamiltonians H and K. Now, using successively (2.76) and (2.77) we have

$$fg^{-1} = f_1^H(f_1^K)^{-1} = f_1^{H\#\bar{K}}$$

hence $fg^{-1} \in$ Ham(n). That Ham(n) is a normal subgroup of Symp(n) immediately follows from formula (2.73) in Proposition 2.50: if g is a symplectomorphism and $f \in \mathrm{Ham}(n)$ then

$$f_1^{H \circ g} = g^{-1} f_1^H g \in \mathrm{Ham}(n) \tag{2.79}$$

so we are done. □

The result above motivates the following definition:

Definition 2.56. The set $\mathrm{Ham}(n)$ of all Hamiltonian symplectomorphisms equipped with the law $fg = f \circ g$ is called the group of Hamiltonian symplectomorphisms of the standard symplectic space $(\mathbb{R}_z^{2n}, \sigma)$.

The topology of $\mathrm{Symp}(n)$ is defined by specifying the convergent sequences: we will say that $\lim_{j\to\infty} f_j = f$ in $\mathrm{Symp}(n)$ if and only if for every compact set $\mathcal{K}$ in $\mathbb{R}_z^{2n}$ the sequences $(f_{j|\mathcal{K}})$ and $(D(f_{j|\mathcal{K}}))$ converge uniformly towards $f_{|\mathcal{K}}$ and $D(f_{|\mathcal{K}})$, respectively. The topology of $\mathrm{Ham}(n)$ is the topology induced by $\mathrm{Symp}(n)$.

We are now going to prove a deep and beautiful result due to Banyaga [6]. It essentially says that a path of time-one Hamiltonian symplectomorphisms passing through the identity at time zero is itself Hamiltonian. It will follow that $\mathrm{Ham}(n)$ is a connected group.

Let $t \longmapsto f_t$ be a path in $\mathrm{Ham}(n)$, defined for $0 \leq t \leq 1$ and starting at the identity: $f_0 = I$. We will call such a path a *one-parameter family of Hamiltonian symplectomorphisms*. Thus, each f_t is equal to some symplectomorphism $f_1^{H_t}$. A striking – and not immediately obvious! – fact is that each path $t \longmapsto f_t$ is itself the flow of a Hamiltonian function!

Theorem 2.57. *Let (f_t) be a one-parameter family in $\mathrm{Ham}(n)$. Then $(f_t) = (f_t^H)$ where the Hamilton function H is given by*

$$H(z,t) = -\int_0^1 \sigma(X(uz,t)du \quad \textit{with} \quad X = (\tfrac{d}{dt}f_t) \circ (f_t)^{-1}. \tag{2.80}$$

Proof. By definition of X we have $\frac{d}{dt}f_t = Xf_t$ so that all we have to do is to prove that X is a (time-dependent) Hamiltonian field. For this it suffices to show that the contraction $i_X\sigma$ of the symplectic form with X is an exact differential one-form, for then $i_X\sigma = -dH$ where H is given by (2.80). The f_t being symplectomorphisms, they preserve the symplectic form σ and hence $\mathcal{L}_X\sigma = 0$. In view of Cartan's homotopy formula we have

$$\mathcal{L}_X\sigma = i_X d\sigma + d(i_X\sigma) = d(i_X\sigma) = 0$$

so that $i_X\sigma$ is closed; by Poincaré's lemma it is also exact. □

Exercise 2.58. Let $(f_t^H)_{0\leq t\leq 1}$ and $(f_t^K)_{0\leq t\leq 1}$ be two arbitrary paths in $\mathrm{Ham}(n)$. The paths $(f_t^H f_t^K)_{0\leq t\leq 1}$ and $(f_t)_{1\leq t\leq 1}$ where

$$f_t = \begin{cases} f_{2t}^K & \text{when } 0 \leq t \leq \frac{1}{2}, \\ f_{2t-1}^H f_1^K & \text{when } \frac{1}{2} \leq t \leq 1 \end{cases}$$

are homotopic with fixed endpoints.

2.3.4 Hamiltonian periodic orbits

Let $H \in C^\infty(\mathbb{R}^{2n}_z)$ be a time-independent Hamiltonian function, and (f_t^H) the flow determined by the associated vector field $X_H = J\partial_z$.

Definition 2.59. Let $z_0 \in \mathbb{R}^{2n}_z$; the mapping

$$\gamma : \mathbb{R}_t \longrightarrow \mathbb{R}^{2n}_z \quad , \quad \gamma(t) = f_t^H(z_0)$$

is called "(Hamiltonian) orbit through z_0".

If there exists $T > 0$ such that $f_{t+T}^H(z_0) = f_t^H(z_0)$ for all $t \in \mathbb{R}$, one says that the orbit γ through z_0 is "periodic with period T". [The smallest possible period is called a "primitive period".]

The following properties are obvious:

- Let γ, γ' be two orbits of H. Then the ranges $\operatorname{Im}\gamma$ and $\operatorname{Im}\gamma'$ are either disjoint or identical.
- The value of H along any orbit is constant ("theorem of conservation of energy").

The first property follows from the uniqueness of the solutions of Hamilton's equations, and the second from the chain rule, setting $\gamma(t) = (x(t), p(t))$:

$$\begin{aligned}\frac{d}{dt}H(\gamma(t)) &= \langle \partial_x H(\gamma(t)), \dot{x}(t)\rangle + \langle \partial_p H(\gamma(t)), \dot{p}(t)\rangle \\ &= -\langle \dot{p}(t), \dot{x}(t)\rangle + \langle \dot{x}(t), \dot{p}(t)\rangle \\ &= 0\end{aligned}$$

where we have taken into account Hamilton's equations.

Assume now that γ is a periodic orbit through z_0. We will use the notation $S_t(z_0) = Df_t^H(z_0)$.

Definition 2.60. Let $\gamma : t \longmapsto f_t^H(z_0)$ be a periodic orbit with period T. The symplectic matrix $S_T(z_0) = Df_T^H(z_0)$ is called a "monodromy matrix". The eigenvalues of $S_T(z_0)$ are called the "*Floquet multipliers*" of γ.

The following property is well known in Floquet theory:

Lemma 2.61.

(i) *Let $S_T(z_0)$ be the monodromy matrix of the periodic orbit γ. We have*

$$S_{t+T}(z_0) = S_t(z_0)S_T(z_0) \tag{2.81}$$

for all $t \in \mathbb{R}$. In particular $S_{NT}(z_0) = S_T(z_0)^N$ for every integer N.

(ii) *Monodromy matrices corresponding to the choice of different origins on the periodic orbit are conjugate of each other in* $\operatorname{Sp}(n)$, *hence the Floquet multipliers do not depend on the choice of origin of the periodic orbit;*

(iii) *Each periodic orbit has an even number > 0 of Floquet multipliers equal to* 1.

Proof. (i) The mappings f_t^H form a group, hence, taking into account the equality $f_T^H(z_0) = z_0$:

$$f_{t+T}^H(z_0) = f_t^H(f_T^H(z_0))$$

so that by the chain rule,

$$Df_{t+T}^H(z_0) = Df_t^H(f_T^H(z_0))Df_T^H(z_0),$$

that is (2.81) since $f_T^H(z_0) = z_0$.

(ii) Evidently the orbit through any point $z(t)$ of the periodic orbit γ is also periodic. We begin by noting that if z_0 and z_1 are points on the same orbit γ, then there exists t_0 such that $z_0 = f_{t_0}^H(z_1)$. We have

$$f_t^H(f_{t_0}^H(z_1)) = f_{t_0}^H(f_t^H(z_1))$$

hence, applying the chain rule of both sides of this equality,

$$Df_t^H(f_{t_0}^H(z_1))Df_{t_0}^H(z_1) = Df_{t_0}^H(f_t^H(z_1))Df_t^H(z_1).$$

Choosing $t = T$ we have $f_t^H(z_0) = z_0$ and hence

$$S_T(f_{t_0}^H(z_1))S_{t_0}(z_1) = S_{t_0}(z_1)S_T(z_1),$$

that is, since $f_{t_0}^H(z_1) = z_0$,

$$S_T(z_0)S_{t_0}(z_1) = S_{t_0}(z_1)S_T(z_1).$$

It follows that the monodromy matrices $S_T(z_0)$ and $S_T(z_1)$ are conjugate and thus have the same eigenvalues.

(iii) We have, using the chain rule together with the relation $f_t^H \circ f_{t'}^H = f_{t+t'}^H$,

$$\frac{d}{dt'} f_t^H(f_{t'}^H(z_0))\Big|_{t'=0} = Df_t(z_0)X_H(z_0) = X_H(f_t^H(z_0)),$$

hence $S_{T_0}(z_0)X_H(z_0) = X_H(z_0)$ setting $t = T$; $X_H(z_0)$ is thus an eigenvector of $S_T(z_0)$ with eigenvalue 1; the lemma follows the eigenvalues of a symplectic matrix occurring in quadruples $(\lambda, 1/\lambda, \bar{\lambda}, 1/\bar{\lambda})$. □

The following theorem is essentially due to Poincaré (see Abraham and Marsden [1] for a proof):

Theorem 2.62. *Let $E_0 = H(z_0)$ be the value of H along a periodic orbit γ_0. Assume that γ_0 has exactly two Floquet multipliers equal to 1. Then there exists a unique smooth 1-parameter family (γ_E) of periodic orbits of E with period T parametrized by the energy E, and each γ_E is isolated on the hypersurface $\Sigma_E = \{z : H(z) = E\}$ among those periodic orbits having periods close to the period T_0 of γ_0 Moreover* $\lim_{E\to E_0} T = T_0$.

One shows, using "normal form" techniques that when the conditions of the theorem above are fulfilled, the monodromy matrix of γ_0 can be written as

$$S_T(z_0) = S_0^T \begin{bmatrix} U & 0 \\ 0 & \tilde{S}(z_0) \end{bmatrix} S_0$$

with $S_0 \in \mathrm{Sp}(n)$, $\tilde{S}(z_0) \in \mathrm{Sp}(n-1)$ and U is of the type $\begin{bmatrix} 1 & \beta \\ 0 & 1 \end{bmatrix}$ for some real number β; the $2(n-1) \times 2(n-1)$ symplectic matrix $\tilde{S}(z_0)$ is called the stability matrix of the isolated periodic orbit γ_0. It plays a fundamental role not only in the study of periodic orbits, but also in semiclassical mechanics ("Gutzwiller's formula" [86], A.V. Sobolev [155]). We will return to the topic when we discuss the Conley–Zehnder index in Chapter 4.

Chapter 3

Multi-Oriented Symplectic Geometry

Multi-oriented symplectic geometry, also called q-symplectic geometry, is a topic which has not been studied as it deserves in the mathematical literature; see however Leray [107], Dazord [28], de Gosson [57, 61]; also [54, 55]. The idea is the following: one begins by observing that since symplectic matrices have determinant 1, the action of $\mathrm{Sp}(n)$ on a Lagrangian plane preserves the orientation of that Lagrangian plane. Thus, ordinary symplectic geometry is not only the study of the action

$$\mathrm{Sp}(n) \times \mathrm{Lag}(n) \longrightarrow \mathrm{Lag}(n)$$

but it is actually the study of the action

$$\mathrm{Sp}(n) \times \mathrm{Lag}_2(n) \longrightarrow \mathrm{Lag}_2(n)$$

where $\mathrm{Lag}_2(n)$ is the double covering of $\mathrm{Lag}(n)$. More generally, q-symplectic geometry will be the study of the action

$$\mathrm{Sp}_q(n) \times \mathrm{Lag}_{2q}(n) \longrightarrow \mathrm{Lag}_{2q}(n)$$

where $\mathrm{Sp}_q(n)$ is the qth order covering of $\mathrm{Sp}(n)$ and $\mathrm{Lag}_{2q}(n)$ is the $2q$th order covering of $\mathrm{Lag}(n)$. In the case $q = 2$ this action highlights the geometrical role of the Maslov indices on the metaplectic group, which we will study in Chapter 7.

The study of q-symplectic geometry makes use of an important generalization of the Maslov index, which we call the Arnol'd–Leray–Maslov index. That index plays a crucial role in at least two other areas of mathematics and mathematical physics:

- It is instrumental in giving the correct phase shifts through caustics in semi-classical quantization (Leray [107], de Gosson [60, 61, 62, 64]) because it allows one to define the argument of the square root of a de Rham form on a Lagrangian manifold;

- It allows a simple and elegant calculation of Maslov indices of both Lagrangian and symplectic paths; these indices play an essential role in the study of "spectral flow" properties related to the theory of the Morse index (see Piccione and his collaborators [130, 131] and Booss–Bavnbek and Wojciechowski [15]).

3.1 Souriau Mapping and Maslov Index

The locution *Maslov index* has become a household name in mathematics. It is actually a collective denomination for a whole constellation of discrete-valued functions defined on loops (or, more generally, on paths) in $\mathrm{Lag}(n)$ or $\mathrm{Sp}(n)$, and which can be viewed as describing the number of times a given loop (or path) intersects some particular locus in $\mathrm{Lag}(n)$ or $\mathrm{Sp}(n)$ known under the omnibus name of "caustic".

In this section we will only deal with the simplest notion of Maslov index, that of loops in $\mathrm{Lag}(n)$, whose definition is due to Maslov and Arnol'd. We will generalize the notion to paths in both $\mathrm{Lag}(n)$ and $\mathrm{Sp}(n)$ when we deal with semi-classical mechanics.

There are several different (but equivalent) ways of introducing the Maslov index on $\mathrm{Lag}(n)$. The simplest (especially for explicit calculation) makes use of the fact that we can identify the Lagrangian Grassmannian $\mathrm{Lag}(n)$ with a set of matrices, using the so-called "Souriau mapping". This will provide us not only with a simple way of defining correctly the Maslov index of Lagrangian loops, but will also allow us to construct the "Maslov bundle" in Chapter 3 (Subsection 3.2.2).

3.1.1 The Souriau mapping

Recall that the mapping

$$A+iB \longmapsto \begin{bmatrix} A & -B \\ B & A \end{bmatrix}$$

identifies $\mathrm{U}(n,\mathbb{C})$ with a subgroup $\mathrm{U}(n)$ of $\mathrm{Sp}(n)$; that subgroup consists of all

$$U = \begin{bmatrix} A & -B \\ B & A \end{bmatrix}$$

where A and B satisfy the conditions

$$A^TA + B^TB = I \ , \ A^TB = B^TA, \tag{3.1a}$$

$$AA^T + BB^T = I \ , \ AB^T = BA^T. \tag{3.1b}$$

Also recall that the unitary group $\mathrm{U}(n,\mathbb{C})$ acts transitively on the Lagrangian Grassmannian $\mathrm{Lag}(n)$ by the law $u\ell = U\ell$ where $U \in \mathrm{U}(n)$ is associated to $u \in \mathrm{U}(n,\mathbb{C})$.

We denote by $\mathrm{W}(n,\mathbb{C})$ the set of all *symmetric* unitary matrices:

$$\mathrm{W}(n,\mathbb{C}) = \{w \in \mathrm{U}(n,\mathbb{C}) : w = w^T\}$$

and by W the image of $w \in \mathrm{W}(n,\mathbb{C})$ in $\mathrm{U}(n) \subset \mathrm{Sp}(n)$. The set of all such matrices W is denoted by $\mathrm{W}(n)$. Applying the conditions (3.1) we thus have

$$W = \begin{bmatrix} A & -B \\ B & A \end{bmatrix} \in \mathrm{W}(n) \Longleftrightarrow \begin{cases} A^2 + B^2 = I, \\ AB = BA, \\ A = A^T \ , \ B = B^T. \end{cases}$$

Observe that neither $\mathrm{W}(n,\mathbb{C})$ or $\mathrm{W}(n)$ are groups: the product of two symmetric matrices is not in general symmetric.

Interestingly enough, the set $\mathrm{W}(n,\mathbb{C})$ is closed under the operation of taking square roots:

Lemma 3.1. *For every $w \in \mathrm{W}(n,\mathbb{C})$ [respectively $W \in \mathrm{W}(n)$] there exists $u \in \mathrm{W}(n,\mathbb{C})$ [resp. $U \in \mathrm{W}(n)$] such that $w = u^2$ [respectively $W = U^2$].*

Proof. Let $w = A+iB$. The condition $ww^* = I$ implies that $AB = BA$. It follows that the symmetric matrices A and B can be diagonalized simultaneously: there exists $R \in \mathrm{O}(n,\mathbb{R})$ such that $G = RAR^T$ and $H = RBR^T$are diagonal. Let g_j and h_j $(1 \leq j \leq n)$ be the eigenvalues of G and H, respectively. Since $A^2 + B^2 = I$ we have $g_j^2 + h_j^2 = 1$ for every j. Choose now real numbers x_j, y_j such that $x_j^2 - y_j^2 = g_j$ and $2x_jy_j = h_j$ for $1 \leq j \leq n$ and let X and Y be the diagonal matrices whose entries are these numbers x_j, y_j. Then $(X+iY)^2 = G+iH$, and $u = R^T(X+iY)R$ is such that $u^2 = w$. □

We are now going prove that $\mathrm{Lag}(n)$ can be identified with $\mathrm{W}(n,\mathbb{C})$ (and hence with $\mathrm{W}(n)$). Let us begin with a preparatory remark:

Remark 3.2. Suppose that $u\ell_P = \ell_P$; writing $u = A + iB$ this implies $B = 0$, and hence $u \in \mathrm{O}(n,\mathbb{R})$. This is immediately seen by noting that the condition $u\ell_P = \ell_P$ can be written in matrix form as: for every p there exists p' such that

$$\begin{bmatrix} A & -B \\ B & A \end{bmatrix} \begin{bmatrix} 0 \\ p \end{bmatrix} = \begin{bmatrix} 0 \\ p' \end{bmatrix}.$$

Theorem 3.3.

(i) *For $\ell \in \mathrm{Lag}(n)$ and $u \in \mathrm{U}(n,\mathbb{C})$ such that $\ell = u\ell_P$ the product $w = uu^T$ only depends on ℓ and not on the choice of u; the correspondence $\ell \longrightarrow uu^T$ is thus the mapping*

$$w(\cdot) : \mathrm{Lag}(n) \longrightarrow \mathrm{W}(n,\mathbb{C}) \quad , \quad w(\ell) = w = uu^T. \tag{3.2}$$

(ii) *That mapping is a bijection, and satisfies*

$$w(u\ell) = uw(\ell)u^T \tag{3.3}$$

for every $u \in \mathrm{U}(n, \mathbb{C})$.

Proof. (i) Let us show that if two unitary matrices u and u' are both such that $u\ell_P = u'\ell_P$ then $uu^T = u'(u'^T)$. This will prove the first statement. The condition $u\ell_P = u'\ell_P$ is equivalent to $u^{-1}u'\ell_P = \ell_P$. In view of the preparatory remark above, this implies that we have $u^{-1}u' = h$ for some $h \in \mathrm{O}(n, \mathbb{R})$. Writing $u' = uh$ we have

$$u'(u\prime)^T = (uh)(uh)^T = u(hh^T)u^T = uu^T$$

since $hh^T = I$.

(ii) Let us show that the mapping $w(\cdot)$ is surjective. In view of Lemma 3.1, for every $w \in \mathrm{W}(n, \mathbb{C})$ there exists $u \in \mathrm{W}(n, \mathbb{C})$ such that $w = u^2 = uu^T$ (since u is symmetric); the Lagrangian plane $\ell = u\ell_P$ is then given by $w(\ell) = w$, hence the surjectivity. To show that $w(\cdot)$ is injective it suffices to show that if $uu^T = u'u'^T$, then $u\ell_P = u'\ell_P$, or equivalently, that $(u')^{-1}u \in \mathrm{O}(n, \mathbb{R})$. Now, the condition $uu^T = u'u'^T$ implies that $(u')^{-1}u = u'^T(u^T)^{-1}$ and hence

$$(u')^{-1}u\left((u')^{-1}u\right)^T = (u')^{-1}u\left(u'^T(u^T)^{-1}\right)^T = I,$$

that is $(u')^{-1}u \in \mathrm{O}(n, \mathbb{R})$ as claimed. There remains to prove formula (3.3). Assume that $\ell = u'\ell_P$; then $u\ell = uu'\ell_P$ and hence

$$w(u\ell) = (uu')(uu')^T = u(u'(u'^T))u^T = u'(u'^T)$$

as claimed. □

The Souriau mapping is a very useful tool when one wants to study transversality properties for Lagrangian planes; for instance

$$\ell \cap \ell' = 0 \Longleftrightarrow \det(w(\ell) - w(\ell')) \neq 0. \tag{3.4}$$

This is immediately seen by noticing that the condition $\det(w(\ell) - w(\ell')) \neq 0$ is equivalent to saying that $w(\ell)(w(\ell'))^{-1}$ does not have $+1$ as an eigenvalue. The equivalence (3.4) is in fact a particular case of the more general result. We denote by $W(\ell)$ the image of $w(\ell)$ in $\mathrm{U}(n)$:

$$w(\ell) = X + iY \Longleftrightarrow W(\ell) = \begin{bmatrix} X & -Y \\ Y & X \end{bmatrix}. \tag{3.5}$$

Proposition 3.4. *For any two Lagrangian planes* ℓ *and* ℓ' *in we have*

$$\operatorname{rank}(W(\ell) - W(\ell')) = 2(n - \dim(\ell \cap \ell')). \tag{3.6}$$

Proof. Since $\mathrm{Sp}(n)$ acts transitively on $\mathrm{Lag}(n)$ it suffices to consider the case $\ell' = \ell_P$, in which case formula (3.6) reduces to

$$\mathrm{rank}(w(\ell) - I) = 2(n - \dim(\ell \cap \ell_P)).$$

Let $w(\ell) = uu^T$ where $u = A + iB$; then, using the relations (3.1),

$$w(\ell) - I = -2(B^T B - iA^T B) = -2B^T(B - iA)$$

hence, with notation (3.5).

$$W(\ell) - I = -2\begin{bmatrix} B^T & 0 \\ 0 & B^T \end{bmatrix}\begin{bmatrix} B & A \\ -A & B \end{bmatrix}.$$

It follows that

$$\mathrm{rank}(W(\ell) - I) = 2\,\mathrm{rank}\, B;$$

this is equivalent to (3.6). □

The Souriau mapping $w(\cdot)$ can also be expressed in terms of projection operators on Lagrangian planes. Let ℓ be a Lagrangian plane and denote by P_ℓ the orthogonal projection in $\mathbb{R}^{2n}_z$ on ℓ:

$$P_\ell^2 = P_\ell \ , \ \mathrm{Ker}(P_\ell) = J\ell \ , \ (P_\ell)^T = P_\ell.$$

We have:

Proposition 3.5. *The image $W(\ell)$ of $w(\ell)$ in $\mathrm{U}(n)$ is given by*

$$W(\ell) = (I - 2P_\ell)C \tag{3.7}$$

where P_ℓ is the orthogonal projection in $\mathbb{R}^{2n}_z$ on ℓ and $C = \begin{bmatrix} I & 0 \\ 0 & -I \end{bmatrix}$ is the "conjugation matrix".

Proof. Since $\mathrm{U}(n)$ acts transitively on $\mathrm{Lag}(n)$, there exists $U \in \mathrm{U}(n)$ such that $\ell = U\ell_P$. Writing

$$U = \begin{bmatrix} A & -B \\ B & A \end{bmatrix}$$

it follows from the relations (3.1) that the vector (Ax, Bx) is orthogonal to ℓ; one immediately checks that the projection operator on ℓ has matrix

$$P_\ell = \begin{bmatrix} BB^T & -AB^T \\ -BA^T & AA^T \end{bmatrix}$$

and hence

$$(I - 2P_\ell)C = \begin{bmatrix} AA^T - BB^T & -2AB^T \\ 2BA^T & BB^T - AA^T \end{bmatrix},$$

that is:

$$(I - 2P_\ell)C = \begin{bmatrix} A & -B \\ B & A \end{bmatrix}\begin{bmatrix} A^T & -B^T \\ B^T & A^T \end{bmatrix} = W(\ell).$$

□

3.1.2 Definition of the Maslov index

We are going to use the Souriau mapping to show that $\pi_1[\mathrm{Lag}(n)]$ is isomorphic to the integer group $(\mathbb{Z},+)$; this will also allow us to define the Maslov index of a loop in $\mathrm{Lag}(n)$.

Let us begin with a preliminary result, interesting by itself. It is a "folk theorem" that the Poincaré group $\pi_1[\mathrm{U}(n,\mathbb{C})]$ is isomorphic to the integer group $(\mathbb{Z},+)$. Let us give a detailed proof of this property; this will at the same time give an explicit isomorphism we will use to define the Maslov index. We recall that the special unitary group $\mathrm{SU}(n,\mathbb{C})$ is connected and simply connected (see, *e.g.*, Leray [107], Ch. I, §2,3).

Lemma 3.6. *The mapping* $\pi_1[\mathrm{U}(n,\mathbb{C})] \longrightarrow \mathbb{Z}$ *defined by*

$$\gamma \longmapsto \frac{1}{2\pi i}\int_\gamma \frac{d(\det u)}{\det u} \tag{3.8}$$

is an isomorphism, and hence $\pi_1[\mathrm{U}(n,\mathbb{C})] \cong (\mathbb{Z},+)$.

Proof. The kernel of the epimorphism $u \longmapsto \det u$ is $\mathrm{SU}(n,\mathbb{C})$ so that we have a fibration $\mathrm{U}(n,\mathbb{C})/\mathrm{SU}(n,\mathbb{C}) = S^1$. The homotopy sequence of that fibration contains the exact sequence

$$\pi_1[\mathrm{SU}(n,\mathbb{C})] \xrightarrow{i} \pi_1[\mathrm{U}(n,\mathbb{C})] \xrightarrow{f} \pi_1[S^1] \longrightarrow \pi_0[\mathrm{SU}(n,\mathbb{C})]$$

where f is induced by $\mathrm{U}(n,\mathbb{C})/\mathrm{SU}(n,\mathbb{C}) = S^1$. Since $\mathrm{SU}(n,\mathbb{C})$ is both connected and simply connected, $\pi_0[\mathrm{SU}(n,\mathbb{C})$ and $\pi_1[\mathrm{SU}(n,\mathbb{C})]$ are trivial, and the sequence above reduces to

$$0 \longrightarrow \pi_1[\mathrm{U}(n,\mathbb{C})] \xrightarrow{f} \pi_1[S^1] \longrightarrow 0$$

hence f is an isomorphism. The result now follows from the fact that the mapping $\pi_1[S^1] \longrightarrow \mathbb{Z}$ defined by

$$\alpha \longmapsto \frac{1}{2\pi i}\int_\alpha \frac{dz}{z}$$

is an isomorphism $\pi_1[S^1] \cong \mathbb{Z}$. □

The next result is important; it shows among other things that the fundamental group of the Lagrangian Grassmannian $\mathrm{Lag}(n)$ is isomorphic to the integer group $(\mathbb{Z},+)$; that isomorphism $\mathrm{Lag}(n) \cong (\mathbb{Z},+)$ is, by definition, the *Maslov index*:

Theorem 3.7.

(i) *The mapping*

$$\pi_1[\mathrm{W}(n,\mathbb{C})] \ni \gamma_W \longmapsto \frac{1}{2\pi i}\int_{\gamma_W} \frac{d(\det w)}{\det w} \in \mathbb{Z} \tag{3.9}$$

is an isomorphism $\pi_1[\mathrm{W}(n,\mathbb{C})] \cong (\mathbb{Z},+)$.

(ii) *The composition of this isomorphism with the isomorphism* $\pi_1[\mathrm{Lag}(n)] \cong \pi_1[\mathrm{W}(n,\mathbb{C})]$ *induced by the Souriau mapping is an isomorphism*

$$m_{\mathrm{Lag}} : \pi_1[\mathrm{Lag}(n)] \cong (\mathbb{Z},+). \tag{3.10}$$

(iii) *In fact* $\pi_1[\mathrm{Lag}(n)]$ *has a generator* β *such that* $m_{\mathrm{Lag}}(\beta^r) = r$ *for every* $r \in \mathbb{Z}$.

Proof. The statement (ii) is an obvious consequence of the statement (i). Let us prove (i). Since $\mathrm{W}(n,\mathbb{C}) \subset \mathrm{U}(n,\mathbb{C})$ It follows from Lemma 3.6 that

$$\frac{1}{2\pi i}\int_{\gamma_W} \frac{d(\det w)}{\det w} \in \mathbb{Z}$$

for every $\gamma_W \in \pi_1[\mathrm{W}(n,\mathbb{C})]$ and that the homomorphism (3.9) is injective. Let us show that this homomorphism is also surjective; it suffices for that to exhibit the generator β in (iii). Writing $(x,p) = (x_1,p_1;\ldots;x_n,p_n)$ the direct sum $\mathrm{Lag}(1) \oplus \cdots \oplus \mathrm{Lag}(1)$ (n terms) is a subset of $\mathrm{Lag}(n)$. Consider the loop $\beta_{(1)} : t \longmapsto e^{2\pi i t}$, $0 \leq t \leq 1$, in $W(1,\mathbb{C}) \equiv \mathrm{Lag}(1)$. Set now $\beta = \beta_{(1)} \oplus I_{2n-2}$ where I_{2n-2} is the identity in $W(n-1,\mathbb{C})$. We have $\beta^r = \beta_{(1)}^r \oplus I_{2n-2}$ and

$$m_{\mathrm{Lag}}(\beta^r) = \frac{1}{2\pi i}\int_0^1 \frac{d(e^{2\pi i r t})}{e^{2\pi i r t}} = r$$

which was to be proven. □

The isomorphism $\pi_1[\mathrm{Lag}(n)] \cong (\mathbb{Z},+)$ constructed in Theorem 3.7 is precisely the Maslov index of the title of this section:

Definition 3.8.

(i) The mapping $m_{\mathrm{Lag}}^{(n)}$ which to every loop γ in $\mathrm{Lag}(n)$ associates the integer

$$m_{\mathrm{Lag}}^{(n)}(\gamma) = \frac{1}{2\pi i}\oint_\gamma \frac{d(\det w)}{\det w} \tag{3.11}$$

is called the "Maslov index" on $\mathrm{Lag}(n)$; when the dimension n is understood we denote it by m_{Lag} as in Theorem 3.7.

(ii) The loop $\beta = \beta_{(1)} \oplus I_{2n-2}$ in $\mathrm{Lag}(n)$ is called the generator of $\pi_1[\mathrm{Lag}(n)]$ whose natural image in $\mathbb{Z}$ is $+1$.

We will extend the definition of the Maslov index in Chapter 5 to loops on Lagrangian manifolds (*i.e.*, submanifolds of $\mathbb{R}_z^{2n}$ whose tangent spaces are Lagrangian planes). This will lead us to the so-called Maslov semiclassical quantization of these manifolds, which is a mathematically rigorous generalization of the physicists' "EBK quantization". Let us study the main properties of the Maslov index. Another much less trivial extension will be constructed in Chapter 3 under the name of "*ALM* index" (*ALM* is an acronym for *Arnol'd–Leray–Maslov*); it will lead us to the definition of quite general Lagrangian intersection indices.

3.1.3 Properties of the Maslov index

If γ and γ' are loops with the same origin, then we denote by $\gamma * \gamma'$ their concatenation, that is the loop γ followed by the loop γ':

$$\gamma * \gamma'(t) = \begin{cases} \gamma(2t) & \text{for } 0 \leq t \leq 1/2, \\ \gamma(2t-1) & \text{for } 1/2 \leq t \leq 1. \end{cases}$$

The following result characterizes the Maslov index:

Proposition 3.9. *The family $(m_{\mathrm{Lag}}^{(n)})_{n\in\mathbb{N}}$ is the only family of mappings $m_{\mathrm{Lag}}^{(n)} : \mathrm{Lag}(n) \longrightarrow \mathbb{Z}$ having the following properties:*

(i) *Homotopy: two loops γ and γ' in $\mathrm{Lag}(n)$ are homotopic if and only if $m_{\mathrm{Lag}}^{(n)}(\gamma) = m_{\mathrm{Lag}}^{(n)}(\gamma')$;*

(ii) *Additivity under concatenation: for all loops γ and γ' in $\mathrm{Lag}(n)$ with the same origin,*

$$m_{\mathrm{Lag}}^{(n)}(\gamma * \gamma') = m_{\mathrm{Lag}}^{(n)}(\gamma) + m_{\mathrm{Lag}}^{(n)}(\gamma');$$

(iii) *Normalization: the generator β of $\pi_1[\mathrm{Lag}(n)]$ has Maslov index $m_{\mathrm{Lag}}^{(n)}(\beta) = +1$.*

(iv) *Dimensional additivity: identifying $\mathrm{Lag}(n_1) \oplus \mathrm{Lag}(n_2)$ with a subset of $\mathrm{Lag}(n)$, $n = n_1 + n_2$, we have*

$$m_{\mathrm{Lag}}^{(n)}(\gamma_1 \oplus \gamma_2) = m_{\mathrm{Lag}}^{(n_1)}(\gamma_1) + m_{\mathrm{Lag}}^{(n_2)}(\gamma_2)$$

if γ_j is a loop in $\mathrm{Lag}(n_j)$, $j = 1, 2$.

Proof. The additivity properties (ii) and (iv) are obvious and so is the normalization property (iii) using formula (3.11). That $m_{\mathrm{Lag}}^{(n)}(\gamma)$ only depends on the homotopy class of the loop γ is clear from the definition of the Maslov index as being a mapping $\pi_1[\mathrm{Lag}(n)] \longrightarrow (\mathbb{Z}, +)$ and that $m_{\mathrm{Lag}}^{(n)}(\gamma) = m_{\mathrm{Lag}}^{(n)}(\gamma')$ implies that γ and γ' are homotopic follows from the injectivity of that mapping. Let us finally prove the uniqueness of $(m_{\mathrm{Lag}}^{(n)})_{n\in\mathbb{N}}$. Suppose there is another family of mappings $\mathrm{Lag}(n) \longrightarrow \mathbb{Z}$ having the same property; then the difference $(\delta_{\mathrm{Lag}}^{(n)})_{n\in\mathbb{N}}$ has the properties (i), (ii), (iv) and (iii) is replaced by $\delta_{\mathrm{Lag}}^{(1)}(\beta_{(1)}) = 0$. Every loop γ in $\mathrm{Lag}(n)$ being homotopic to β^r for some $r \in \mathbb{Z}$, it follows from the concatenation property (ii) that $\delta_{\mathrm{Lag}}^{(n)}(\gamma) = \delta_{\mathrm{Lag}}^{(n)}(\beta^r) = 0$. □

Remark 3.10. Notice that we did not use in the proof of uniqueness in Proposition 3.9 the dimensional additivity property: properties (i), (ii), and (iii) thus characterize the Maslov index.

3.1.4 The Maslov index on $\mathrm{Sp}(n)$

Let $\gamma : [0,1] \longrightarrow \mathrm{Sp}(n)$ be a loop of symplectic matrices: $\gamma(t) \in \mathrm{Sp}(n)$ and $\gamma(0) = \gamma(1)$. The orthogonal part of the polar decomposition $\gamma(t) = U(t)e^{X(t)}$ is given by the formula

$$U(t) = \gamma(t)(\gamma(t)^T\gamma(t))^{-1/2}. \tag{3.12}$$

Definition 3.11. The Maslov index of the symplectic loop γ is the integer

$$m_{\mathrm{Sp}}(\gamma) = \theta(1) - \theta(0)$$

where θ is the continuous function $[0,1] \longrightarrow \mathbb{R}$ defined by $\det u(t) = e^{2\pi i\theta(t)}$ where $u(t)$ is the image in $\mathrm{U}(n,\mathbb{C})$ of the matrix $U(t) \in \mathrm{U}(n)$ defined by (3.12).

Let us exhibit a particular generator of $\pi_1[\mathrm{Sp}(n)]$. (In addition to the fact that it allows easy calculations of the Maslov index it will be useful in the study of general symplectic intersection indices in Chapter 3).

Let us rearrange the coordinates in $\mathbb{R}^{2n}_z$ and identify (x,p) with the vector $(x_1, p_1, \ldots, x_n, p_n)$; denoting by $\mathrm{Sp}(1)$ the symplectic group acting on pairs (x_j, p_j) the direct sum

$$\mathrm{Sp}(1) \oplus \mathrm{Sp}(1) \oplus \cdots \oplus \mathrm{Sp}(1) \quad (n \text{ terms})$$

is identified with a subgroup of $\mathrm{Sp}(n)$ in the obvious way. We will denote by J_1 the standard 2×2 symplectic matrix:

$$J_1 = \begin{bmatrix} 0 & 1 \\ -1 & 0 \end{bmatrix}.$$

With the notation we have (*cf.* the proof of Theorem 3.7):

Proposition 3.12.

(i) *The fundamental group $\pi_1[\mathrm{Sp}(n)]$ is generated by the loop*

$$\alpha : t \longmapsto e^{2\pi t J_1} \oplus I_{n-2} \quad , \quad 0 \leq t \leq 1 \tag{3.13}$$

where I_{2n-2} is the identity on $\mathbb{R}^{2n-2}$.

(ii) *The Maslov index of any symplectic loop γ is $m_{\mathrm{Sp}}(\gamma) = r$ where the integer r is defined by the condition: "γ is homotopic to α^r".*

Proof. (i) Clearly $J_1 \in \mathfrak{sp}(1)$ hence $\alpha(t) \in \mathrm{Sp}(n)$; since

$$e^{2\pi t J_1} = \begin{bmatrix} \cos 2\pi t & -\sin 2\pi t \\ \sin 2\pi t & \cos 2\pi t \end{bmatrix}$$

we have $\alpha(0) = \alpha(1)$. Now $\alpha(t)^T\alpha(t)$ is the identity, hence

$$\alpha(t)(\alpha(t)^T\alpha(t))^{-1/2} = \alpha(t) = \begin{bmatrix} A(t) & -B(t) \\ B(t) & A(t) \end{bmatrix}$$

where $A(t)$ and $B(t)$ are the diagonal matrices

$$A(t) = \operatorname{diag}[\cos(2\pi t), 1, \dots, 1],$$
$$B(t) = \operatorname{diag}[\sin(2\pi t), 0, \dots, 0].$$

Let Δ be the mapping $\operatorname{Sp}(n) \longrightarrow S^1$ defined in Proposition 2.26 above; we have

$$\Delta(\alpha(t)) = \det(A(t) + iB(t)) = e^{2\pi i t} \tag{3.14}$$

hence $t \mapsto \Delta(\alpha(t))$ is the generator of $\pi_1[S^1]$. The result follows.

(ii) It suffices to show that $m_{\operatorname{Sp}}(\alpha) = 1$. But this follows from formula (3.14). □

Definition 3.13.

(i) The loop α defined by (3.13) will be called the "generator of $\pi_1[\operatorname{Sp}(n)]$ whose image in $\mathbb{Z}$ is $+1$".

(ii) Let γ be an arbitrary loop in $\operatorname{Sp}(n)$; the integer r such that γ is homotopic to α^r is called the "Maslov index of γ".

The following result is the analogue of Proposition 3.9; its proof being quite similar it is left to the reader as an exercise:

Proposition 3.14. *The family $(m_{\operatorname{Sp}}^{(n)})_{n\in\mathbb{N}}$ is the only family of mappings $m_{\operatorname{Sp}}^{(n)}$: $\operatorname{Sp}(n) \longrightarrow \mathbb{Z}$ having the following properties:*

(i) *Homotopy: two loops γ and γ' in $\operatorname{Sp}(n)$ are homotopic if and only if $m_{\operatorname{Sp}}^{(n)}(\gamma) = m_{\operatorname{Sp}}^{(n)}(\gamma')$;*

(ii) *Additivity under concatenation: for all loops γ and γ' in $\operatorname{Sp}(n)$ with the same origin*

$$m_{\operatorname{Sp}}^{(n)}(\gamma * \gamma') = m_{\operatorname{Sp}}^{(n)}(\gamma) + m_{\operatorname{Sp}}^{(n)}(\gamma');$$

(iii) *Normalization: the generator α of $\pi_1[\operatorname{Sp}(n)]$ has Maslov index $m_{\operatorname{Sp}}^{(n)}(\beta) = +1$.*

(iv) *Dimensional additivity: identifying $\operatorname{Sp}(n_1) \oplus \operatorname{Sp}(n_2)$ with a subset of $\operatorname{Lag}(n)$, $n = n_1 + n_2$ we have*

$$m_{\operatorname{Sp}}^{(n)}(\gamma_1 \oplus \gamma_2) = m_{\operatorname{Sp}}^{(n_1)}(\gamma_1) + m_{\operatorname{Sp}}^{(n_2)}(\gamma_2)$$

if γ_j is a loop in $\operatorname{Sp}(n_j)$, $j = 1, 2$.

3.2 The Arnol'd–Leray–Maslov Index

Following ideas of Maslov [119] and Arnol'd [4] Leray constructed in [107, 108, 109] an index m such that

$$m(\ell_\infty, \ell'_\infty) - m(\ell_\infty, \ell''_\infty) + m(\ell'_\infty, \ell''_\infty) = \operatorname{Inert}(\ell, \ell', \ell'')$$

for all triples $(\ell_\infty, \ell'_\infty, \ell''_\infty)$ with pairwise transversal projections:

$$\ell \cap \ell' = \ell' \cap \ell'' = \ell'' \cap \ell = 0. \tag{3.15}$$

The integer $\mathrm{Inert}(\ell, \ell', \ell'')$ is the index of inertia of the triple (ℓ, ℓ', ℓ'') (Leray [107]); it is defined as follows: the conditions

$$(z, z', z'') \in \ell \times \ell' \times \ell'' \quad , \quad z + z' + z'' = 0$$

define three isomorphisms $z \longmapsto z'$, $z' \longmapsto z''$, $z'' \longmapsto z$ whose product is the identity. It follows that

$$\sigma(z, z') = \sigma(z', z'') = \sigma(z'', z)$$

is the value of a quadratic form at $z \in \ell$ (or $z' \in \ell'$, or $z'' \in \ell''$); these quadratic forms have the same index of inertia, denoted by $\mathrm{Inert}(\ell, \ell', \ell'')$.

Since Leray's index of inertia $\mathrm{Inert}(\ell, \ell', \ell'')$ is defined in terms of quadratic forms which only exist when the transversality conditions (3.15) are satisfied, it is not immediately obvious how to extend $m(\ell_\infty, \ell'_\infty)$ to arbitrary pairs $(\ell_\infty, \ell'_\infty)$. The extension presented in this chapter is due to the author; it first appeared in [54] and was then detailed in [57]. (Dazord has constructed in [28] a similar index using methods from algebraic topology, for a different approach see Leray [109]). The main idea is to use the signature $\tau(\ell, \ell', \ell'')$ instead of $\mathrm{Inert}(\ell, \ell', \ell'')$; this idea probably goes back to Lion and Vergne [111], albeit in a somewhat incomplete form: see the remarks in de Gosson [54, 55]. For a very detailed study of the *ALM* and related indices see the paper [22] by Cappell *et al.*

The theory of the Arnol'd–Leray–Maslov index – which we will call for short the *ALM index* – is a beautiful generalization of the theory of the Maslov index of Lagrangian loops. It is a very useful mathematical object, which can be used to express various other indices: Lagrangian and symplectic path intersection indices, and, as we will see, the Conley–Zehnder index.

3.2.1 The problem

Recall from Chapter 1 that the Wall–Kashiwara signature associates to every triple (ℓ, ℓ', ℓ'') of Lagrangian planes in $(\mathbb{R}^{2n}_z, \sigma)$ an integer $\tau(\ell, \ell', \ell'')$ which is the signature of the quadratic form

$$(z, z', z'') \longmapsto \sigma(z, z') + \sigma(z', z'') + \sigma(z'', z)$$

on $\ell \oplus \ell' \oplus \ell''$. Besides being antisymmetric and $\mathrm{Sp}(n)$-invariant, τ is a cocycle, that is:

$$\tau(\ell_1, \ell_2, \ell_3) - \tau(\ell_2, \ell_3, \ell_4) + \tau(\ell_1, \ell_3, \ell_4) - \tau(\ell_1, \ell_2, \ell_4) = 0. \tag{3.16}$$

As briefly mentioned in the statement of Theorem 1.32 this property can be expressed in terms of the coboundary operator ∂ (see the section devoted to the notations in the preface) in the abbreviated form

$$\partial\tau(\ell_1, \ell_2, \ell_3, \ell_4) = 0.$$

Let us look for primitives of the cocycle τ; by primitive we mean a "1-cochain"

$$\mu : \mathrm{Lag}(n) \times \mathrm{Lag}(n) \longrightarrow \mathbb{Z}$$

such that

$$\partial\mu(\ell_1, \ell_2, \ell_3) = \tau(\ell_1, \ell_2, \ell_3)$$

where ∂ is the usual "coboundary operator". That primitives exist is easy to see: for instance, for every fixed Lagrangian plane ℓ the cochain μ_ℓ defined by

$$\mu_\ell(\ell_1, \ell_2) = \tau(\ell, \ell_1, \ell_2) \tag{3.17}$$

satisfies, in view of (3.16),

$$\begin{aligned}\mu_\ell(\ell_1, \ell_2) - \mu_\ell(\ell_1, \ell_3) + \mu_\ell(\ell_2, \ell_3) &= \tau(\ell, \ell_1, \ell_2) - \tau(\ell, \ell_1, \ell_3) + \tau(\ell, \ell_2, \ell_3) \\ &= \tau(\ell_1, \ell_2, \ell_3)\end{aligned}$$

and hence $\partial\mu_\ell = \tau$. We however want the primitive we are looking for to satisfy, in addition, topological properties consistent with those of the signature τ. Recall that we showed that $\tau(\ell_1, \ell_2, \ell_3)$ remains constant when the triple (ℓ_1, ℓ_2, ℓ_3) moves continuously in such a way that the dimensions of the intersections $\dim(\ell_1, \ell_2)$, $\dim(\ell_1, \ell_3)$, $\dim(\ell_2, \ell_3)$ do not change. It is therefore reasonable to demand that the primitive μ also is locally constant on all pairs (ℓ_1, ℓ_2) such that $\dim(\ell_1, \ell_2)$ is fixed. It is easy to see why the cochain (3.17) does not satisfy this property: assume, for instance, that the pair (ℓ_1, ℓ_2) moves continuously while remaining transversal: $\ell_1 \cap \ell_2 = 0$. Then, $\tau(\ell, \ell_1, \ell_2)$ would – if the desired condition is satisfied – remain constant. This is however not the case, since the signature of a triple of Lagrangian planes changes when we change the relative positions of the involved planes (see Subsection 1.4.1). It turns out that we will actually *never* be able to find a cochain μ on $\mathrm{Lag}(n)$ which is both a primitive of τ *and* satisfies the topological condition above: to construct such an object we have to pass to the universal covering $\mathrm{Lag}_\infty(n)$ ("Maslov bundle") of $\mathrm{Lag}(n)$.

Let $\pi : \mathrm{Lag}_\infty(n) \longrightarrow \mathrm{Lag}(n)$ be the universal covering of the Lagrangian Grassmannian $\mathrm{Lag}(n)$. We will write $\ell = \pi(\ell_\infty)$.

Definition 3.15. The ALM (= Arnol'd–Leray–Maslov) index is the unique mapping

$$\mu : (\mathrm{Lag}_\infty(n))^2 \longrightarrow \mathbb{Z}$$

having the two following properties:

(i) μ is locally constant on the set $\{(\ell_\infty, \ell'_\infty) : \ell \cap \ell' = 0\}$;
(ii) $\partial\mu : (\mathrm{Lag}_\infty(n))^3 \longrightarrow \mathbb{Z}$ descends to $(\mathrm{Lag}(n))^3$ and is equal to τ:

$$\mu(\ell_\infty, \ell'_\infty) - \mu(\ell_\infty, \ell''_\infty) + \mu(\ell'_\infty, \ell''_\infty) = \tau(\ell, \ell', \ell''). \tag{3.18}$$

Notice that property (3.18) implies, together with the antisymmetry of the signature, that if the *ALM* index exists then it must satisfy

$$\mu(\ell_\infty, \ell'_\infty) = -\mu(\ell'_\infty, \ell_\infty) \tag{3.19}$$

for all pairs $(\ell_\infty, \ell'_\infty)$.

Admittedly, the definition above is not very constructive. And, by the way, why is μ (provided that it exists) *unique*? This question is at least easily answered. Suppose there are two mappings μ and μ' satisfying the same conditions as above: for all triples $(\ell_\infty, \ell'_\infty, \ell''_\infty)$,

$$\mu(\ell_\infty, \ell'_\infty) - \mu(\ell_\infty, \ell''_\infty) + \mu(\ell'_\infty, \ell''_\infty) = \tau(\ell, \ell', \ell''),$$
$$\mu'(\ell_\infty, \ell'_\infty) - \mu'(\ell_\infty, \ell''_\infty) + \mu'(\ell'_\infty, \ell''_\infty) = \tau(\ell, \ell', \ell'').$$

It follows that $\delta = \mu - \mu'$ is such that

$$\delta(\ell_\infty, \ell'_\infty) = \delta(\ell_\infty, \ell''_\infty) - \delta(\ell'_\infty, \ell''_\infty); \tag{3.20}$$

since μ and μ' are locally constant on $\{(\ell_\infty, \ell'_\infty) : \ell \cap \ell' = 0\}$ the same is true of δ. Choosing ℓ'' such that $\ell'' \cap \ell = \ell'' \cap \ell' = 0$ we see that in fact δ is locally constant on all of $(\mathrm{Lag}_\infty(n))^2$. Now $\mathrm{Lag}_\infty(n)$, and hence $(\mathrm{Lag}_\infty(n))^2$, is connected so that δ is actually constant. Its constant value is

$$\delta(\ell_\infty, \ell_\infty) = \delta(\ell_\infty, \ell''_\infty) - \delta(\ell_\infty, \ell''_\infty) = 0$$

hence $\mu = \mu'$ and the Arnol'd–Leray–Maslov index (if it exists) is thus indeed unique.

We will see that the action of fundamental group of $\mathrm{Lag}(n)$ on $\mathrm{Lag}_\infty(n)$ is reflected on the *ALM* index by the formula

$$\mu(\beta^r \ell_\infty, \beta^{r'} \ell'_\infty) = \mu(\ell_\infty, \ell'_\infty) + 2(r - r') \tag{3.21}$$

where β is the generator of $\pi_1[\mathrm{Lag}(n)] \cong (\mathbb{Z}, +)$ whose image in $\mathbb{Z}$ is $+1$. This formula shows that the *ALM* index is effectively defined on $(\mathrm{Lag}_\infty(n))^2$ (*i.e.*, it is "multi-valued" on $(\mathrm{Lag}(n))^2$); it also shows why we could not expect to find a function having similar properties on $\mathrm{Lag}(n)$ itself: if such a function μ' existed, we could "lift" it to a function on $(\mathrm{Lag}_\infty(n))^2$ in an obvious way by the formula $\mu'(\ell_\infty, \ell'_\infty) = \mu'(\ell, \ell')$; but the uniqueness of the *ALM* index would then imply that $\mu' = \mu$ which is impossible since μ' cannot satisfy (3.21).

An important consequence of this uniqueness is the invariance of the *ALM* index under the action of the universal covering group $\mathrm{Sp}_\infty(n)$ of $\mathrm{Sp}(n)$:

Proposition 3.16. *For all* $(S_\infty, \ell_\infty, \ell'_\infty) \in \mathrm{Sp}_\infty(n) \times (\mathrm{Lag}_\infty(n))^2$ *we have*

$$\mu(S_\infty \ell_\infty, S_\infty \ell'_\infty) = \mu(\ell_\infty, \ell'_\infty). \tag{3.22}$$

Proof. Set, for fixed $S_\infty \in \mathrm{Sp}_\infty(n)$, $\mu'(\ell_\infty, \ell'_\infty) = \mu(S_\infty \ell_\infty, S_\infty \ell'_\infty)$. We have

$$\mu'(\ell_\infty, \ell'_\infty) - \mu'(\ell_\infty, \ell''_\infty) + \mu'(\ell'_\infty, \ell''_\infty) = \tau(S\ell, S\ell', S\ell'')$$

where $S \in \mathrm{Sp}(n)$ is the projection of S_∞. In view of the symplectic invariance of the Wall–Kashiwara signature we have $\tau(S\ell, S\ell', S\ell'')$ and hence

$$\mu'(\ell_\infty, \ell'_\infty) - \mu'(\ell_\infty, \ell''_\infty) + \mu'(\ell'_\infty, \ell''_\infty) = \tau(\ell, \ell', \ell'').$$

Since on the other hand $S\ell \cap S\ell' = 0$ if and only if $\ell \cap \ell' = 0$, the index μ' is locally constant on $\{(\ell_\infty, \ell'_\infty) : \ell \cap \ell' = 0\}$ and must thus be equal to μ, that is (3.22). □

There now remains the hard part of the work, namely the explicit construction of the ALM index. Let us show how this can be done in the case $n = 1$. The general case will definitely require more work. The Lagrangian Grassmannian Lag(1) consists of all straight lines through the origin in the symplectic plane $(\mathbb{R}^2_z, -\det)$. Let $\ell = \ell$ and $\ell' = \ell'$ be the lines with equations

$$x \cos\alpha + p \sin\alpha = 0 \quad , \quad x \cos\alpha' + p \sin\alpha' = 0$$

and identify ℓ_∞ and ℓ'_∞ with $\theta = 2\alpha$ and $\theta' = 2\alpha'$. Denoting by $[r]$ the integer part of $r \in \mathbb{R}$ we then have

$$\mu(\theta, \theta') = \begin{cases} 2\left[\frac{\theta - \theta'}{2\pi}\right] + 1 \text{ if } \theta - \theta' \notin \pi\mathbb{Z}, \\ k \text{ if } \theta - \theta' = k\pi. \end{cases} \tag{3.23}$$

Introducing the antisymmetric integer part function

$$[r]_{\text{anti}} = \frac{1}{2}([r] - [-r]) = \begin{cases} [r] + \frac{1}{2} \text{ if } r \notin \mathbb{Z}, \\ r \text{ if } r \in \mathbb{Z}, \end{cases}$$

definition (3.23) can be rewritten in compact form as

$$\mu(\theta, \theta') = 2\left[\frac{\theta - \theta'}{2\pi}\right]_{\text{anti}}. \tag{3.24}$$

The coboundary $\partial\mu$ is the function

$$\partial\mu(\theta, \theta', \theta'') = 2\left[\frac{\theta - \theta'}{2\pi}\right]_{\text{anti}} - 2\left[\frac{\theta - \theta''}{2\pi}\right]_{\text{anti}} + 2\left[\frac{\theta' - \theta''}{2\pi}\right]_{\text{anti}}$$

and this is precisely the signature $\tau(\ell, \ell', \ell'')$ in view of formula (1.21) in Section 1.4.

To generalize this construction to arbitrary n we need a precise "numerical" description of the universal covering of Lag(n).

3.2.2 The Maslov bundle

The Maslov bundle is, by definition, the universal covering manifold $\mathrm{Lag}_\infty(n)$ of the Lagrangian Grassmannian $\mathrm{Lag}(n)$.

The homomorphism

$$\pi_1\left[\mathrm{U}(n,\mathbb{C})\right] \ni \gamma \longmapsto k_\gamma \in \mathbb{Z}$$

defined by

$$k_\gamma = \frac{1}{2\pi i}\int_\gamma \frac{d(\det u)}{\det u}$$

is an isomorphism $\pi_1\left[\mathrm{U}(n,\mathbb{C})\right] \cong (\mathbb{Z},+)$. Set now

$$\mathrm{U}_\infty(n,\mathbb{C}) = \left\{(u,\theta) : u \in \mathrm{U}(n,\mathbb{C}), \det u = e^{i\theta}\right\}$$

and equip this set with the topology induced by the product $\mathrm{U}(n,\mathbb{C}) \times \mathbb{R}$. Define a projection $\pi_\infty : \mathrm{U}_\infty(n,\mathbb{C}) \longrightarrow \mathrm{U}(n,\mathbb{C})$ by $\pi_\infty(U,\theta) = W$, and let the group $\pi_1\left[\mathrm{U}(n,\mathbb{C})\right]$ act on $\mathrm{U}_\infty(n,\mathbb{C})$ by the law

$$\gamma(u,\theta) = (u, \theta + 2k_\gamma\pi).$$

That action is clearly transitive, hence $\mathrm{U}_\infty(n,\mathbb{C})$ is the universal covering group of $\mathrm{U}(n,\mathbb{C})$, the group structure being given by

$$(U,\theta)(U',\theta') = (UU', \theta+\theta').$$

Let us now identify the Maslov bundle with a subset of $\mathrm{U}_\infty(n,\mathbb{C})$:

Proposition 3.17. *The universal covering of* $\mathrm{Lag}(n) \equiv \mathrm{W}(n,\mathbb{C})$ *is the set*

$$\mathrm{W}_\infty(n,\mathbb{C}) = \left\{(w,\theta) : w \in \mathrm{W}(n,\mathbb{C}), \det w = e^{i\theta}\right\}$$

equipped with the topology induced by $\mathrm{U}_\infty(n)$, *together with the projection* π_∞ : $\mathrm{W}_\infty(n,\mathbb{C}) \longrightarrow \mathrm{W}(n,\mathbb{C})$ *defined by* $\pi_\infty(w,\theta) = w$.

Proof. In view of Theorem 3.7 of Subsection 3.1.2 where the fundamental group of $\mathrm{W}(n,\mathbb{C})$ is identified with $(\mathbb{Z},+)$ it is sufficient to check that $\mathrm{W}_\infty(n,\mathbb{C})$ is connected because it will then indeed be the universal covering of $\mathrm{W}(n,\mathbb{C})$. Let $\mathrm{U}_\infty(n,\mathbb{C})$ act on $\mathrm{W}_\infty(n,\mathbb{C})$ via the law

$$(u,\varphi)(w,\theta) = (uwu^T, \theta + 2\varphi). \tag{3.25}$$

The stabilizer of $(I,0)$ in $\mathrm{U}_\infty(n)$ under this action is the subgroup of $\mathrm{U}_\infty(n)$ consisting of all pairs (U,φ) such that $UU^T = I$ and $\varphi = 0$ (and hence $\det U = 1$); it can thus be identified with the rotation group $\mathrm{SO}(n)$ and hence

$$\mathrm{W}_\infty(n,\mathbb{C}) = \mathrm{U}_\infty(n,\mathbb{C})/\,\mathrm{SO}(n,\mathbb{R}).$$

Since $\mathrm{U}_\infty(n,\mathbb{C})$ is connected, so is $\mathrm{W}_\infty(n,\mathbb{C})$. □

The Maslov bundle $\mathrm{Lag}_\infty(n)$ is the universal covering of the Lagrangian Grassmannian. Quite abstractly, it is constructed as follows (the construction is not specific of $\mathrm{Lag}_\infty(n)$, it is the way one constructs the universal covering of any topological space: see the Appendix B). Choose a "base point" ℓ_0 in $\mathrm{Lag}(n)$: it is any fixed Lagrangian plane; let ℓ be an arbitrary element of $\mathrm{Lag}(n)$. Since $\mathrm{Lag}(n)$ is path-connected, there exists at least one continuous path $\lambda : [0,1] \longmapsto \mathrm{Lag}(n)$ going from ℓ_0 to ℓ: $\lambda(0) = \ell_0$ and $\lambda(1) = \ell$. We say that two such paths λ and λ' are 'homotopic with fixed endpoints' if one of them can be continuously deformed into the other while keeping its origin ℓ_0 and its endpoint ℓ fixed. Homotopy with fixed endpoints is an equivalence relation; denote the equivalence class of the path λ by ℓ_∞. The universal covering of $\mathrm{Lag}_\infty(n)$ is the set of all the equivalence classes ℓ_∞ as ℓ ranges over $\mathrm{Lag}(n)$; the mapping $\pi_\infty : \mathrm{Lag}_\infty(n) \longrightarrow \mathrm{Lag}(n)$ which to ℓ_∞ associates the endpoint ℓ of a path λ in ℓ_∞ is called a 'covering projection'. One shows that it is possible to endow the set $\mathrm{Lag}_\infty(n)$ with a topology for which it is both connected and simply connected, and such that every $\ell \in \mathrm{Lag}(n)$ has an open neighborhood $\mathcal{U}_\ell$ such that $\pi_\infty^{-1}(\mathcal{U}_\ell)$ is the disjoint union of open neighborhoods $\mathcal{U}_\ell^{(1)}, \ldots, \mathcal{U}_\ell^{(k)}, \ldots$ of the points of $\pi_\infty^{-1}(\ell)$, and the restriction of π_∞ to each $\mathcal{U}_\ell^{(k)}$ is a homeomorphism $\mathcal{U}_\ell^{(k)} \longrightarrow \mathcal{U}_\ell$.

3.2.3 Explicit construction of the ALM index

Let us identify ℓ_∞ with (w, θ), w being the image of ℓ in $\mathrm{W}(n, \mathbb{C})$ by the Souriau mapping and $\det w = e^{i\theta}$. We are going to prove that:

- The *ALM* index exists and is given by $\mu(\ell_\infty, \ell'_\infty) = \frac{1}{2} m(\ell_\infty, \ell'_\infty)$, that is
$$\mu(\ell_\infty, \ell'_\infty) = \frac{1}{\pi}\left[\theta - \theta' + i \operatorname{Tr}\log(-w(w')^{-1}\right] \tag{3.26}$$
when $\ell \cap \ell' \neq 0$;
- When $\ell \cap \ell'$ has arbitrary dimension, one chooses ℓ'' such that $\ell \cap \ell'' = \ell' \cap \ell'' = 0$ and one then calculates $\mu(\ell_\infty, \ell'_\infty)$ using the property
$$\mu(\ell_\infty, \ell'_\infty) = -\mu(\ell'_\infty, \ell''_\infty) - \mu(\ell''_\infty, \ell_\infty) + \tau(\ell, \ell', \ell'') \tag{3.27}$$
and the expressions for $\mu(\ell_\infty, \ell''_\infty)$ and $\mu(\ell'_\infty, \ell''_\infty)$ given by (3.26).

Exercise 3.18. Check, using the cocycle property of the signature τ, that the left-hand side of (3.27) does not depend on the choice of ℓ'' such that $\ell \cap \ell'' = \ell' \cap \ell'' = 0$.

Let us begin by showing that $\mu(\ell_\infty, \ell'_\infty)$ defined by (3.26)–(3.27) always is an integer:

Proposition 3.19. *We have*
$$\mu(\ell_\infty, \ell'_\infty) \equiv n \mod 2 \quad \textit{when } \ell \cap \ell' = 0; \tag{3.28}$$

more generally

$$\mu(\ell_\infty, \ell'_\infty) \equiv n + \dim(\ell \cap \ell') \mod 2. \tag{3.29}$$

Proof. Setting $\mu = \mu(\ell_\infty, \ell'_\infty)$ we have

$$\begin{aligned} e^{i\pi\mu} &= \exp[i(\theta - \theta')](\exp[\operatorname{Tr}\log(-w(w')^{-1}])^{-1} \\ &= \exp[i(\theta - \theta')](\det(-w(w')^{-1}])^{-1} \\ &= \exp[i(\theta - \theta')](-1)^n \exp[-i(\theta - \theta')] \\ &= (-1)^n \end{aligned}$$

hence (3.28). Formula (3.29) follows, using formula (3.27) together with the value modulo 2 of the index τ given by formula (1.31) in Section 1.4). □

Let us now prove the main result of this subsection, namely the existence of the *ALM* index:

Theorem 3.20. *The ALM index $\mu(\ell_\infty, \ell'_\infty)$ exists and is calculated as follows:*

(i) *If $\ell \cap \ell' = 0$ then*

$$\mu(\ell_\infty, \ell'_\infty) = \frac{1}{\pi}\left[\theta - \theta' + i \operatorname{Tr}\operatorname{Log}(-w(w')^{-1}\right];$$

(ii) *In the general case choose ℓ'' such that $\ell \cap \ell'' = \ell' \cap \ell'' = 0$ and calculate $\mu(\ell_\infty, \ell'_\infty)$ using the formula:*

$$\mu(\ell_\infty, \ell'_\infty) = \mu(\ell_\infty, \ell''_\infty) - \mu(\ell'_\infty, \ell''_\infty) + \tau(\ell, \ell', \ell'')$$

[the right-hand side is independent of the choice of ℓ''].

Proof. It is clear that μ defined by (3.26) is locally constant on the set of all (ℓ, ℓ') such that $\ell \cap \ell' = 0$. Let us prove that

$$\mu(\ell_\infty, \ell'_\infty) - \mu(\ell_\infty, \ell''_\infty) + \mu(\ell'_\infty, \ell''_\infty) = \tau(\ell, \ell', \ell'')$$

when

$$\ell \cap \ell' = \ell \cap \ell'' = \ell' \cap \ell'' = 0;$$

the formula will then hold in the general case as well in view of (3.27). We are going to proceed along the lines in [84], p. 126. Since μ is locally constant on its domain. It follows that the composed mapping $S_\infty \longmapsto \mu(S_\infty\ell_\infty, S_\infty\ell'_\infty)$ is (for fixed $(\ell_\infty, \ell'_\infty)$ such that $\ell \cap \ell' = 0$) is locally constant on $\operatorname{Sp}_\infty(n)$; since $\operatorname{Sp}(n)$ is connected this mapping is in fact constant so we have $\mu(S_\infty\ell_\infty, S_\infty\ell'_\infty) = \mu(\ell_\infty, \ell'_\infty)$ and it is thus sufficient to show that

$$\mu(S_\infty\ell_\infty, S_\infty\ell'_\infty) - \mu(S_\infty\ell_\infty, S_\infty\ell''_\infty) + \mu(S_\infty\ell'_\infty, S_\infty\ell''_\infty) = \tau(\ell, \ell', \ell'')$$

for some convenient $S_\infty \in \operatorname{Sp}(n)$. Since $\operatorname{Sp}(n)$ acts transitively on pairs of Lagrangian planes, there exists $S \in \operatorname{Sp}(n)$ such that $S(\ell, \ell'') = (\ell_P, \ell_X)$ where

$\ell_X = \mathbb{R}^n \times 0$ and $\ell_P = 0 \times \mathbb{R}^n$. The transversality condition $\ell' \cap \ell = \ell' \cap \ell'' = 0$ then implies that

$$\ell' = \{(x,p) : p = Ax\} = \ell_A$$

for some symmetric matrix A with $\det A \neq 0$. We have thus reduced the proof to the case $(\ell, \ell', \ell'') = (\ell_P, \ell_A, \ell_X)$ and we have to show that

$$\mu(\ell_{P,\infty}, \ell_{A,\infty}) - \mu(\ell_{P,\infty}, \ell_{X,\infty}) + \mu(\ell_{A,\infty}, \ell_{X,\infty}) = \tau(\ell_P, \ell_A, \ell_X). \tag{3.30}$$

Now, in view of formula (1.24) (Corollary 1.31, Section 1.4) we have

$$\tau(\ell_P, \ell_A, \ell_X) = \operatorname{sign}(A) = p - q$$

where p (resp. q) is the number of > 0 (resp. < 0) eigenvalues of the symmetric matrix A. Let us next calculate $\mu(\ell_\infty, \ell''_\infty) = \mu(\ell_{P,\infty}, \ell_{X,\infty})$. Identifying $\mathrm{Lag}_\infty(n)$ with $\mathrm{W}_\infty(n)$ there exist integers k and k' such that

$$\ell_{P,\infty} = (I, 2k\pi) \quad and \quad \ell_{X,\infty} = (-I, (2k'+n)\pi)$$

and hence

$$\mu(\ell_{P,\infty}, \ell_{X,\infty}) = \frac{1}{\pi}(2k\pi - (2k'+n)\pi + i \operatorname{Tr} \operatorname{Log} I) = 2(k-k') - n.$$

Let us now calculate $\mu(\ell_{P,\infty}, \ell_{A,\infty})$. Recall that V_{-A} and M_L ($\det L \neq 0$) denote the symplectic matrices defined by (2.50) in Subsection 2.2.3:

$$V_{-A} = \begin{bmatrix} I & 0 \\ A & I \end{bmatrix} \quad , \quad M_L = \begin{bmatrix} L^{-1} & 0 \\ 0 & L^T \end{bmatrix}.$$

We begin by noting that we have $\ell_A = V_{-A}\ell_X$ hence $M_L \ell_A = V_{A'}\ell_X$, using the intertwining formula

$$M_L V_{-A} = V_{-A'} M_L \quad , \quad A' = L^T A L.$$

We may thus assume, replacing ℓ_A by $M_L^{-1}\ell_A$ and A by $L^T A L$ where L diagonalizes A, that

$$A = \operatorname{diag}[+1, \ldots, +1, -1, \ldots, -1]$$

with p plus signs and $q = n - p$ minus signs. Let now $\mathcal{B} = \{e_1, \ldots, e_n; f_1, \ldots, f_n\}$ be the canonical symplectic basis of $(\mathbb{R}^{2n}_z, \sigma)$. The n vectors

$$g_i = \frac{1}{\sqrt{2}}(e_i + f_i) \quad , \quad 1 \leq i \leq p,$$

$$g_j = \frac{1}{\sqrt{2}}(e_j + f_j) \quad , \quad p+1 \leq j \leq n$$

(with obvious conventions if $p = n$ or $q = n$) form an orthonormal basis of $\ell_A = V_{-A}\ell_X$. We thus have $\ell_A = U\ell_P$ where

$$UR = \frac{1}{\sqrt{2}}\begin{bmatrix} A & I \\ -I & A \end{bmatrix} \in \mathrm{U}(n).$$

The identification of U with $u = \frac{1}{\sqrt{2}}(A - iI)$ in $\mathrm{U}(n, \mathbb{C})$ identifies ℓ_A with $uu^T = -iA$. We have $\det(-iA) = i^{q-p}$, hence

$$\ell_{A,\infty} \equiv (iA, \tfrac{1}{2}(q-p)\pi + 2r\pi)$$

for some $r \in \mathbb{Z}$. To calculate $\mu(\ell_{P,\infty}, \ell_{A,\infty})$ we need to know

$$\mathrm{Log}(-I(-iA)^{-1} = \mathrm{Log}(-iA);$$

the choice of the logarithm being the one which is obtained by analytic continuation from the positive axis we have

$$\begin{aligned}
\mathrm{Log}(-iA) &= \mathrm{Log}(-i\,\mathrm{diag}[+1, \ldots, +1, -1, \ldots, -1]) \\
&= \mathrm{Log}\,\mathrm{diag}[-i, \ldots, -i, +i, \ldots, +i] \\
&= \mathrm{diag}[\tfrac{1}{2}\pi(-i, \ldots, -i, +i, \ldots, +i)]
\end{aligned}$$

(p plus signs and q minus signs) hence, by definition (3.26) of μ,

$$\begin{aligned}
\mu(\ell_{P,\infty}, \ell_{A,\infty}) &= \frac{1}{\pi}\left[2k\pi - (\tfrac{1}{2}(q-p)\pi + 2r\pi) + i\,\mathrm{Tr}\,\mathrm{Log}(-iA)\right] \\
&= \frac{1}{\pi}\left[2k\pi - (\tfrac{1}{2}(q-p)\pi + 2r\pi) + i(\tfrac{1}{2}\pi(q-p)i)\right] \\
&= 2(k-r) + p - q.
\end{aligned}$$

Similarly

$$\begin{aligned}
\mu(\ell_{A,\infty}, \ell_{X,\infty}) &= -\mu(\ell_{X,\infty}, \ell_{A,\infty}) \\
&= -\frac{1}{\pi}\left[(2k' + n)\pi - (\tfrac{1}{2}(q-p)\pi + 2r\pi) + i\,\mathrm{Tr}\,\mathrm{Log}(iA)\right] \\
&= 2(r - k') - n,
\end{aligned}$$

hence

$$\mu(\ell_{P,\infty}, \ell_{A,\infty}) - \mu(\ell_{P,\infty}, \ell_{X,\infty}) + \mu(\ell_{A,\infty}, \ell_{X,\infty}) = p - q$$

which ends the proof since $p - q = \tau(\ell_P, \ell_A, \ell_X)$. □

In the following exercise the reader is encouraged to find an explicit expression for the ALM index when $n = 1$:

Exercise 3.21. Using the formula

$$\operatorname{Log} e^{i\varphi} = i\left(\varphi - 2\pi\left[\frac{\varphi+\pi}{2\pi}\right]\right) \quad \text{for } \varphi \notin \pi\mathbb{Z}$$

calculate $\mu(\ell_\infty, \ell'_\infty)$ when $n = 1$.

The following consequence of the theorem above describes the action of $\pi_1[\mathrm{Lag}(n)] \cong (\mathbb{Z}, +)$ on the *ALM* index, and shows why the *ALM* index is an extension of the usual Maslov index defined and studied in Chapter 5, Section 3.1:

Corollary 3.22. *Let β be the generator of $\pi_1[\mathrm{Lag}(n)]$ whose natural image in $\mathbb{Z}$ is $+1$. We have*

$$\mu(\beta^r\ell_\infty, \beta^{r'}\ell'_\infty) = \mu(\ell_\infty, \ell'_\infty) + 2(r - r') \tag{3.31}$$

for all $(\ell_\infty, \ell'_\infty) \in (\mathrm{Lag}(n))^2$ and $(r, r') \in \mathbb{Z}^2$ and hence

$$\mu(\gamma\ell_\infty, \ell'_\infty) - \mu(\ell_\infty, \ell'_\infty) = 2m(\gamma) \tag{3.32}$$

for every loop γ in $\mathrm{Lag}(n)$ ($m(\gamma)$ the Maslov index of γ).

Proof. Formula (3.32) follows from formula (3.31) since every loop γ is homotopic to β^r for some $r \in \mathbb{Z}$. Let us first prove (3.31) when $\ell \cap \ell' = 0$. Assume that $\ell_\infty \equiv (w, \theta)$ and $\ell'_\infty \equiv (w', \theta')$ with $w, w' \in \mathrm{W}(n, \mathbb{C})$, $\det w = e^{i\theta}$, and $\det w' = e^{i\theta'}$. Then

$$\beta^r\ell_\infty \equiv (w, \theta + 2r\pi) \quad , \quad \beta^{r'}\ell'_\infty \equiv (w', \theta' + 2r'\pi)$$

and hence, by definition (3.26)

$$\begin{aligned}\mu(\beta^r\ell_\infty, \beta^{r'}\ell'_\infty) &= \frac{1}{\pi}\left[\theta + 2r - \theta' - 2r' + i \operatorname{Tr}\operatorname{Log}(-w(w')^{-1}\right] \\ &= \mu(\ell_\infty, \ell'_\infty) + 2(r - r').\end{aligned}$$

The general case immediately follows using formula (3.27) and the fact that $\beta^r\ell_\infty$ and $\beta^{r'}\ell'_\infty$ have projections ℓ and ℓ' on $\mathrm{Lag}(n)$. □

3.3 q-Symplectic Geometry

Now we can – at last! – study the central topic of this chapter, the action of $\mathrm{Sp}_q(n)$ on $\mathrm{Lag}_{2q}(n)$. Due to the properties of the fundamental groups of $\mathrm{Sp}(n)$ and $\mathrm{Lag}(n)$ the general case will easily follow from the case $q = +\infty$. We begin by identifying $\mathrm{Lag}_\infty(n)$ with $\mathrm{Lag}(n) \times \mathbb{Z}$ and $\mathrm{Sp}_\infty(n)$ with a subgroup of $\mathrm{Sp}(n) \times \mathbb{Z}$ equipped with a particular group structure.

3.3.1 The identification $\mathrm{Lag}_\infty(n) = \mathrm{Lag}(n) \times \mathbb{Z}$

The title of this subsection is at first sight provocative: how can we identify the Maslov bundle $\mathrm{Lag}_\infty(n)$, which is a connected manifold, with a Cartesian product where one of the factors is a discrete space? The answer is that we will identify $\mathrm{Lag}_\infty(n)$ and $\mathrm{Lag}(n) \times \mathbb{Z}$ as *sets*, not as topological spaces, and equip $\mathrm{Lag}(n) \times \mathbb{Z}$ with the transported topology (which is of course not the product topology).

Let us justify this in detail.

We denote by $\partial \dim$ the coboundary of the 1-cochain $\dim(\ell, \ell') = \dim \ell \cap \ell'$ on $\mathrm{Lag}(n)$. It is explicitly given by

$$\partial \dim(\ell, \ell', \ell'') = \dim \ell \cap \ell' - \dim \ell \cap \ell'' + \dim \ell' \cap \ell''$$

(see definition (9) in the Preface).

Definition 3.23.

(i) The function $m : \mathrm{Lag}_\infty(n) \longrightarrow \mathbb{Z}$ defined by

$$m(\ell_\infty, \ell'_\infty) = \frac{1}{2}(\mu(\ell_\infty, \ell'_\infty) + n + \dim \ell \cap \ell') \tag{3.33}$$

is called the "reduced *ALM index*" *on* $\mathrm{Lag}_\infty(n)$.

(ii) The function $(\mathrm{Lag}(n))^3 \longrightarrow \mathbb{Z}$ defined by

$$\mathrm{Inert}(\ell, \ell', \ell'') = \frac{1}{2}(\tau(\ell, \ell', \ell'') + n + \partial \dim(\ell, \ell', \ell''))$$

where τ is the signature is called the "index of inertia" of (ℓ, ℓ', ℓ'').

That $m(\ell_\infty, \ell'_\infty)$ is an integer follows from the congruence (3.29) in Proposition 3.19 (Subsection 3.2.3). That $\mathrm{Inert}(\ell, \ell', \ell'')$ also is an integer follows from the congruence (1.31) in Proposition 1.34 (Subsection 1.4.3). These congruences, together with the antisymmetry of the signature τ moreover imply that

$$m(\ell_\infty, \ell'_\infty) + m(\ell'_\infty, \ell_\infty) = n + \dim \ell \cap \ell' \tag{3.34}$$

for all $(\ell'_\infty, \ell_\infty) \in (\mathrm{Lag}_\infty(n))^2$.

Proposition 3.24. *The reduced ALM index has the following properties:*

(i) *For all* $(\ell_\infty, \ell'_\infty, \ell''_\infty) \in (\mathrm{Lag}_\infty(n))^3$,

$$m(\ell_\infty, \ell'_\infty) - m(\ell_\infty, \ell''_\infty) + m(\ell'_\infty, \ell''_\infty) = \mathrm{Inert}(\ell, \ell', \ell''); \tag{3.35}$$

(ii) *Let* β *be the generator of* $\pi_1[\mathrm{Lag}(n)]$ *whose natural image in* $\mathbb{Z}$ *is* $+1$*; then*

$$m(\beta^r \ell_\infty, \beta^{r'} \ell'_\infty) = m(\ell_\infty, \ell'_\infty) + r - r' \tag{3.36}$$

for all $(r, r') \in \mathbb{Z}^2$.

Proof. Formula (3.35) is equivalent to the property (3.18) of the *ALM* index. Formula (3.36) follows from (3.31) in Corollary 3.22. □

Remark 3.25. Formula (3.36) shows that the range of the mapping

$$(\ell_\infty, \ell'_\infty) \longmapsto m(\ell_\infty, \ell'_\infty)$$

is all of $\mathbb{Z}$.

Exercise 3.26. Check that $\mathrm{Inert}(\ell, \ell', \ell'')$ coincides with Leray's index of inertia defined at the beginning of Section 3.2 when $\ell \cap \ell' = \ell' \cap \ell'' = \ell'' \cap \ell = 0$.

Let us state and prove the main result of this subsection:

Theorem 3.27. *Let $\ell_{\alpha,\infty}$ be an arbitrary element of $\mathrm{Lag}_\infty(n)$ and define a mapping*

$$\Phi_\alpha : \mathrm{Lag}_\infty(n) \longrightarrow \mathrm{Lag}(n) \times \mathbb{Z}$$

by the formula

$$\Phi_\alpha(\ell_\infty) = (\ell, m(\ell_\infty, \ell_{\alpha,\infty})) \quad , \quad \ell = \pi^{\mathrm{Lag}}(\ell_\infty).$$

(i) *The mapping Φ_α is a bijection whose restriction to the subset $\{\ell_\infty : \ell \cap \ell_\alpha = 0\}$ of $\mathrm{Lag}_\infty(n)$ is a homeomorphism onto $\{\ell : \ell \cap \ell_\alpha = 0\} \times \mathbb{Z}$.*

(ii) *The set of all bijections $(\Phi_\alpha)_{\ell_{\alpha,\infty}}$ form a system of local charts of $\mathrm{Lag}_\infty(n)$ whose transitions $\Phi_{\alpha\beta} = \Phi_\alpha \Phi_\beta^{-1}$ are the functions*

$$\Phi_{\alpha\beta}(\ell, \lambda) = (\ell, \lambda + \mathrm{Inert}(\ell, \ell_\alpha, \ell_\beta) - m(\ell_{\alpha,\infty}, \ell_{\beta,\infty})). \tag{3.37}$$

Proof. (i) Assume that $\Phi_\alpha(\ell_\infty) = \Phi_\alpha(\ell'_\infty)$; then $\ell = \ell'$ and $m(\ell_\infty, \ell_{\alpha,\infty}) = m(\ell'_\infty, \ell_{\alpha,\infty})$. Let $r \in \mathbb{Z}$ be such that $\ell'_\infty = \beta^r \ell_\infty$ (β the generator of $\pi_1[\mathrm{Lag}(n)]$); in view of formula (3.36) we have

$$m(\ell'_\infty, \ell_{\alpha,\infty}) = m(\beta^r \ell_\infty, \ell_{\alpha,\infty}) = m(\ell_\infty, \ell_{\alpha,\infty}) + r$$

hence $r = 0$ and $\ell'_\infty = \ell_\infty$, so that Φ_α is injective. Let us show it is surjective. For $(\ell, k) \in \mathrm{Lag}(n) \times \mathbb{Z}$ choose $\ell_\infty \in \mathrm{Lag}_\infty(n)$ such that $\ell = \pi^{\mathrm{Lag}}(\ell_\infty)$. If $m(\ell_\infty, \ell_{\alpha,\infty}) = k$ we are done. If $m(\ell_\infty, \ell_{\alpha,\infty}) \neq k$ replace ℓ_∞ by $\beta^r \ell_\infty$ such that $m(\ell_\infty, \ell_{\alpha,\infty}) + r = k$ (*cf.* Remark 3.25). The *ALM* index μ is locally constant on the set

$$\{(\ell_\infty, \ell'_\infty) : \ell \cap \ell' = 0\} \subset (\mathrm{Lag}_\infty(n))^2$$

hence so is m; it follows that the restriction of Φ_α to $\{\ell_\infty : \ell \cap \ell_\alpha = 0\}$ indeed is a homeomorphism onto its image $\{\ell : \ell \cap \ell_\alpha = 0\} \times \mathbb{Z}$.

(ii) The mapping $\Phi_{\alpha\beta} = \Phi_\alpha \Phi_\beta^{-1}$ takes $(\ell, \lambda) = (\ell, m(\ell_\infty, \ell_{\beta,\infty}))$ to $(\ell', \lambda') = (\ell, m(\ell_\infty, \ell_{\alpha,\infty}))$ hence

$$\Phi_{\alpha\beta}(\ell_\infty) = (\ell, \lambda + m(\ell_\infty, \ell_{\alpha,\infty}) - m(\ell_\infty, \ell_{\beta,\infty}))$$

which is the same thing as (3.37) in view of formula (3.35). □

We are going to perform a similar identification for the universal covering of the symplectic group; this will allow us to exhibit precise formulas for q-symplectic geometry.

3.3.2 The universal covering $\mathrm{Sp}_\infty(n)$

Recall that the *ALM* index is $\mathrm{Sp}_\infty(n)$-invariant:

$$\mu(S_\infty\ell_\infty, S_\infty\ell'_\infty) = \mu(\ell_\infty, \ell'_\infty)$$

for all $(S_\infty, \ell_\infty, \ell'_\infty) \in \mathrm{Sp}_\infty(n) \times \mathrm{Lag}^2_\infty(n)$ (Proposition 3.16, Subsection 3.2.1).

Definition 3.28. Let $\ell \in \mathrm{Lag}(n)$; the "Maslov index" on $\mathrm{Sp}_\infty(n)$ relative to ℓ is the mapping $\mu_\ell : \mathrm{Sp}_\infty(n) \longrightarrow \mathbb{Z}$ defined by

$$\mu_\ell(S_\infty) = \mu(S_\infty\ell_\infty, \ell_\infty) \tag{3.38}$$

where ℓ_∞ is an arbitrary element of $\mathrm{Lag}_\infty(n)$ with projection $\pi^{\mathrm{Lag}}(\ell_\infty) = \ell$.

This definition makes sense in view of the following observation: suppose that we change ℓ_∞ into another element ℓ'_∞ with the same projection ℓ. Then there exists an integer m such that $\ell_\infty = \beta^m\ell'_\infty$ and

$$\begin{aligned}
\mu(S_\infty\ell_\infty, \ell_\infty) &= \mu(S_\infty(\beta^r\ell'_\infty), \beta^r\ell'_\infty) \\
&= \mu(S_\infty(\beta^r\ell'_\infty), \ell'_\infty) - 2r \\
&= \mu(\beta^r\ell'_\infty, S_\infty^{-1}\ell'_\infty) - 2r \\
&= \mu(\ell'_\infty, S_\infty^{-1}\ell'_\infty) + 2r - 2r \\
&= \mu(S_\infty\ell'_\infty, \ell'_\infty)
\end{aligned}$$

where we have used successively (3.31), (3.50), again (3.31), and finally the $\mathrm{Sp}_\infty(n)$-invariance (3.38) of the *ALM* index.

Here are a few properties which immediately follow from those of the *ALM* index:

- In view of property (3.29) (Proposition 3.19) of the *ALM* index we have

  $$\mu_\ell(S_\infty) \equiv n - \dim(S\ell \cap \ell) \mod 2 \tag{3.39}$$

  for all $S_\infty \in \mathrm{Sp}_\infty(n)$.
- The antisymmetry (3.19) of the *ALM* index implies that we have

  $$\mu_\ell(S_\infty^{-1}) = -\mu_\ell(S_\infty) \quad , \quad \mu_\ell(I_\infty) = 0 \tag{3.40}$$

  (I_∞ the unit of $\mathrm{Sp}_\infty(n)$).
- Let α be the generator of $\pi_1[\mathrm{Sp}(n)] \cong (\mathbb{Z}, +)$ whose image in $\mathbb{Z}$ is $+1$; then

  $$\mu_\ell(\alpha^r S_\infty) = \mu_\ell(S_\infty) + 4r \tag{3.41}$$

  for every $S_\infty \in \mathrm{Sp}(n)$ and $r \in \mathbb{Z}$: this immediately follows from formula (3.31) for the action of $\pi_1[\mathrm{Lag}(n)]$ on the *ALM* index.

The following properties of μ_ℓ are immediate consequences of the characteristic properties of the *ALM* index:

Proposition 3.29.

(i) *For all* S_∞, S'_∞ *in* $\mathrm{Sp}_\infty(n)$,

$$\mu_\ell(S_\infty S'_\infty) = \mu_\ell(S_\infty) + \mu_\ell(S'_\infty) + \tau_\ell(S, S') \tag{3.42}$$

where $\tau_\ell : (\mathrm{Sp}(n))^2 \longrightarrow \mathbb{Z}$ *is defined by*

$$\tau_\ell(S, S') = \tau(\ell, S\ell, SS'\ell). \tag{3.43}$$

(ii) *The function* $(S_\infty, \ell, \ell') \longmapsto \mu_\ell(S) - \tau(S\ell, \ell', \ell'')$ *is locally constant on the set*

$$\{(S_\infty, \ell, \ell') : S\ell \cap \ell'' = \ell \cap \ell'' = 0\} \subset \mathrm{Sp}_\infty(n) \times (\mathrm{Lag}(n))^2;$$

in particular μ_ℓ *is locally constant on* $\{S_\infty : \dim(S\ell \cap \ell) = 0\}$.

(iii) *We have*

$$\mu_\ell(S_\infty) - \mu_{\ell'}(S_\infty) = \tau(S\ell, \ell, \ell') - \tau(S\ell, S\ell', \ell') \tag{3.44}$$

for every $S_\infty \in \mathrm{Sp}(n)$ *and* $(\ell, \ell') \in (\mathrm{Lag}(n))^2$.

Proof. (i) By definition of μ_ℓ,

$$\begin{aligned}\mu_\ell(S_\infty S'_\infty) &- \mu_\ell(S_\infty) - \mu_\ell(S'_\infty)\\ &= \mu(S_\infty S'_\infty \ell_\infty, \ell_\infty) - \mu(S_\infty \ell_\infty, \ell_\infty) - \mu(S'_\infty \ell_\infty, \ell_\infty)\end{aligned}$$

that is, using the $\mathrm{Sp}_\infty(n)$-invariance and the antisymmetry of μ:

$$\begin{aligned}\mu_\ell(S_\infty S'_\infty) - \mu_\ell(S_\infty) - \mu_\ell(S'_\infty) &= \mu(S_\infty S'_\infty \ell_\infty, \ell_\infty) + \mu(\ell_\infty, S_\infty \ell_\infty)\\ &\quad - \mu(S_\infty S'_\infty \ell_\infty, S_\infty \ell_\infty).\end{aligned}$$

In view of the cocycle property $\partial\mu = \pi^*\tau$ of the *ALM* index the right-hand side of this equality is equal to

$$\tau(SS'\ell, \ell, S\ell) = \tau(\ell, S\ell, SS'\ell) = \tau_\ell(S, S'),$$

hence (3.42).

Property (ii) immediately follows from the two following observations: the *ALM* index is locally constant on

$$\{(\ell_\infty, \ell'_\infty) : \ell \cap \ell' = 0\} \subset (\mathrm{Lag}(n))^2$$

and the signature $\tau(\ell, \ell', \ell'')$ is locally constant on

$$\{(\ell, \ell', \ell'') : \ell \cap \ell' = \ell' \cap \ell'' = \ell'' \cap \ell = 0\} \subset (\mathrm{Lag}(n))^3.$$

(iii) Using again the property $\partial\mu = \pi^*\tau$ and the $\mathrm{Sp}_\infty(n)$-invariance of μ we have

$$\begin{aligned}\mu(S_\infty \ell_\infty, \ell_\infty) - \mu(S_\infty \ell_\infty, \ell'_\infty) + \mu(S_\infty \ell_\infty, S_\infty \ell'_\infty) &= \tau(S\ell, \ell, \ell'),\\ \mu(S_\infty \ell_\infty, S_\infty \ell'_\infty) - \mu(S_\infty \ell_\infty, \ell'_\infty) + \mu(S_\infty \ell'_\infty, \ell'_\infty) &= \tau(S\ell, S\ell', \ell')\end{aligned}$$

which yields (3.44) subtracting the first identity from the second. □

The practical calculation of $\mu_\ell(S_\infty)$ does not always require the determination of an ALM index; formula (3.42) can often be used with profit. Here is an example:

Example 3.30. Assume that $n = 1$. Let $(-I)_\infty$ be the homotopy class of the symplectic path $t \longmapsto e^{\pi t J}$, $0 \leq t \leq 1$, joining I to $-I$ in Sp(1). We have $(-I)^2_\infty = \alpha$ (the generator of $\pi_1[\mathrm{Sp}(n)]$), hence

$$\mu_\ell((-I)^2_\infty) = \mu_\ell(\alpha) = 4.$$

But (3.42) implies that

$$\mu_\ell((-I)^2_\infty) = 2\mu_\ell((-I)_\infty) + \tau(\ell,\ell,\ell) = 2\mu_\ell((-I)_\infty),$$

hence $\mu_\ell((-I)_\infty) = 2$.

It turns out that the properties (i) and (ii) of the Maslov index μ_ℓ listed in Proposition 3.29 characterize that index. More precisely:

Proposition 3.31. *Assume that* $\mu'_\ell : \mathrm{Sp}_\infty(n) \longrightarrow \mathbb{Z}$ *is locally constant on* $\{S_\infty : \dim(S\ell \cap \ell) = 0\}$ *and satisfies*

$$\mu_\ell(S_\infty S'_\infty) = \mu_\ell(S_\infty) + \mu_\ell(S'_\infty) + \tau_\ell(S, S') \tag{3.45}$$

for all S_∞, S'_∞ *in* $\mathrm{Sp}_\infty(n)$ $(S = \pi^{\mathrm{Sp}}(S_\infty)$, $S' = \pi^{\mathrm{Sp}}(S'_\infty))$. *Then* $\mu'_\ell = \mu_\ell$.

Proof. The function $\delta_\ell = \mu_\ell - \mu'_\ell$ satisfies

$$\delta_\ell(S_\infty S'_\infty) = \delta_\ell(S_\infty) + \delta_\ell(S'_\infty)$$

and is locally constant on $\{S_\infty : \dim(S\ell \cap \ell) = 0\}$. In view of Proposition 2.36 of Chapter 2 every $S \in \mathrm{Sp}(n)$ can be factorized as $S = S_1S_2$ with $S_1\ell_0 \cap \ell_0 = S_2\ell_0 \cap \ell_0 = 0$; since

$$\delta_\ell(S_\infty) = \delta_\ell(S_{1,\infty}) + \delta_\ell(S_{2,\infty})$$

it follows that δ_ℓ is actually constant on $\mathrm{Sp}_\infty(n)$; taking $S_\infty = S'_\infty$ we thus have $\delta_\ell(S_\infty) = 0$ for all S_∞ hence $\mu'_\ell = \mu_\ell$. □

Exercise 3.32. Use the uniqueness property above to prove the conjugation formula

$$\mu_\ell((S')^{-1}S_\infty S') = \mu_{S'\ell}(S_\infty)$$

where $(S')^{-1}S_\infty S'$ denotes the homotopy class of the path $t \longmapsto (S')^{-1}S(t)S'$ if S_∞ is the homotopy class of a $t \longmapsto S(t)$, $0 \leq t \leq 1$ in Sp(n).

Let us mention the following result which will be proven in Chapter 7 (Lemma 7.24) in connection with the study of the metaplectic group: assume that $\ell = \ell_P$

and that S, S' are free symplectic matrices. Writing S, S', and $S'' = SS'$ in block matrix form

$$S = \begin{bmatrix} A & B \\ C & D \end{bmatrix} \quad , \quad S' = \begin{bmatrix} A' & B' \\ C' & D' \end{bmatrix} \quad , \quad S'' = \begin{bmatrix} A'' & B'' \\ C'' & D'' \end{bmatrix}$$

we have $\det(BB') \neq 0$ and

$$\tau(\ell_P, S\ell_P, SS'\ell_P) = \operatorname{sign}(B^{-1}B''(B')^{-1}) = \tau^+ - \tau^- \tag{3.46}$$

where $\tau^{\pm}$ is the number of positive (resp. negative) eigenvalues of the symmetric matrix $B^{-1}B''(B')^{-1}$.

For $\ell \in \operatorname{Lag}(n)$ define a function $m_\ell : \operatorname{Sp}_\infty(n) \longrightarrow \mathbb{Z}$ by

$$m_\ell(S_\infty) = \frac{1}{2}(\mu_\ell(S_\infty) + n + \dim(S\ell \cap \ell)), \tag{3.47}$$

that is

$$m_\ell(S_\infty) = m(S_\infty \ell_\infty, \ell_\infty)$$

where ℓ_∞ has projection $\pi^{\operatorname{Lag}}(\ell_\infty) = \ell$. Since $m(\ell_\infty, \ell'_\infty) \in \mathbb{Z}$ for all $(\ell_\infty, \ell'_\infty) \in (\operatorname{Lag}_\infty(n))^2$ it follows that $m_\ell(S_\infty) \in \operatorname{Sp}_\infty(n)$.

Definition 3.33. The mapping $m_\ell : \operatorname{Sp}_\infty(n) \longrightarrow \mathbb{Z}$ defined by (3.47) is called the reduced Maslov index on $\operatorname{Sp}_\infty(n)$ relatively to $\ell \in \operatorname{Lag}(n)$.

The properties of the reduced Maslov index immediately follow from those of μ_ℓ; in particular (3.41) implies that

$$m_\ell(\alpha^r S_\infty) = m_\ell(S_\infty) + 2r \tag{3.48}$$

for every integer r (α being the generator of $\pi_1[\operatorname{Sp}(n)]$). An immediate consequence of (3.48) is that the value modulo 2 of $m_\ell(S_\infty)$ only depends on the projection $S = \pi^{\operatorname{Sp}}(S_\infty)$. We will denote by $m_\ell(S)$ the corresponding equivalence class:

$$m \in m_\ell(S) \Longleftrightarrow m \equiv m_\ell(S_\infty) \mod 2.$$

The following result is the symplectic equivalent of Theorem 3.27; it identifies $\operatorname{Sp}_\infty(n)$ with a subset of $\operatorname{Sp}(n) \times \mathbb{Z}$:

Theorem 3.34. *For $\ell_\alpha \in \operatorname{Lag}(n)$ define a mapping*

$$\Psi_\alpha : \operatorname{Sp}_\infty(n) \longrightarrow \operatorname{Sp}(n) \times \mathbb{Z}$$

by the formula

$$\Psi_\alpha(S_\infty) = (S, m_{\ell_\alpha}(S_\infty)).$$

(i) *The mapping Ψ_α is a bijection*

$$\operatorname{Sp}_\infty(n) \longrightarrow \{(S, m) : S \in \operatorname{Sp}(n),\ m \in m_{\ell_\alpha}(S)\}$$

whose restriction to the subset $\{S_\infty : S\ell_\alpha \cap \ell_\alpha = 0\}$ is a homeomorphism onto

$$\operatorname{Sp}_{\ell_\alpha}(n) = \{(S, m) : S \in \operatorname{Sp}(n),\ S\ell_\alpha \cap \ell_\alpha = 0\ ,\ m \in m_\ell(S)\}.$$

(ii) *The set of all bijections* $(\Psi_\alpha)_{\ell_\alpha}$ *form a system of local charts of* $\mathrm{Sp}_\infty(n)$ *whose transitions* $\Psi_\alpha\Psi_\beta^{-1}$ *are the functions*

$$\Psi_{\alpha\beta}(S,m) = (S, m + \mathrm{Inert}(S\ell_\alpha, \ell_\alpha, \ell_\beta) - \mathrm{Inert}(S\ell_\alpha, S\ell_\beta, \ell_\beta).$$

Proof. (i) By definition of $m_\ell(S)$ the range of Ψ_α consists of all pairs (S,m) with $m \in m_\ell(S)$. Assume that $(S, m_{\ell_\alpha}(S_\infty)) = (S', m_{\ell_\alpha}(S'_\infty))$; then $S = S'$ and $S'_\infty = \alpha^r S_\infty$ for some $r \in \mathbb{Z}$ (*cf.* the proof of (i) in Theorem 3.27). In view of (3.48) we must have $r = 0$ and hence $S_\infty = S'_\infty$ so that Ψ_α is injective.

(ii) It is identical to that of the corresponding properties in Theorem 3.27 and is therefore left to the reader as an exercise. □

The theorem above allows us to describe in a precise way the composition law of the universal covering group $\mathrm{Sp}_\infty(n)$:

Corollary 3.35. *Let* $\ell_\alpha \in \mathrm{Lag}(n)$. *Identifying* $\mathrm{Sp}_\infty(n)$ *with the subset*

$$\{(S,m) : S \in \mathrm{Sp}(n), m \in m_{\ell_\alpha}(S)\}$$

of $\mathrm{Sp}(n) \times \mathbb{Z}$ *the composition law of* $\mathrm{Sp}_\infty(n)$ *is given by the formula*

$$(S,m) *_{\ell_\alpha} (S',m') = (SS', m + m' + \mathrm{Inert}(\ell_\alpha, S\ell_\alpha, SS'\ell_\alpha). \tag{3.49}$$

Proof. This is obvious since we have

$$m_{\ell_\alpha}(S_\infty S'_\infty) = m_{\ell_\alpha}(S_\infty) + m_{\ell_\alpha}(S'_\infty) + \mathrm{Inert}(\ell_\alpha, S\ell_\alpha, SS'\ell_\alpha)$$

in view of property (3.42) of μ_ℓ and definition (3.47) of m_ℓ. □

Let us now proceed to prove the main results of this section.

3.3.3 The action of $\mathrm{Sp}_q(n)$ on $\mathrm{Lag}_{2q}(n)$

Let $\mathrm{St}(\ell_P)$ be the isotropy subgroup of $\ell_P = 0 \times \mathbb{R}^n_p$ in $\mathrm{Sp}(n)$: $S \in \mathrm{St}(\ell_P)$ if and only if $S \in \mathrm{Sp}(n)$ and $S\ell_P = \ell_P$. The fibration

$$\mathrm{Sp}(n)/\mathrm{St}(\ell_P) = \mathrm{Lag}(n)$$

defines an isomorphism

$$\mathbb{Z} \cong \pi_1[\mathrm{Sp}(n)] \longrightarrow \pi_1[\mathrm{Lag}(n)] \cong \mathbb{Z}$$

which is multiplication by 2 on $\mathbb{Z}$. It follows that the action of $\mathrm{Sp}(n)$ on $\mathrm{Lag}(n)$ can be lifted to a transitive action of the universal covering $\mathrm{Sp}_\infty(n)$ on the Maslov bundle $\mathrm{Lag}_\infty(n)$ such that

$$(\alpha S_\infty)\ell_\infty = \beta^2(S_\infty \ell_\infty) = S_\infty(\beta^2\ell_\infty) \tag{3.50}$$

for all $(S_\infty, \ell_\infty) \in \mathrm{Sp}_\infty(n) \times \mathrm{Lag}_\infty(n)$; as previously α (resp. β) is the generator of $\pi_1[\mathrm{Sp}(n)]$ (resp. $\pi_1[\mathrm{Lag}(n)]$) whose natural image in $\mathbb{Z}$ is $+1$.

The following theorem describes ∞-symplectic geometry:

Theorem 3.36. *Let $\ell_\alpha \in \mathrm{Lag}(n)$. Identifying $\mathrm{Sp}_\infty(n)$ with the subset*

$$\{(S,m) : S \in \mathrm{Sp}(n), m \in m_{\ell_\alpha}(S)\}$$

of $\mathrm{Sp}(n)\times\mathbb{Z}$ defined in Theorem 3.34 and $\mathrm{Lag}_\infty(n)$ with $\mathrm{Lag}(n)\times\mathbb{Z}$ as in Theorem 3.27, the action of $\mathrm{Sp}_\infty(n)$ on $\mathrm{Lag}_\infty(n)$ is given by the formula

$$(S,m)\cdot_{\ell_\alpha}(\ell,\lambda) = (S\ell, m+\lambda-\mathrm{Inert}(S\ell, S\ell_\alpha, \ell_\alpha)). \tag{3.51}$$

Proof. We have $\lambda = m(\ell_\infty,\ell_{\alpha,\infty})$ for some ℓ_∞ covering ℓ, and $m = m(S_\infty\ell_{\alpha,\infty},\ell_{\alpha,\infty})$ for some S_∞ covering S. Let us define the integer ∂ by the condition

$$m+\lambda+\delta = m(S_\infty\ell_\infty,\ell_{\alpha,\infty}),$$

that is

$$\delta = m(S_\infty\ell_\infty,\ell_{\alpha,\infty}) - m(\ell_\infty,\ell_{\alpha,\infty}) - m(S_\infty\ell_{\alpha,\infty},\ell_{\alpha,\infty}).$$

We have to show that

$$\delta = -\,\mathrm{Inert}(S\ell, S\ell_\alpha, \ell_\alpha). \tag{3.52}$$

In view of the $\mathrm{Sp}_\infty(n)$-invariance of the reduced *ALM* index we have $m(\ell_\infty,\ell_{\alpha,\infty}) = m(S_\infty\ell_\infty, S_\infty\ell_{\alpha,\infty})$ and hence

$$\delta = m(S_\infty\ell_\infty,\ell_{\alpha,\infty}) - m(S_\infty\ell_\infty,S_\infty\ell_{\alpha,\infty}) - m(S_\infty\ell_{\alpha,\infty},\ell_{\alpha,\infty});$$

on the other hand

$$m(S_\infty\ell_{\alpha,\infty},\ell_{\alpha,\infty}) + m(\ell_{\alpha,\infty},S_\infty\ell_{\alpha,\infty}) = n + \dim(S\ell_\alpha\cap\ell_\alpha)$$

(formula (3.34)) so that

$$\begin{aligned}\delta = m(S_\infty\ell_\infty,\ell_{\alpha,\infty}) - m(S_\infty\ell_\infty,S_\infty\ell_{\alpha,\infty}) + m(\ell_{\alpha,\infty},S_\infty\ell_{\alpha,\infty})\\ - n - \dim(S\ell_\alpha\cap\ell_\alpha).\end{aligned}$$

Using property (3.35) of m this can be rewritten

$$\delta = \mathrm{Inert}(S\ell,\ell_\alpha,S\ell_\alpha) - n - \dim(S\ell_\alpha\cap\ell_\alpha).$$

The equality (3.52), follows noting that by definition of the index of inertia and the antisymmetry of τ

$$\mathrm{Inert}(S\ell,\ell_\alpha,S\ell_\alpha) - n - \dim(S\ell_\alpha\cap\ell_\alpha) = -\,\mathrm{Inert}(S\ell,S\ell_\alpha,\ell_\alpha). \qquad \square$$

Recall that there is an isomorphism

$$\mathbb{Z}\cong\pi_1[\mathrm{Sp}(n)] \longrightarrow \pi_1[\mathrm{Lag}(n)]\cong\mathbb{Z}$$

which is multiplication by 2 on $\mathbb{Z}$; in fact (formula (3.50))

$$(\alpha S_\infty)\ell_\infty = \beta^2(S_\infty \ell_\infty) = S_\infty(\beta^2 \ell_\infty)$$

for all $(S_\infty, \ell_\infty) \in \mathrm{Sp}_\infty(n) \times \mathrm{Lag}_\infty(n)$. Also recall that the *ALM* and Maslov indices satisfy

$$\mu(\beta^r \ell_\infty, \beta^{r'} \ell'_\infty) = \mu(\ell_\infty, \ell'_\infty) + 4(r - r')$$

(formula (3.31) and

$$\mu_\ell(\alpha^r S_\infty) = \mu_\ell(S_\infty) + 2r$$

(formula (3.41) for all integers r and r'.

Let now q be an integer, $q \geq 1$. We have

$$\pi_1[\mathrm{Lag}(n)] = \{\beta^k : k \in \mathbb{Z}\}$$

hence (see Appendix B)

$$\mathrm{Lag}_q(n) = \mathrm{Lag}(n)/\{\beta^{qk} : k \in \mathbb{Z}\}. \tag{3.53}$$

Similarly, since

$$\pi_1[\mathrm{Sp}(n)] = \{\alpha^k : k \in \mathbb{Z}\}$$

we have

$$\mathrm{Sp}_q(n) = \mathrm{Sp}(n)/\{\alpha^{qk} : k \in \mathbb{Z}\}. \tag{3.54}$$

Let us now identify $\pi_1[\mathrm{Lag}(n)]$ with $\mathbb{Z}$; recalling that the natural homomorphism $\pi_1[\mathrm{Sp}(n)] \longrightarrow \pi_1[\mathrm{Lag}(n)]$ is multiplication by 2 in $\mathbb{Z}$ (*cf.* formula (3.50) $\pi_1[\mathrm{Sp}(n)]$ is then identified with $2\mathbb{Z}$. This leads us, taking (3.53) and (3.54) into account, to the identifications

$$\mathrm{Lag}_q(n) \equiv \mathrm{Lag}_\infty(n)/q\mathbb{Z} \quad , \quad \mathrm{Sp}_q(n) \equiv \mathrm{Sp}_\infty(n)/2q\mathbb{Z}. \tag{3.55}$$

The *ALM* index on $\mathrm{Lag}_q(n)$ is now defined as being the function

$$[\mu]_q : (\mathrm{Lag}_q(n))^2 \longrightarrow \mathbb{Z}_q = \mathbb{Z}/q\mathbb{Z}$$

given by

$$[\mu]_q(\ell_{(q)}, \ell'_{(q)}) = \mu(\ell_\infty, \ell'_\infty) \mod q$$

if $(\ell_\infty, \ell'_\infty) \in (\mathrm{Lag}_\infty(n))^2$ covers $(\ell_{(q)}, \ell'_{(q)}) \in \mathrm{Lag}_q(n))^2$; similarly the Maslov index relative to $\ell \in \mathrm{Lag}(n)$ on $\mathrm{Sp}_q(n)$ is the function

$$[\mu_\ell]_{2q} : \mathrm{Sp}_q(n) \longrightarrow \mathbb{Z}_{2q} = \mathbb{Z}/2q\mathbb{Z}$$

defined by

$$[\mu_\ell]_{2q}(S_{(q)}) = \mu_\ell(S_\infty) \mod 2q$$

if $S_\infty \in \mathrm{Sp}_\infty(n)$ covers $S_{(q)} \in \mathrm{Sp}_q(n)$.

Exactly as was the case for ∞-symplectic geometry the study of q-symplectic geometry requires the use of *reduced indices*:

Definition 3.37. The reduced ALM index on $\mathrm{Lag}_{2q}(n)$ is defined by

$$[m]_{2q}(\ell_{(q)}, \ell'_{(q)}) = m(\ell_\infty, \ell'_\infty) \mod 2q$$

and the reduced Maslov index $[m_\ell]_{2q}$ on $\mathrm{Sp}_q(n)$ by

$$[m_\ell]_{2q}(S_{(q)}) = m_\ell(S_\infty) \mod 2q.$$

Let us denote by $[r]_{2q}$ the equivalence class modulo $2q$ of $r \in \mathbb{Z}$. Corollary 3.35 and Theorem 3.36 immediately imply that:

Corollary 3.38. *Let $\ell_\alpha \in \mathrm{Lag}(n)$ and identify $\mathrm{Sp}_q(n)$ with the subset*

$$\{(S, m) : S \in \mathrm{Sp}(n), m \subset m_{\ell_\alpha}(S)\}$$

of $\mathrm{Sp}(n) \times \mathbb{Z}_{2q}$ equipped with the composition law

$$(S, [m]_{2q}) *_{\ell_\alpha} (S', [m']_{2q}) = (SS', [m + m' + \mathrm{Inert}(\ell_\alpha, S\ell_\alpha, SS'\ell_\alpha)]_{2q})$$

and $\mathrm{Lag}_{2q}(n)$ with $\mathrm{Lag}(n) \times \mathbb{Z}_{2q}$. The action of $\mathrm{Sp}_q(n)$ on $\mathrm{Lag}_{2q}(n)$ is then given by the formula

$$(S, [m]_{2q}) \cdot_{\ell_\alpha} (\ell, [\lambda]_{2q}) = (S\ell, [m + \lambda - \mathrm{Inert}(S\ell, \ell, \ell_\alpha)]_{2q}).$$

We have now achieved our goal which was to describe the algebraic structure of q-symplectic geometry.

A related interesting notion, to which we will come back later, is that of *q-orientation* of a Lagrangian plane. Recall that we mentioned in the beginning of this chapter that the action of $\mathrm{Sp}(n)$ on $\mathrm{Lag}(n)$ automatically induces an action

$$\mathrm{Sp}(n) \times \mathrm{Lag}_2(n) \longrightarrow \mathrm{Lag}_2(n)$$

("1-symplectic geometry") since linear symplectic transformations have determinant 1 and are thus preserving the orientation of Lagrangian planes. If we view the datum of an element $\ell_{(\pm)}$ of $\mathrm{Lag}_2(n)$ as the choice of an orientation of the Lagrangian plane $\ell \in \mathrm{Lag}(n)$ it covers, the following definition makes sense:

Definition 3.39. Let $\ell \in \mathrm{Lag}(n)$. A q-orientation of ℓ is the datum of an element $\ell_{(q)}$ of $\mathrm{Lag}_{2q}(n)$ covering ℓ. (Every Lagrangian plane thus has exactly $2q$ q-orientations).

This definition is at first sight rather artificial. It is however not a useless extension of the notion of orientation; it will play an important role in the understanding of the Maslov quantization of Lagrangian manifold.

Notice that the action of $\mathrm{Sp}_q(n)$ on a q-oriented Lagrangian plane preserves its q-orientation: this is one of the main interests of q-symplectic geometry, and justifies *a posteriori* the title of this chapter.

Chapter 4

Intersection Indices in $\mathrm{Lag}(n)$ and $\mathrm{Sp}(n)$

In this chapter we generalize the notion of Maslov index to arbitrary paths (not just loops) in $\mathrm{Lag}(n)$ and $\mathrm{Sp}(n)$. We will study two (related) constructions of these "intersection indices": the Lagrangian and symplectic "Maslov indices", which extend the usual notion of Maslov index for loops to arbitrary paths in $\mathrm{Lag}(n)$ and $\mathrm{Sp}(n)$, and which are directly related to the notion of "spectral flow", and the Conley–Zehnder index, which plays a crucial role in Morse theory, and its applications to mathematical physics (we will apply the latter to a precise study of metaplectic operators in Chapter 7).

The results in the two first sections are taken from de Gosson [63] and de Gosson and de Gosson [67]. For applications to Morse theory and to functional analysis, see Booss–Bavnbek and Furutani [14], Javaloyes and Piccione [130], Nostre Marques *et al.* [131]

4.1 Lagrangian Paths

A Lagrangian path is a continuous mapping $\Lambda:[0,1]\longrightarrow\mathrm{Sp}(n)$; the vocation of a Lagrangian intersection index is to keep a precise account of the way that path intersects a given locus (or "Maslov cycle") in $\mathrm{Lag}(n)$. It can be viewed as a generalization of the usual Maslov index in $\mathrm{Lag}(n)$, to which it reduces (up to the factor two) when Λ is a loop. We begin by briefly discussing the notion of stratum in $\mathrm{Lag}(n)$.

4.1.1 The strata of $\mathrm{Lag}(n)$

Let $\mathbb{M}^m$ be a m-dimensional topological manifold. A *stratification* of $\mathbb{M}^m$ is a partition of $\mathbb{M}^m$ in a family $\{\mathbb{M}^k_\alpha\}_{\alpha\in A}$ of connected submanifolds ("strata") of dimension $k \leq m$ such that:

- The family $\{\mathbb{M}^k_\alpha\}_{\alpha\in A}$ is a locally finite partition of $\mathbb{M}^m$;
- If $\mathbb{M}^k_\alpha \cap \overline{\mathbb{M}^{k'}_{\alpha'}} \neq \emptyset$ for $\alpha \neq \alpha'$ then $\mathbb{M}^k_\alpha \subset \mathbb{M}^{k'}_{\alpha'}$ and $k \leq k'$;
- $\overline{\mathbb{M}^k_\alpha} \setminus \mathbb{M}^k_\alpha$ is a disjoint union of strata of dimension $< k$.

It turns out that to every Lagrangian plane ℓ we can associate a natural stratification of Lag(n):

For $\ell \in \mathrm{Lag}(n)$ and $0 \leq k \leq n$ set

$$\mathrm{Lag}_\ell(n,k) = \{\ell' \in \mathrm{Lag}(n) : \dim \ell \cap \ell' = k\}.$$

We will call $\mathrm{Lag}_\ell(n;k)$ the *stratum of* Lag(n) *of order* k, *relative to the Lagrangian plane* ℓ. Clearly

$$\mathrm{Lag}_\ell(n;k) \cap \mathrm{Lag}_\ell(n;k') = \emptyset \quad \textit{if} \quad k \neq k'$$

and

$$\mathrm{Lag}(n) = \cup_{0\leq k\leq n} \mathrm{Lag}_\ell(n;k).$$

One proves (see, *e.g.*, Trèves [164]; also see Robbin and Salamon [135]) that the sets $\mathrm{Lag}_\ell(n,k)$ form a stratification of Lag(n); moreover $\mathrm{Lag}_\ell(n;0)$ is an open subset of Lag(n) and the sets $\mathrm{Lag}_\ell(n;k)$ are, for $0 \leq k \leq n$, *connected* submanifolds of Lag(n) with codimension $k(k+1)/2$.

Definition 4.1. The closed set

$$\Sigma_\ell = \mathrm{Lag}(n) \setminus \mathrm{Lag}_\ell(n;0) = \overline{\mathrm{Lag}_\ell(n,1)}$$

is called the "*Maslov cycle relative to* ℓ": it is the set of Lagrangians that are not transverse to ℓ. When $\ell = \ell_P$ we call Σ_ℓ simply the "*Maslov cycle*", and denote it by Σ.

Let us now enunciate a system of "reasonable" axioms that should be satisfied by a generalization of the Maslov index for loops.

4.1.2 The Lagrangian intersection index

The definition we give here is sightly more general than those of, for instance, Robbin and Salamon [135]. We do not in particular impose from the beginning any "dimensional additivity"; this property is however satisfied by the explicit indices we construct in the next subsection.

Let $\mathcal{C}(\mathrm{Lag}(n))$ be the set of continuous paths $\Lambda : [0,1] \longrightarrow \mathrm{Lag}(n)$. If Λ and Λ' are two consecutive paths (*i.e.*, if $\Lambda(1) = \Lambda'(0)$) we shall denote by $\Lambda * \Lambda'$ the concatenation of Λ and Λ', that is, the path Λ followed by the path Λ':

$$\Lambda * \Lambda'(t) = \begin{cases} \Lambda(2t) & \textit{if} \quad 0 \leq t \leq \frac{1}{2}, \\ \Lambda'(2t-1) & \textit{if} \quad \frac{1}{2} \leq t \leq 1. \end{cases}$$

We will use the notation Λ^o for the inverse of the path Λ:

$$\Lambda^o(t) = \Lambda(1-t) \quad , \quad 0 \leq t \leq 1.$$

Finally, we shall write $\Lambda \sim \Lambda'$ when the paths Λ and Λ' are homotopic with *fixed* endpoints.

Definition 4.2. A "Lagrangian intersection index" is a mapping

$$\begin{aligned} \mu_{\text{Lag}} : \mathcal{C}(\text{Lag}(n)) \times \text{Lag}(n) &\longrightarrow \mathbb{Z}, \\ (\Lambda, \ell) &\longmapsto \mu_{\text{Lag}}(\Lambda, \ell) \end{aligned}$$

having the following four properties:

(L_1) Homotopy invariance: *If the paths Λ and Λ' have the same endpoints, then* $\mu_{\text{Lag}}(\Lambda, \ell) = \mu_{\text{Lag}}(\Lambda', \ell)$ if and only if $\Lambda \sim \Lambda'$;

(L_2) Additivity under composition: if Λ and Λ' are two consecutive paths, then

$$\mu_{\text{Lag}}(\Lambda * \Lambda', \ell) = \mu_{\text{Lag}}(\Lambda, \ell) + \mu_{\text{Lag}}(\Lambda', \ell)$$

for every $\ell \in \text{Lag}(n)$;

(L_3) Zero in strata: if the path Λ remains in the same stratum $\text{Lag}_\ell(n; k)$, then $\mu_{\text{Lag}}(\Lambda, \ell)$ is zero:

$$\dim(\Lambda(t) \cap \ell) = k \quad (0 \leq t \leq 1) \Longrightarrow \mu_{\text{Lag}}(\Lambda, \ell) = 0;$$

(L_4) Restriction to loops: if γ is a loop in $\text{Lag}(n)$ then

$$\mu_{\text{Lag}}(\gamma, \ell) = 2m(\gamma)$$

($m(\gamma)$ the Maslov index of γ) for every $\ell \in \text{Lag}\, g(n)$.

We note in particular that the axioms (L_2) and (L_4) imply that an intersection index is antisymmetric in the sense that

$$\mu_{\text{Lag}}(\Lambda^o, \ell) = -\mu_{\text{Lag}}(\Lambda, \ell). \tag{4.1}$$

Indeed, by (L_2) we have

$$\mu_{\text{Lag}}(\Lambda * \Lambda^o, \ell) = \mu_{\text{Lag}}(\Lambda, \ell) + \mu_{\text{Lag}}(\Lambda^o, \ell)$$

and since the loop $\Lambda * \Lambda^o = \gamma$ is homotopic to a point (L_4) implies that

$$\mu_{\text{Lag}}(\Lambda * \Lambda^o, \ell) = 2m(\gamma) = 0.$$

The system of axioms (L_1)–(L_4) is in fact equivalent to the system of axioms obtained by replacing (L_1) by the apparently stronger condition (4.2) below. Let us first define the notion of "homotopy in strata":

Definition 4.3. Two Lagrangian paths Λ and Λ' are said to be "homotopic in the strata relative to ℓ" (denoted $\Lambda \approx_\ell \Lambda'$) if there exist a continuous mapping $h : [0,1] \times [0,1] \longrightarrow \mathrm{Lag}(n)$ such that

$$h(t,0) = \Lambda(t) \quad , \; h(t,1) = \Lambda'(t) \;\; \mathit{for} \;\; 0 \leq t \leq 1$$

and two integers k_0, k_1 $(0 \leq k_0, k_1 \leq n)$ such that

$$h(0,s) \in \mathrm{Lag}_\ell(n;k_0) \;\; \mathit{and} \;\; h(1,s) \in \mathrm{Lag}_\ell(n;k_1) \;\; \mathit{for} \;\; 0 \leq s \leq 1.$$

Intuitively $\Lambda \approx_\ell \Lambda$ means that Λ and Λ' are homotopic in the usual sense and that the endpoints $\Lambda(0)$ and $\Lambda'(0)$ (resp. $\Lambda(1)$ and $\Lambda'(1)$) remain in the same stratum during the homotopy taking Λ to Λ'.

The intersection indices μ_{Lag} have the following property that strengthens (L_1):

Proposition 4.4. *If the paths Λ and Λ' are homotopic in strata relative to ℓ, then $\mu_{\mathrm{Lag}}(\Lambda,\ell) = \mu_{\mathrm{Lag}}(\Lambda',\ell)$:*

$$\Lambda \approx_\ell \Lambda' \Longrightarrow \mu_{\mathrm{Lag}}(\Lambda,\ell) = \mu_{\mathrm{Lag}}(\Lambda',\ell). \tag{4.2}$$

Proof. Suppose that $\Lambda \approx_\ell \Lambda'$ and define the paths ε_0 and ε_1 joining $\Lambda'(0)$ to $\Lambda(0)$ and $\Lambda(1)$ to $\Lambda'(1)$, respectively, by $\varepsilon_0(s) = h(0,1-s)$ and $\varepsilon_1(s) = h(1,s)$ $(0 \leq s \leq 1)$. Then $\Lambda * \varepsilon_1 * \Lambda'^{-1} * \varepsilon_0$ is homotopic to a point, and hence, in view of (L_2) and (L_4):

$$\mu_{\mathrm{Lag}}(\Lambda,\ell) + \mu_{\mathrm{Lag}}(\varepsilon_1,\ell) + \mu_{\mathrm{Lag}}(\Lambda'^{-1},\ell) + \mu_{\mathrm{Lag}}(\varepsilon_0,\ell) = 0.$$

But, in view of (L_3),

$$\mu_{\mathrm{Lag}}(\varepsilon_1,\ell) = \mu_{\mathrm{Lag}}(\varepsilon_0,\ell) = 0$$

and thus

$$\mu_{\mathrm{Lag}}(\Lambda,\ell) + \mu_{\mathrm{Lag}}(\Lambda'^{-1},\ell) = 0;$$

the conclusion now follows from the antisymmetry property (4.1). □

We will not discuss the uniqueness of an index defined by the axioms above; the interested reader is referred to Serge de Gosson's thesis [74] for a study of this question.

Let us next construct explicitly a Lagrangian intersection index using the properties of the *ALM* index studied in the previous chapter.

4.1.3 Explicit construction of a Lagrangian intersection index

Our approach is purely topological, and does not appeal to any differentiability conditions for the involved paths, as opposed to the construction given in, for instance, Robbin and Salamon [134].

Theorem 4.5. *For* $(\Lambda_{12}, \ell) \in \mathcal{C}(\mathrm{Lag}(n)) \times \mathrm{Lag}(n)$ *let us define* ℓ_∞, $\ell_{1,\infty}$ *and* $\ell_{2,\infty}$ *in the following way:*

(i) ℓ_∞ *is an arbitrary element of* $\mathrm{Lag}_\infty(n)$ *covering* ℓ*;*

(ii) $\ell_{1,\infty}$ *is the equivalence class of an arbitrary path* $\Lambda_{01} \in \mathcal{C}(\mathrm{Lag}(n))$ *joining* ℓ_0 *to* ℓ_1*;*

(iii) $\ell_{2,\infty}$ *is the equivalence class of* $\Lambda_{02} = \Lambda_{01} * \Lambda_{12}$.

Then the formula

$$\mu_{\mathrm{Lag}}(\Lambda_{12}, \ell) = \mu(\ell_{2,\infty}, \ell_\infty) - \mu(\ell_{1,\infty}, \ell_\infty) \tag{4.3}$$

defines an intersection index on $\mathrm{Lag}(n)$.

Proof. Let us first show that $\mu_{\mathrm{Lag}}(\Lambda_{12}, \ell)$ is independent of the choice of the element ℓ_∞ of $\mathrm{Lag}_\infty(n)$ covering ℓ. Assume in fact that

$$\pi^{\mathrm{Lag}}(\ell'_\infty) = \pi^{\mathrm{Lag}}(\ell_\infty) = \ell;$$

then there exists $r \in \mathbb{Z}$ such that $\ell'_\infty = \beta^r \ell_\infty$ (β is as usual the generator of $\pi_1[\mathrm{Lag}(n)]$) and hence

$$\mu(\ell_{2,\infty}, \ell'_\infty) = \mu(\ell_{2,\infty}, \ell_\infty) - 2r \quad , \quad \mu(\ell_{1,\infty}, \ell'_\infty) = \mu(\ell_{1,\infty}, \ell_\infty) - 2r$$

in view of property (3.31) of the *ALM* index; it follows that

$$\mu(\ell_{2,\infty}, \ell'_\infty) - \mu(\ell_{1,\infty}, \ell'_{1,\infty}) = \mu(\ell_{2,\infty}, \ell_\infty) - \mu(\ell_{1,\infty}, \ell_\infty).$$

Let us next show that $\mu_{\mathrm{Lag}}(\Lambda_{12}, \ell)$ is also independent of the choice of Λ_{01} and hence of the choice of the element $\ell_{1,\infty}$ such that $\pi^{\mathrm{Lag}}(\ell_{1,\infty}) = \ell_1$. Let us replace Λ_{01} by a path Λ'_{01} with same attributes and such that $\ell'_{1,\infty} = \beta^r \ell_{1,\infty}$; $\ell_{2,\infty}$ will thus be replaced by $\ell'_{2,\infty} = \beta^r \ell_{2,\infty}$. Using again (3.31) we have

$$\mu(\ell'_{2,\infty}, \ell_\infty) - \mu(\ell'_{1,\infty}, \ell_\infty) = \mu(\ell_{2,\infty}, \ell_\infty) - \mu(\ell_{1,\infty}, \ell_\infty),$$

hence our claim. It remains to prove that the function μ_{Lag} defined by (4.3) satisfies the axioms (L_1)–(L_4). *Axiom* L_1. Let us replace the path Λ_{12} by any path Λ'_{12} homotopic (with fixed endpoints) to Λ_{12}. Then $\Lambda_{02} = \Lambda_{01} * \Lambda_{12}$ is replaced by a homotopic path $\Lambda'_{02} = \Lambda_{01} * \Lambda'_{12}$ and the homotopy class $\ell_{2,\infty}$ does not change. Consequently, $\mu_{\mathrm{Lag}}(\Lambda'_{12}, \ell) = \mu_{\mathrm{Lag}}(\Lambda_{12}, \ell)$. *Axiom* L_2. Consider two consecutive paths Λ_{12} and Λ_{23}. By definition

$$\mu_{\mathrm{Lag}}(\Lambda_{23}, \ell) = \mu(\ell_{3,\infty}, \ell_\infty) - \mu(\ell'_{2,\infty}, \ell_\infty)$$

where $\ell'_{2,\infty}$ is the homotopy class of an arbitrary path Λ'_{02} and $\ell_{3,\infty}$ that of $\Lambda'_{02} * \Lambda_{23}$. Let us choose $\Lambda'_{02} = \Lambda_{02}$. Then $\ell'_{2,\infty} = \ell_{2,\infty}$ and

$$\begin{aligned}\mu_{\mathrm{Lag}}(\Lambda_{12}, \ell) + \mu_{\mathrm{Lag}}(\Lambda_{23}, \ell) = \mu(\ell_{2,\infty}, \ell_\infty) - \mu(\ell_{1,\infty}, \ell_\infty)& \\ + \mu(\ell_{3,\infty}, \ell_\infty) - \mu(\ell_{2,\infty}, \ell_\infty)&,\end{aligned}$$

that is

$$\mu_{\mathrm{Lag}}(\Lambda_{12},\ell)+\mu_{\mathrm{Lag}}(\Lambda_{23},\ell)=\mu_{\mathrm{Lag}}(\Lambda_{13},\ell)$$

which we set out to prove. *Axiom* L_3. Let Λ_{12} be a path in the stratum $\mathrm{Lag}_\ell(n;k)$ and denote by $\ell_\infty(t)$ the equivalence class of $\Lambda_{01}*\Lambda_{12}(t)$ for $0\leq t\leq 1$. The mapping $t\longmapsto \ell_\infty(t)$ being continuous, the composition mapping $t\longmapsto \mu(\ell_\infty(t),\ell_\infty)$ is locally constant on the interval $[0,1]$. It follows that it is constant on that interval since $\mathrm{Lag}_\ell(n;k)$ is connected; its value is

$$\mu(\ell_\infty(0),\ell_\infty)=\mu(\ell_\infty(1),\ell_\infty)$$

hence $\mu(\Lambda_{12},\ell)=0$. *Axiom* L_4. Let $\gamma\in\pi_1[\mathrm{Lag}(n),\ell_0)]$. In view of formula (3.32) in Corollary 3.22 we have the equality

$$\mu_{\mathrm{Lag}}(\gamma,\ell)-\mu(\gamma\ell_{0,\infty},\ell_\infty)\quad\mu(\ell_{0,\infty},\ell_\infty)-2m(\gamma)$$

which concludes the proof. □

Let us now proceed to the study of symplectic intersection indices.

4.2 Symplectic Intersection Indices

The theory of symplectic intersection indices is analogue to the Lagrangian case; in fact each theory can be deduced from the other. For the sake of clarity we however treat the symplectic case independently. For different points of view and deep applications to the theory of Hamiltonian periodic orbits see the monographs by Ekeland [39] and Long [113].

4.2.1 The strata of Sp(n)

Similar definitions are easy to give for the symplectic group $\mathrm{Sp}(n)$. For $\ell\in\mathrm{Lag}(n)$ and k an integer we call the set

$$\mathrm{Sp}_\ell(n;k)=\{S\in\mathrm{Sp}(n):\dim S\ell\cap\ell=k\}$$

the *stratum of* $\mathrm{Sp}(n)$ *of order* k*, relative to the Lagrangian plane* ℓ. The sets $\mathrm{Sp}_\ell(n;k)$ indeed form a stratification of the Lie group $\mathrm{Sp}(n)$; clearly $\mathrm{Sp}_\ell(n;k)$ is empty for $k<0$ or $k>n$, and we have

$$\mathrm{Sp}_\ell(n;0)=\mathrm{St}(\ell)$$

(the stabilizer of ℓ in $\mathrm{Sp}(n)$). We have of course

$$\mathrm{Sp}(n)=\cup_{0\leq k\leq n}\mathrm{Sp}_\ell(n;k).$$

The strata $\mathrm{Sp}_\ell(n;k)$ are not in general connected:

Exercise 4.6. Show directly that $\mathrm{Sp}_\ell(n;0)$ has two connected components.

Exercise 4.7. Show that $\mathrm{Sp}_\ell(n;k)$ is a submanifold of $\mathrm{Sp}(n)$ with codimension $k(k+1)/2$. [Hint: it is sufficient to prove this for $\ell=\ell_P$; then use block matrices.]

4.2.2 Construction of a symplectic intersection index

Let us now define symplectic intersection indices. We denote by $\mathcal{C}(\mathrm{Sp}(n))$ the set of continuous paths $[0,1] \longrightarrow \mathrm{Sp}(n)$:

$$\mathcal{C}(\mathrm{Sp}(n)) = C^0([0,1], \mathrm{Sp}(n)).$$

Definition 4.8. A symplectic intersection index is a mapping

$$\mu_{\mathrm{Sp}} : \mathcal{C}(\mathrm{Sp}(n)) \times \mathrm{Lag}(n) \longrightarrow \mathbb{Z},$$
$$(\Sigma, \ell) \longmapsto \mu_{\mathrm{Sp}}(\Sigma, \ell)$$

satisfying the following four axioms:

(S_1) Homotopy invariance: if the symplectic paths Σ and Σ' are homotopic with fixed endpoints, then $\mu_{\mathrm{Sp}}(\Sigma, \ell) = \mu_{\mathrm{Sp}}(\Sigma', \ell)$ for all $\ell \in \mathrm{Lag}(n)$.

(S_2) Additivity under concatenation: if Σ and Σ' are two consecutive symplectic paths, then

$$\mu_{\mathrm{Sp}}(\Sigma * \Sigma', \ell) = \mu_{\mathrm{Sp}}(\Sigma, \ell) + \mu_{\mathrm{Sp}}(\Sigma', \ell)$$

for all $\ell \in \mathrm{Lag}(n)$.

(S_3) Zero in strata: if Σ and ℓ are such that $\mathrm{Im}(\Sigma\ell) \subset \mathrm{Lag}_\ell(n)$, then $\mu_{\mathrm{Sp}}(\Sigma, \ell) = 0$.

(S_4) Restriction to loops: is ψ is a loop in $\mathrm{Sp}(n)$, then

$$\mu_{\mathrm{Sp}}(\psi, \ell) = 2m(\psi\ell)$$

for all $\ell \in \mathrm{Lag}(n)$; $m(\psi\ell)$ is the Maslov index of the loop $t \longmapsto \psi(t)\ell$ in $\mathrm{Lag}(n)$.

The following exercise proposes a symplectic version of Proposition 4.4:

Exercise 4.9. Show that if the symplectic paths Σ and Σ' are such that $\Sigma\ell$ and $\Sigma'\ell$ are homotopic in strata relative to ℓ, then $\mu_{\mathrm{Sp}}(\Sigma, \ell) = \mu_{\mathrm{Sp}}(\Sigma', \ell)$.

The data of an intersection index on $\mathrm{Lag}(n)$ is equivalent to that of an intersection index on $\mathrm{Sp}(n)$. Indeed, let μ_{Lag} be an intersection index on $\mathrm{Lag}(n)$ and let $\Sigma \in \mathcal{C}(\mathrm{Sp}(n))$ be a symplectic path. Then the function

$$\mathcal{C}(\mathrm{Sp}(n)) \times \mathrm{Lag}(n) \ni (\Sigma, \ell) \longmapsto \mu(\Sigma\ell, \ell) \in \mathbb{Z} \tag{4.4}$$

($\Sigma\ell$ being the path $t \longmapsto \Sigma(t)\ell$) is an intersection index on $\mathrm{Sp}(n)$.

Conversely, to each intersection index μ_{Sp} we may associate an intersection index μ_{Lag} on $\mathrm{Lag}(n)$ in the following way. For each $\ell \in \mathrm{Lag}(n)$ we have a fibration

$$\mathrm{Sp}(n) \longrightarrow \mathrm{Sp}(n)/\,\mathrm{St}(\ell) = \mathrm{Lag}(n)$$

($\mathrm{St}(\ell)$ the stabilizer of ℓ in $\mathrm{Sp}(n)$), hence, every path $\Lambda \in \mathcal{C}(\mathrm{Lag}(n))$ can be lifted to a path $\Sigma_\Lambda \in \mathcal{C}(\mathrm{Sp}(n))$ such that $\Lambda = \Sigma_A\ell$. One verifies that the mapping

$$\mu_{\mathrm{Lag}} : \mathcal{C}(\mathrm{Lag}(n)) \times \mathrm{Lag}(n) \longrightarrow \mathbb{Z}$$

defined by

$$(\Lambda, \ell) \longmapsto \mu_{\mathrm{Sp}}(\Sigma_\Lambda, \ell)$$

is an intersection index on Lag(n).

In Theorem 4.5 we expressed a Lagrangian intersection index as the difference between two values of the *ALM* index. A similar result holds for symplectic intersection indices:

Proposition 4.10. *Let $\Sigma_{12} \in \mathcal{C}(\mathrm{Sp}(n))$ be a symplectic path joining S_1 to S_2 in* Sp(n)*. Let $S_{1,\infty}$ be an arbitrary element of $\mathrm{Sp}_\infty(n)$ covering S_1 and $S_{2,\infty}$ the homotopy class of $\Sigma_{01} * \Sigma_{12}$ (Σ_{01} a representative of $S_{1,\infty}$). The function $\mu_{\mathrm{Sp}} : \mathcal{C}(\mathrm{Sp}(n)) \longrightarrow \mathbb{Z}$ defined by*

$$\mu_{\mathrm{Sp}}(\Sigma_{12}, \ell) = \mu_\ell(S_{2,\infty}) - \mu_\ell(S_{1,\infty}) \tag{4.5}$$

is an intersection index on Sp(n).

Proof. Consider ℓ_∞ to be the homotopy class of an arbitrary path Λ joining ℓ_0 (the base point of $\mathrm{Lag}(n)_\infty$) to ℓ. We have

$$S_{1,\infty}\ell_\infty = \mathrm{class}\,[t \longmapsto \Sigma_{01}(t)\Lambda(t),\ 0 \leq t \leq 1]$$

(where "class" means "equivalence class of") and

$$S_{2,\infty}\ell_\infty = \mathrm{class}\left[t \longmapsto \begin{cases} \Sigma_{01}(2t)\Lambda(2t),\ 0 \leq t \leq \frac{1}{2} \\ \Sigma_{12}(2t-1)\Lambda(2t-1),\ \frac{1}{2} \leq t \leq 1 \end{cases}\right]$$

hence, using (4.4) and (4.3),

$$\mu_{\mathrm{Sp}}(\Sigma_{12}, \ell) = \mu_{\mathrm{Lag}(n)}(\Sigma_{12}\ell, \ell) = \mu_\infty(S_{2,\infty}\ell_\infty, \ell_\infty) - \mu_\infty(S_{1,\infty}\ell_\infty, \ell_\infty)$$

which is (4.5). □

Let us illustrate the notions studied above on a simple example.

4.2.3 Example: spectral flows

Here is a simple application of the constructions above. Let $(A(t))_{0\leq t\leq 1}$ be a family of real symmetric matrices of order n depending continuously on $t \in [0, 1]$. By definition the "spectral flow" of $(A(t))_{0\leq t\leq 1}$ is the integer

$$\mathrm{SF}(A(t))_{0\leq t\leq 1} = \mathrm{sign}\, A(1) - \mathrm{sign}\, A(0) \tag{4.6}$$

where $\mathrm{sign}\, A(t)$ is the difference between the number of eigenvalues > 0 and the number of eigenvalues < 0 of $A(t)$.

We have the following result, which has been established in a particular case by Duistermaat [34], and which relates the spectral flow to the notions of Lagrangian and symplectic intersection indices:

Proposition 4.11. *Let* Λ *be the Lagrangian path associated to the family* $(A(t))_{0\leq t\leq 1}$ *by*

$$\Lambda(t) = \{(x, A(t)x) : x \in \mathbb{R}^n\} \quad , \quad 0 \leq t \leq 1. \tag{4.7}$$

Then the spectral flow of $(A(t))_{0\leq t\leq 1}$ *is given by*

$$\mathrm{SF}(A(t))_{0\leq t\leq 1} = \mu_{\mathrm{Lag}}(\Lambda, \ell_X) \tag{4.8}$$

or, equivalently

$$\mathrm{SF}(A(t))_{0\leq t\leq 1} = \mu_{\mathrm{Sp}}(\Sigma, \ell_X) \tag{4.9}$$

where Σ_A *the symplectic path defined by*

$$\Sigma(t) = V_{-A(t)} = \begin{bmatrix} I & 0 \\ A(t) & I \end{bmatrix}$$

for $0 \leq t \leq 1$.

Proof. Formula (4.9) follows immediately from formula (4.8) observing that

$$\begin{bmatrix} I & 0 \\ A(t) & I \end{bmatrix} \begin{bmatrix} x \\ 0 \end{bmatrix} = \begin{bmatrix} x \\ A(t)x \end{bmatrix}.$$

To prove formula (4.8) we begin by noting that $\dim(\Lambda(t) \cap \ell_P) = n$ for $0 \leq t \leq 1$, and hence

$$\bar{\mu}_{\mathrm{Lag}}(\Lambda, \ell_P) = 0 \tag{4.10}$$

in view of the axiom (L_3) of nullity in the strata. By definition of μ_{Lag} we have, with obvious notation,

$$\mu_{\mathrm{Lag}}(\Lambda, \ell_X) = \mu(\Lambda(1)_\infty, \ell_{X,\infty}) - \mu(\Lambda(0)_\infty, \ell_{X,\infty}).$$

In view of the property $\partial\mu = \tau$ of the ALM index,

$$\mu(\Lambda(t)_\infty, \ell_{X,\infty}) - \mu(\Lambda(t)_\infty, \ell_{P,\infty}) = -\mu(\ell_{X,\infty}, \ell_{P,\infty}) + \tau(\Lambda(t), \ell_X, \ell_P)$$

for $0 \leq t \leq 1$, and hence, taking (4.10) into account,

$$\begin{aligned} \mu_{\mathrm{Lag}}(\Lambda, \ell_X) &= \tau(\Lambda(1), \ell_X, \ell_P) - \tau(\Lambda(0), \ell_X, \ell_P) \\ &= \tau(\ell_P, \Lambda(1), \ell_X) - (\ell_P, \Lambda(1), \ell_X). \end{aligned}$$

In view of formula (1.24) in Corollary 1.31 (Subsection 1.4.1) we have

$$\tau(\ell_P, \Lambda(t), \ell_X) = \mathrm{sign}(A(t))$$

and formula (4.8) follows. □

The result above is rather trivial in the sense that the spectral flow (4.6) depends only on the extreme values $A(1)$ and $A(0)$. The situation is however far more complicated in the case of infinite-dimensional symplectic spaces and its analysis requires elaborated functional analytical techniques (see for instance Booss–Bavnbek and Furutani [14]).

4.3 The Conley–Zehnder Index

Here is another type of symplectic intersection index, due to Conley and Zehnder [25] (also see Hofer *et al.* [90]). We will use it in the study of the Weyl representation of the metaplectic group in Chapter 7 (Subsection 7.4.2). It also plays an important role in the Gutzwiller theory [86] of semiclassical quantization of classically chaotic Hamiltonian systems and in the theory of periodic Hamiltonian orbits and related topics (such as Morse theory and Floer homology).

4.3.1 Definition of the Conley–Zehnder index

Let Σ be a continuous path $[0,1] \longrightarrow \mathrm{Sp}(n)$ such that $\Sigma(0) = I$ and $\det(\Sigma(1)-I) \neq 0$. Loosely speaking, the Conley–Zehnder index [25, 90] counts algebraically the number of times this path crosses the locus

$$\mathrm{Sp}_0(n) = \{S : \det(S - I) = 0\}.$$

To give a more precise definition we need some additional notation. Let us define

$$\begin{aligned} \mathrm{Sp}^+(n) &= \{S : \det(S - I) > 0\}, \\ \mathrm{Sp}^-(n) &= \{S : \det(S - I) < 0\}. \end{aligned}$$

These sets partition $\mathrm{Sp}(n)$, and $\mathrm{Sp}^+(n)$ and $\mathrm{Sp}^-(n)$ are moreover arcwise connected (this is proven in [25]); the symplectic matrices $S^+ = -I$ and

$$S^- = \begin{bmatrix} L & 0 \\ 0 & L^{-1} \end{bmatrix} \quad , \quad L = \mathrm{diag}[2, -1, \dots, -1]$$

belong to $\mathrm{Sp}^+(n)$ and $\mathrm{Sp}^-(n)$, respectively.

Let us denote by ρ the mapping $\mathrm{Sp}(n) \longrightarrow S^1$ defined as follows:

$$S \in \mathrm{Sp}(n) \longmapsto U = S(S^TS)^{-1/2} \in \mathrm{U}(n) \longmapsto \det_{\mathbb{C}} U \in S^1$$

where

$$\det_{\mathbb{C}} U = \det(A + iB) \;\text{ if }\; U = \begin{bmatrix} A & -B \\ B & A \end{bmatrix}.$$

We obviously have $\rho(S^+) = (-1)^n$ and $\rho(S^-) = (-1)^{n-1}$.

We now have all we need to define the Conley–Zehnder index. Let us denote by $C^{\pm}(2n, \mathbb{R})$ the space of all paths $\Sigma : [0,1] \longrightarrow \mathrm{Sp}(n)$ with $\Sigma(0) = I$ and $\Sigma(1) \in \mathrm{Sp}^{\pm}(n)$. Any such path can be extended into a path $\widetilde{\Sigma} : [0,2] \longrightarrow \mathrm{Sp}(n)$ such that $\widetilde{\Sigma}(t) \in \mathrm{Sp}^{\pm}(n)$ for $1 \leq t \leq 2$ and $\widetilde{\Sigma}(2) = S^+$ or $\widetilde{\Sigma}(2) = S^-$. The orthogonal part of the polar decomposition of $\widetilde{\Sigma}(t)$ is given by the formula

$$U(t) = \widetilde{\Sigma}(t)(\widetilde{\Sigma}(t)^T\widetilde{\Sigma}(t))^{-1/2}$$

(*cf.* the definition of the Maslov index on $\mathrm{Sp}(n)$, formula (3.12)). When t varies from 0 to 2 the complex number $\det_{\mathbb{C}} U(t) = e^{i\theta(t)}$ varies from $e^{i\theta(0)} = 1$ to $e^{i\theta(2)} = \pm 1$ so that $\theta(2) \in \pi\mathbb{Z}$.

Definition 4.12. The mapping $i_{\mathrm{CZ}} : C^{\pm}(2n,\mathbb{R}) \longrightarrow \mathbb{Z}$ defined by

$$i_{\mathrm{CZ}}(\Sigma) = \frac{\theta(2)}{\pi}$$

is called the Conley–Zehnder index on $C^{\pm}(2n,\mathbb{R})$.

It turns out that $i_{\mathrm{CZ}}(\Sigma)$ is invariant under homotopy as long as the endpoint $S = \Sigma(1)$ remains in $\mathrm{Sp}^{\pm}(n)$; in particular it does not change under homotopies with fixed endpoints so we may view i_{CZ} as defined on the subset

$$\mathrm{Sp}^*_\infty(n) = \{S_\infty : \det(S - I) \neq 0\}$$

of the universal covering group $\mathrm{Sp}_\infty(n)$. With this convention one proves (see [90]) that the Conley–Zehnder index is the unique mapping $i_{\mathrm{CZ}} : \mathrm{Sp}^*_\infty(n) \longrightarrow \mathbb{Z}$ having the following properties:

(CZ_1) *Antisymmetry*: For every S_∞ we have

$$i_{\mathrm{CZ}}(S^{-1}_\infty) = -i_{\mathrm{CZ}}(S_\infty) \tag{4.11}$$

where S^{-1}_∞ is the homotopy class of the path $t \longmapsto S^{-1}_t$;

(CZ_2) *Continuity*: Let Σ be a symplectic path representing S_∞ and Σ' a path joining S to an element S' belonging to the same component $\mathrm{Sp}^{\pm}(n)$ as S. Let S'_∞ be the homotopy class of $\Sigma * \Sigma'$. We have

$$i_{\mathrm{CZ}}(S_\infty) = i_{\mathrm{CZ}}(S'_\infty); \tag{4.12}$$

(CZ_3) *Action of* $\pi_1[\mathrm{Sp}(n)]$:

$$i_{\mathrm{CZ}}(\alpha^r S_\infty) = i_{\mathrm{CZ}}(S_\infty) + 2r \tag{4.13}$$

for every $r \in \mathbb{Z}$.

The uniqueness of a mapping $\mathrm{Sp}^*_\infty(n) \longrightarrow \mathbb{Z}$ satisfying these properties is actually rather obvious: suppose $i'_{\mathrm{CZ}} : \mathrm{Sp}^*_\infty(n) \longrightarrow \mathbb{Z}$ has the same properties and set $\delta = i_{\mathrm{CZ}} - i'_{\mathrm{CZ}}$. In view of ($\mathrm{CZ}_3$) we have $\delta(\alpha^r S_\infty) = \delta(S_\infty)$ for all $r \in \mathbb{Z}$ hence δ is defined on $\mathrm{Sp}^*(n) = \mathrm{Sp}^+(n) \cup \mathrm{Sp}^-(n)$ so that $\delta(S_\infty) = \delta(S)$ where $S = S_1$, the endpoint of the path $t \longmapsto S_t$. Property (CZ_2) implies that this function $\mathrm{Sp}^*(n) \longrightarrow \mathbb{Z}$ is constant on both $\mathrm{Sp}^+(n)$ and $\mathrm{Sp}^-(n)$. We next observe that since $\det S = 1$ we have $\det(S^{-1} - I) = \det(S - I)$ so that S and S^{-1} always belong to the same set $\mathrm{Sp}^+(n)$ or $\mathrm{Sp}^-(n)$ if $\det(S - I) \neq 0$. Property (CZ_1) then implies that δ must be zero on both $\mathrm{Sp}^+(n)$ or $\mathrm{Sp}^-(n)$.

One proves that the Conley–Zehnder in addition satisfies:

(CZ_4) *Normalization*: Let J_1 be the standard symplectic matrix in $Sp(1)$. If S_1 is the path $t \longrightarrow e^{\pi t J_1}$ $(0 \leq t \leq 1)$ joining I to $-I$ in Sp(1), then $i_{CZ,1}(S_{1,\infty}) = 1$ ($i_{CZ,1}$ the Conley–Zehnder index on Sp(1));

(CZ_5) *Dimensional additivity*: if $S_{1,\infty} \in \mathrm{Sp}^*_\infty(n_1)$, $S_{2,\infty} \in \mathrm{Sp}^*_\infty(n_2)$, $n_1 + n_2 = n$, then

$$i_{CZ}(S_{1,\infty} \oplus S_{2,\infty}) = i_{CZ,1}(S_{1,\infty}) + i_{CZ,2}(S_{2,\infty})$$

where $i_{CZ,j}$ is the Conley–Zehnder index on Sp(n_j), $j = 1, 2$.

These properties will actually easily follow from the properties of the extended index we will construct. Let us first introduce a useful notion of Cayley transform for symplectic matrices.

4.3.2 The symplectic Cayley transform

Our extension of the index i_{CZ} requires a notion of Cayley transform for symplectic matrices.

Definition 4.13. If $S \in \mathrm{Sp}(n)$, $\det(S - I) \neq 0$, we call the matrix

$$M_S = \tfrac{1}{2}J(S + I)(S - I)^{-1} \tag{4.14}$$

the "symplectic Cayley transform of S". Equivalently:

$$M_S = \tfrac{1}{2}J + J(S - I)^{-1}. \tag{4.15}$$

It is straightforward to check that M_S always is a symmetric matrix: $M_S = M_S^T$. In fact:

$$M_S^T = -\tfrac{1}{2}J - (S^T - I)^{-1}J = -\tfrac{1}{2}J + (JS^T - J)^{-1},$$

that is, since $JS^T = S^{-1}J$ and $(S^{-1} - I)^{-1} = (I - S)^{-1}S$:

$$M_S^T = -\tfrac{1}{2}J + (S^{-1}J - J)^{-1} = -\tfrac{1}{2}J - J(I - S)^{-1}S.$$

Noting the trivial identity $(I - S)^{-1}S = -I + (I - S)^{-1}$ we finally obtain

$$M_S^T = \tfrac{1}{2}J - J(I - S)^{-1} = M_S.$$

The symplectic Cayley transform has the following properties:

Lemma 4.14.

(i) *We have*

$$(M_S + M_{S'})^{-1} = -(S' - I)(SS' - I)^{-1}(S - I)J \tag{4.16}$$

and the symplectic Cayley transform of the product SS' is (when defined) given by the formula

$$M_{SS'} = M_S + (S^T - I)^{-1}J(M_S + M_{S'})^{-1}J(S - I)^{-1}. \tag{4.17}$$

(ii) *The symplectic Cayley transform of S and S^{-1} are related by*

$$M_{S^{-1}} = -M_S. \tag{4.18}$$

Proof. (i) We begin by noting that (4.15) implies that

$$M_S + M_{S'} = J(I + (S - I)^{-1} + (S' - I)^{-1}), \tag{4.19}$$

hence the identity (4.16). In fact, writing $SS' - I = S(S' - I) + S - I$, we have

$$\begin{aligned}(S' - I)(SS' - I)^{-1}(S - I) &= (S' - I)(S(S' - I) + S - I)^{-1}(S - I)\\ &= ((S - I)^{-1}S(S' - I)(S' - I)^{-1} + (S' - I)^{-1})^{-1}\\ &= ((S - I)^{-1}S + (S' - I)^{-1})\\ &= I + (S - I)^{-1} + (S' - I)^{-1};\end{aligned}$$

the equality (4.16) follows in view of (4.19). Let us prove (4.17); equivalently

$$M_S + M = M_{SS'} \tag{4.20}$$

where M is the matrix defined by

$$M = (S^T - I)^{-1}J(M_S + M_{S'})^{-1}J(S - I)^{-1}$$

that is, in view of (4.16),

$$M = (S^T - I)^{-1}J(S' - I)(SS' - I)^{-1}.$$

Using the obvious relations $S^T = -JS^{-1}J$ and $(-S^{-1} + I)^{-1} = S(S - I)^{-1}$ we have

$$\begin{aligned}M &= (S^T - I)^{-1}J(S' - I)(SS' - I)^{-1}\\ &= -J(-S^{-1} + I)^{-1}(S' - I)(SS' - I)^{-1}\\ &= -JS(S - I)^{-1}(S' - I)(SS' - I)^{-1}\end{aligned}$$

that is, writing $S = S - I + I$,

$$M = -J(S' - I)(SS' - I)^{-1} - J(S - I)^{-1}(S' - I)(SS' - I)^{-1}.$$

Replacing M_S by its value (4.15) we have

$$\begin{aligned}&M_S + M\\ &= J(\tfrac{1}{2}I + (S - I)^{-1} - (S' - I)(SS' - I)^{-1} - (S - I)^{-1}(S' - I)(SS' - I)^{-1});\end{aligned}$$

noting that

$$\begin{aligned}(S-I)^{-1}&-(S-I)^{-1}(S'-I)(SS'-I)^{-1}\\&=(S-I)^{-1}(SS'-I-S'+I)(SS'-I)^{-1})\end{aligned}$$

that is

$$\begin{aligned}(S-I)^{-1}-(S-I)^{-1}(S'-I)(SS'-I)^{-1}&=(S-I)^{-1}(SS'-S')(SS'-I)^{-1}\\&=S'(SS'-I)^{-1})\end{aligned}$$

we get

$$\begin{aligned}M_S+M&=J(\tfrac{1}{2}I-(S'-I)(SS'-I)^{-1}+S'(SS'-I)^{-1})\\&=J(\tfrac{1}{2}I+(SS'-I)^{-1})\\&=M_{SS'}\end{aligned}$$

which we set out to prove.

(ii) Formula (4.18) follows from the sequence of equalities

$$\begin{aligned}M_{S^{-1}}&=\tfrac{1}{2}J+J(S^{-1}-I)^{-1}\\&=\tfrac{1}{2}J-JS(S-I)^{-1}\\&=\tfrac{1}{2}J-J(S-I+I)(S-I)^{-1}\\&=-\tfrac{1}{2}J-J(S-I)^{-1}\\&=-M_S.\end{aligned}$$

□

4.3.3 Definition and properties of $\nu(S_\infty)$

We define on $\mathbb{R}^{2n}\oplus\mathbb{R}^{2n}$ a symplectic form $\sigma^\ominus$ by

$$\sigma^\ominus(z_1,z_2;z_1',z_2')=\sigma(z_1,z_1')-\sigma(z_2,z_2')$$

and denote by $\mathrm{Sp}^\ominus(2n)$ and $\mathrm{Lag}^\ominus(2n)$ the corresponding symplectic group and Lagrangian Grassmannian. Let $\mu^\ominus$ be the Leray index on $\mathrm{Lag}^\ominus_\infty(2n)$ and $\mu_L^\ominus$ the Maslov index on $\mathrm{Sp}^\ominus_\infty(2n)$ relative to $L\in\mathrm{Lag}^\ominus(2n)$.

For $S_\infty\in\mathrm{Sp}_\infty(n)$ we define

$$\nu(S_\infty)=\frac{1}{2}\mu^\ominus((I\oplus S)_\infty\Delta_\infty,\Delta_\infty) \tag{4.21}$$

where $(I\oplus S)_\infty$ is the homotopy class in $\mathrm{Sp}^\ominus(2n)$ of the path

$$t\longmapsto\{(z,S_tz):z\in\mathbb{R}^{2n}\}\ ,\ \ 0\le t\le 1$$

and $\Delta = \{(z, z) : z \in \mathbb{R}^{2n}\}$ the diagonal of $\mathbb{R}^{2n} \oplus \mathbb{R}^{2n}$. Setting $S_t^{\ominus} = I \oplus S_t$ we have $S_t^{\ominus} \in \mathrm{Sp}^{\ominus}(2n)$ hence formulae (4.21) is equivalent to

$$\nu(S_\infty) = \frac{1}{2}\mu_\Delta^{\ominus}(S_\infty^{\ominus}) \tag{4.22}$$

where $\mu_\Delta^{\ominus}$ is the relative Maslov index on $\mathrm{Sp}_\infty^{\ominus}(2n)$ corresponding to the choice $\Delta \in \mathrm{Lag}^{\ominus}(2n)$.

Note that replacing n by $2n$ in the congruence (3.28) (Proposition 3.19) we have

$$\begin{aligned}\mu^{\ominus}((I \oplus S)_\infty \Delta_\infty, \Delta_\infty) &\equiv \dim((I \oplus S)\Delta \cap \Delta) \mod 2 \\ &\equiv \dim \mathrm{Ker}(S - I) \mod 2\end{aligned}$$

and hence

$$\nu(S_\infty) \equiv \frac{1}{2} \dim \mathrm{Ker}(S - I) \mod 1.$$

Since the eigenvalue 1 of S has even multiplicity, $\nu(S_\infty)$ is thus always an integer.

The index ν has the following three important properties; the third is essential for the calculation of the index of repeated periodic orbits (it clearly shows that ν is not in general additive):

Proposition 4.15.

(i) *For all $S_\infty \in \mathrm{Sp}_\infty(n)$ we have*

$$\nu(S_\infty^{-1}) = -\nu(S_\infty) \quad , \quad \nu(I_\infty) = 0 \tag{4.23}$$

(I_∞ the identity of the group $\mathrm{Sp}_\infty(n)$).

(ii) *For all $r \in \mathbb{Z}$ we have*

$$\nu(\alpha^r S_\infty) = \nu(S_\infty) + 2r \quad , \quad \nu(\alpha^r) = 2r. \tag{4.24}$$

(iii) *Let S_∞ be the homotopy class of a path Σ in $\mathrm{Sp}(n)$ joining the identity to $S \in \mathrm{Sp}^*(n)$, and let $S' \in \mathrm{Sp}(n)$ be in the same connected component $\mathrm{Sp}^{\pm}(n)$ as S. Then $\nu(S'_\infty) = \nu(S_\infty)$ where S'_∞ is the homotopy class in $\mathrm{Sp}(n)$ of the concatenation of Σ and a path joining S to S' in $\mathrm{Sp}_0(n)$.*

Proof. (i) Formulae (4.23) immediately follows from the equality $(S_\infty^{\ominus})^{-1} = (I \oplus S^{-1})_\infty$ and the antisymmetry of $\mu_\Delta^{\ominus}$.

(ii) The second formula (4.24) follows from the first using (4.23). To prove the first formula (4.24) it suffices to observe that to the generator α of $\pi_1[\mathrm{Sp}(n)]$ corresponds the generator $I_\infty \oplus \alpha$ of $\pi_1[\mathrm{Sp}^{\ominus}(2n)]$; in view of property (3.41) of the Maslov indices it follows that

$$\begin{aligned}\nu(\alpha^r S_\infty) &= \tfrac{1}{2}\mu_\Delta^{\ominus}((I_\infty \oplus \alpha)^r S_\infty^{\ominus}) \\ &= \tfrac{1}{2}(\mu_\Delta^{\ominus}(S_\infty^{\ominus}) + 4r) \\ &= \nu(S_\infty) + 2r.\end{aligned}$$

(iii) Assume in fact that S and S' belong to, say, $\mathrm{Sp}^+(n)$. Let S_∞ be the homotopy class of the path Σ, and Σ' a path joining S to S' in $\mathrm{Sp}^+(n)$ (we parametrize both paths by $t \in [0,1]$). Let $\Sigma'_{t'}$ be the restriction of Σ' to the interval $[0,t']$, $t' \leq t$ and $S_\infty(t')$ the homotopy class of the concatenation $\Sigma * \Sigma'_{t'}$. We have $\det(S(t)-I) > 0$ for all $t \in [0,t']$, hence $S^\ominus_\infty(t)\Delta \cap \Delta \neq 0$ as t varies from 0 to 1. It follows from the fact that the $\mu^\ominus_\Delta$ is locally constant on the set $\{S^\ominus_\infty : S^\ominus_\infty \Delta \cap \Delta = 0\}$ (property (ii) in Proposition 3.29) that the function $t \longmapsto \mu^\ominus_\Delta(S^\ominus_\infty(t))$ is constant, and hence

$$\begin{aligned}\mu^\ominus_\Delta(S^\ominus_\infty) &= \mu^\ominus_\Delta(S^\ominus_\infty(0)) \\ &= \mu^\ominus_\Delta(S^\ominus_\infty(1)) \\ &= \mu^\ominus_\Delta(S'^\ominus_\infty)\end{aligned}$$

which was to be proven. □

The following consequence of the result above shows that the indices ν and i_{CZ} coincide on their common domain of definition:

Corollary 4.16. *The restriction of the index ν to $\mathrm{Sp}^*(n)$ is the Conley–Zehnder index:*

$$\nu(S_\infty) = i_{CZ}(S_\infty) \quad \text{if} \quad \det(S-I) \neq 0.$$

Proof. The restriction of ν to $\mathrm{Sp}^*(n)$ satisfies the properties (CZ_1), (CZ_2), and (CZ_3) of the Conley–Zehnder index listed in §4.3.1; we showed that these properties uniquely characterize i_{CZ}. □

Let us prove a formula for the index of the product of two paths:

Proposition 4.17. *If S_∞, S'_∞, and $S_\infty S'_\infty$ are such that* $\det(S-I) \neq 0$, $\det(S'-I) \neq 0$, *and* $\det(SS'-I) \neq 0$, *then*

$$\nu(S_\infty S'_\infty) = \nu(S_\infty) + \nu(S'_\infty) + \tfrac{1}{2}\operatorname{sign}(M_S + M_{S'}) \tag{4.25}$$

where M_S is the symplectic Cayley transform of S; in particular

$$\nu(S^r_\infty) = r\nu(S_\infty) + \tfrac{1}{2}(r-1)\operatorname{sign} M_S \tag{4.26}$$

for every integer r.

Proof. In view of (4.22) and the product property (3.42) of the Maslov index (Proposition 3.29) we have

$$\begin{aligned}\nu(S_\infty S'_\infty) &= \nu(S_\infty) + \nu(S'_\infty) + \tfrac{1}{2}\tau^\ominus(\Delta, S^\ominus\Delta, S^\ominus S'^\ominus\Delta) \\ &= \nu(S_\infty) + \nu(S'_\infty) - \tfrac{1}{2}\tau^\ominus(S^\ominus S'^\ominus\Delta, S^\ominus\Delta, \Delta)\end{aligned}$$

where $S^\ominus = I \oplus S$, $S'^\ominus = I \oplus S'$ and $\tau^\ominus$ is the signature on the symplectic space $(\mathbb{R}^{2n} \oplus \mathbb{R}^{2n}, \sigma^\ominus)$. The condition $\det(SS'-I) \neq 0$ is equivalent to $S^\ominus S'^\ominus\Delta \cap \Delta = 0$

hence we can apply Proposition 1.29 with $\ell = S^{\ominus}S'^{\ominus}\Delta$, $\ell' = S^{\ominus}\Delta$, and $\ell'' = \Delta$. The projection operator onto $S^{\ominus}S'^{\ominus}\Delta$ along Δ is easily seen to be

$$\Pr_{S^{\ominus}S'^{\ominus}\Delta,\Delta} = \begin{bmatrix} (I - SS')^{-1} & -(I - SS')^{-1} \\ SS'(I - SS')^{-1} & -SS'(I - SS')^{-1} \end{bmatrix},$$

hence $\tau^{\ominus}(S^{\ominus}S'^{\ominus}\Delta, S^{\ominus}\Delta, \Delta)$ is the signature of the quadratic form

$$Q(z) = \sigma^{\ominus}(\Pr_{S^{\ominus}S'^{\ominus}\Delta,\Delta}(z, Sz); (z, Sz)),$$

that is, since $\sigma^{\ominus} = \sigma \ominus \sigma$:

$$\begin{aligned} Q(z) &= \sigma((I - SS')^{-1}(I - S)z, z)) - \sigma(SS'(I - SS')^{-1}(I - S)z, Sz)) \\ &= \sigma((I - SS')^{-1}(I - S)z, z)) - \sigma(S'(I - SS')^{-1}(I - S)z, z)) \\ &= \sigma((I - S')(I - SS')^{-1}(I - S)z, z)). \end{aligned}$$

In view of formula (4.16) in Lemma 4.14 we have

$$(I - S')(SS' - I)^{-1}(I - S) = (M_S + M_{S'})^{-1}J,$$

hence

$$Q(z) = -\left\langle (M_S + M_{S'})^{-1}Jz, Jz \right\rangle$$

and the signature of Q is thus the same as that of the quadratic form

$$Q'(z) = -\left\langle (M_S + M_{S'})^{-1}z, z \right\rangle,$$

that is $-\operatorname{sign}(M_S + M_{S'})$ proving formula (4.25). Formula (4.26) follows by a straightforward induction on the integer r. □

It is not immediately obvious that the index μ_γ of the periodic orbit γ is independent of the choice of the origin of the orbit. Let us prove that this is in fact the case:

Proposition 4.18. *Let (f_t) be the flow determined by a (time-independent) Hamiltonian function on $\mathbb{R}^{2n}$ and $z \neq 0$ such that $f_T(z) = z$ for some $T > 0$. Let $z' = f_{t'}(z)$ for some t' and denote by $S_T(z) = Df_T(z)$ and $S_T(z') = Df_T(z')$ the corresponding monodromy matrices. Let $S_T(z)_\infty$ and $S_T(z')_\infty$ be the homotopy classes of the paths $t \longmapsto S_t(z) = Df_t(z)$ and $t \longmapsto S_t(z') = Df_t(z')$, $0 \leq t \leq T$. We have $\nu(S_T(z)_\infty) = \nu(S_T(z')_\infty)$.*

Proof. We have proven in Lemma 2.61 that monodromy matrices $S_T(z)$ and $S_T(z')$ are conjugate of each other. Since we will need to let t' vary we write $S_T(z') = S_T(z', t')$ so that

$$S_T(z', t') = S_{t'}(z')S_T(z)S_{t'}(z')^{-1}.$$

The paths

$$t \longmapsto S_t(z') \quad \text{and} \quad t \longmapsto S_{t'}(z')S_t(z)S_{t'}(z')^{-1}$$

being homotopic with fixed endpoints $S_T(z',t')_\infty$ is also the homotopy class of the path $t \longmapsto S_{t'}(z')S_t(z)S_{t'}(z')^{-1}$. We thus have, by definition (4.21) of ν,

$$\nu(S_T(z',t')_\infty) = \frac{1}{2}\mu^{\ominus}_{\Delta_{t'}}(S^{\ominus}_T(z)_\infty)$$

where we have set

$$\Delta_{t'} = (I \oplus S_{t'}(z')^{-1})\Delta \quad\text{and}\quad S^{\ominus}_T(z)_\infty = I_\infty \oplus S_T(z)_\infty.$$

Consider now the mapping $t' \longmapsto \mu^{\ominus}_{\Delta_{t'}}(S^{\ominus}_T(z)_\infty)$; we have

$$S^{\ominus}_T(z)\Delta_{t'} \cap \Delta_{t'} = \{z : Sz = z\},$$

hence the dimension of the intersection $S^{\ominus}_T(z)\Delta_{t'} \cap \Delta_{t'}$ remains constant as t' varies; in view of the topological property of the relative Maslov index the mapping $t' \longmapsto \mu^{\ominus}_{\Delta_{t'}}(S^{\ominus}_T(z)_\infty)$ is thus constant and hence

$$\nu(S_T(z',t')_\infty) = \nu(S_T(z',0)_\infty) = \nu(S_T(z)_\infty)$$

which concludes the proof. □

4.3.4 Relation between ν and μ_{ℓ_P}

The index ν can be expressed in a simple – and useful – way in terms of the Maslov index μ_{ℓ_P} on $\mathrm{Sp}_\infty(n)$. The following technical result will be helpful in establishing this relation. Recall that $S \in \mathrm{Sp}(n)$ is free if $S\ell_P \cap \ell_P = 0$ and that this condition is equivalent to $\det B \neq 0$ when S is identified with the matrix

$$S = \begin{bmatrix} A & B \\ C & D \end{bmatrix} \tag{4.27}$$

in the canonical basis; the set of all free symplectic matrices is dense in $\mathrm{Sp}(n)$. The quadratic form W on $\mathbb{R}^n_x \times \mathbb{R}^n_x$ defined by

$$W(x,x') = \tfrac{1}{2}\langle Px,x\rangle - \langle Lx,x'\rangle + \tfrac{1}{2}\langle Qx',x'\rangle$$

where

$$P = DB^{-1},\ L = B^{-1},\ Q = B^{-1}A \tag{4.28}$$

then generates S in the sense that

$$(x,p) = S(x',p') \Longleftrightarrow p = \partial_x W(x,x')\ ,\ p' = -\partial_{x'}W(x,x')$$

(observe that P and Q are symmetric). We have:

Lemma 4.19. *Let $S = S_W \in \mathrm{Sp}(n)$ be given by* (4.27). *We have*

$$\det(S_W - I) = (-1)^n \det B \det(B^{-1}A + DB^{-1} - B^{-1} - (B^T)^{-1}), \tag{4.29}$$

that is:

$$\det(S_W - I) = (-1)^n \det(L^{-1}) \det(P + Q - L - L^T).$$

In particular the symmetric matrix

$$P + Q - L - L^T = DB^{-1} + B^{-1}A - B^{-1} - (B^T)^{-1}$$

is invertible.

Proof. Since B is invertible we can factorize $S - I$ as

$$\begin{bmatrix} A-I & B \\ C & D-I \end{bmatrix} = \begin{bmatrix} 0 & B \\ I & D-I \end{bmatrix} \begin{bmatrix} C-(D-I)B^{-1}(A-I) & 0 \\ B^{-1}(A-I) & I \end{bmatrix}$$

and hence

$$\begin{aligned} \det(S_W - I) &= \det(-B)\det(C - (D-I)B^{-1}(A-I)) \\ &= (-1)^n \det B \det(C - (D-I)B^{-1}(A-I)). \end{aligned}$$

Since S is symplectic we have $C - DB^{-1}A = -(B^T)^{-1}$ and hence

$$C - (D-I)B^{-1}(A-I)) = B^{-1}A + DB^{-1} - B^{-1} - (B^T)^{-1};$$

the lemma follows. □

Let us now introduce the notion of index of concavity of a Hamiltonian periodic orbit γ, defined for $0 \leq t \leq T$, with $\gamma(0) = \gamma(T) = z_0$. As t goes from 0 to T the linearized part $D\gamma(t) = S_t(z_0)$ goes from the identity to $S_T(z_0)$ (the monodromy matrix) in $\mathrm{Sp}(n)$. We assume that $S_T(z_0)$ is free and that $\det(S_T(z_0) - I) \neq 0$. Writing

$$S_t(z_0) = \begin{bmatrix} A(t) & B(t) \\ C(t) & D(t) \end{bmatrix}$$

we thus have $\det B(t) \neq 0$ in a neighborhood $[T-\varepsilon, T+\varepsilon]$ of the time T. The generating function

$$W(x, x', t) = \frac{1}{2}\langle P(t)x, x\rangle - \langle L(t)x, x'\rangle + \frac{1}{2}\langle Q(t)x', x'\rangle$$

(with $P(t)$, $Q(t)$, $L(t)$ defined by (4.28) thus exists for $T - \varepsilon \leq t \leq T + \varepsilon$. By definition Morse's index of concavity [126] of the periodic orbit γ is the index of inertia,

$$\mathrm{Inert}\, W''_{xx} = \mathrm{Inert}(P + Q - L - L^T)$$

of W''_{xx}, the matrix of second derivatives of the function $x \longmapsto W(x, x; T)$ (we have set $P = P(T)$, $Q = Q(T)$, $L = L(T)$).

Let us now prove the following essential result; recall that m_ℓ denotes the reduced Maslov index associated to μ_ℓ:

Proposition 4.20. *Let $t \longmapsto S_t$ be a symplectic path, $0 \leq t \leq 1$. Let $S_\infty \in \mathrm{Sp}_\infty(n)$ be the homotopy class of that path and set $S = S_1$. If $\det(S-I) \neq 0$ and $S\ell_P \cap \ell_P = 0$, then*

$$\nu(S_\infty) = \frac{1}{2}(\mu_{\ell_P}(S_\infty) + \operatorname{sign} W''_{xx}) = m_{\ell_P}(S_\infty) - \operatorname{Inert} W''_{xx} \tag{4.30}$$

where $\operatorname{Inert} W''_{xx}$ *is the index of concavity corresponding to the endpoint S of the path $t \longmapsto S_t$.*

Proof. We will divide the proof in three steps. *Step 1.* Let $L \in \mathrm{Lag}^\ominus(4n)$. Using successively formulae (4.22) and (3.44) we have

$$\nu(S_\infty) = \frac{1}{2}(\mu_L^\ominus(S_\infty^\ominus) + \tau^\ominus(S^\ominus\Delta, \Delta, L) - \tau^\ominus(S^\ominus\Delta, S^\ominus L, L)). \tag{4.31}$$

Choosing in particular $L = L_0 = \ell_P \oplus \ell_P$ we get

$$\begin{aligned}
\mu_{L_0}^\ominus(S_\infty^\ominus) &= \mu^\ominus((I \oplus S)_\infty(\ell_P \oplus \ell_P), (\ell_P \oplus \ell_P)) \\
&= \mu(\ell_{P,\infty}, \ell_{P,\infty}) - \mu(\ell_{P,\infty}, S_\infty \ell_{P,\infty}) \\
&= -\mu(\ell_{P,\infty}, S_\infty \ell_{P,\infty}) \\
&= \mu_{\ell_P}(S_\infty)
\end{aligned}$$

so that there remains to prove that

$$\tau^\ominus(S^\ominus\Delta, \Delta, L_0) - \tau^\ominus(S^\ominus\Delta, S^\ominus L_0, L_0) = -2 \operatorname{sign} W''_{xx}.$$

Step 2. We are going to show that

$$\tau^\ominus(S^\ominus\Delta, S^\ominus L_0, L_0) = 0;$$

in view of the symplectic invariance and the antisymmetry of $\tau^\ominus$ this is equivalent to

$$\tau^\ominus(L_0, \Delta, L_0, (S^\ominus)^{-1} L_0) = 0. \tag{4.32}$$

We have

$$\Delta \cap L_0 = \{(0, p; 0, p) : p \in \mathbb{R}^n\}$$

and $(S^\ominus)^{-1} L_0 \cap L_0$ consists of all $(0, p', S^{-1}(0, p''))$ with $S^{-1}(0, p'') = (0, p')$; since S (and hence S^{-1}) is free we must have $p' = p'' = 0$ so that

$$(S^\ominus)^{-1} L_0 \cap L_0 = \{(0, p; 0, 0) : p \in \mathbb{R}^n\}.$$

It follows that we have

$$L_0 = \Delta \cap L_0 + (S^\ominus)^{-1} L_0 \cap L_0,$$

hence (4.32) in view of Proposition 1.30. *Step 3.* Let us finally prove that.

$$\tau^\ominus(S^\ominus\Delta, \Delta, L_0) = -2 \operatorname{sign} W''_{xx};$$

this will complete the proof of the proposition. The condition $\det(S-I) \neq 0$ is equivalent to $S^{\ominus}\Delta \cap \Delta = 0$ hence, using Proposition 1.29, the number

$$\tau^{\ominus}(S^{\ominus}\Delta, \Delta, L_0) = -\tau^{\ominus}(S^{\ominus}\Delta, L_0, \Delta)$$

is the signature of the quadratic form Q on L_0 defined by

$$Q(0,p,0,p') = -\sigma^{\ominus}(\mathrm{Pr}_{S^{\ominus}\Delta,\Delta}(0,p,0,p'); 0,p,0,p')$$

where

$$\mathrm{Pr}_{S^{\ominus}\Delta,\Delta} = \begin{bmatrix} (S-I)^{-1} & -(S-I)^{-1} \\ S(S-I)^{-1} & -S(S-I)^{-1} \end{bmatrix}$$

is the projection on $S^{\ominus}\Delta$ along Δ in $\mathbb{R}^{2n} \oplus \mathbb{R}^{2n}$. It follows that the quadratic form Q is given by

$$Q(0,p,0,p') = -\sigma^{\ominus}((I-S)^{-1}(0,p''), S(I-S)^{-1}(0,p''); 0,p,0,p')$$

where we have set $p'' = p - p'$; by definition of $\sigma^{\ominus}$ this is

$$Q(0,p,0,p') = -\sigma((I-S)^{-1}(0,p''),(0,p)) + \sigma(S(I-S)^{-1}(0,p''),(0,p')).$$

Let now M_S be the symplectic Cayley transform (4.14) of S; we have

$$(I-S)^{-1} = JM_S + \tfrac{1}{2}I \quad , \quad S(I-S)^{-1} = JM_S - \tfrac{1}{2}I$$

and hence

$$\begin{aligned} Q(0,p,0,p') &= -\sigma((JM_S + \tfrac{1}{2}I)(0,p''),(0,p)) + \sigma((JM_S - \tfrac{1}{2}I)(0,p''),(0,p')) \\ &= -\sigma(JM_S(0,p''),(0,p)) + \sigma(JM_S(0,p''),(0,p')) \\ &= \sigma(JM_S(0,p''),(0,p'')) \\ &= -\langle M_S(0,p''),(0,p'')\rangle . \end{aligned}$$

Let us calculate explicitly M_S. Writing S in usual block-form we have

$$S - I = \begin{bmatrix} 0 & B \\ I & D-I \end{bmatrix} \begin{bmatrix} C-(D-I)B^{-1}(A-I) & 0 \\ B^{-1}(A-I) & I \end{bmatrix},$$

that is

$$S - I = \begin{bmatrix} 0 & B \\ I & D-I \end{bmatrix} \begin{bmatrix} W''_{xx} & 0 \\ B^{-1}(A-I) & I \end{bmatrix}$$

where we have used the identity

$$C-(D-I)B^{-1}(A-I)) = B^{-1}A + DB^{-1} - B^{-1} - (B^T)^{-1}$$

which follows from the relation $C-DB^{-1}A=-(B^T)^{-1}$ (the latter is a rephrasing of the equalities $D^TA-B^TC=I$ and $D^TB=B^TD$, which follow from the fact that $S^TJS=S^TJS$ since $S\in\mathrm{Sp}(n)$). It follows that

$$(S-I)^{-1}=\begin{bmatrix}(W''_{xx})^{-1} & 0\\ B^{-1}(I-A)(W''_{xx})^{-1} & I\end{bmatrix}\begin{bmatrix}(I-D)B^{-1} & I\\ B^{-1} & 0\end{bmatrix}$$
$$=\begin{bmatrix}(W''_{xx})^{-1}(I-D)B^{-1} & (W''_{xx})^{-1}\\ B^{-1}(I-A)(W''_{xx})^{-1}(I-D)B^{-1}+B^{-1} & B^{-1}(I-A)(W''_{xx})^{-1}\end{bmatrix}$$

and hence

$$M_S=\begin{bmatrix}B^{-1}(I-A)(W''_{xx})^{-1}(I-D)B^{-1}+B^{-1} & \frac{1}{2}I+B^{-1}(I-A)(W''_{xx})^{-1}\\ -\frac{1}{2}I-(W''_{xx})^{-1}(I-D)B^{-1} & -(W''_{xx})^{-1}\end{bmatrix}$$

from which follows that

$$\begin{aligned}Q(0,p,0,p')&=\langle (W''_{xx})^{-1}p'',p''\rangle\\ &=\langle (W''_{xx})^{-1}(p-p'),(p-p')\rangle\,.\end{aligned}$$

The matrix of the quadratic form Q is thus

$$2\begin{bmatrix}(W''_{xx})^{-1} & -(W''_{xx})^{-1}\\ -(W''_{xx})^{-1} & (W''_{xx})^{-1}\end{bmatrix}$$

and this matrix has signature $2\,\mathrm{sign}(W''_{xx})^{-1}=2\,\mathrm{sign}\,W''_{xx}$, proving the first equality (4.30); the second equality follows because $\mu_{\ell_P}(S_\infty)=2m_{\ell_P}(S_\infty)-n$ since $S\ell_P\cap\ell_P=0$ and the fact that W''_{xx} has rank n in view of Lemma 4.19. □

Remark 4.21. Lemma 4.19 above shows that if S is free then we have

$$\begin{aligned}\frac{1}{\pi}\arg\det(S-I)&\equiv n+\arg\det B+\arg\det W''_{xx}\mod 2\\ &\equiv n-\arg\det B+\arg\det W''_{xx}\mod 2.\end{aligned}$$

The reduced Maslov index $m_{\ell_P}(S_\infty)$ corresponds to a choice of $\arg\det B$ modulo 4; Proposition 4.20 thus justifies the following definition of the argument of $\det(S-I)$ modulo 4:

$$\frac{1}{\pi}\arg\det(S-I)\equiv n-\nu(S_\infty)\mod 4.$$

Let us finish with an example. Consider first the one-dimensional harmonic oscillator with Hamiltonian function

$$H=\frac{\omega}{2}(p^2+x^2);$$

all the orbits are periodic with period $2\pi/\omega$. The monodromy matrix is simply the identity: $\Sigma_T = I$ where

$$\Sigma_t = \begin{bmatrix} \cos\omega t & \sin\omega t \\ -\sin\omega t & \cos\omega t \end{bmatrix}.$$

Let us calculate the corresponding index $\nu(\Sigma_\infty)$. The homotopy class of path $t \longmapsto \Sigma_t$ as t goes from 0 to $T = 2\pi/\omega$ is just the inverse of α, the generator of $\pi_1[\mathrm{Sp}(1)]$ hence $\nu(\Sigma_\infty) = -2$ in view of (4.24). If we had considered r repetitions of the orbit we would likewise have obtained $\nu(\Sigma_\infty) = -2r$.

Consider next a two-dimensional harmonic oscillator with Hamiltonian function

$$H = \frac{\omega_x}{2}(p_x^2 + x^2) + \frac{\omega_y}{2}(p_y^2 + y^2);$$

we assume that the frequencies ω_y, ω_x are incommensurate, so that the only periodic orbits are librations along the x and y axes. Let us focus on the orbit γ_x along the x axis; its prime period is $T = 2\pi/\omega_x$ and the corresponding monodromy matrix is

$$S_1 = \begin{bmatrix} 1 & 0 & 0 & 0 \\ 0 & \cos\chi & 0 & \sin\chi \\ 0 & 0 & 1 & 0 \\ 0 & -\sin\chi & 0 & \cos\chi \end{bmatrix} \quad , \quad \chi = 2\pi\frac{\omega_y}{\omega_x};$$

it is the endpoint of the symplectic path $t \longmapsto S_t$, $0 \leq t \leq 1$, consisting of the matrices

$$S_t = \begin{bmatrix} \cos 2\pi t & 0 & \sin 2\pi t & 0 \\ 0 & \cos\chi t & 0 & \sin\chi t \\ -\sin 2\pi t & 0 & \cos 2\pi t & 0 \\ 0 & -\sin\chi t & 0 & \cos\chi t \end{bmatrix}.$$

In Gutzwiller's trace formula [86] the sum is taken over periodic orbits, including their repetitions; we are thus led to calculate the Conley–Zehnder index of the path $t \longmapsto S_t$ with $0 \leq t \leq r$ where the integer r indicates the number of repetitions of the orbit. Let us calculate the Conley–Zehnder index $\nu(\tilde{S}_{r,\infty})$ of this path. We have $S_t = \Sigma_t \oplus \tilde{S}_t$ where

$$\Sigma_t = \begin{bmatrix} \cos 2\pi t & \sin 2\pi t \\ -\sin 2\pi t & \cos 2\pi t \end{bmatrix} \quad , \quad \tilde{S}_t = \begin{bmatrix} \cos\chi t & \sin\chi t \\ -\sin\chi t & \cos\chi t \end{bmatrix};$$

in view of the additivity property of the relative Maslov index we thus have

$$\nu(S_{r,\infty}) = \nu(\Sigma_{r,\infty}) + \nu(\tilde{S}_{r,\infty})$$

where the first term is just

$$\nu(\Sigma_{r,\infty}) = -2r$$

in view of the calculation we made in the one-dimensional case with a different parametrization. Let us next calculate $\nu(\tilde{S}_{r,\infty})$. We will use formula (4.30) relating the index ν to the Maslov index via the index of concavity, so we begin by calculating the relative Maslov index

$$m_{\ell_P}(\tilde{S}_{r,\infty}) = m(\tilde{S}_{r,\infty}\ell_{P,\infty}, \ell_{P,\infty}).$$

Here is a direct argument; in more complicated cases the formulas we proved in [68] are useful. When t goes from 0 to r the line $\tilde{S}_t\ell_P$ describes a loop in Lag(1) going from ℓ_P to $\tilde{S}_r\ell_P$. We have $\tilde{S}_t \in U(1)$; its image in $U(1,\mathbb{C})$ is $e^{-i\chi t}$ hence the Souriau mapping identifies $\tilde{S}_t\ell_P$ with $e^{-2i\chi t}$. It follows, using formula (3.26), that

$$\begin{aligned} m_{\ell_P}(\tilde{S}_{r,\infty}) &= \frac{1}{2\pi}\left(-2r\chi + i\operatorname{Log}(-e^{-2ir\chi})\right) + \frac{1}{2} \\ &= \frac{1}{2\pi}\left(-2r\chi + i\operatorname{Log}(e^{i(-2r\chi+\pi)})\right) + \frac{1}{2}. \end{aligned}$$

The logarithm is calculated as follows: for $\theta \neq (2k+1)\pi$ ($k \in \mathbb{Z}$),

$$\operatorname{Log} e^{i\theta} = i\theta - 2\pi i \left[\frac{\theta+\pi}{2\pi}\right]$$

and hence

$$\operatorname{Log}(e^{i(-2r\chi+\pi)}) = -i(2r\chi + \pi + 2\pi\left[\frac{r\chi}{\pi}\right]);$$

it follows that the Maslov index is

$$m_{\ell_P}(\tilde{S}_{r,\infty}) = -\left[\frac{r\chi}{\pi}\right]. \tag{4.33}$$

To obtain $\nu(\tilde{S}_{r,\infty})$ we note that by (4.30)

$$\nu(\tilde{S}_{r,\infty}) = m_{\ell_P}(\tilde{S}_{1,\infty}) - \operatorname{Inert} W''_{xx}$$

where $\operatorname{Inert} W''_{xx}$ is the concavity index corresponding to the generating function of $\tilde{S}_t$; the latter is

$$W(x,x',t) = \frac{1}{2\sin\chi t}((x^2+x'^2)\cos\chi t - 2xx'),$$

hence $W''_{xx} = -\tan(\chi t/2)$. We thus have, taking (4.33) into account,

$$\nu(\tilde{S}_{r,\infty}) = -\left[\frac{r\chi}{\pi}\right] - \operatorname{Inert}\left(-\tan\frac{r\chi}{2}\right);$$

a straightforward induction on r shows that this can be rewritten more conveniently as

$$\nu(\tilde{S}_{r,\infty}) = -1 - 2\left[\frac{r\chi}{2\pi}\right].$$

Summarizing, we have

$$\begin{aligned} \nu(S_{r,\infty}) &= \nu(\Sigma_{r,\infty}) + \nu(\tilde{S}_{r,\infty}) \\ &= -2r - 1 - 2\left[\frac{r\chi}{2\pi}\right], \end{aligned}$$

hence the index in Gutzwiller's formula corresponding to the rth repetition is

$$\mu_{x,r} = -\nu(S_{r,\infty}) = 1 + 2r + 2\left[\frac{r\chi}{2\pi}\right]$$

that is, by definition of χ,

$$\mu_{x,r} = 1 + 2r + 2\left[r\frac{\omega_y}{\omega_x}\right].$$

Remark 4.22. The calculations above are valid when the frequencies are incommensurate. If, say, $\omega_x = \omega_y$, the calculations are much simpler: in this case the homotopy class of the loop $t \longmapsto S_t$, $0 \leq t \leq 1$, is $\alpha^{-1} \oplus \alpha^{-1}$ and by the second formula (4.24),

$$\mu_{x,r} = -\nu(S_{r,\infty}) = 4r$$

which is zero modulo 4.

Part II

Heisenberg Group, Weyl Calculus, and Metaplectic Representation

Chapter 5

Lagrangian Manifolds and Quantization

Lagrangian manifolds are (immersed) submanifolds of the standard symplectic space $\mathbb{R}_z^{2n}$ whose tangent space at every point is a Lagrangian plane. What makes Lagrangian manifolds interesting are that they are perfect candidates for the (semi-classical) quantization of integrable Hamiltonian systems. Moreover, every Lagrangian manifold has a "phase", which is defined on its universal covering; this notion will ultimately lead us, using only classical arguments, to the Heisenberg–Weyl operators, which are the first step towards quantum mechanics. In the next chapter they will be used to define the notion of Weyl pseudo-differential operator.

In his *Bulletin* review paper[1] [177] Weinstein sustains that "*Everything is a Lagrangian manifold!*". Weinstein has made a point here, because mathematically and physically interesting examples of Lagrangian manifolds abound in the literature on both classical and quantum mechanics.

5.1 Lagrangian Manifolds and Phase

In the first subsection we state the main definitions and properties of (immersed) Lagrangian manifolds; for more on this topic the reader is invited to consult the existing literature (for instance Maslov [119], Mischenko et al. [124], Vaisman [168]) We thereafter proceed to study the important notion of phase of a Lagrangian manifold, as defined by Leray [107] (also see de Gosson [59, 62, 70] for additional results).

[1] It also contains an interesting review of the state of symplectic geometry in the beginning of the 1980s.

5.1.1 Definition and examples

All manifolds are assumed to be C^∞. We begin by defining the notion of immersed Lagrangian manifold; unless otherwise specified we will assume that all involved manifolds are connected.

Definition 5.1. Let $\mathbb{V}^n$ be an n-dimensional manifold and $\iota : \mathbb{V}^n \longrightarrow \mathbb{R}^{2n}_z$ an immersion (*i.e.*, ι is a differentiable mapping such that $d\iota$ is injective at every point). We will say that $\mathbb{V}^n$ is an "immersed Lagrangian manifold" if $\iota^*\sigma = 0$. The tangent space $T_z\mathbb{V}^n$ at z will be denoted $\ell(z)$.

An immersed manifold $\mathbb{V}^n$ is thus Lagrangian if and only if we have

$$\sigma(d\iota(z)X(z), d\iota(z)X'(z)) = 0$$

for every pair $(X(z), X'(z))$ of tangent vectors to $\mathbb{V}^n$ (at every point z); in intrinsic notation:

$$\iota^*\sigma = 0$$

where $\iota^*\sigma$ is the pull-back of the symplectic form by the immersion $\iota : \mathbb{V}^n \longrightarrow \mathbb{R}^{2n}_z$.

Example 5.2. Any smooth curve in the symplectic plane $(\mathbb{R}^2, -\det)$ is a Lagrangian manifold (every line is a Lagrangian plane in $(\mathbb{R}^2, -\det)$. The helix

$$x(\theta) = R\cos\theta,\ x(\theta) = R\cos\theta\ \ ,\ \ \theta = \theta$$

is an immersed Lagrangian submanifold of $(\mathbb{R}^2, -\det)$.

Proposition 5.3. *Let $\mathbb{V}^n$ be a Lagrangian submanifold of $(\mathbb{R}^{2n}_z, \sigma)$ and f a symplectomorphism $\mathbb{R}^{2n}_z \longrightarrow \mathbb{R}^{2n}_z$ defined on a neighborhood of $\mathbb{V}^n$. The manifold $f(\mathbb{V}^n)$ is also Lagrangian.*

Proof. It is clear that $f(\mathbb{V}^n)$ is a differentiable manifold since a symplectomorphism is a diffeomorphism. The tangent plane to $f(\mathbb{V}^n)$ at a point $f(z)$ is $Df(z)\ell(z)$; the result follows since $Df(z) \in \mathrm{Sp}(n)$ by definition of a symplectomorphism and using the fact that the image of a Lagrangian plane by a symplectic linear mapping is a Lagrangian plane. □

It follows, in particular, that the image by a Lagrangian plane by a symplectic diffeomorphism is a Lagrangian manifold.

A basic (but not generic) example of Lagrangian manifold in $(\mathbb{R}^{2n}_z, \sigma)$ is that of the graph of the gradient of a function:

Proposition 5.4. *Let $\Phi \in C^\infty(U, \mathbb{R})$ where U is an open subset of $\mathbb{R}^n$.*

(i) *The graph*

$$\mathbb{V}^n_\Phi = \{(x, \partial_x\Phi(x)) : x \in U\}$$

is a Lagrangian submanifold of $(\mathbb{R}^{2n}_z, \sigma)$.

(ii) *The orthogonal projection on $\ell_X = \mathbb{R}^n_x$ is a diffeomorphism $\mathbb{V}^n_\Phi \longrightarrow U$, and thus a global chart.*

Proof. (i) It suffices to notice that an equation for the tangent space $\ell(z_0)$ to $\mathbb{V}^n_\Phi$ at $z_0 = (x_0, p_0)$ is

$$p = [D(\partial_x \Phi)](x_0)x = [D^2\Phi(x_0)]x$$

where $D^2\Phi(x_0)$ is the Hessian matrix of Φ calculated at x_0. The latter being symmetric, $\ell(z_0)$ is a Lagrangian plane in view of Corollary 1.23 in Chapter 1, Section 1.3.

Property (ii) is obvious. □

Lagrangian manifolds $\mathbb{V}^n$ having the property that there exists a Lagrangian plane ℓ such that the projection $\pi_\ell : \mathbb{V}^n \to \ell$ is a diffeomorphism are called *exact Lagrangian manifolds.* It turns out that every Lagrangian submanifold of $(\mathbb{R}^{2n}_z, \sigma)$ is *locally* exact. Let us make this statement precise. We denote by I an arbitrary subset of $\{1, 2, \ldots, n\}$ and by $\bar{I}$ its complement in $\{1, 2, \ldots, n\}$; if $I = \{i_1, \ldots, i_k\}$ we write $x_I = (x_{i_1}, \ldots, x_{i_k})$ and use similar conventions for $x_{\bar{I}}$, p_I, and $p_{\bar{I}}$.

Proposition 5.5. *Let $\mathbb{V}^n$ be a submanifold of $(\mathbb{R}^{2n}_z, \sigma)$. $\mathbb{V}^n$ is Lagrangian if and only every $z_0 \in \mathbb{V}^n$ has a neighborhood $\mathcal{U}_{z_0}$ in $\mathbb{V}^n$ defined in $\mathbb{R}^{2n}_z$ by equations $x_{\bar{I}} = x_{\bar{I}}(x_I, p_{\bar{I}})$, $p_I = p_I(x_I, p_{\bar{I}})$ such that*

$$\left(\frac{\partial x_{\bar{I}}}{\partial p_{\bar{I}}}\right)^T = \frac{\partial x_{\bar{I}}}{\partial p_{\bar{I}}}\ , \quad \left(\frac{\partial p_I}{\partial x_I}\right)^T = \frac{\partial p_I}{\partial x_I}$$
$$\left(\frac{\partial p_I}{\partial p_{\bar{I}}}\right)^T = -\frac{\partial x_{\bar{I}}}{\partial x_I}.$$

Proof. The proof of this result relies on Proposition 1.25 on canonical coordinates for a Lagrangian plane; since we will not use it in the rest of this chapter we omit its proof and refer to, for instance Vaisman [168], §3.3, or Mischenko *et al.* [124]. □

5.1.2 The phase of a Lagrangian manifold

Consider now an arbitrary Lagrangian submanifold $\mathbb{V}^n$ of $(\mathbb{R}^{2n}_z, \sigma)$ equipped with a "base point" z_0; we denote by $\pi_1[\mathbb{V}^n]$ the fundamental group $\pi_1[\mathbb{V}^n, z_0]$. Let us denote by $\check{\mathbb{V}}^n$ the set of all homotopy classes $\check{z}$ of paths $\gamma(z_0, z)$ starting at z_0 and ending at z, and by

$$\pi : \check{\mathbb{V}}^n \longrightarrow \mathbb{V}^n$$

the mapping which to $\check{z}$ associates the endpoint z of any of its representatives $\gamma(z_0, z)$. It is a classical result from elementary algebraic topology (see for instance Seifert–Threlfall [148] or Singer–Thorpe [154]) that the set $\check{\mathbb{V}}^n$ can be equipped with a differential structure having the following properties:

- $\check{\mathbb{V}}^n$ is a simply connected and connected C^∞ manifold;
- π is a covering mapping: every $z \in \mathbb{V}^n$ has an open neighborhood U such that $\pi^{-1}(U)$ is the disjoint union of a sequence of open sets $\check{U}_1, \check{U}_2, \ldots$, in $\check{\mathbb{V}}^n$ and the restriction of π to each of the $\check{U}_j$ is a diffeomorphism $\check{U}_j \longrightarrow U$.

- The tangent mapping
$$d_{\check{z}}\pi : T_{\check{z}}\check{\mathbb{V}}^n \longrightarrow T_z\mathbb{V}^n$$
has maximal rank n at every point $\check{z} \in \check{\mathbb{V}}^n$.

With that topology and projection, $\check{\mathbb{V}}^n$ is the *universal covering manifold* of $\mathbb{V}^n$.

We claim that:

Lemma 5.6. *The universal covering manifold $\pi : \check{\mathbb{V}}^n \longrightarrow \mathbb{V}$ of a Lagrangian submanifold $\mathbb{V}^n$ of $(\mathbb{R}^{2n}_z, \sigma)$ is an immersed Lagrangian manifold.*

Proof. Since the tangent mapping $d_{\check{z}}\pi$ has maximal rank n at every point, π is an immersion. Set now $\ell(\check{z}) = T_{\check{z}}\check{\mathbb{V}}^n$; we have

$$d\pi(\ell(\check{z})) = \ell(z) \in \mathrm{Lag}(n)$$

hence $\check{\mathbb{V}}^n$ is a Lagrangian manifold, as claimed. □

Let $\mathbb{V}^n$ be an exact Lagrangian submanifold of $(\mathbb{R}^{2n}_z, \sigma)$: it is thus defined by the equation $p = \partial_x\Phi(x)$. Setting $\varphi(z) = \Phi(x)$ we obviously have

$$d\varphi(z) = d\Phi(x) = \langle \partial_x\Phi(x), dx\rangle = pdx.$$

We will call the function $\varphi : \mathbb{V}^n \longrightarrow \mathbb{R}$ a *phase* of $\mathbb{V}^n$. More generally:

Definition 5.7. Let $\iota : \mathbb{V}^n \longrightarrow \mathbb{R}^{2n}_z$ be an immersed Lagrangian manifold. Let

$$\lambda = pdx = p_1dx_1 + \cdots + p_ndx_n$$

be the "action form"[2] in $\mathbb{R}^{2n}_z$. Any smooth function $\varphi : \mathbb{V}^n \longrightarrow \mathbb{R}$ such that

$$d\varphi = \iota^*\lambda$$

is called a "phase" of $\mathbb{V}^n$.

The differential form $\iota^*\lambda$ on $\mathbb{V}^n$ is closed: since the manifold $\mathbb{V}^n$ is Lagrangian, we have

$$d\iota^*(pdx) = \iota^*(dp \wedge dx) = \iota^*\sigma = 0.$$

In view of "Poincaré's relative lemma" (see Vaisman [168], Weinstein [178]) the phase is thus always locally defined on $\mathbb{V}^n$; if $\mathbb{V}^n$ is simply connected it is even globally defined. In the general case one immediately encounters the usual cohomological obstructions for the global existence of a phase. Here is a simple but typical example:

[2]it is sometimes also called the "Liouville form".

Example 5.8. Let $S^1(R) : x^2 + p^2 = R^2$ be a circle in the symplectic plane. Setting $x = R\cos\theta$ and $p = R\sin\theta$ the condition $d\varphi(z) = pdx$ becomes

$$d\varphi(\theta) = -R^2 \sin^2\theta d\theta$$

which, when integrated, yields the following expression for φ:

$$\varphi(\theta) = \frac{R^2}{2}(\cos\theta\sin\theta - \theta). \tag{5.1}$$

The rub in this example comes from the fact that φ is not defined on the circle itself, because $\varphi(\theta+2\pi) = \varphi(\theta) - \pi R^2 \neq \varphi(\theta)$. There is, however, a way out: we can view $\varphi(\theta)$ as defined on the universal covering of $S^1(R)$, identified with the real line $\mathbb{R}_\theta$, the projection $\pi : \mathbb{R}_\theta \longrightarrow S^1(R)$ being given by $\pi(\theta) = (R\cos\theta, R\sin\theta)$. This trick extends without difficulty to the general case as well:

Theorem 5.9. *Let* $\pi : \check{\mathbb{V}}^n \longrightarrow \mathbb{V}^n$ *be the universal covering of the Lagrangian manifold* $\mathbb{V}^n$.

Proposition 5.10.

(i) *There exists a differentiable function* $\varphi : \check{\mathbb{V}}^n \longrightarrow \mathbb{R}$ *such that*

$$d\varphi(\check{z}) = pdx \quad \textit{if} \quad \pi(\check{z}) = z = (x,p) \tag{5.2}$$

where we are writing pdx *for* $\iota^*\lambda$;

(ii) *That function is given by*

$$\varphi(\check{z}) = \int_{\gamma(z_0,z)} pdx \tag{5.3}$$

where $\gamma(z_0, z)$ *is an arbitrary continuous path in* $\mathbb{V}^n$ *joining* z_0 *to* z.

Proof. It suffices to prove that the right-hand side of (5.3) only depends on the homotopy class in $\mathbb{V}^n$ of the path $\gamma(z_0, z)$ and that $d\varphi(\check{z}) = pdx$. Let $\gamma'(z_0, z)$ be another path joining z_0 to z in $\mathbb{V}^n$ and homotopic to $\gamma(\bar{z}, z)$; the loop $\delta = \gamma(z_0, z) - \gamma'(z_0, z)$ is thus homotopic to a point in $\mathbb{V}^n$. Let $h = h(s,t)$, $0 \leq s,t \leq 1$ be such a homotopy: $h(0,t) = \delta(t)$, $h(1,t) = 0$. As s varies from 0 to 1 the loop δ will sweep out a two-dimensional surface $\mathcal{D}$ with boundary δ contained in $\mathbb{V}^n$. In view of Stokes' theorem we have

$$\int_\delta pdx = \iint_{\mathcal{D}} dp \wedge dx = 0$$

where the last equality follows from the fact that $\mathcal{D}$ is a subset of a Lagrangian manifold. It follows that

$$\int_{\gamma(z_0,z)} pdx = \int_{\gamma'(z_0,z)} pdx$$

hence the integral of pdx along $\gamma(z_0, z)$ only depends on the homotopy class in $\mathbb{V}^n$ of the path joining z_0 to z; it is thus a function of $\check{z} \in \check{\mathbb{V}}^n$. There remains to show that the function

$$\varphi(\check{z}) = \int_{\gamma(z_0,z)} pdx \tag{5.4}$$

is such that $d\varphi(z) = pdx$. The property being local, we can assume that $\mathbb{V}^n$ is simply connected, so that $\check{\mathbb{V}}^n = \mathbb{V}^n$ and write $\varphi(z) = \varphi(\check{z})$. Since $\mathbb{V}^n$ is diffeomorphic to $\ell(z) = T_z\mathbb{V}^n$ in a neighborhood of z, we can reduce the proof to the case where $\mathbb{V}^n$ is a Lagrangian plane ℓ. Let $Ax + Bp = 0$ ($A^TB = BA^T$) be an equation of ℓ, and

$$\gamma(z) : t \longmapsto (-B^Tu(t), A^Tu(t)) \ , \ \ 0 \leq t \leq 1$$

be a differentiable curve starting from 0 and ending at $z = (-B^Tu(1), A^Tu(1))$. We have

$$\varphi(z) = -\int_0^1 \langle A^Tu(t), B^T\dot{u}(t)\rangle \, dt = -\int_0^1 \langle BA^Tu(t), \dot{u}(t)\rangle \, dt$$

and hence, since BA^T is symmetric:

$$\varphi(z) = -\tfrac{1}{2}BA^Tu(1)^2$$

that is

$$d\varphi(z) = -BA^Tu(1)du(1) = pdx$$

which was to be proven. □

Definition 5.11. A function $\varphi : \check{\mathbb{V}}^n \longrightarrow \mathbb{R}$ such that

$$d\varphi(\check{z}) = pdx \ \text{ if } \ \pi(\check{z}) = z = (x, p)$$

is called *"a phase" of* $\mathbb{V}^n$.

Notice that we can always fix a phase by imposing a given value at some point of $\check{\mathbb{V}}^n$; for instance we can choose $\varphi(z_0) = 0$ where z_0 is identified with the (homotopy class of) the constant loop $\gamma(z_0, z)$.

As already observed above we are slightly abusing language by calling φ a "phase of $\mathbb{V}^n$" since φ is multi-valued on $\mathbb{V}^n$. This multi-valuedness is made explicit by studying the action of $\pi_1[\mathbb{V}^n]$ on $\check{\mathbb{V}}^n$, which is defined as follows: let γ be a loop in $\mathbb{V}^n$ with origin the base point z_0 and $\check{\gamma} \in \pi_1[\mathbb{V}^n]$ its homotopy class. Then $\check{\gamma}\check{z}$ is the homotopy class of the loop γ followed by the path $\gamma(z)$ representing $\check{z}$. From the definition of the phase φ follows that

$$\varphi(\check{\gamma}\check{z}) = \varphi(\check{z}) + \oint_\gamma pdx. \tag{5.5}$$

The phase is thus defined on $\mathbb{V}^n$ itself if and only if $\int_\gamma pdx = 0$ for all loops in $\mathbb{V}^n$; this is the case if $\mathbb{V}^n$ is simply connected.

Remark 5.12. Gromov has proved in [81] (also see [91]) that if $\mathbb{V}^n$ is a closed Lagrangian manifold (*i.e.*, compact and without boundary) then we cannot have $\oint_\gamma pdx = 0$ for all loops γ in $\mathbb{V}^n$; to construct the phase of such a manifold we thus have to use the procedure above.

5.1.3 The local expression of a phase

Recall that a Lagrangian manifold which can be represented by an equation $p = \partial_x \Phi(x)$ is called an "exact Lagrangian manifold". It turns out that Lagrangian manifolds are (locally) exact outside their caustic set, and this is most easily described in terms of the phase defined above.

Definition 5.13. A point $z = (x, p)$ of a Lagrangian manifold $\mathbb{V}^n$ is called a "caustic point" if z has no neighborhood in $\mathbb{V}^n$ for which the restriction of the projection $(x,p) \longmapsto x$ is a diffeomorphism. The set of all caustic points of $\mathbb{V}^n$ is called the "*caustic*" *of* $\mathbb{V}^n$ and is denoted by $\Sigma_{\mathbb{V}^n}$.

For instance the caustic of the circle $x^2 + p_x^2 = R^2$ in the symplectic plane consists of the points $(\pm 1, 0)$.

Of course, caustics have no intrinsic meaning, whatsoever: there are just artefacts coming from the choice of a privileged n-dimensional plane (*e.g.*, the position space $\mathbb{R}_x^n$) on which one projects the motion.

Let U be an open subset of $\mathbb{V}^n$ which contains no caustic points: $U \cap \Sigma_{\mathbb{V}^n} = \emptyset$. Then the restriction χ_U to U of the projection $\chi : (x,p) \longmapsto x$ is a local diffeomorphism of U onto its image $\chi_U(U)$; choosing U sufficiently small we can thus assume that (U, χ_U) is a local chart of $\mathbb{V}^n$ and that the fiber $\pi^{-1}(U)$ is the disjoint union of a family of open sets $\check{U}$ in the universal covering of $\mathbb{V}^n$ such that the restriction $\pi_{\check{U}}$ to $\check{U}$ of the projection $\pi : \check{\mathbb{V}}^n \longrightarrow \mathbb{V}^n$ is a local diffeomorphism $\check{U} \longrightarrow U$. It follows that we can always assume that $(\check{U}, \chi_U \circ \pi_{\check{U}})$ is a local chart of $\mathbb{V}^n$.

Proposition 5.14. *Let Φ be the local expression of the phase φ in any of the local charts $(\check{U}, \chi_U \circ \pi_{\check{U}})$ such that $U \cap \Sigma_{\mathbb{V}^n} = \emptyset$:*

$$\Phi(x) = \varphi((\chi_U \circ \pi_{\check{U}})^{-1}(x)). \tag{5.6}$$

The Lagrangian submanifold U is exact and can be represented by the equation

$$p = \partial_x \Phi(x) = \partial_x \varphi((\chi_U \circ \pi_{\check{U}})^{-1}(x)). \tag{5.7}$$

Proof. Let us first show that the equation (5.7) remains unchanged if we replace $(\check{U}, \chi_U \circ \pi_U)$ by a chart $(\check{U}', \chi_{U'} \circ \pi_{\check{U}'})$ such that $\pi(\check{U}') = \pi(\check{U})$. There exists $\gamma \in \pi_1[\mathbb{V}^n]$ such that $\check{U}' = \gamma\check{U}$ hence, by (5.5), the restrictions $\varphi_{\check{U}'}$ and $\varphi_{\check{U}}$ differ by the constant

$$C(\gamma) = \oint_\gamma pdx.$$

It follows that

$$\partial_x\varphi((\chi_{U'} \circ \pi_{U'})^{-1}(x)) = \partial_x\varphi((\chi_U \circ \pi_U)^{-1}(x))$$

and hence the right-hand side of the identity (5.7) does not depend on the choice of local chart $(\check{U}, \chi_U \circ \pi_U)$. Set now $(\chi_U \circ \pi_U)^{-1}(x) = (p(x), x)$; we have, for $x \in \chi_U \circ \pi_U(U)$,

$$d\Phi(x) = d\varphi(p(x), x) = p(x)dx$$

hence (5.7). □

Exercise 5.15. For $S \in \mathrm{Sp}(n)$ set $(x_S, p_S) = S(x, p)$.

(i) Show that

$$p_S dx_S - x_S dp_S = pdx - xdp. \tag{5.8}$$

(ii) Define the differentiable function $\varphi_S : \mathbb{V}^n \longrightarrow \mathbb{R}$ by the formula

$$\varphi_S(\check{z}) = \varphi(\check{z}) + \tfrac{1}{2}(\langle p_S, x_S\rangle - \langle p, x\rangle). \tag{5.9}$$

Show that

$$d\varphi_S(\check{z}) = p_S dx_S \text{ if } \pi(\check{z}) = (x, p). \tag{5.10}$$

5.2 Hamiltonian Motions and Phase

In this section we investigate the action of Hamiltonian flows on the phase of a Lagrangian manifold; it will lead us to the definition of the Heisenberg–Weyl operators in Section 5.5.

We begin by studying the properties of an important integral invariant, the Poincaré–Cartan differential form. The study of integral invariants was initiated in a systematic way by the mathematician E. Cartan in his *Leçons sur les invariants intégraux* in 1922. There are many excellent books on the topic; see for instance Abraham–Marsden [1], Choquet-Bruhat and DeWitt–Morette [23], Libermann and Marle [110], or Godbillon [50].

5.2.1 The Poincaré–Cartan Invariant

Let H be a Hamiltonian function (possibly time-dependent).

Definition 5.16. The Poincaré–Cartan form associated to H is the differential 1-form

$$\alpha_H = pdx - Hdt \tag{5.11}$$

on extended phase space $\mathbb{R}^{2n}_z \times \mathbb{R}_t$ where

$$pdx = p_1dx_1 + \cdots + p_ndx_n$$

is the action form (or "Liouville form").

The interest of the Poincaré–Cartan form comes from the fact that it is a *relative integral invariant*. This means that the contraction $i_{\tilde{X}_H} d\alpha_H$ of the exterior derivative

$$d\alpha_H = dp \wedge dx - dH \wedge dt$$

of α_H with the suspended Hamilton vector field

$$\tilde{X}_H = (X_H, 1) = (J\partial_z H, 1)$$

is zero:

$$i_{\tilde{X}_H} d\alpha_H = 0.$$

There actually is a whole family of closely related invariants having the same property. These invariants are defined, for $\lambda \in \mathbb{R}$, by

$$\alpha_H^{(\lambda)} = \lambda p dx + (\lambda - 1) x dp - H dt; \tag{5.12}$$

of course $\alpha_H^{(1)} = \alpha_H$. Since we have

$$\begin{aligned} d(\lambda p dx + (\lambda - 1) x dp) &= \lambda dp \wedge dx + (\lambda - 1) dx \wedge dp \\ &= \lambda dp \wedge dx - (\lambda - 1) dp \wedge dx \\ &= dp \wedge dx. \end{aligned}$$

it follows that

$$d\alpha_H^{(\lambda)} = d\alpha_H \quad \text{for every } \lambda \in \mathbb{R}.$$

Example 5.17. The case $\lambda = 1/2$. The corresponding form can be formally written as

$$\alpha_H^{(1/2)} = \frac{1}{2}\sigma(z, dz) - H dt;$$

we will see in Chapter 10 that this form is particularly convenient in the sense that it leads to a Schrödinger equation in phase space where the variables x and p are placed on similar footing, very much as in Hamilton's equations.

Let us prove the main result of this subsection, namely the relative invariance of the forms $\alpha_H^{(\lambda)}$:

Proposition 5.18. *The forms* $\alpha_H^{(\lambda)}$ *defined by* (5.12) *satisfy, for every* $\lambda \in \mathbb{R}$,

$$d\alpha_H^{(\lambda)}(\tilde{X}_H(z,t), \tilde{Y}(z,t)) = 0 \tag{5.13}$$

for every vector $\tilde{Y}(z,t) = (Y(z,t), \alpha(t))$ *in* $\mathbb{R}_z^{2n} \times \mathbb{R}_t$ *originating at a point* (z,t).

Proof. Since

$$d\alpha_H^{(\lambda)} = \sigma - dH \wedge dt$$

we have

$$d\alpha_H^{(\lambda)}(\tilde{X}_H(z,t),\tilde{Y}(z,t)) = \\ \sigma(X_H(x,t),Y(z,t)) - (dH\wedge dt)(\tilde{X}_H(z,t),\tilde{Y}(z,t)).$$

Writing, for short,

$$\begin{aligned}\tilde{X}_H(z,t) &= (X_H,1) = (\partial_p H, -\partial_x H, 1),\\ \tilde{Y}(z,t) &= (Y,\alpha) = (Y_x, Y_p, \alpha),\end{aligned}$$

we have to show that

$$\sigma(X_H,Y) - (dH\wedge dt)(X_H,1;Y,\alpha) = 0. \tag{5.14}$$

By definition of σ we have

$$\sigma(X_H,Y) = -\langle \partial_x H, Y_x\rangle - \langle \partial_p H, Y_p\rangle\,;$$

on the other hand

$$dH\wedge dt = \partial_x H(dx\wedge dt) + \partial_p H(dp\wedge dt)$$

and

$$\begin{aligned}(dx\wedge dt)(X_H,1;Y,\alpha) &= \alpha\partial_p H - Y_x,\\ (dp\wedge dt)(X_H,1;Y,\alpha) &= -\alpha\partial_x H - Y_p,\end{aligned}$$

so that

$$(dH\wedge dt)(X_H,1;Y,\alpha) = -\langle \partial_x H, Y_x\rangle - \langle \partial_p H, Y_p\rangle = \sigma(X_H,Y)$$

which is the equality (5.14) we set out to prove. □

The relative invariance of the forms $\alpha_H^{(\lambda)}$ has the following important consequence: let $\tilde{\gamma} : [0,1] \longrightarrow \mathbb{R}_z^{2n}\times\mathbb{R}_t$ be a smooth curve in extended phase space on which we let the suspended flow $\tilde{f}_t^H$ act; as time varies, $\tilde{\gamma}$ will sweep out a two-dimensional surface Σ_t whose boundary $\partial\Sigma_t$ consists of $\tilde{\gamma}$, $\tilde{f}_t^H(\tilde{\gamma})$, and two arcs of phase-space trajectory, $\tilde{\gamma}_0$ and $\tilde{\gamma}_1$: $\tilde{\gamma}_0$ is the trajectory of the origin $\tilde{\gamma}(0)$ of $\tilde{\gamma}$, and $\tilde{\gamma}_1$ that of its endpoint $\tilde{\gamma}(1)$. We have

$$\int_{\partial\Sigma_t} \alpha_H^{(\lambda)} = 0. \tag{5.15}$$

Here is a sketch of the proof (for details, and more on invariant forms in general, see Libermann and Marle [110]). Using the multi-dimensional Stokes formula we have

$$\int_{\partial\Sigma_t} \alpha_H^{(\lambda)} = \int_{\Sigma_t} d\alpha_H^{(\lambda)} = \int_{\Sigma_t} d\alpha_H.$$

Since the surface Σ_t consists of flow lines of $\tilde{X}_H$ each pair $(\tilde{X}, \tilde{Y})$ of tangent vectors at a point (z, t) can be written as a linear combination of two independent vectors, and one of these vectors can be chosen as $\tilde{X}_H$. It follows that $d\alpha_H(\tilde{X}, \tilde{Y})$ is a sum of terms of the type $d\alpha_H(\tilde{X}_H, \tilde{Y})$, which are equal to zero in view of (5.13). We thus have

$$\int_{\Sigma_t} d\alpha_H = 0$$

whence (5.15).

5.2.2 Hamilton–Jacobi theory

Here is one method that can be used (at least theoretically) to integrate Hamilton's equations; historically it is one of the first known resolution schemes[3]. For a very interesting discussion of diverse related questions such as calculus of variations and Bohmian mechanics see Butterfield's paper [20]. a complete rigorous treatment is to be found in, for instance, Abraham and Marsden [1].

Given an arbitrary Hamiltonian function $H \in C^\infty(\mathbb{R}^{2n+1}_{z,t}, \mathbb{R})$ the associated Hamilton–Jacobi equation is the (usually non-linear) partial differential equation with unknown Φ:

$$\frac{\partial \Phi}{\partial t} + H(x, \partial_x \Phi, t) = 0. \tag{5.16}$$

The interest of this equation comes from the fact that the knowledge of a sufficiently general solution Φ yields the solutions of Hamilton's equations for H. (At first sight it may seem strange that one replaces a system of ordinary differential equations by a non-linear partial differential equation, but this procedure is often the only available method; see the examples in Goldstein [53].)

Proposition 5.19. *Let $\Phi = \Phi(x, t, \alpha)$ be a solution of*

$$\frac{\partial \Phi}{\partial t} + H(x, \partial_x \Phi, t) = 0 \tag{5.17}$$

depending on n non-additive constants of integration $\alpha_1, \ldots, \alpha_n$, and such that

$$\det D^2_{x,\alpha}\Phi(x, t, \alpha) \neq 0. \tag{5.18}$$

Let $\beta_1, \ldots, \beta_n$ be constants; the functions $t \longmapsto x(t)$ and $t \longmapsto p(t)$ determined by the implicit equations

$$\partial_\alpha \Phi(x, t, \alpha) = \beta \quad , \quad p = \partial_x \Phi(x, t, \alpha) \tag{5.19}$$

are solutions of Hamilton's equations for H.

[3] Butterfield [20] tells us that Whittaker reports that what we call Hamilton–Jacobi theory was actually already developed by Pfaff and Cauchy using earlier results of Lagrange and Monge, well before Hamilton and Jacobi's work.

Proof. We assume $n = 1$ for notational simplicity; the proof extends to the general case without difficulty. Condition (5.18) implies, in view of the implicit function theorem, that the equation $\partial_\alpha \Phi(x, t, \alpha) = \beta$ has a unique solution $x(t)$ for each t; this defines a function $t \longmapsto x(t)$. Inserting $x(t)$ in the formula $p = \partial_x \Phi(x, t, \alpha)$ we also get a function $t \longmapsto p(t) = \partial_x \Phi(x(t), t, \alpha)$. Let us show that $t \longmapsto (x(t), p(t))$ is a solution of Hamilton's equations for H. Differentiating the equation (5.17) with respect to α yields, using the chain rule,

$$\frac{\partial^2 \Phi}{\partial \alpha \partial t} + \frac{\partial H}{\partial p} \frac{\partial^2 \Phi}{\partial \alpha \partial x} = 0; \tag{5.20}$$

differentiating the first equation (5.19) with respect to t yields

$$\frac{\partial^2 \Phi}{\partial x \partial \alpha} x + \frac{\partial^2 \Phi}{\partial t \partial \alpha} = 0; \tag{5.21}$$

subtracting (5.21) from (5.20) we get

$$\frac{\partial^2 \Phi}{\partial x \partial \alpha} \left(\frac{\partial H}{\partial p} - \dot{x} \right) = 0,$$

hence we have proven that $\dot{x} = \partial_p H$ since $\partial^2 \Phi / \partial x \partial \alpha$ is assumed to be non-singular. To show that $\dot{p} = -\partial_x H$ we differentiate (5.17) with respect to x:

$$\frac{\partial^2 \Phi}{\partial x \partial t} + \frac{\partial H}{\partial x} + \frac{\partial H}{\partial p} \frac{\partial^2 \Phi}{\partial x^2} = 0 \tag{5.22}$$

and $p = \partial_x \Phi$ with respect to t:

$$\dot{p} = \frac{\partial^2 \Phi}{\partial t \partial x} + \frac{\partial^2 \Phi}{\partial x^2} \dot{x}. \tag{5.23}$$

Inserting the value of $\partial^2 \Phi / \partial x \partial t$ given by (5.23) in (5.22) yields

$$\frac{\partial H}{\partial x} + \frac{\partial^2 \Phi}{\partial x^2} \dot{x} - \frac{\partial H}{\partial p} \frac{\partial^2 \Phi}{\partial x^2} + \dot{p} = 0,$$

hence $\dot{p} = -\partial_x H$ since $\dot{x} = \partial_p H$. □

When the Hamiltonian is time-independent, the Hamilton–Jacobi equation is separable: inserting $\Phi = \Phi_0 - Et$ in (5.17) we get the 'reduced Hamilton–Jacobi equation':

$$H(x, \partial_x \Phi_0, t) = E \tag{5.24}$$

which is often easier to solve in practice; the energy E can be taken as a constant of integration.

Exercise 5.20.
(i) Let $H = \frac{1}{2m}p^2$ be the Hamiltonian of a particle with mass m moving freely along the x-axis. Use (5.24) to find a complete family of solutions of the time-dependent Hamilton–Jacobi equation for H.
(ii) Do the same with the harmonic oscillator Hamiltonian $H = \frac{1}{2m}(p^2+m^2\omega^2x^2)$.

It turns out that for a wide class of physically interesting Hamiltonians the Hamilton–Jacobi equation can be explicitly solved using the notion of free generating function defined in Chapter 2, Subsection 2.2.3.

Proposition 5.21. *Suppose that there exists ε such that for $0 < |t| < \varepsilon$ the mappings f_t^H are free symplectomorphisms when defined. The Cauchy problem*

$$\frac{\partial \Phi}{\partial t} + H\left(x, \partial_x \Phi, t\right) = 0 \ , \ \Phi(x,0) = \Phi_0(x) \tag{5.25}$$

has a solution Φ, defined for $0 < |t| < \varepsilon$, and given the formula

$$\Phi(x,t) = \Phi_0(x') + W(x,x';t) \tag{5.26}$$

where x' is defined by the condition

$$(x,p) = f_t^H(x', \partial_x \Phi_0(x')) \tag{5.27}$$

and W is the generating function

$$W(x,x';t) = \int_{x',0}^{x,t} pdx - Hdt$$

of the symplectomorphism f_t^H.

Proof. As in the proof of Proposition 5.19 we assume that $n = 1$; the generalization to arbitrary n is straightforward. We first note that formula (5.27) uniquely defines x' for small t. In fact, writing $x = (x', \partial_x \Phi_0(x'), t)$ we have

$$\frac{dx}{dx'} = \frac{\partial x}{\partial x'} + \frac{\partial x}{\partial p'}\frac{\partial^2 \Phi_0}{\partial x'^2};$$

since the limit of the Jacobian

$$D_t^H(z') = \begin{bmatrix} \frac{\partial x}{\partial x'} & \frac{\partial x}{\partial p'} \\ \frac{\partial p}{\partial x'} & \frac{\partial p}{\partial p'} \end{bmatrix}$$

is the identity as $t \to 0$, it follows that dx/dx' is different from zero in some interval $[-\alpha, \alpha]$, $\alpha > 0$, hence $x' \longrightarrow f_t^H(x', \partial_x \Phi_0(x'))$ is a local diffeomorphism for each $t \in [-\alpha, \alpha]$. Obviously $\lim_{t \to 0} \Phi(x,t) = \Phi_0(x)$ since $x' \to x$ as $t \to 0$, so that the

Cauchy condition is satisfied. To prove that Φ is a solution of Hamilton–Jacobi's equation one notes that

$$\Phi(x+\Delta x, t+\Delta t) - \Phi(x,t) = \int_L pdx - Hdt$$

where L is the line segment joining (x,p,t) to $(x+\Delta x, p+\Delta p, t+\Delta t)$; p and $p+\Delta p$ are determined by the relations $p = \partial_x W(x,x';t)$ and

$$p+\Delta p = \partial_x W(x+\Delta x, x'+\Delta x'; t+\Delta t)$$

where $\Delta x' = x'(x+\Delta x) - x'(x)$. Thus,

$$\Phi(x+\Delta x, t+\Delta t) - \Phi(x,t) = p\Delta x + \frac{1}{2}\Delta p \Delta x \\ - \Delta t \int_0^1 H(x+s\Delta x, p+s\Delta p, t+s\Delta t)ds$$

and hence

$$\frac{\Phi(x,t+\Delta t) - \Phi(x,t)}{\Delta t} = -\int_0^1 H(x,p+s\Delta p, t+s\Delta t)ds$$

from which follows that

$$\frac{\partial \Phi}{\partial t}(x,t) = -H(x,p,t) \tag{5.28}$$

since $\Delta p \to 0$ when $\Delta t \to 0$. Similarly,

$$\Phi(x+\Delta x, t) - \Phi(x,t) = p\Delta x + \frac{1}{2}\Delta p \Delta x$$

and $\Delta p \to 0$ as $\Delta x \to 0$ so that

$$\frac{\partial \Phi}{\partial x}(x,t) = p. \tag{5.29}$$

Combining (5.28) and (5.29) shows that Φ satisfies Hamilton–Jacobi's equation. □

5.2.3 The Hamiltonian phase

Let $\mathbb{V}^n$ be a Lagrangian manifold on which a base point z_0 is chosen. Let us set $\mathbb{V}^n_t = f^H_t(\mathbb{V}^n)$ and $z_t = f^H_t(z_0)$ where (f^H_t) is the flow determined by some Hamiltonian H. Since Hamiltonian flows consist of symplectomorphisms each $\mathbb{V}_t$ is a Lagrangian manifold (Proposition 5.3), and the function $\varphi_t : \check{\mathbb{V}}^n_t \longrightarrow \mathbb{R}$ defined by

$$\varphi_t(\check{z}(t)) = \int_{\gamma(z_t, z(t))} pdx \tag{5.30}$$

($\check{z}(t)$ being the homotopy class in $\mathbb{V}^n_t$ of a path $\gamma(z_t, z(t))$) obviously is a phase of $\mathbb{V}^n_t$ when $z_t = f^H_t(z_0)$ is chosen as base point in $\mathbb{V}^n_t$. The following result relates φ_t to the phase φ_0 of the initial manifold $\mathbb{V}^n = \mathbb{V}^n_0$:

Lemma 5.22. *Let $\check{z}$ be a point in $\check{\mathbb{V}}^n = \check{\mathbb{V}}_0^n$ and $\check{z}(t)$ its image in $\check{\mathbb{V}}_t^n$ by f_t^H (i.e., $\check{z}(t)$ is the homotopy class in $\mathbb{V}_t^n$ of the image by f_t^H of a path representing $\check{z}$). We have*

$$\varphi_t(\check{z}(t)) - \varphi(\check{z}) = \int_z^{z(t)} \alpha_H - \int_{z_0}^{z_t} \alpha_H. \tag{5.31}$$

Proof. Let Σ_t be the closed piecewise smooth curve

$$\Sigma_t = [z_0, z_t] + \gamma(z_t, z(t)) - [z, z(t)] - \gamma(z_0, z)$$

where $[z_0, z_t]$ (resp. $[z, z(t)]$) is the Hamiltonian trajectory joining z_0 to z_t (resp. z to $z(t)$). In view of the property (5.15) of the Poincaré–Cartan form we have

$$\int_{\Sigma_t} \alpha_H = 0. \tag{5.32}$$

Since $dt = 0$ along both paths $\gamma(z_t, z(t))$ and $\gamma(z_0, z)$ we have

$$\int_{\gamma(z_t,z(t))} \alpha_H = \int_{\gamma(z_t,z(t))} pdx \quad , \quad \int_{\gamma(z_0,z)} \alpha_H = \int_{\gamma(z_0,z)} pdx$$

and hence (5.32) is equivalent to

$$\int_{\gamma(z_0,z)} pdx + \int_z^{z(t)} \alpha_H - \int_{\gamma(z_t,z(t))} pdx - \int_{z_0}^{z_t} \alpha_H = 0$$

that is to (5.31). □

Lemma 5.22 has the following important consequence:

Proposition 5.23. *Let $\check{z} \in \check{\mathbb{V}}_t^n$ and define $\check{z}' \in \check{\mathbb{V}}^n$ by the condition $\check{z} = f_t^H(\check{z}')$. The function $\varphi(\cdot, t) : \check{\mathbb{V}}_t^n \longrightarrow \mathbb{R}$ defined by*

$$\varphi(\check{z}, t) = \varphi(\check{z}') + \int_{z',0}^{z,t} \alpha_H \tag{5.33}$$

is a phase of $\mathbb{V}_t^n$: for fixed t we have

$$d\varphi(\check{z}, t) = pdx \quad \textit{if} \quad \pi_t(\check{z}) = z = (x, p) \tag{5.34}$$

where π_t is the projection $\check{\mathbb{V}}_t^n \longrightarrow \mathbb{V}_t^n$.

Proof. In view of Lemma 5.22 we have

$$\varphi(\check{z}, t) = \int_{z_0,0}^{z_t,t} \alpha_H + \int_{\gamma(z_t,z)} pdx \tag{5.35}$$

where α_H is integrated along the arc of extended phase space trajectory leading from z_0 at time $t = 0$ to z_t at time t and $\gamma(z_t, z)$ is the image in $\check{\mathbb{V}}_t^n$ of a path joining z_0 to z' in $\check{\mathbb{V}}^n$ and belonging to the homotopy class $\check{z}'$. Differentiating (5.35) for fixed t we get (5.34). □

Definition 5.24. We will call the function $\varphi(\cdot,t) : \check{\mathbb{V}}^n \longrightarrow \mathbb{R}$ defined by (5.33), (5.35) the "Hamiltonian *phase*" of $\mathbb{V}_t^n$ (as opposed to the phase (5.30).

Here is an interesting particular case of the result above:

Exercise 5.25. Let H be a Hamiltonian which is quadratic and homogeneous in the position and momentum variables; its flow thus consists of symplectic matrices S_t^H.

(i) Show that the Hamiltonian phase of $S_t^H(\mathbb{V})$ is

$$\varphi(\check{z},t) = \varphi(\check{z}') + \tfrac{1}{2}(\langle p_t, x_t\rangle - \langle p, x\rangle) \tag{5.36}$$

where $(x_t, p_t) = S_t^H(x,p)$.

(ii) Explicit this condition when $\mathbb{V}$ is a graph: $p = \partial_x\Phi(x)$ and $S_t^H(\mathbb{V})$ is itself a graph.

An interesting particular case of Proposition 5.23 is when the Lagrangian manifold $\mathbb{V}^n$ is invariant under the Hamiltonian flow; this situation typically occurs when one has a completely integrable system and $\mathbb{V}^n = \mathbb{T}^n$ is an associated Lagrangian torus:

Corollary 5.26. *Let H be a time-independent Hamiltonian, (f_t^H) its flow, and assume that $f_t^H(\mathbb{V}^n) = \mathbb{V}^n$ for all t. If $\check{z}$ is the homotopy class in $\mathbb{V}^n$ of a path $\gamma(z_0,z)$ and $\gamma(z,z(t))$ is the piece of Hamiltonian trajectory joining z to $z(t)$, then*

$$\varphi(\check{z},t) = \varphi(\check{z}(t)) - Et \tag{5.37}$$

where E is the (constant) value of H on $\mathbb{V}^n$ and $\check{z}(t)$ the homotopy class of the path $\gamma(z_0,z) + \gamma(z,z(t))$ in $\mathbb{V}^n$.

Proof. The trajectory $s \longmapsto z_s = f_s^H(z)$ is a path $\gamma(z,z(t))$ in $\mathbb{V}^n$ joining z to $z(t)$, hence

$$\int_{\gamma(z_0,z)} pdx + \int_z^{z(t)} \alpha_H = \int_{\gamma(z_0,z(t))} pdx - Et$$

where $\gamma(z_0,z(t)) = \gamma(z_0,z) + \gamma(z,z(t))$. The result follows since the first integral in the right-hand side of this equality is by definition $\varphi(\check{z}(t))$. □

Proposition 5.23 links the notion of phase of a Lagrangian manifold to the standard Hamilton–Jacobi theory:

Proposition 5.27. *Let $z \in \mathbb{V}^n$ have a neighborhood U in $\mathbb{V}^n$ projecting diffeomorphically on $\mathbb{R}_x^n$.*

(i) *There exists $\varepsilon > 0$ such that the local expression $\Phi = \Phi(x,t)$ of the phase φ is defined for $|t| < \varepsilon$;*

(ii) *Φ satisfies the Hamilton–Jacobi equation*

$$\frac{\partial\Phi}{\partial t} + H(x, \partial_x\Phi) = 0$$

for x in the projection of U and $|t| < \varepsilon$.

Proof. The first part (i) is an immediate consequence of Proposition 5.14 (the existence of ε follows from the fact that the caustic is a closed subset of $\mathbb{V}^n$). To prove (ii) we observe that

$$\Phi(x,t) = \Phi(x',0) + \int_{z',0}^{z,t} pdx - Hdt$$

in view of formula (5.33) in Proposition 5.23; now we can parametrize the arc joining $z',0$ to z,t by x and t hence

$$\Phi(x,t) = \Phi(x',0) + \int_{x',0}^{x,t} pdx - Hdt$$

which is precisely the solution of Hamilton–Jacobi's equation with initial datum Φ at time $t = 0$. □

5.3 Integrable Systems and Lagrangian Tori

Completely integrable systems (sometimes also called Liouville integrable systems) are exceptions rather than the rule. They play however a privileged role in Hamiltonian mechanics, because the associated Hamilton equations are explicitly solvable by passing to the so-called "angle-action variables"; they are also historically the first to have been rigorously semiclassically quantized (the Keller–Maslov theory [102, 119, 120, 124, 128] which we will discuss in the last section of this chapter). The reason for which we introduce them in this chapter is that the phase-space curves corresponding to the solutions of Hamilton's equations lie on Lagrangian manifolds of a particular type.

5.3.1 Poisson brackets

Let F and G be two continuously differentiable functions on $\mathbb{R}^{2n}_z$. By definition, the Poisson bracket of F and G is

$$\{F,G\} = \sum_{j=1}^{n} \frac{\partial F}{\partial x_j}\frac{\partial G}{\partial p_j} - \frac{\partial F}{\partial p_j}\frac{\partial G}{\partial x_j} \tag{5.38}$$

(in some texts the opposite sign convention is chosen), that is

$$\{F,G\} = \langle \partial_x F, \partial_p G\rangle - \langle \partial_p F, \partial_x G\rangle . \tag{5.39}$$

The Poisson bracket is related in an obvious way to the symplectic structure on $\mathbb{R}^{2n}_z$: defining the Hamiltonian vector fields

$$X_F = J\partial_z F \quad , \quad X_G = J\partial_z G$$

we have $\sigma(X_F, X_G) = \sigma(\partial_z F, \partial_z G)$ hence formula (5.39) can be rewritten for short as

$$\{F, G\} = -\sigma(X_F, X_G). \tag{5.40}$$

Exercise 5.28. Show that $X_{\{F,G\}} = [X_F, X_G]$ where $[X_F, X_G] = X_F X_G - X_G X_F$ is the commutator of the vector fields X_F and X_G.

Exercise 5.29. Hamilton's equations can be rewritten in terms of the Poisson bracket as

$$\dot{x}_j = \{x_j, H\} \quad , \quad \dot{p}_j = \{p_j, H\}.$$

Indeed, $\{x_j, H\} = \partial H / \partial p_j$ and $\{p_j, H\} = -\partial H / \partial x_j$.

When $\{F, G\} = 0$ we will say that the functions F and G are in *involution* (one also says that they are *Poisson commuting).*

The Poisson bracket has the following obvious properties:

$$\begin{aligned}
&\{\lambda F, G\} = \lambda \{F, G\} \quad \text{for } \lambda \in \mathbb{R}, \\
&\{F, G + H\} = \{F, G\} + \{F, H\}, \\
&\{F, G\} = -\{G, F\}, \\
&\{\{F, G\}, H\} + \{\{G, H\}, F\} + \{\{H, F\}, G\} = 0;
\end{aligned}$$

the last formula is called "Jacobi's identity". All these relations are of course immediate consequences of the definition of the Poisson bracket.

Definition 5.30. Let H and F be smooth functions on $\mathbb{R}^{2n}_z$. Viewing H as a Hamiltonian defining a "motion" we will say that F is a "constant of the motion" if its Poisson commutes with H, that is if

$$\{F, H\} = 0. \tag{5.41}$$

The terminology comes from the following remark: let $t \longmapsto z(t) = (x(t), p(t))$ be a solution of Hamilton's equations $\dot{z} = J\partial_z H(z)$. Then

$$\begin{aligned}
\frac{d}{dt} F(z(t)) &= \langle \partial_x F, \dot{x} \rangle + \langle \partial_p F, \dot{p} \rangle \\
&= \langle \partial_x F, \partial_p H \rangle - \langle \partial_p F, \partial_x H \rangle \\
&= 0
\end{aligned}$$

hence the function $t \longmapsto F(z(t))$ is constant "along the motion". In particular $\{H, H\} = 0$ hence the Hamiltonian itself is a constant of the motion: this is the theorem of conservation of energy for time-independent Hamiltonian systems.

The notion of constant of the motion makes sense for time-dependent Hamiltonians as well: a constant of the motion for a time-dependent Hamiltonian H is a function $F : \mathbb{R}^{2n+1} \longrightarrow \mathbb{R}$ which is constant along the curves $t \longmapsto (z(t), t)$ in extended phase space. Notice that if F is defined along $t \longmapsto (z(t), t)$ then, by the chain rule

$$\frac{dF}{dt} = \frac{\partial F}{\partial t} + \{F, H\}. \tag{5.42}$$

Thus $F : \mathbb{R}^{2n+1} \longrightarrow \mathbb{R}$ is a constant of the motion if and only if it satisfies *Liouville's equation*

$$\frac{\partial F}{\partial t} + \{F, H\} = 0. \tag{5.43}$$

Since $\{H, H\} = 0$ we see that in particular the energy is a constant of the motion if and only if $\partial H/\partial t = 0$ that is if and only if H is time-independent.

Exercise 5.31. Let H be the one-dimensional harmonic oscillator Hamiltonian

$$H = \frac{\omega(t)}{2}(p^2 + x^2)$$

and $E(t) = H(z(t), t)$ the energy along a solution $t \longmapsto z(t)$. In view of (5.42) we have $\dot{E}(t) = (\dot{\omega}(t)/\omega(t))E(t)$, hence the energy and the frequency are proportional: $E(t) = k\omega(t)$ with $k = E(0)/\omega(0)$.

For a harmonic oscillator with n degrees of freedom it is not true in general that energy is proportional to the frequencies. Suppose however H is a Hamiltonian of the type

$$H = \sum_{j=1}^{n} \frac{\omega_j(t)}{2}(p_j^2 + x_j^2)$$

and $t \longmapsto z(t)$ is a solution of Hamilton's equations for H. If there exists a real number k such that $\omega(t) = k\omega(t')$ ($\omega = (\omega_1, \ldots, \omega_n)$) for some instants t and t', then $E(t)/\omega_j(t) = E(t')/\omega_j(t')$.

5.3.2 Angle-action variables

From now on, and until the end of this section, we will assume that all Hamiltonians are time-independent.

Let us begin by discussing a simple example. Consider the n-dimensional harmonic oscillator Hamiltonian in normal form

$$H = \sum_{j=1}^{n} \frac{\omega_j}{2}(p_j^2 + x_j^2)$$

and define new variables (ϕ, I) by

$$x_j = \sqrt{2\mathrm{I}_j}\cos\phi_j \quad , \quad p_j = \sqrt{2\mathrm{I}_j}\sin\phi_j \tag{5.44}$$

for $1 \leq j \leq n$; we assume $\mathrm{I}_j \geq 0$ and the angles ϕ_j are chosen such that $0 \leq \phi_j < 2\pi$. In these new variables the Hamiltonian H takes the simple form

$$K(\mathrm{I}) = \sum_{j=1}^{n} \omega_j \mathrm{I}_j \tag{5.45}$$

(observe that K does not contain ϕ). The transformation $f : (x,p) \longmapsto (\phi, \mathrm{I})$ is a symplectomorphism outside the origin of $\mathbb{R}_z^{2n}$; it suffices to verify this in the case $n = 1$. We have

$$\phi = \tan^{-1}\left(\frac{x}{p}\right) \quad , \quad \mathrm{I} = \frac{1}{2}(x^2 + p^2)$$

hence the Jacobian matrix of the mapping $f : z = (x,p) \longmapsto (\phi, \mathrm{I})$ is

$$Df(z) = \begin{bmatrix} \frac{x}{x^2+p^2} & \frac{-p}{x^2+p^2} \\ p & x \end{bmatrix}$$

which obviously is a symplectic matrix (it has determinant equal to 1). It suffices now to solve Hamilton's equations

$$\dot{\phi}_j = \partial_{\mathrm{I}_j} K(\mathrm{I}) = \omega_j \quad , \quad \dot{I}_j = -\partial_{\phi_j} K(\mathrm{I}) = 0 \tag{5.46}$$

for K and then to return to the original variables (x,p) using the inverse change of variables. The equations (5.46) have the obvious solutions

$$\phi_j(t) = \omega_j t + \phi_j(0) \quad , \quad \mathrm{I}_j(t) = \mathrm{I}_j(0).$$

Inserting these values in (5.44) we get

$$x_j(t) = \sqrt{2\mathrm{I}_j(0)} \cos(\omega_j t + \phi_j(0)),$$
$$p_j(t) = \sqrt{2\mathrm{I}_j(0)} \sin(\omega_j t + \phi_j(0))$$

for $1 \leq j \leq n$.

This example leads us quite naturally to the following general definition:

Definition 5.32. A time-independent Hamiltonian H on $\mathbb{R}_z^{2n}$ is *"completely integrable" (for short: "integrable")* if there exists a symplectomorphism $f : (x,p) \longmapsto (\phi, \mathrm{I})$ (in general not globally defined) such that the composed function $K = H \circ f$ only depends explicitly on the action variables $\mathrm{I} = (\mathrm{I}_1, \ldots, \mathrm{I}_n)$:

$$H(z) = H(f(\phi, \mathrm{I})) = K(\mathrm{I}). \tag{5.47}$$

The numbers $\omega_j(\mathrm{I}) = \omega_j(\mathrm{I}_1, \ldots, \mathrm{I}_n)$ defined by

$$\omega_j(\mathrm{I}) = \frac{\partial K}{\partial \mathrm{I}_j}(\mathrm{I}) \tag{5.48}$$

are called the "frequencies of the motion".

The Hamilton equations for K are

$$\dot{\phi} = \partial_{\mathrm{I}} K(\mathrm{I}) \quad , \quad \dot{I} = -\partial_{\phi} K(\mathrm{I}); \tag{5.49}$$

since $\partial_{\phi} K = 0$, these can be explicitly solved "by quadratures", yielding the solutions

$$\phi_j(t) = \omega_j(\mathrm{I}(0))t + \phi(0) \quad , \quad \mathrm{I}_j(t) = \mathrm{I}_j(0). \tag{5.50}$$

The flow (f_t^K) determined by K is thus given by

$$f_t^K(\phi, \mathrm{I}) = (\omega(\mathrm{I})t + \phi, \mathrm{I}) \quad , \quad \omega = (\omega_1, \ldots, \omega_n); \tag{5.51}$$

it is called a "Kronecker flow'; ω is the "frequency vector".

We have already seen one example of a completely integrable Hamiltonian system, namely the n-dimensional harmonic oscillator discussed in the beginning of this subsection. More generally every Hamiltonian function which is a positive definite quadratic form in the position and momentum coordinates is integrable, and even admits global angle-action variables. Namely choose a symplectic matrix S such that $H \circ S$ has the normal form

$$H(Sz) = \sum_{j=1}^{n} \frac{\omega_j}{2}(p_j^2 + x_j^2) \tag{5.52}$$

(that this is possible is a consequence of Williamson's theorem, which we will discuss in Chapter 8). Setting $(X, P) = S(x, p)$ and $X_j = \sqrt{2\mathrm{I}_j} \cos\phi_j$, $P_j = \sqrt{2\mathrm{I}_j} \sin\phi_j$, the symplectomorphism $(x, p) \longmapsto (\phi_j, \mathrm{I}_j)$ again brings H into the form

$$K = \sum_{j=1}^{n} \omega_j \mathrm{I}_j. \tag{5.53}$$

5.3.3 Lagrangian tori

Let us now study the notion of Lagrangian torus attached to a completely integrable Hamiltonian system.

Consider again the Hamiltonian (5.53); the solutions of the associated Hamilton equations are

$$\dot{\phi}_j(t) = \omega_j t + \dot{\phi}_j(0) \quad , \quad \mathrm{I}_j(t) = \mathrm{I}_j(0) \quad (1 \leq j \leq n).$$

Let us now state a condition ensuring us the existence of angle-action variables and invariant tori. Here is an important classical result:

Theorem 5.33. *Assume that $F_1 = H, F_2, \ldots, F_n$ are n constants of the motion in involution on an open dense subset of $\mathbb{R}_z^{2n}$. Set $F = (F_1, F_2, \ldots, F_n)$ and assume that $F^{-1}(z_0)$ is a compact and connected n-dimensional manifold. Then:*

(i) *there exists a neighborhood $\mathcal{V}$ of $F^{-1}(0)$ in $\mathbb{R}^n$, an open subset $\mathcal{U}$ of $\mathbb{R}^n$ and a diffeomorphism*

$$f : (\mathbb{R}/2\pi\mathbb{Z})^n \times \mathcal{V} \longrightarrow \mathcal{U}$$

such that if $(x,p) = f(\phi, \mathrm{I})$, then

$$dp \wedge dx = d\mathrm{I} \wedge d\phi;$$

The diffeomorphism f is thus canonical.

(ii) *In the (ϕ, I) variables the Hamiltonian becomes $K(\mathrm{I}) = H(x,p)$ and the motion thus takes place on an n-dimensional submanifold $\mathbb{V}^n$ of $\mathbb{R}^{2n}_z$ such that*

$$f(\mathbb{V}^n) = \{\mathrm{I}_0\} \times (\mathbb{R}/2\pi\mathbb{Z})^n.$$

(*The motion thus takes place on a torus in (ϕ, I) phase space.*)

The manifold $F^{-1}(z_0)$ described in this theorem is topologically an n-dimensional torus $\mathbb{T}^n = (\mathbb{R}/2\pi\mathbb{Z})^n$ invariant under the flow determined by the Hamiltonian, and hence compact. It is a Lagrangian manifold. This can be seen directly, without reference to the angle-action variables as follows. Viewing each of the constants of the motion F_j as a Hamiltonian function in its own right, we denote by X_j the associated Hamiltonian vector field $J\partial_z F_j$, and note that we obviously have, since $\sigma(z,z') = \langle Jz, z'\rangle$:

$$\{F_j(z), F_k(z)\} = \sigma(X_j(z), X_k(z))$$

at every $z \in \mathbb{V}^n$. It follows that the involution conditions $\{F_j, F_k\} = 0$ are equivalent to

$$\sigma(X_j(z), X_k(z)) = 0$$

for every $z \in V$. Since the functions F_j are independent, the vector fields X_j span $\mathbb{V}^n$. It follows that for all pairs $Y(z)$, $Y'(z)$ of tangent vectors at $z \in \mathbb{V}^n$, the skew-product $\sigma(Y(z), Y'(z))$ can be expressed as a linear combination of the terms $\sigma(X_j(z), X_k(z))$ and hence $\sigma(Y(z), Y'(z)) = 0$, so that $\mathbb{V}^n$ is indeed a Lagrangian manifold.

The proof of Theorem 5.33 is long and technical; it can be found for instance in the references [1, 3, 27, 91]. The last statement in (ii) follows from formula (5.51) due to the identification of I-coordinates modulo 2π. The action variables I are constructed as follows: let γ_j form a basis for the 1-cycles on $\mathbb{T}^n$ and set

$$\mathrm{I}_j = \frac{1}{2\pi}\oint_{\gamma_j} p_j dx_j$$

for $1 \leq j \leq n$.

Note that Theorem 5.33 is in a sense a local statement, because it only guarantees the existence of angle-action angles in a neighborhood of $F^{-1}(z_0)$: the underlying phase space in the (ϕ, I) variables is $(\mathbb{R}/2\pi\mathbb{Z})^n \times \mathcal{U}$ where the neighborhood $\mathcal{U}$ is determined by the neighborhood $\mathcal{U}$ of $F^{-1}(z_0)$. It turns out that there are major obstructions for the construction of global angle-action variables (see the discussion in [27], Appendix D.2, and the references therein).

Remark 5.34. The energy being a constant of the motion, it follows that all systems with one degree of freedom are completely integrable.

5.4 Quantization of Lagrangian Manifolds

Physics describes the real world, and its domain of competence does this in two modes: *classical* (including Hamiltonian mechanics and relativity theory), and *quantum*. A classical problem (which might not be so well posed) is that of "quantization". Quantization refers to a variety of procedures of which two of the most important are operator quantization, and semi-classical quantization. While we will study in detail operator quantization in the forthcoming chapters, semi-classical quantization is more than just an approximate "poor man's quantum mechanics"; it is in fact a mathematical theory, or rather a collection of mathematical theories interesting in their own right. The number of books devoted to quantum mechanics defies the imagination; here are a few basic sources (in alphabetical order): Bohm [12] (from the point of view of one the great physicists of last century), Isham [96], Merzbacher [122], Messiah [123] (a classic, very readable by mathematicians), Park [129]. The rigorous mathematical foundations of quantum mechanics were pioneered by von Neumann [171] and Weyl [179]; see in this context the book [18] on the Schrödinger equation by Berezin and Shubin; another valuable book is Schechter [139]. A classical historical review is Jammer's book [97] (very complete). I also recommend the book [48] by Giacchetta *et al.* which contains an advanced treatment of up-to-date mathematical tools to be used in studying quantum problems.

We begin by discussing the Maslov quantization rules (Arnol'd [3], Leray [107], Maslov [119], Maslov–Fedoriuk [120]) for completely integrable Hamiltonian systems. The idea of these rules goes back to the pioneering work of Keller [102], elaborating on an idea of Einstein [37] (see Bergia and Navarro [8] and Gutzwiller [86] for a historical discussion of Einstein's idea which goes back to... 1917!).

5.4.1 The Keller–Maslov quantization conditions

The passage from classical to semiclassical mechanics consists in imposing selection rules on the Lagrangian manifolds $\mathbb{V}^n$; these rules are the Keller–Maslov quantization conditions:

$$\frac{1}{2\pi\hbar}\oint_\gamma pdx - \frac{1}{4}m(\gamma) \in \mathbb{Z} \text{ for all one-cycles } \gamma \text{ on } \mathbb{V}^n \tag{5.54}$$

where $m(\gamma)$ is the Maslov index. In the physical literature these conditions are often called the *EBK* (Einstein–Brillouin–Keller) quantum conditions, or the Bohr–Sommerfeld–Maslov conditions. Since the integral of the action form pdx only depends on the homotopy class of γ in $\mathbb{V}^n$ and the Maslov index is a homotopy invariant, the same is true of conditions (5.54).

Definition 5.35. When the conditions (5.54) hold, the Lagrangian manifold $\mathbb{V}^n$ is said to be "quantized" (in the sense of Keller and Maslov)

We will not justify the conditions (5.54) here; we will in fact see in Subsection 5.4.3 that they are equivalent to the existence of "waveforms" on the Lagrangian manifold $\mathbb{V}^n$. In Chapter 8 we will derive (5.54) using a topological argument involving the notion of symplectic capacity.

Remark 5.36. Souriau [156] has proven that if the Lagrangian manifold $\mathbb{V}^n$ is oriented then $m(\gamma)$ is even. We will re-derive this result in the more general context of q-oriented Lagrangian manifold in Subsection 5.4.2.

The semiclassical values of the energy corresponding to the Keller–Maslov conditions are obtained as follows: let $\mathrm{I} = (\mathrm{I}_1, \ldots, \mathrm{I}_n)$ be the action variables corresponding to the basic one-cycles $\gamma^1, \ldots, \gamma^n$ on $\mathbb{V}^n$. These are defined as follows: let $\bar{\gamma}^1, \ldots, \bar{\gamma}^n$ be the loops in $\mathbb{T}^n(R_1, \ldots, R_n)$ defined, for $0 \leq t \leq 2\pi$, by

$$\begin{aligned}
\bar{\gamma}^1(t) &= R_1(\cos t, 0, \ldots, 0; \sin t, 0, \ldots, 0),\\
\bar{\gamma}^2(t) &= R_2(0, \cos t, \ldots, 0; 0, \sin t, \ldots, 0),\\
&\cdots\cdots\cdots\cdots\cdots\cdots\cdots\cdots\\
\bar{\gamma}^n(t) &= R_n(0, \ldots, 0, \cos t; 0, \ldots, 0, \sin t).
\end{aligned}$$

The basic one-cycles $\gamma^1, \ldots, \gamma^1$ of $\mathbb{V}^n$ are then just

$$\gamma^1 = f^{-1}(\bar{\gamma}^1), \ldots, \gamma^n = f^{-1}(\bar{\gamma}^n).$$

The action variables being given by

$$\mathrm{I}_j = \frac{1}{2\pi} \oint_{\gamma^j} pdx \quad , \quad 1 \leq j \leq n,$$

the quantization conditions (5.54) imply that we must have

$$I_j = (N_j + \tfrac{1}{4} m(\gamma^j))\hbar \quad \text{for} \quad 1 \leq j \leq n, \tag{5.55}$$

each N_j being an integer ≥ 0. Writing $H(x,p) = K(I)$ the semiclassical energy levels are then given by the formula

$$E_{N_1,\ldots,N_n} = K((N_1 + \tfrac{1}{4} m(\gamma^1))\hbar, \ldots, (N_n + \tfrac{1}{4} m(\gamma^1))\hbar) \tag{5.56}$$

where $N_1, \ldots, N_n$ range over all *non-negative* integers; they correspond to the physical "quantum states" labeled by the sequence of integers $(N_1, \ldots, N_n)$.

Remark 5.37. We do not discuss here the ambiguity that might arise in calculation of the energy because of the non-uniqueness of the angle action coordinates; that ambiguity actually disappears if one requires that the system under consideration is non-degenerate, that is $\partial^2 K(\mathrm{I}) \neq 0$. (See Arnol'd [3]), Ch. 10, §52.)

Applying the Keller–Maslov quantization conditions (5.54) to a system with quadratic Hamiltonian, or even an atom placed in a constant magnetic field, gives the same energy levels as those predicted by quantum mechanics. However in general these conditions only lead to approximate values of observable quantities. This reflects the fact that they are semi-classical quantization conditions, obtained by imposing the quantum selection rule on a Lagrangian torus, which is an object associated to a classical system. We notice that Giacchetta *et al.* give in [47] a geometric quantization scheme for completely integrable Hamiltonian systems in the action-angle variables around an invariant torus with respect to polarizations spanned by almost-Hamiltonian vector fields of angle variables; also see these authors' recent book [48].

5.4.2 The case of q-oriented Lagrangian manifolds

Recall that we defined in Chapter 3, at the end of Section 3.3 devoted to q-symplectic geometry, the notion of a q-oriented Lagrangian plane: it is the datum of a pair $(\ell, [\lambda]_{2q})$ where $\ell \in \mathrm{Lag}(n)$ and $[\lambda]_{2q} \in \mathbb{Z}_{2q}$. For instance, if $q = 1$, to every $\ell \in \mathrm{Lag}(n)$ we can associate two pairs $(\ell, +)$, $(\ell, -)$ corresponding to the choices of a "positive" and of a "negative" orientation.

Let $\mathbb{V}^n$ be a (connected) Lagrangian manifold; we denote by $z \longmapsto \ell(z)$ the continuous mapping $\mathbb{V}^n \longrightarrow \mathrm{Lag}(n)$ which to every $z \in \mathbb{V}^n$ associates the tangent plane $\ell(z) \in \mathrm{Lag}(n)$ to $\mathbb{V}^n$ at z:

$$\ell(z) = T_z\mathbb{V}^n \quad , \quad z \in \mathbb{V}^n.$$

The notion of q-orientation makes sense for $\mathbb{V}^n$. Let

$$\pi^{\mathrm{Lag}_{2q}} : \mathrm{Lag}_{2q}(n) \longrightarrow \mathrm{Lag}(n)$$

be the covering of order $2q$ of $\mathrm{Lag}(n)$.

Definition 5.38. We say that the connected Lagrangian manifold $\mathbb{V}^n$ is "q-oriented" $(1 \leq q \leq \infty)$ if the mapping

$$\ell(\cdot) : \mathbb{V}^n \longrightarrow \mathrm{Lag}(n) \quad , \quad z \longmapsto \ell(z)$$

has a continuous lift

$$\ell_{2q}(\cdot) : \mathbb{V}^n \longrightarrow \mathrm{Lag}_{2q}(n) \quad , \quad z \longmapsto \ell_{2q}(z)$$

(*i.e.*, $\pi^{\mathrm{Lag}_{2q}}(\ell_{2q}(z)) = \ell(z)$ for every $z \in \mathbb{V}^n$). The datum of such a lift $\ell_{2q}(\cdot)$ is called a "q-orientation" of $\mathbb{V}^n$.

For instance every oriented Lagrangian manifold has exactly two 1-orientations, namely $z \longmapsto (\ell(z), +)$ and $z \longmapsto (\ell(z), -)$ where "+" and "−" correspond to the positive or negative orientation of all the tangent planes.

Example 5.39. The circle S^1 is oriented and hence 1-oriented. The same is true of the n-torus $\mathbb{T}^n = (S^1)^n$. $\mathbb{T}^n$ is not q-oriented if $q > 1$.

If the Lagrangian manifold $\mathbb{V}^n$ is simply connected, then it is ∞-orientable (and hence orientable). The converse is not true: in [64], *p.* 175–177 we have constructed, in relation with the quantization of the "plane rotator", an ∞-orientable Lagrangian manifold which is not simply connected; that manifold is the submanifold of $\mathbb{R}^4_{x,y,p_x,p_y}$ defined by the equations

$$x^2 + y^2 = r^2 \quad , \quad xp_y - yp_x = \mathcal{L}_0.$$

The following result provides us with a very useful test for deciding whether a given Lagrangian manifold is q-oriented for some integer q (or $+\infty$). It is also the key to the understanding of Maslov quantization for other Lagrangian manifolds than tori.

Theorem 5.40. *A connected Lagrangian manifold $\mathbb{V}^n$ is q-oriented" $(1 \leq q < \infty)$ if and only if the Maslov index $m_{\mathbb{V}^n}$ is such that*

$$m(\gamma) \equiv 0 \mod 2q \tag{5.57}$$

for every loop γ in $\mathbb{V}^n$. It is ∞-oriented if and only if $m(\gamma) = 0$ for every γ.

Proof. Choose a base point z_0 in $\mathbb{V}^n$ and set $\ell_0 = \ell(z_0)$. Let

$$\ell_* : \pi_1[\mathbb{V}^n, z_0] \longrightarrow \pi_1[\mathrm{Lag}(n), \ell_0]$$

be the group homomorphism induced by the continuous $\ell(\cdot) : \mathbb{V}^n \longrightarrow \mathrm{Lag}(n)$. Let $\ell_{0,2q} \in \mathrm{Lag}_{2q}(n)$ be such that $\pi^{\mathrm{Lag}_{2q}}(\ell_{0,2q}) = \ell_0$ and

$$\pi_*^{\mathrm{Lag}_{2q}} : \pi_1[\mathrm{Lag}_{2q}(n), \ell_{0,2q}] \longrightarrow \pi_1[\mathrm{Lag}(n), \ell_0]$$

is the homomorphism induced by $\pi^{\mathrm{Lag}_{2q}}$. The mapping $\ell(\cdot) : \mathbb{V}^n \longrightarrow \mathrm{Lag}(n)$ can be lifted to a mapping $\ell_{2q}(\cdot) : \mathbb{V}^n \longrightarrow \mathrm{Lag}_{2q}(n)$ if and only we have the inclusion

$$\ell_*(\pi_1[\mathbb{V}^n, z_0]) \subset \pi_*^{\mathrm{Lag}_{2q}}(\pi_1[\mathrm{Lag}_{2q}(n), \ell_{0,2q}]).$$

Let β_0 be the generator of $\pi_1[\mathrm{Lag}(n), \ell_0]$ whose image in $\mathbb{Z}$ is $+1$; we have

$$\pi_1[\mathrm{Lag}_{2q}(n), \ell_{0,2q}] = \{\beta_0^{2kq} : k \in \mathbb{Z}\},$$

hence the inclusion above is equivalent to saying that for every $\gamma \in \pi_1[\mathbb{V}^n, z_0]$ there exists $k \in \mathbb{Z}$ such that $m(\gamma) = 2kq$, that is (5.57). □

Specializing to the case $q = 1$ we immediately recover Souriau's result (Remark 5.36):

Corollary 5.41. *Assume that $\mathbb{V}^n$ is an orientable Lagrangian manifold; then $m(\gamma)$ is even.*

Proof. Immediate since orientability is equivalent to 1-orientability. □

5.4.3 Waveforms on a Lagrangian Manifold

The machinery developed above can be used to define a notion of phase space wavefunction for quantum systems associated to general Lagrangian manifolds, and generalizing Leray's [107] "Lagrangian functions". We will only sketch the approach here and refer to de Gosson [60, 62, 66, 64] for details.

We begin by reviewing the notion of de Rham forms and their square roots.

Let $\mathbb{V}^n$ be a connected manifold; we do not assume for the moment that it is Lagrangian. In a neighborhood $U \subset \mathbb{V}^n$ of every point z of $\mathbb{V}^n$ there are two orientations; using them we construct the two-sheeted oriented double covering of $\mathbb{V}^n$. Let $(U_\alpha, f_\alpha)_\alpha$ be a maximal atlas of $\mathbb{V}^n$; for α and β such that $U_\alpha \cap U_\beta \neq \emptyset$ we define a locally constant mapping

$$g_{\alpha\beta} : U_\alpha \cap U_\beta \longrightarrow \{-1, +1\}$$

by the formula

$$g_{\alpha\beta}(z) = \operatorname{sign}(\det D(f_\alpha f_\beta^{-1}))(f_\beta(z))$$

where $\operatorname{sign} u = u/|u|$ if $u \in \mathbb{R}$, $u \neq 0$. The mappings $g_{\alpha\beta}$ obviously satisfy the cocycle relation

$$g_{\alpha\beta}(z)g_{\beta\gamma}(z) = g_{\alpha\gamma}(z) \quad \text{for} \quad z \in U_\alpha \cap U_\beta \cap U_\gamma$$

and therefore define a two-sheeted covering $\pi^o : (\mathbb{V}^n)^o \longrightarrow \mathbb{V}^n$ with trivializations

$$\Phi_\alpha : (\pi^o)^{-1}(U_\alpha) \longrightarrow U_\alpha \times \{-1, +1\}$$

such that

$$\Phi_\alpha \Phi_\beta^{-1}(z, \pm 1) = (z, \pm g_{\alpha\beta}(z)) \quad \text{for} \quad z \in U_\alpha \cap U_\beta.$$

One proves (see for instance Godbillon [51], Ch. X, §5, or Abraham *et al.* [2], Ch. 7) that $(\mathbb{V}^n)^o$ is an orientable manifold and that:

- $\mathbb{V}^n$ is oriented if and only if $(\mathbb{V}^n)^o$ is trivial, that is $(\mathbb{V}^n)^o = \mathbb{V}^n \times \{-1, +1\}$;
- $(\mathbb{V}^n)^o$ is connected if and only if $\mathbb{V}^n$ is non-orientable.

These properties motivate the following definition:

Definition 5.42.

(i) The manifold $(\mathbb{V}^n)^o$ is called the "two-sheeted orientable covering" of $\mathbb{V}^n$.
(ii) A "de Rham form" (or: "twisted form") on $\mathbb{V}^n$ is a differential form on $(\mathbb{V}^n)^o$.
(iii) A "de Rham (or: twisted) volume form" is a volume form on $(\mathbb{V}^n)^o$.

Locally, in a neighborhood $U \subset \mathbb{V}^n$ of $z \in \mathbb{V}^n$, a de Rham form μ on $\mathbb{V}^n$ can be written $\mathrm{Orient}(U)\mu$ where $\mathrm{Orient}(U) = \pm 1$ is an orientation of the set U and μ a differential form on U.

From now on we again assume that $\mathbb{V}^n$ is a Lagrangian manifold coming equipped with a Rham form μ. In [64], Ch. 5, §5.3, we showed that the pull-back of μ to the universal covering $\pi : \check{\mathbb{V}}^n \longrightarrow \mathbb{V}^n$ can be written outside the caustic $\Sigma_{\mathbb{V}^n}$ as

$$\pi^*\mu(\check{z}) = (-1)^{m(\check{z})}\pi^*\rho(\check{z}) = (-1)^{m(\check{z})}\rho(z) \tag{5.58}$$

where ρ is a density on $\mathbb{V}^n$ and

$$m(\check{z}) = m(\ell_{P,\infty}, \ell_\infty(\check{z}))$$

corresponds to a choice of $\ell_{P,\infty}$ such that $\pi^{\mathrm{Lag}}(\ell_{P,\infty}) = \ell_P$ (m being the reduced *ALM* index: Definition 3.23, Chapter 3.2, Section 3.3). The mapping

$$\ell_\infty(\cdot) : \check{\mathbb{V}}^n \longrightarrow \mathrm{Lag}_\infty(n) \ \ , \ \ \check{z} \longrightarrow \ell_\infty(\check{z})$$

is a lift of the tangent mapping

$$\ell(\cdot) : \mathbb{V}^n \longrightarrow \mathrm{Lag}(n) \ \ , \ \ z \longmapsto \ell(z) = T_z\mathbb{V}^n.$$

Formula (5.58) allows us to define the square root of the pulled-back de Rham form outside the caustic by

$$\sqrt{\pi^*\mu}(\check{z}) = i^{m(\check{z})}\sqrt{\rho}(z)$$

where $\sqrt{\rho}$ is a choice of half-density corresponding to the density ρ. This motivates the following definition:

Definition 5.43. A "waveform" on $\check{\mathbb{V}}^n$ is the family $(\Psi_{\ell_{\alpha,\infty}})$ where

$$\Psi_{\ell_{\alpha,\infty}}(\check{z}) = e^{\frac{i}{\hbar}\varphi(\check{z})} i^{m_\alpha(\check{z})}\sqrt{\rho}(z)$$

with $\ell_{\alpha,\infty} \in \mathrm{Lag}_\infty(n)$ and $m_\alpha(\check{z}) = m(\ell_{\alpha,\infty}, \ell_\infty(\check{z}))$, m being the reduced ALM index.

Exercise 5.44. Show that the expressions Ψ_{ℓ_α} and Ψ_{ℓ_β} corresponding to choices $\ell_\alpha, \ell_\beta \in \mathrm{Lag}(n)$ are related by the formula $\Psi_{\ell_\alpha} = i^{m_{\alpha\beta}(\cdot)}\Psi_{\ell_\beta}$ where

$$m_{\alpha\beta}(z) = m(\ell_{\alpha,\infty}, \ell_{\beta,\infty}) - \mathrm{Inert}(\ell_\alpha, \ell_\beta, \ell(z)).$$

Waveforms are defined on the universal covering $\check{\mathbb{V}}^n$ of $\mathbb{V}^n$; to use an older terminology they are "multi-valued" on $\mathbb{V}^n$. They are single-valued if and only if $\mathbb{V}^n$ is a quantized Lagrangian manifold in the sense of Definition 5.35.

Proposition 5.45. *The Lagrangian manifold $\mathbb{V}^n$ satisfies the Keller–Maslov quantization conditions*

$$\frac{1}{2\pi\hbar}\oint_\gamma pdx - \frac{1}{4}m(\gamma) \in \mathbb{Z} \tag{5.59}$$

for every $\gamma \in \pi_1[\mathbb{V}^n, z_0]$ (z_0 an arbitrary point of $\check{\mathbb{V}}^n$) if and only if

$$\Psi_{\ell_\alpha}(\gamma\check{z}) = \Psi_{\ell_\alpha}(\check{z})$$

for every γ, that is, if and only if Ψ_{ℓ_α} is defined on $\mathbb{V}^n$.

Proof. By definition of Ψ_{ℓ_α} we have

$$\Psi_{\ell_\alpha}(\gamma\check{z}) = e^{\frac{i}{\hbar}\varphi(\gamma\check{z})} i^{m_\alpha(\gamma\check{z})}\sqrt{\mu}(z).$$

In view of formula (5.5) in Subsection 5.1.2 we have

$$\varphi(\gamma\check{z}) = \varphi(\check{z}) + \oint_\gamma pdx.$$

On the other hand, using property (3.36) of the reduced ALM index (Proposition 3.24, Subsection 3.3.1) and the fact that $\ell_\infty(\gamma\check{z}) = \beta^{m(\gamma)}\ell_\infty(\check{z})$, we have

$$\begin{aligned} m_\alpha(\gamma\check{z}) &= m(\ell_{\alpha,\infty}, \ell_\infty(\gamma\check{z})) \\ &= m(\ell_{\alpha,\infty}, \beta^{m(\gamma)}\ell_\infty(\check{z})) \\ &= m(\ell_{\alpha,\infty}, \ell_\infty(\check{z})) - m(\gamma), \end{aligned}$$

hence $\Psi_{\ell_\alpha}(\gamma\check{z}) = \Psi_{\ell_\alpha}(\check{z})$ is equivalent to the quantization condition (5.59). □

The time-evolution of a waveform under the action of Hamiltonian flows is defined as follows: let H be a Hamiltonian function (possibly time-dependent), and (f_t^H) the flow determined by the associated Hamiltonian equations. Then

$$f_t^H\Psi_{\ell_\alpha}(\hat{z}) = \Psi_{\ell_\alpha}(\hat{z}, t) \tag{5.60}$$

where

$$\Psi_{\ell_\alpha}(\hat{z}, t) = e^{\frac{i}{\hbar}\varphi(\check{z},t)} i^{m_\alpha(\check{z},t)}\sqrt{(f_t^H)_*\rho(z)}$$

and:

- $\varphi(\check{z}, t)$ is the phase of the Lagrangian manifold $f_t^H(\mathbb{V}^n)$:

$$\varphi(\check{z}, t) = \varphi(\check{z}') + \int_{z',0}^{z,t} pdx - Hdt \tag{5.61}$$

  with $z = f_t^H(z')$ (formula (5.33));

- $m_\alpha(\check{z}, t)$ is the integer

$$m_\alpha(\check{z}, t) = m(\ell_{\alpha,\infty}, S_t(z)_\infty \ell_\infty(\check{z})) \tag{5.62}$$

where $S_t(z)_\infty$ is the homotopy class in $\mathrm{Sp}(n)$ of the symplectic path

$$t' \longmapsto S_{t'}(z) = D(f_{t'}^H)(z),$$

$0 \leq t' \leq t$;

- $(f_t^H)_*\rho$ is the push-forward of the density ρ by the diffeomorphism f_t^H:

$$(f_t^H)_*\rho(z)(Z_1, \ldots, Z_n) = \rho((f_t^H)^{-1}z)(S_t(z)^{-1}Z_1, \ldots, S_t(z)^{-1}Z_n)$$

for tangent vectors $(Z_1, \ldots, Z_n)$ to $\mathbb{V}^n$ at z.

The motion of waveforms defined above is semi-classical; see de Gosson [60, 62] for a discussion of the relation between waveforms with approximate solutions of Schrödinger's equation.

5.5 Heisenberg–Weyl and Grossmann–Royer Operators

The Weyl–Heisenberg operators are the quantum-mechanical analogues of the phase space translation operators $T(z_0) : z \longmapsto z + z_0$. As such, they translate functions (in fact waveforms), but at the same time they change their phase. Even though their use is quantum-mechanical, their definition only makes use of the notion of phase of a Lagrangian manifold. In the two following subsections we study the most elementary properties of the Heisenberg–Weyl operators; in Subsection 5.5.3 we describe a class of related operators, whose definition apparently goes back to Grossmann [82] and Royer [137] and which were exploited by Gracia-Bondia [77]. Besides the fact that the Grossmann–Royer operators sometimes considerably simplify calculations in Weyl calculus, they have an intrinsic interest in quantization (and dequantization) problems.

5.5.1 Definition of the Heisenberg–Weyl operators

Let $z_a = (x_a, p_a)$ be an arbitrary point[4] of the standard symplectic phase space $(\mathbb{R}_z^{2n}, \sigma)$. The translation operator

$$T(z_a) : z \longmapsto z + z_a$$

can be viewed as the time-one map of a Hamiltonian flow. Consider the Hamiltonian function

$$H(z) = \sigma(z, z_a); \tag{5.63}$$

[4] We will use for a while the subscript a in place of 0 to avoid confusion with the basepoint of the Lagrangian manifold.

the associated Hamilton equations

$$\dot{x} = \partial_p \sigma(x, p; x_a, p_a) \quad , \quad \dot{p} = -\partial_x \sigma(x, p; x_a, p_a)$$

reduce to the simple form $\dot{x} = x_a$, $\dot{p} = p_a$ and the solution $t \longmapsto z(t)$ passing through a point z at time $t = 0$ is thus given by

$$z(t) = f_t^H(z) = z + tz_a.$$

The translation $T(z_a)$ is thus the affine symplectic mapping associating to each z the point $z(1) = f_1^a(z)$ where (f_t^a) denotes the flow determined by the Hamiltonian (5.63).

The following result describes the phase of the Langrangian manifold $T(z_a)(\mathbb{V}^n) = f_1^a(\mathbb{V}^n)$. To simplify notation we assume that $\mathbb{V}^n$ is simply connected so that it coincides with its universal covering $\check{\mathbb{V}}^n$.

Proposition 5.46. *Let (f_t^a) be the flow determined by the Hamiltonian $H^a(z) = \sigma(z, z_a)$.*

(i) *The Hamiltonian phase of $f_t^a(\mathbb{V}^n)$ is*

$$\varphi_{a,t}(z) = \varphi(z - tz_a) + t \langle p_a, x \rangle - \tfrac{1}{2} t^2 \langle p_a, x_a \rangle \tag{5.64}$$

where $z = (x, p)$, hence

(ii) *The phase φ_a of $f_1^a(\mathbb{V}^n) = T(z_a)\mathbb{V}^n$ is*

$$\varphi_a(\check{z}) = \varphi(z - tz_a) + \langle p_a, x \rangle - \tfrac{1}{2} \langle p_a, x_a \rangle . \tag{5.65}$$

Proof. We have, since the energy has constant value $\sigma(z, z_a)$ along the trajectory,

$$\begin{aligned}
\varphi_a(z) &= \varphi(z - tz_a) + \int_0^t \langle p + (s-t)p_a, x_a \rangle \, ds - \int_0^t \sigma(z, z_a) ds \\
&= \varphi(z - tz_a) + t \langle p, x_a \rangle - \tfrac{1}{2} t^2 \langle p_a, x_a \rangle - t(\langle p, x_a \rangle - \langle p_a, x \rangle) \\
&= \varphi(z - tz_a) + t \langle p_a, x \rangle + \tfrac{1}{2} t^2 \langle p_a, x_a \rangle ,
\end{aligned}$$

whence (5.65). □

Exercise 5.47. Let $z_a = (x_a, p_a)$ and $z_b = (x_b, p_b)$ be in $\mathbb{R}_z^{2n}$. Let $\varphi_{a,b}$ be the phase of $T(z_a)(T(z_b)\mathbb{V}^n)$ and φ_{a+b} that of $T(z_a + z_b)\mathbb{V}^n$. Calculate $\varphi_{a,b}(z) - \varphi_{a+b}(z)$ and $\varphi_{a,b}(z) - \varphi_{b,a}(z)$.

Exercise 5.48. Compare the Hamiltonian phases of the identical Lagrangian manifolds $S_t^H(T(z_a)\mathbb{V}^n)$ and $T(S_t^H(z_a))\mathbb{V}^n$.

Let now $\psi_{\ell_{\alpha,\infty}}$ be the expression of a waveform on a Lagrangian manifold $\mathbb{V}^n$; for simplicity we assume again that $\mathbb{V}^n$ is simply connected, so that $\psi_{\ell_{\alpha,\infty}}$ is defined on $\mathbb{V}^n$ itself:

$$\Psi_{\ell_{\alpha,\infty}}(z) = e^{\frac{i}{\hbar}\varphi(z)} i^{m_\alpha(z)} \sqrt{\rho}(z).$$

In view of formulae (5.60), (5.61), (5.62) the action of the flow (f_t^0) determined by the translation Hamiltonian (5.63) is given by

$$f_t^H \Psi_{\ell_{\alpha,\infty}}(z) = e^{\frac{i}{\hbar}\varphi(z,t)} i^{m_\alpha(z,t)} \sqrt{\rho}(z - tz_a)$$

where $m_\alpha(z,t) = m_\alpha(z)$ (because the Jacobian matrix of a translation is the identity) and the Hamiltonian phase $\varphi(\check{z},t)$ is given by

$$\varphi(z,t) = \varphi(z - tz_a) + t\langle p_a, x\rangle - \tfrac{1}{2}t^2\langle p_a, x_a\rangle$$

in view of formula (5.64) in Proposition 5.46. We thus have the simple formula

$$f_t^H \Psi_{\ell_{\alpha,\infty}}(z) - e^{\frac{i}{\hbar}(t\langle p_a,x\rangle - \frac{1}{2}t^2\langle p_a,x_a\rangle)} T(tz_a)\Psi_{\ell_{\alpha,\infty}}(z)$$

which yields, for $t = 1$,

$$f_1^H \Psi_{\ell_{\alpha,\infty}}(z) = e^{\frac{i}{\hbar}(\langle p_a,x\rangle - \frac{1}{2}\langle p_a,x_a\rangle)} T(z_a)\Psi_{\ell_{\alpha,\infty}}(z) \tag{5.66}$$

and motivates the following definition:

Definition 5.49. Let Ψ be an arbitrary function defined on $\mathbb{R}_z^{2n}$; the operator $\widehat{T}(z_a)$ defined by

$$\widehat{T}(z_a)\Psi(z) = e^{\frac{i}{\hbar}(\langle p_a,x\rangle - \frac{1}{2}\langle p_a,x_a\rangle)}(T(z_a)\Psi)(z)$$

is called the Weyl–Heisenberg operator associated to $z_a = (x_a, p_a)$.

Since we obviously have

$$|\widehat{T}(z_a)\Psi(z)| = |\Psi(z - z_a)|,$$

the Heisenberg–Weyl operators are unitary in $L^2(\mathbb{R}_x^n)$:

$$||\widehat{T}(z_a)\Psi||_{L^2(\mathbb{R}_z^{2n})} = ||\Psi||_{L^2(\mathbb{R}_z^{2n})} \text{ for } \psi \in L^2(\mathbb{R}_z^{2n}).$$

We will come back to this phase-space picture of the Heisenberg–Weyl operators when we study the Schrödinger equation in phase space in Chapter 10.

5.5.2 First properties of the operators $\widehat{T}(z)$

While ordinary translation operators obviously form an abelian group,

$$T(z_a)T(z_b) = T(z_b)T(z_a) = T(z_a + z_b),$$

this is not true of the Weyl–Heisenberg operators, because they do not commute. The following important result actually only reflects the properties of the Hamiltonian phase (*cf.* Exercise 5.47):

Proposition 5.50. *The Heisenberg–Weyl operators satisfy the relations*

$$\widehat{T}(z_a)\widehat{T}(z_b) = e^{\frac{i}{\hbar}\sigma(z_a,z_b)}\widehat{T}(z_b)\widehat{T}(z_a), \tag{5.67}$$

$$\widehat{T}(z_a+z_b) = e^{-\frac{i}{2\hbar}\sigma(z_a,z_b)}\widehat{T}(z_a)\widehat{T}(z_b), \tag{5.68}$$

for all $z_a, z_b \in \mathbb{R}^{2n}_z$.

Proof. Let us prove (5.67). We have

$$\begin{aligned}\widehat{T}(z_a)\widehat{T}(z_b) &= \widehat{T}(z_a)(e^{\frac{i}{\hbar}(\langle p_b,x\rangle - \frac{1}{2}\langle p_b,x_b\rangle)}T(z_b)) \\ &= e^{\frac{i}{\hbar}(\langle p_a,x\rangle - \frac{1}{2}\langle p_a,x_a\rangle)}e^{\frac{i}{\hbar}(\langle p_b,x-x_a\rangle - \frac{1}{2}\langle p_b,x_b\rangle)}T(z_a+z_b)\end{aligned}$$

and, similarly

$$\widehat{T}(z_b)\widehat{T}(z_a) = e^{\frac{i}{\hbar}(\langle p_b,x\rangle - \frac{1}{2}\langle p_b,x_b\rangle)}e^{\frac{i}{\hbar}(\langle p_a,x-x_b\rangle - \frac{1}{2}\langle p_a,x_a\rangle)}T(z_a+z_b).$$

Defining the quantities

$$\begin{aligned}\Phi &= \langle p_a,x\rangle - \tfrac{1}{2}\langle p_a,x_a\rangle + \langle p_b,x-x_a\rangle - \tfrac{1}{2}\langle p_b,x_b\rangle, \\ \Phi' &= \langle p_b,x\rangle - \tfrac{1}{2}\langle p_b,x_b\rangle + \langle p_a,x-x_b\rangle - \tfrac{1}{2}\langle p_a,x_a\rangle,\end{aligned}$$

we have

$$\widehat{T}(z_a)\widehat{T}(z_b) = e^{\frac{i}{\hbar}(\Phi-\Phi')}\widehat{T}(z_b)\widehat{T}(z_a)$$

and an immediate calculation yields

$$\Phi - \Phi' = \langle p_a,x_b\rangle - \langle p_b,x_a\rangle = \sigma(z_a,z_b)$$

which proves (5.67). Let us next prove formula (5.68). We have

$$\widehat{T}(z_a+z_b) = e^{\frac{i}{\hbar}\Phi''}T(z_a+z_b)$$

with

$$\Phi'' = \langle p_a+p_b,x\rangle - \tfrac{1}{2}\langle p_a+p_b,x_a+x_b\rangle.$$

On the other hand we have seen above that

$$\widehat{T}(z_a)\widehat{T}(z_b) = e^{\frac{i}{\hbar}\Phi}T(z_a+z_b)$$

so that

$$\widehat{T}(z_a+z_b) = e^{\frac{i}{\hbar}(\Phi''-\Phi)}\widehat{T}(z_a)\widehat{T}(z_b).$$

A straightforward calculation shows that

$$\Phi'' - \Phi = \tfrac{1}{2}\langle p_b,x_a\rangle - \tfrac{1}{2}\langle p_a,x_b\rangle = -\tfrac{1}{2}\sigma(z_a,z_b),$$

hence formula (5.68). □

The Heisenberg–Weyl operators are an easy bridge to quantum mechanics[5] – or, at least, to Schrödinger's equation. Consider the equation

$$i\hbar\frac{\partial\psi}{\partial t} = \widehat{\sigma}(\hat{z}, z_a)\psi \tag{5.69}$$

where the operator $\widehat{\sigma}(\hat{z}, z_a)$ is obtained by replacing formally $z = (x,p)$ by $\hat{z} = (x, -i\hbar\partial_x)$ in $\sigma(z, z_a)$:

$$\widehat{\sigma}(\hat{z}, z_a) = \langle -i\hbar\partial_x, x_a\rangle - \langle p_a, x\rangle\,. \tag{5.70}$$

This is the "Schrödinger equation associated via the Weyl correspondence to the Hamiltonian function $\sigma(z, z_a)$" (the Weyl correspondence will be studied in detail in Chapter 6). Formally, the solution of (5.69) is given by

$$\psi(x,t) = e^{\frac{i}{\hbar}t\sigma(z_a,\hat{z})}\psi_0(x) \quad , \quad \psi(x,0) = \psi_0(x); \tag{5.71}$$

one easily checks by a direct calculation that this solution is explicitly given by the formula

$$\psi(x,t) = \exp\left[\tfrac{i}{\hbar}(t\langle p_a, x\rangle - \tfrac{1}{2}t^2\langle p_a, x_a\rangle)\right]\psi_0(x - tx_a) \tag{5.72}$$

which we can as well rewrite as

$$\psi(x,t) = \widehat{T}(tz_a)\psi(z). \tag{5.73}$$

We will come back to this relationship between the Heisenberg–Weyl operators in the next chapter, where we will begin to let them act – in conformity with tradition! – on functions defined on position space $\mathbb{R}^n_x$, even though the definition we have given (and which is not standard) shows that their true vocation is to act on phase space objects (functions, or waveforms, for instance). This fact will be exploited later in this book to construct a quantum mechanics in phase space.

5.5.3 The Grossmann–Royer operators

For $z_0 = (x_0, p_0) \in \mathbb{R}^{2n}_z$ and a function ψ on $\mathbb{R}^n_x$ we define

$$\widetilde{T}(z_0)\psi(x) = e^{\frac{2i}{\hbar}\langle p_0, x - x_0\rangle}\psi(2x_0 - x). \tag{5.74}$$

The operator $\widetilde{T}(z_0)$ is clearly linear and unitary: if $\psi \in L^2(\mathbb{R}^n_x)$ then

$$||\widetilde{T}(z_0)\psi||_{L^2(\mathbb{R}^n_x)} = ||\psi||_{L^2(\mathbb{R}^n_x)}.$$

Definition 5.51. The unitary operators $\widetilde{T}(z_0)$ on $L^2(\mathbb{R}^n_x)$ are called the Grossmann–Royer operators; note that $\widetilde{T}(0)$ is just the reflection operator $\psi \longmapsto \psi^\vee$ where $\psi^\vee(x) = \psi(-x)$.

[5] Another such bridge is provided by the (related) metaplectic representation of the symplectic group we will study in Chapter 7.

Each of the operators $\widetilde{T}(z_0)$ can be obtained from the reflection operator $\widetilde{T}(0)$ using conjugation by Heisenberg–Weyl operators:

Proposition 5.52. *The intertwining relation*

$$\widetilde{T}(z_0) = \widehat{T}(z_0)\widetilde{T}(0)\widehat{T}(z_0)^{-1} \tag{5.75}$$

holds for every $z_0 \in \mathbb{R}^{2n}_z$.

Proof. Let us establish the equivalent relation $\widetilde{T}(z_0)\widehat{T}(z_0) = \widehat{T}(z_0)\widetilde{T}(0)$. We have

$$\begin{aligned}
\widetilde{T}(z_0)\widehat{T}(z_0)\psi(x) &= \widetilde{T}(z_0)\left[e^{\frac{i}{\hbar}(\langle p_0,x\rangle - \frac{1}{2}\langle p_0,x_0\rangle)}\psi(x-x_0)\right] \\
&= e^{\frac{2i}{\hbar}\langle p_0,x-x_0\rangle}e^{\frac{i}{\hbar}(\langle p_0,2x_0-x\rangle-\frac{1}{2}\langle p_0,x_0\rangle)}\psi(x_0-x) \\
&= e^{\frac{i}{\hbar}(\langle p_0,x\rangle - \frac{1}{2}\langle p_0,x_0\rangle)}\psi(x_0-x) \\
&= \widehat{T}(z_0)\widetilde{T}(0)\psi(x).
\end{aligned}$$

□

The fact that the product of two reflections is a translation has its counterpart in terms of the Grossmann–Royer operators:

Proposition 5.53. *The Grossmann–Royer operators satisfy the product formula*

$$\widetilde{T}(z_a)\widetilde{T}(z_b) = e^{-\frac{2i}{\hbar}\sigma(z_a,z_b)}\widehat{T}(2(z_a-z_b)) \tag{5.76}$$

for all $z_a, z_b \in \mathbb{R}^{2n}_z$; *in particular* $\widetilde{T}(z)$ *is an involution.*

Proof. We have

$$\begin{aligned}
\widetilde{T}(z_a)\widetilde{T}(z_b)\psi(x) &= \widetilde{T}(z_a)\left[e^{\frac{2i}{\hbar}\langle p_b,2x_a-x-x_b\rangle}\psi(2x_b-x)\right] \\
&= e^{\frac{2i}{\hbar}\langle p_a,x-x_a\rangle}e^{\frac{2i}{\hbar}\langle p_b,x-x_b\rangle}\psi(2x_b-(2x_a-x)) \\
&= e^{\frac{i}{\hbar}\Phi}\psi(x-2(x_a-x_b))
\end{aligned}$$

with

$$\Phi = 2((p_a-p_b)x - p_a x_a - p_b x_b + 2p_b x_a).$$

On the other hand

$$\widehat{T}(2(z_a-z_b))\psi(x) = e^{\frac{i}{\hbar}\Phi'}\psi(x-2(x_a-x_b))$$

with

$$\Phi' = 2((p_a-p_b)x - (p_a-p_b)(x_a-x_b)).$$

We have $\Phi - \Phi' = -2\sigma(z_a,z_b)$, hence the result. □

Exercise 5.54. Express the product $\widetilde{T}(z_a)\widetilde{T}(z_b)\widetilde{T}(z_c)$ in terms of the Heisenberg–Weyl operator $\widehat{T}(z_a - z_b + z_c)$.

Let us now proceed to the next chapter, where the study of the Heisenberg–Weyl operators is taken up from an algebraic point of view.

Chapter 6

Heisenberg Group and Weyl Operators

The algebraic approach to the Heisenberg group we outline in the first section goes back to Weyl's work [179] on the applications of group theory to quantum mechanics, a precursor of which is Heisenberg's "matrix mechanics,[1]" which was rigorously presented by Born and Jordan in [13]. It is remarkable that such an *ad hoc* inductive argument has led to one of the most important developments in mathematical physics, with ramifications in many areas of pure mathematics. For explicit applications of the Heisenberg group to various problems in information theory see Binz *et al.* [9, 10], Schempp [140, 141]. Gaveau *et al.* [45] study the symbolic calculus on the Heisenberg group; also see the books by Folland [42] or Stein [158] where interesting material can be found.

The second part of this chapter is devoted to the definition and study of the notion of Weyl pseudo-differential calculus. Due to practical limitations we do not pretend to give a full account of the Weyl pseudo-differential calculus; the interested reader is referred to the (vast) literature on the subject. Some references for Weyl calculus I have in mind are Leray [107], Trèves [164], Wong [181]. For recent advances and a new point of view see Unterberger's book [167]. For modern aspects of pseudo-differential calculus in general, including a thorough study of the fascinating topic of pseudo-differential operators on manifolds with conical singularities the reader is referred to Schulze [144, 145, 146] and Egorov and Schulze [36].

[1] It was during a time dubbed "Hexenmathematik" – which is German for "witch mathematics" – because of its non-commutativity, which was a quite unfamiliar feature for physicists at that time.

6.1 Heisenberg Group and Schrödinger Representation

Born and Jordan's argument mentioned above went as follows: noting that Hamilton's equations can be written in terms of the Poisson brackets as

$$\dot{x}_j = \{x_j, H\} \ , \ \dot{p}_j = \{p_j, H\}, \tag{6.1}$$

we choose matrices

$$X_1, X_2, \ldots, X_n; P_1, P_2, \ldots, P_n$$

satisfying the "canonical commutation relations"

$$[X_i, X_j] = 0 \ , \ [P_i, P_j] = 0 \ , \ [X_i, P_j] = i\hbar\delta_{ij}I. \tag{6.2}$$

Assuming that the "matrix Hamiltonian"

$$\tilde{H} = H(X_1, \ldots, X_n; P_1, \ldots, P_n)$$

is unambiguously defined, one postulates that, in analogy with (6.1), the time-evolution of the matrices X_i, P_j is determined by the equations

$$\dot{X}_j = [X_j, \tilde{H}] \ , \ \dot{P}_j = [P_j, \tilde{H}].$$

It turns out that the relations (6.2), viewed as commutation relations for *operators*, lead to the definition of a Lie algebra, which is called the *Heisenberg algebra* in the literature. We will see that it is the Lie algebra of a group closely related to the Heisenberg–Weyl operators defined at the end of the last chapter.

6.1.1 The Heisenberg algebra and group

The Heisenberg group is a simple mathematical object; its interest in quantization problems comes from the fact that its Lie algebra reproduces in abstract form the canonical commutation relations. We will see that there is a unitary representation of the Heisenberg group in $L^2(\mathbb{R}^n_x)$, commonly called the *Schrödinger representation*, which leads to an important version of pseudo-differential calculus, the *Weyl calculus.* (In Chapter 10 we will study another representation of the Heisenberg group, leading to quantum mechanics in phase space.)

Consider the textbook quantum operators $\widehat{X_j}$, $\widehat{P_j}$ on $\mathcal{S}(\mathbb{R})$ defined, for $\psi \in \mathcal{S}(\mathbb{R}^n_x)$, by

$$\widehat{X}_j\psi = x_j\psi \quad , \quad \widehat{P}_j\psi = -i\hbar\frac{\partial\psi}{\partial x_j}.$$

These operators satisfy the commutation relations $[\widehat{X}_i, \widehat{X}_j] = [\widehat{P}_i, \widehat{P}_j] = 0$ and $[\widehat{X}_i, \widehat{P}_j] = i\hbar\delta_{ij}I$, justifying the following general definition:

Definition 6.1. A "*Heisenberg algebra*" is a Lie algebra $\mathfrak{h}_n$ with a basis $\{\widehat{X}_1,\ldots,\widehat{X}_n; \widehat{P}_1,\ldots,\widehat{P}_n; \widehat{T}\}$ satisfying the "canonical commutation relations"[2]

$$\begin{gathered}[\widehat{X}_i,\widehat{X}_j]=0\ ,\ \ [\widehat{P}_i,\widehat{P}_j]=0\ \ ,\ \ [\widehat{X}_i,\widehat{P}_j]=\delta_{ij}\widehat{T}\\ [\widehat{X}_i,\widehat{T}]=0\ \ ,\ \ [\widehat{P}_i,\widehat{T}]=0\end{gathered} \tag{6.3}$$

for $1\leq i,j\leq n$; $\hbar$ is a constant identified in Physics with Planck's constant h divided by 2π.

In the example above the operator $\widehat{T}$ corresponds to multiplication by $i\hbar$: $\widehat{T}\psi=i\hbar\psi$; it plays the role of *time*.

The canonical commutation relations (6.3) are strongly reminiscent of the formulae (1.9) defining a symplectic basis (see Subsection 1.1.2 of Chapter 1). It is in fact easy to make the link with symplectic geometry: writing $z=(x,p)$ and

$$\widehat{U}=\sum_{i=1}^{n}x_i\widehat{X}_i+p_i\widehat{P}_i+t\widehat{T}\ \ ,\ \ \widehat{U}'=\sum_{i=1}^{n}x_i'\widehat{X}_i+p_i'\widehat{P}_i+t'\widehat{T}$$

the relations (6.3) are immediately seen to be equivalent to

$$[\widehat{U},\widehat{U}']=\sigma(z,z')\widehat{T}. \tag{6.4}$$

Let us now describe the simply connected Lie group $\mathbf{H}_n$ corresponding to the Lie algebra in $\mathfrak{h}_n$. Since all Lie brackets of length superior to 2 vanish, the Baker–Campbell–Hausdorff formula (see Appendix A) shows that if $\widehat{U}$ and $\widehat{U}'$ are sufficiently close to zero, then we have

$$\exp(\widehat{U})\exp(\widehat{U}')=\exp(\widehat{U}+\widehat{U}'+\tfrac{1}{2}[\widehat{U},\widehat{U}']),$$

that is, in view of (6.4),

$$\exp(\widehat{U})\exp(\widehat{U}')=\exp(\widehat{U}+\widehat{U}'+\tfrac{1}{2}\sigma(z,z')\widehat{T}). \tag{6.5}$$

The exponential being a diffeomorphism of a neighborhood $\mathcal{U}$ of zero in $\mathfrak{h}_n$ onto a neighborhood of the identity in $\mathbf{H}_n$, we can identify $\widehat{U},\widehat{U}'$, for small z,z',t,t', with the exponentials $\exp(\widehat{U})$ and $\exp(\widehat{U}')$, and $\exp(\widehat{U})\exp(\widehat{U}')$ with the element $(z,t)\maltese(z',t)$ of $\mathbb{R}_z^{2n}\times\mathbb{R}_t$ defined by

$$(z,t)\maltese(z',t)=(z+z',t+t'+\tfrac{1}{2}\sigma(z,z')). \tag{6.6}$$

Clearly $(0,0)$ is a unit for the composition law $\maltese$, and each (z,t) is invertible with inverse $(-z,-t)$. One verifies that the law $\maltese$ is associative, so it defines a group structure on $\mathbb{R}_z^{2n}\times\mathbb{R}_t$, identifying $\mathbf{H}_n$ with the set $\mathbb{R}_z^{2n}\times\mathbb{R}_t$.

[2]Often abbreviated to "CCR" in the physical literature.

Definition 6.2. The set $\mathbb{R}^{2n}_z \times \mathbb{R}_t$ equipped with the group law

$$(z,t)\maltese(z',t) = (z+z', t+t'+\tfrac{1}{2}\sigma(z,z'))$$

is called the $(2n+1)$-dimensional Heisenberg group[3] $\mathbf{H}_n$.

At this point we remark that some authors define the Heisenberg group as being the set $\mathbb{R}^{2n}_z \times S^1$ equipped with the law

$$(z,u)(z',u') = (zz', uu'e^{\frac{i}{2}\sigma(z,z')}). \tag{6.7}$$

It is immediate to check that $\mathbf{H}_n$ is the universal covering group of this "exponentiated" version $\mathbf{H}^{\exp}_n$ of $\mathbf{H}_n$; since the covering groups of a Lie group all have the same Lie algebra as the group itself, the Lie algebra of $\mathbf{H}^{\exp}_n$ indeed is $\mathfrak{h}_n$.

There is a useful identification of $\mathbf{H}_n$ with a subgroup $\mathbf{H}^{\text{pol}}_n$ of $\mathrm{GL}(2n+2,\mathbb{R})$. That group (the "polarized Heisenberg group") consists of all $(2n+2)\times(2n+2)$ upper-triangular matrices of the type

$$M(z,t) = \begin{bmatrix} 1 & p_1 & \cdots & p_n & t \\ 0 & 1 & \cdots & 0 & x_1 \\ \vdots & \vdots & \vdots & \vdots & \vdots \\ 0 & 0 & \cdots & 1 & x_n \\ 0 & 0 & \cdots & 0 & 1 \end{bmatrix}$$

(the entries of the principal diagonal are all equal to 1); we will write these matrices for short as

$$M(z,t) = \begin{bmatrix} 1 & p^T & t \\ 0 & 1 & x \\ 0 & 0 & 1 \end{bmatrix}.$$

A simple calculation shows that the determinant of $M(z,t)$ is 1 and that its inverse is

$$M(z,t)^{-1} = \begin{bmatrix} 1 & -p^T & -t+\langle p,x\rangle \\ 0 & 1 & -x \\ 0 & 0 & 1 \end{bmatrix};$$

moreover

$$M(z,t)M(z',t') = M(z+z', t+t'+\langle p,x'\rangle). \tag{6.8}$$

Exercise 6.3.

(i) Show that the mapping $\phi : \mathbf{H}^{\text{pol}}_n \longrightarrow \mathbf{H}_n$ defined by

$$\phi(M(z,t)) = (z, t-\tfrac{1}{2}\langle p,x\rangle) \tag{6.9}$$

is a group isomorphism.

[3]It is sometimes also called the *Weyl group*, especially in the physical literature.

(ii) Show that the Lie algebra $\mathfrak{h}_n^{\mathrm{pol}}$ of $\mathbf{H}_n^{\mathrm{pol}}$ consists of all matrices

$$X^{\mathrm{pol}}(z,t) = \begin{bmatrix} 0 & p^T & t - \frac{1}{2}\langle p,x\rangle \\ 0 & 0 & x \\ 0 & 0 & 0 \end{bmatrix}.$$

We are next going to represent the elements of the Heisenberg group as unitary operators acting on the Hilbert space $L^2(\mathbb{R}^n_x)$ of square-integrable functions on position space; doing this we will recover the Heisenberg–Weyl operators defined at the end of the last chapter. This representation is the key to the Weyl calculus that will be developed in Section 6.2.

6.1.2 The Schrödinger representation of $\mathbf{H}_n$

Let us recall some terminology and basic facts from representation theory (see for instance Schempp [140], Varadarajan [170], or Wallach [175]). Let G be a topological group and $\mathcal{H}$ a Hilbert space. A unitary representation $(T,\mathcal{H})$ of G is a homomorphism $T: G \longrightarrow \mathcal{U}(\mathcal{H})$ where $\mathcal{U}(\mathcal{H})$ is the group of all unitary bijections $\mathcal{H} \longrightarrow \mathcal{H}$ equipped with the strong topology.

- Two unitary representations $(T,\mathcal{H})$ and $(T',\mathcal{H}')$ are *equivalent* if there exists a unitary bijection $U: \mathcal{H} \longrightarrow \mathcal{H}'$ such that

$$U \circ T(g) = T'(g) \circ U \quad \text{for all } g \in G;$$

- The unitary representation $(T,\mathcal{H})$ is *irreducible* if the only closed subspaces of $\mathcal{H}$ which are invariant under all $T(g)$, $g \in G$, are $\{0\}$ or $\mathcal{H}$ itself. [One sometimes says *topologically irreducible*].

A theorem of Schur (see Folland [42], Appendix B, for a proof) says that:

- The unitary representation $(T,\mathcal{H})$ is irreducible if and only if all bounded operators $A: \mathcal{H} \longrightarrow \mathcal{H}$ such that

$$A \circ T(g) = T(g) \circ A \quad \text{for all } g \in G$$

  are of the type λI, $\lambda \in \mathbb{C}$.

Recall that the Heisenberg–Weyl operators were already defined in Chapter 5, Subsection 5.5, by the formula

$$\widehat{T}(z_0)\psi(z) = e^{\frac{i}{\hbar}(\langle p_0,x\rangle - \frac{1}{2}\langle p_0,x_0\rangle)}\psi(x-x_0).$$

We also saw that these operators do not commute, and satisfy the commutation relations (5.67)–(5.68), which we restate here:

$$\widehat{T}(z_0)\widehat{T}(z_1) = e^{\frac{i}{\hbar}\sigma(z_0,z_1)}\widehat{T}(z_1)\widehat{T}(z_0), \tag{6.10}$$

$$\widehat{T}(z_0+z_1) = e^{-\frac{i}{2\hbar}\sigma(z_0,z_1)}\widehat{T}(z_0)\widehat{T}(z_1). \tag{6.11}$$

We will need the following classical result from harmonic analysis (see, *e.g.*, Stein and Weiss [159], Chapter 1, §3):

Lemma 6.4. *Let A be a bounded operator $L^2(\mathbb{R}^n_x) \longrightarrow L^2(\mathbb{R}^n_x)$. If A commutes with the translations $x \longmapsto x + x_0$, then there exists a function $a \in L^\infty(\mathbb{R}^n_x)$ such that $F(A\psi) = aF\psi$ for all $\psi \in \mathcal{S}(\mathbb{R}^n_x)$.*

Proof. We will give the proof of this property in the special case where A is an integral operator

$$A\psi(x) = \int K(x,y)\psi(y)d^n y$$

with kernel $K \in L^1(\mathbb{R}^n_x \times \mathbb{R}^n_x)$ satisfying

$$\sup_y \int |K(x,y)|\, dx \leq C \quad , \quad \sup_x \int |K(x,y)|\, dx \leq C.$$

In view of a classical lemma of Schur (see for instance Hörmander [92], *p.* 74) A is then continuous $L^2(\mathbb{R}^n_x) \to L^2(\mathbb{R}^n_x)$ and has norm $\leq C$. Let $T(x_0)$ be the translation $x \longmapsto x + x_0$; the relation $T(x_0)A = AT(x_0)$ is easily seen to be equivalent to $K(x - x_0, y) = K(x, y + x_0)$. For $\psi, \phi \in \mathcal{S}(\mathbb{R}^n_x)$ we have

$$\int F\phi(y)\psi(y)d^n y = \int \phi(\eta)F\psi(\eta)d^n\eta$$

where F is the usual Fourier transform

$$F\psi(\xi) = \left(\tfrac{1}{2\pi}\right)^{n/2} \int e^{-i\langle \xi, x\rangle}\psi(x)d^n_x.$$

We have

$$A\psi(x) \int (F^{-1}K)(x,\eta)F\psi(\eta)d^n\eta$$

where F^{-1} is the inverse Fourier transform acting on the second variable. We can rewrite this formula as

$$A\psi(x) = \left(\tfrac{1}{2\pi}\right)^{n/2} \int e^{i\langle x,\eta\rangle} a(x,\eta)F\psi(\eta)d^n\eta$$

where the symbol a is given by

$$a(x,\eta) = (2\pi)^{n/2} e^{-i\langle x,\eta\rangle} F^{-1}K(x,\eta),$$

that is

$$a(x,\eta) = \int e^{i\langle -x+y,\eta\rangle} K(x,y)d^n y.$$

Setting $y' = -x + y$ this equality becomes

$$a(x,\eta) = \int e^{i\langle y',\eta\rangle} K(x, y' + x)dy' = \int e^{i\langle y',\eta\rangle} K(0, y')dy'$$

so that a is independent of $x : a(x,\eta) = a(\eta)$. Summarizing we have shown that

$$A\psi(x) = \left(\tfrac{1}{2\pi}\right)^{n/2} \int e^{i\langle x,\eta\rangle} a(\eta) F\psi(\eta) d^n\eta,$$

hence $F(A\psi) = aF\psi$ using the Fourier inversion formula. □

Let us now state and prove the main result of this subsection, namely the existence of an irreducible representation of the Heisenberg group $\mathbf{H}_n$ compatible (in a sense we will make precise) with the Schrödinger equation. (For different approaches see Wallach [175], or Folland [42]; the latter gives a proof related to the Stone–von Neumann theorem which we will discuss below); there are also relations with Kirillov theory [175] which we do not discuss here because of lack of space.

Theorem 6.5. *The mapping* $\widehat{T}(z_0, t_0) : \mathbf{H}_n \longrightarrow \mathcal{U}(L^2(\mathbb{R}^n_x))$ *defined by*

$$\widehat{T}(z_0, t_0)\psi(x) = e^{\frac{i}{\hbar}t_0}\widehat{T}(z_0)\psi(x)$$

is a unitary and irreducible representation of $\mathbf{H}_n$.

Proof. It is clear that each operator $T(z_0, t_0)$ is unitary:

$$||\widehat{T}(z_0, t_0)\psi||_{L^2} = ||\psi||_{L^2} \text{ for all } \psi \in L^2(\mathbb{R}^n_x).$$

On the other hand, by formula (6.11),

$$\widehat{T}(z_0, t_0)\widehat{T}(z_1, t_1) = e^{\frac{i}{2\hbar}\sigma(z_0,z_1)}\widehat{T}(z_0 + z_1, t_0 + t_1),$$

that is

$$\widehat{T}(z_0, t_0)\widehat{T}(z_1, t_1) = \widehat{T}(z_0 + z_1, t_0 + t_1 + \tfrac{1}{2}\sigma(z_0, z_1)), \tag{6.12}$$

hence $\widehat{T}$ is a continuous homomorphism $\mathbf{H}_n \longrightarrow \mathcal{U}(L^2(\mathbb{R}^n_x))$ showing that $\widehat{T}$ indeed is a unitary representation of $\mathbf{H}_n$. Let us now prove the non-trivial part of the theorem, namely the irreducibility of that representation. We must show that if $\widehat{A}$ is a bounded operator $L^2(\mathbb{R}^n_x) \longrightarrow L^2(\mathbb{R}^n_x)$ such that

$$\widehat{A} \circ \widehat{T}(z_0, t_0) = \widehat{T}(z_0, t_0) \circ \widehat{A}$$

for all $z_0 \in \mathbb{R}^{2n}_z$, then $\widehat{A} = \lambda I$ for some complex constant λ. Choosing $p_0 = 0$ and $t_0 = 0$ we have in particular

$$\widehat{A} \circ T(x_0) = T(x_0) \circ \widehat{A}$$

where the operator $T(x_0)$ is defined by $T(x_0)\psi(x) = \psi(x - x_0)$; it now suffices to apply Lemma 6.4 above. □

A natural question that arises at this stage is whether there are other irreducible unitary representations of $\mathbf{H}_n$ in the Hilbert space $L^2(\mathbb{R}^n_x)$. A famous result, the Stone–von Neumann theorem, asserts that the Schrödinger representation of it is, up to unitary equivalences, the only irreducible representation of $\mathbf{H}_n$ in *that* Hilbert space (we are emphasizing the word "that" because the Stone–von Neumann theorem does not prevent us from constructing non-trivial irreducible representations of $\mathbf{H}_n$ in other Hilbert spaces than $L^2(\mathbb{R}^n_x)$. We will return to this important question in Chapter 10 when we discuss the Schrödinger equation in phase space; we will see that it is perfectly possible to construct (infinitely many) representations of the Heisenberg group on closed subspaces of $L^2(\mathbb{R}^{2n}_z)$.

6.2 Weyl Operators

The first mathematically rigorous treatment of the Weyl calculus is probably the paper by Grossmann *et al.* [83], which was later taken up by Hörmander in [93]. Weyl calculus is often called by quantum physicists and chemists the *WWMG formalism*; the acronym "WWMG" stands for Weyl–Wigner–Moyal–Groenewold, thus doing justice to the three other major contributors, Wigner [180], Moyal [127], and Groenewold [80], who are often forgotten in the mathematical literature. Here are a few references for pseudo-differential calculi in general, and Weyl operators in particular. These theories have passed through so many stages since the foundational work of Kohn and Nirenberg [104] that it is impossible to even sketch a complete bibliography for this vast subject. Two classical references are however the treatises of Hörmander [92] or Trèves [164]; Folland also gives a very readable short account of pseudo-differential operators in general and Weyl operators in particular[4]. A detailed treatment of the Weyl calculus is Chapter III in Leray's book [107]; it is however written without concession for the reader and a beginner might therefore find it somewhat difficult to digest. Toft [161] studies various regularity properties of Weyl operators in various Sobolev and Besov spaces, and applies his results to Toeplitz operators. Voros [172, 173] has given very interesting asymptotic results related to the metaplectic group; his work certainly deserves to be taken up. The reading of this chapter could be fruitfully complemented by that of Wong's [181] book (especially Chapters 4, 9, 29, 30). We also wish to draw the reader's attention to the deep and brilliant paper [77] by Gracia-Bondia where the Weyl formalism is analyzed in detail from the point of view of a class of simple operators earlier defined by Grossmann [82] and Royer [137]. This approach allows the author to discuss in a pertinent way the problem of dequantization.

Weyl operators are of course only one possible choice for the operators arising in quantum mechanics; our choice has been dictated by the very agreeable symplectic (or rather metaplectic) covariance properties of these operators. There are

[4]Folland insists on putting 2π's in the exponentials: his normalizations are typical of mathematicians working in harmonic analysis, while we are using the normalizations common among people working in partial differential equations.

of course other options, corresponding to different "ordering" procedures (see for instance the treatise by Nazaikiinskii, Schulze and Sternin [128], which in addition contains many topics we have not been able to include in this book, in particular an up-to-date treatment of Maslov's canonical operator method).

6.2.1 Basic definitions and properties

Let us begin by introducing a notion of Fourier transform for functions (or distributions) defined on the symplectic phase space $(\mathbb{R}_z^{2n}, \sigma)$.

Definition 6.6. Let $f \in \mathcal{S}(\mathbb{R}_z^{2n})$. The "symplectic Fourier transform" of f is the function $f_\sigma = \mathcal{F}_\sigma f$ defined by

$$f_\sigma(z_0) = \left(\tfrac{1}{2\pi\hbar}\right)^n \int e^{-\frac{i}{\hbar}\sigma(z_0,z)} f(z) d^{2n}z \tag{6.13}$$

(σ the standard symplectic form on $\mathbb{R}_z^{2n}$).

Let $\mathcal{F}$ be the usual Fourier transform on $\mathcal{S}(\mathbb{R}_z^{2n})$ defined, for $f \in \mathcal{S}(\mathbb{R}_z^{2n})$, by

$$\mathcal{F}f(z) = \left(\tfrac{1}{2\pi\hbar}\right)^n \int e^{-\frac{i}{\hbar}\langle z,z'\rangle} f(z') d^{2n}z'.$$

The symplectic Fourier transform is related to $\mathcal{F}$ by the obvious formula

$$\mathcal{F}_\sigma f(Jz) = \mathcal{F}f(z). \tag{6.14}$$

Since $\mathcal{F}$ extends into a (unitary) operator $L^2(\mathbb{R}_z^{2n}) \longrightarrow L^2(\mathbb{R}_z^{2n})$ and, by duality, into an operator $\mathcal{S}'(\mathbb{R}_z^{2n}) \longrightarrow \mathcal{S}'(\mathbb{R}_z^{2n})$ so does $\mathcal{F}_\sigma$. Summarizing:

> *The symplectic Fourier transform $\mathcal{F}_\sigma$ is a unitary operator on $L^2(\mathbb{R}_z^{2n})$ which extends into an automorphism of the space $\mathcal{S}'(\mathbb{R}_z^{2n})$ of tempered distributions.*

It follows from formula (6.14) that $\mathcal{F}_\sigma$ is an involution:

$$\mathcal{F}_\sigma^{-1} = \mathcal{F}_\sigma \quad (\text{or } \mathcal{F}_\sigma^2 = I). \tag{6.15}$$

Exercise 6.7. Prove the formula above using the relation (6.14) and the properties of the ordinary Fourier transform. The following symplectic covariance property generalizes formula (6.14):

Proposition 6.8. *For every $S \in \mathrm{Sp}(n)$ we have*

$$\mathcal{F}_\sigma(f \circ S) = (\mathcal{F}_\sigma f) \circ S \quad \textit{for every} \quad S \in \mathrm{Sp}(n). \tag{6.16}$$

Proof. Since $\sigma(Sz, z') = \sigma(z, S^{-1}z')$ we have

$$\mathcal{F}_\sigma f(Sz) = \left(\tfrac{1}{2\pi\hbar}\right)^n \int e^{-\frac{i}{\hbar}\sigma(z,S^{-1}z')} f(z') d^{2n}z'$$

that is, performing the change of variable $z' = Sz''$ and taking into account the fact that the determinant of a symplectic matrix is 1,

$$\begin{aligned}\mathcal{F}_\sigma f(Sz) &= \left(\tfrac{1}{2\pi\hbar}\right)^n \int e^{-\frac{i}{\hbar}\sigma(z,z'')} f(Sz'')|\det S| d^{2n}z'' \\ &= \left(\tfrac{1}{2\pi\hbar}\right)^n \int e^{-\frac{i}{\hbar}\sigma(z,z'')} f(Sz'') d^{2n}z'',\end{aligned}$$

hence (6.16). □

Let us now define the Weyl operator associated to a symbol. We will see below that our definition just represents the Fourier decomposition of a pseudo-differential operator of a particular type.

Definition 6.9. Let $a \in \mathcal{S}(\mathbb{R}^{2n}_z)$.

(i) The "Weyl operator associated to the symbol a" is the operator

$$\widehat{A} : \mathcal{S}(\mathbb{R}^n_x) \longrightarrow \mathcal{S}(\mathbb{R}^n_x)$$

defined by

$$\widehat{A}\psi(x) = \left(\tfrac{1}{2\pi\hbar}\right)^n \int a_\sigma(z_0)\widehat{T}(z_0)\psi(x) d^{2n}z_0. \tag{6.17}$$

(ii) The symplectic Fourier transform $a_\sigma = \mathcal{F}_\sigma a$ of the symbol a is called the "twisted symbol of $\widehat{A}$". We will write $\widehat{A} \overset{\text{Weyl}}{\longleftrightarrow} a$ or $a \overset{\text{Weyl}}{\longleftrightarrow} \widehat{A}$ (the "Weyl correspondence").

We will often write (6.17) in the shorter form

$$\widehat{A} = \left(\tfrac{1}{2\pi\hbar}\right)^n \int a_\sigma(z)\widehat{T}(z) d^{2n}z \tag{6.18}$$

where the right-hand side should be interpreted as a "Bochner integral", *i.e.*, an integral with value in a Banach space.

It follows from the inversion formula (6.15) that the ordinary Weyl symbol a and its twisted version a_σ are explicitly related by the simple formulae

$$a_\sigma(z) = \left(\tfrac{1}{2\pi\hbar}\right)^n \int e^{-\frac{i}{\hbar}\sigma(z,z')} a(z') d^{2n}z', \tag{6.19}$$

$$a(z) = \left(\tfrac{1}{2\pi\hbar}\right)^n \int e^{-\frac{i}{\hbar}\sigma(z,z')} a_\sigma(z') d^{2n}z'. \tag{6.20}$$

In view of the definition of the Heisenberg–Weyl operator $\widehat{T}(z_0)$, definition (6.17) of $\widehat{A}$ can be rewritten in a slightly more explicit form as

$$\widehat{A}\psi(x) = \left(\tfrac{1}{2\pi\hbar}\right)^n \int a_\sigma(z_0) e^{\frac{i}{\hbar}(\langle p_0,x\rangle - \frac{1}{2}\langle p_0,x_0\rangle)} \psi(x-x_0) d^{2n}z_0. \tag{6.21}$$

This formula shows that, indeed, $\widehat{A}\psi$ is well defined under the assumptions on a and ψ made in Definition 6.9: the integrand is rapidly decreasing as $|z| \to \infty$ so the integral is absolutely convergent. In fact, a modest amount of work involving repeated use of Leibniz's formula for the differentiation of products shows that $\widehat{A}\psi \in \mathcal{S}(\mathbb{R}^n_x)$, and even that $\widehat{A}$ is continuous for the topology of $\mathcal{S}(\mathbb{R}^n_x)$. We leave these trivial (but lengthy) verifications to the reader as a useful but boring exercise, and rather focus on the question whether formula (6.17) could be given a reasonable meaning for larger classes of symbols than $\mathcal{S}(\mathbb{R}^{2n}_z)$, namely those having a well-defined symplectic Fourier transform (even in the distributional sense). This is more than an academic question, since in the applications to quantum mechanics the symbol a represents "observables" which can be quite general functions of position and momentum (for instance energy) that have no reason to be rapidly decreasing at infinity.

Let us determine the Weyl symbol and the twisted symbol of the identity operator:

Proposition 6.10. *Let $\widehat{I}$ be the identity operator in $\mathcal{S}(\mathbb{R}^n_x)$ and a its Weyl symbol. We have*

$$a(z) = 1 \quad \text{and} \quad a_\sigma(z) = (2\pi\hbar)^n \delta(z).$$

Proof. We obviously have

$$\begin{aligned}\psi(x) &= \iint \delta(z_0)\psi(x - x_0)d^n p_0 d^n x_0 \\ &= \iint \delta(z_0)\widehat{T}(z_0)\psi(x)d^n p_0 d^n x_0,\end{aligned}$$

hence $a_\sigma(z) = (2\pi\hbar)^n\delta(z)$ as claimed. Since on the other hand we have, in view of the inversion formula (6.15),

$$a(z) = \mathcal{F}_\sigma a_\sigma(z) = \int e^{-\frac{i}{\hbar}\sigma(z,z')}\delta(z')d^{2n}z' = 1,$$

this proves the proposition. □

Here is another example:

Example 6.11. Let us next consider the case where the symbol is the Dirac distribution δ on $\mathbb{R}^{2n}_z$. We have here $a_\sigma(z_0) = (2\pi\hbar)^{-n}$ and hence

$$\begin{aligned}\widehat{A}\psi(x) &= \left(\tfrac{1}{2\pi\hbar}\right)^{2n} \int \left(\int e^{\frac{i}{\hbar}\langle p_0, x - \frac{1}{2}x_0\rangle} d^n p_0 \right) \psi(x - x_0)d^n x_0 \\ &= \left(\tfrac{1}{2\pi\hbar}\right)^{n} \int \delta(x - \tfrac{1}{2}x_0)\psi(x - x_0)d^n x_0 \\ &= \left(\tfrac{1}{\pi\hbar}\right)^{n} \psi(-x)\end{aligned}$$

so that the operator $\widehat{A}$ is, up to the factor $(\pi\hbar)^{-n}$, a reflection operator.

6.2.2 Relation with ordinary pseudo-differential calculus

The two examples above show that one can indeed expect to define Weyl operators for quite general symbols. To see how this can be done it is a good idea to look at the kernel of the Weyl operator with symbol $a \in \mathcal{S}(\mathbb{R}_z^{2n})$. We recall the following theorem from functional analysis (for a proof see, *e.g.*, Trèves [165]):

Schwartz's kernel theorem: *The continuous linear transforms*

$$\widehat{A} : \mathcal{S}(\mathbb{R}_x^n) \longrightarrow \mathcal{S}'(\mathbb{R}_x^n)$$

are precisely the operators with kernel $K_{\widehat{A}} \in \mathcal{S}'(\mathbb{R}_x^n \times \mathbb{R}_x^n)$.

We will write the action of such an operator on a function $\psi \in \mathcal{S}(\mathbb{R}_x^n)$ as

$$\widehat{A}\psi(x) = \int K_{\widehat{A}}(x,y)\psi(y)d^n y \tag{6.22}$$

where the integral should be interpreted, for fixed x, as the distributional bracket

$$\widehat{A}\psi(x) = \langle K_{\widehat{A}}(x,\cdot), \psi(\cdot)\rangle \,.$$

The following theorem shows that Weyl operators are just pseudo-differential operators of a special kind:

Theorem 6.12. *Let* $a \in \mathcal{S}(\mathbb{R}_z^{2n})$ *and* $\widehat{A} \overset{\text{Weyl}}{\longleftrightarrow} a$.

(i) *The kernel of* $\widehat{A}$ *is given by*

$$K_{\widehat{A}}(x,y) = \left(\tfrac{1}{2\pi\hbar}\right)^n \int e^{\frac{i}{\hbar}\langle p, x-y\rangle} a(\tfrac{1}{2}(x+y),p)d^n p \tag{6.23}$$

and hence

$$\widehat{A}\psi(x) = \left(\tfrac{1}{2\pi\hbar}\right)^n \iint e^{\frac{i}{\hbar}\langle p, x-y\rangle} a(\tfrac{1}{2}(x+y),p)\psi(y)d^n p d^n y \tag{6.24}$$

for $\psi \in \mathcal{S}(\mathbb{R}_x^n)$.

(ii) *Conversely*

$$a(x,p) = \int e^{-\frac{i}{\hbar}\langle p,y\rangle} K_{\widehat{A}}(x+\tfrac{1}{2}y, x-\tfrac{1}{2}y)d^n y. \tag{6.25}$$

Proof. (i) We have

$$\widehat{A}\psi(x) = \left(\tfrac{1}{2\pi\hbar}\right)^n \int a_\sigma(z_0) e^{\frac{i}{\hbar}(\langle p_0, x\rangle - \frac{1}{2}\langle p_0, x_0\rangle)}\psi(x-x_0)d^{2n}z_0,$$

that is, setting $y = x - x_0$:

$$\widehat{A}\psi(x) = \left(\tfrac{1}{2\pi\hbar}\right)^n \iint e^{\frac{i}{2\hbar}\langle p_0, x+y\rangle} a_\sigma(x-y,p_0)\psi(y)d^n y d^n p_0.$$

We thus have

$$\widehat{A}\psi(x) = \int K_{\widehat{A}}(x,y)\psi(y)d^n y$$

where the kernel $K_{\widehat{A}}$ is given by

$$K_{\widehat{A}}(x,y) = \left(\tfrac{1}{2\pi\hbar}\right)^n \int e^{\frac{i}{2\hbar}\langle p_0, x+y\rangle} a_\sigma(x-y, p_0)d^n p_0.$$

Since $a_\sigma(z) = \mathcal{F}a(Jz)$ (formula (6.14)) this can be rewritten as

$$\begin{aligned}
K_{\widehat{A}}(x,y) &= \left(\tfrac{1}{2\pi\hbar}\right)^n \int e^{\frac{i}{2\hbar}\langle p_0, x+y\rangle} Fa(p_0, y-x)d^n p_0 \\
&= \left(\tfrac{1}{2\pi\hbar}\right)^{2n} \iint e^{\frac{i}{2\hbar}\langle p_0, x+y\rangle} e^{-\frac{i}{\hbar}\sigma(x-y,p_0;z')} a(z')d^n p_0 d^{2n}z' \\
&= \left(\tfrac{1}{2\pi\hbar}\right)^{2n} \int e^{\frac{i}{\hbar}\langle p', x-y\rangle} \left(\int e^{\frac{i}{\hbar}\langle p_0, \frac{1}{2}(x+y)-x'\rangle} d^n p_0\right) a(z')d^{2n}z' \\
&= \left(\tfrac{1}{2\pi\hbar}\right)^n \int e^{\frac{i}{\hbar}\langle p', x-y\rangle} \delta(\tfrac{1}{2}(x+y)-x')a(z')d^{2n}z' \\
&= \left(\tfrac{1}{2\pi\hbar}\right)^n \int e^{\frac{i}{\hbar}\langle p', x-y\rangle} a(\tfrac{1}{2}(x+y), p')d^n p'
\end{aligned}$$

which is (6.23). Formula (6.24) follows.
To prove (ii) it suffices to note that (6.23) implies that

$$K_{\widehat{A}}(x+\tfrac{1}{2}y, x-\tfrac{1}{2}y) = \left(\tfrac{1}{2\pi\hbar}\right)^n \int e^{-\frac{i}{\hbar}\langle p,y\rangle} a(x,p)d^n p; \tag{6.26}$$

fixing x, formula (6.25) follows from the Fourier inversion formula. □

Part (i) of the theorem above shows that Weyl operators can be expressed in an amazingly simple way in terms of the Grossmann–Royer operators

$$\widetilde{T}(z_0)\psi(x) = e^{\frac{2i}{\hbar}\langle p_0, x-x_0\rangle}\psi(2x_0 - x)$$

defined in Section 5.5 of Chapter 5:

Corollary 6.13. *Let $a \in \mathcal{S}(\mathbb{R}^{2n}_z)$ and $\widehat{A} \overset{\text{Weyl}}{\longleftrightarrow} a$; we have*

$$\widehat{A} = \left(\tfrac{1}{\pi\hbar}\right)^n \int a(z)\widetilde{T}(z)d^{2n}z \tag{6.27}$$

where the integral is interpreted in the sense of Bochner; in particular we have the "reproducing property"

$$\left(\tfrac{1}{\pi\hbar}\right)^n \int \widetilde{T}(z)d^{2n}z = \widehat{I} \tag{6.28}$$

($\widehat{I}$ the identity operator).

Proof. Let $\psi \in \mathcal{S}(\mathbb{R}^n_x)$; setting $x' = \frac{1}{2}(x+y)$ in (6.24) we get

$$\widehat{A}\psi(x) = \left(\tfrac{1}{\pi\hbar}\right)^n \iint e^{\frac{2i}{\hbar}\langle p, x-x'\rangle} a(x',p)\psi(2x'-x)d^np d^nx'$$

which is precisely (6.27). □

Notice that if we interpret formula (6.24) in the distributional sense we recover the fact that the identity operator has symbol $a = 1$. In fact it suffices to observe that in view of the Fourier inversion formula,

$$\left(\tfrac{1}{2\pi\hbar}\right)^n \iint e^{\frac{i}{\hbar}\langle p, x-y\rangle}\psi(y)d^ny d^np = \psi(x). \tag{6.29}$$

In fact, Schwartz's kernel theorem implies:

Corollary 6.14. *For every continuous operator $\widehat{A} : \mathcal{S}(\mathbb{R}^n_x) \longrightarrow \mathcal{S}'(\mathbb{R}^n_x)$ there exists $a \in \mathcal{S}'(\mathbb{R}^{2n}_z)$ such that the kernel of $\widehat{A}$ is given by formula* (6.23) *interpreted in the distributional sense:*

$$K_{\widehat{A}}(x,y) = \left\langle e^{\frac{i}{\hbar}\langle \cdot, x-y\rangle}, a(\tfrac{1}{2}(x+y),\cdot)\right\rangle. \tag{6.30}$$

Proof. Formulae (6.23) and (6.25) in Theorem 6.12 show that the linear mapping is an automorphism $\mathcal{S}(\mathbb{R}^n_x) \longrightarrow \mathcal{S}(\mathbb{R}^n_x)$. It is easy to verify that this automorphism is continuous and hence extends to an automorphism $\mathcal{S}'(\mathbb{R}^n_x) \longrightarrow \mathcal{S}'(\mathbb{R}^n_x)$. In view of Schwartz's kernel theorem every continuous operator $\widehat{A} : \mathcal{S}(\mathbb{R}^n_x) \longrightarrow \mathcal{S}'(\mathbb{R}^n_x)$ can be written as

$$\widehat{A}\psi(x) = \langle K_{\widehat{A}}(x,\cdot), \psi(\cdot)\rangle$$

where $K_{\widehat{A}} \in \mathcal{S}'(\mathbb{R}^{2n}_z)$. The corollary follows. □

Here is one basic example, useful in quantum mechanics:

Example 6.15. Let $a(z) = x_jp_j$. Then $\widehat{A}$ is the operator

$$\widehat{A} = \tfrac{1}{2}(\widehat{X_j}\widehat{P_j} + \widehat{P_j}\widehat{X_j}). \tag{6.31}$$

To see this we notice that

$$\begin{aligned}
\widehat{A}\psi(x) &= \left(\tfrac{1}{2\pi\hbar}\right)^n \iint e^{\frac{i}{\hbar}\langle p, x-y\rangle}\tfrac{1}{2}(x_j+y_j)p_j\psi(y)d^ny d^np \\
&= \left(\tfrac{1}{2\pi\hbar}\right)^n \tfrac{1}{2}x_j \iint e^{\frac{i}{\hbar}\langle p, x-y\rangle}p_j\psi(y)d^ny d^np \\
&+ \left(\tfrac{1}{2\pi\hbar}\right)^n \tfrac{1}{2} \iint e^{\frac{i}{\hbar}\langle p, x-y\rangle}p_jy_j\psi(y)d^ny d^np;
\end{aligned}$$

in view of the obvious equalities

$$\left(\tfrac{1}{2\pi\hbar}\right)^n \tfrac{1}{2}x_j \iint e^{\frac{i}{\hbar}\langle p,x-y\rangle} p_j\psi(y)d^n y d^n p = \tfrac{1}{2}x_j \widehat{P_j}\psi(x),$$

$$\left(\tfrac{1}{2\pi\hbar}\right)^n \tfrac{1}{2} \iint e^{\frac{i}{\hbar}\langle p,x-y\rangle} p_j y_j\psi(y)d^n y d^n p = \tfrac{1}{2}\widehat{P_j}(x_j\psi)(x),$$

formula (6.31) follows.

Theorem 6.12 shows us the way for the definition of the Weyl correspondence for more general symbols. For instance, we can notice that for all $\psi, \phi \in \mathcal{S}(\mathbb{R}^n_x)$ we have

$$\left\langle \widehat{A}\psi, \phi \right\rangle = \left(\tfrac{1}{2\pi\hbar}\right)^n \iiint e^{\frac{i}{\hbar}\langle p,x-y\rangle} a(\tfrac{1}{2}(x+y),p)\psi(y)\phi(x)d^n y d^n p d^n x;$$

this formula actually defines $\widehat{A}\psi$ in the weak (= distributional) sense for $a \in \mathcal{S}'(\mathbb{R}^n_x)$. (Wong [181] uses the Riesz–Thorin interpolation theorem to define $\widehat{A}$ for arbitrary symbols $a \in \mathcal{S}'(\mathbb{R}^n_x)$; in particular he defines $\widehat{A}$ for $a \in L^p(\mathbb{R}^{2n}_z)$, $1 \leq p \leq \infty$; see in this context Simon [151]). We will come back to these questions of extension in the next subsection. Let us first emphasize that one major advantage of the Weyl correspondence on traditional "classical" pseudo-differential calculus (see Appendix C) is that if the symbol a is *real*, then the operator $\widehat{A}$ is *symmetric* (or *essentially self-adjoint*). This property is of paramount importance in applications to quantum mechanics, where one wants to associate a self-adjoint operator to a real "observable":

Proposition 6.16. *The following properties hold:*

(i) *If* $a \overset{\text{Weyl}}{\longleftrightarrow} \widehat{A}$, *then* $\overline{a} \overset{\text{Weyl}}{\longleftrightarrow} \widehat{A}^*$.

(ii) *The operator* $\widehat{A} \overset{\text{Weyl}}{\longleftrightarrow} a$ *is self-adjoint:* $\widehat{A}^* = \widehat{A}$ *if and only if* a *is a real symbol.*

Proof. (i) The adjoint $\widehat{A}^*$ of $\widehat{A}$ is defined by

$$(\widehat{A}^*\psi, \phi)_{L^2(\mathbb{R}^n_x)} = (\psi, \widehat{A}\phi)_{L^2(\mathbb{R}^n_x)}$$

for $\psi, \phi \in \mathcal{S}(\mathbb{R}^n_x)$; this means that

$$\int \left(\int K^*_{\widehat{A}}(x,y)\psi(y)d^n y \right) \phi(x)d^n x = \int \overline{\left(\int K_{\widehat{A}}(y,x)\phi(x)d^n x \right)} \psi(y)d^n y$$

which is equivalent to the condition

$$K^*_{\widehat{A}}(x,y) = \overline{K_{\widehat{A}}(y,x)}. \tag{6.32}$$

The adjoint of $\widehat{A}$ is obtained by replacing a by its complex conjugate.

(ii) We have $\widehat{A} = \widehat{A}^*$ if and only if $K_{\widehat{A}}(x,y) = \overline{K^*_{\widehat{A}}(y,x)}$, that is, by formula (6.23), to

$$\int e^{\frac{i}{\hbar}\langle p,x-y\rangle} a(\tfrac{1}{2}(x+y),p)d^n p = \int e^{\frac{i}{\hbar}\langle p,x-y\rangle} \overline{a(\tfrac{1}{2}(x+y),p)}d^n p$$

which is satisfied if and only if a is real-valued. ∎

6.3 Continuity and Composition

In this section we begin by stating and proving a few continuity properties of Weyl operators; because of lack of space we limit ourselves to the cases of interest to us in the applications to quantum mechanics. We then study in detail the composition formula for Weyl operators.

6.3.1 Continuity properties of Weyl operators

Let us begin by the mathematically "traditional" case, and assume that the symbol a belongs to one of the standard classes $S^m_{1,0}$ (see Appendix C for a definition of the classes $S^m_{\rho,\delta}$).

Theorem 6.17. *Assume that $a \in S^m_{1,0}(\mathbb{R}^{2n}_z)$. Then the operator A defined, for $N > m+n$ and $\psi \in \mathcal{S}(\mathbb{R}^n_x)$, by*

$$A\psi(x) = \left(\tfrac{1}{2\pi\hbar}\right)^n \iint e^{\frac{i}{\hbar}\langle p,x-y\rangle} L^N_y a(\tfrac{1}{2}(x+y),p)(1+|p|)^{-N}\psi(y)d^nyd^np \quad (6.33)$$

where $L_y = 1 - \Delta_y$ (Δ_y the Laplace operator in $y = (y_1,\dots,y_n)$) is a continuous operator $\mathcal{S}(\mathbb{R}^n_x) \longrightarrow \mathcal{S}(\mathbb{R}^n_x)$ coinciding with $\widehat{A} \overset{\text{Weyl}}{\longleftrightarrow} a$ when $a \in \mathcal{S}(\mathbb{R}^{2n}_z)$.

Proof. We only sketch the proof of this well-known result here. (We are following Wong [181].) Choose $\theta \in C^\infty_0(\mathbb{R}^n_p)$ such that $\theta(0) = 1$ and set, for $\varepsilon > 0$,

$$\widehat{A}_{\theta,\varepsilon}\psi(x) = \left(\tfrac{1}{2\pi\hbar}\right)^n \iint e^{\frac{i}{\hbar}\langle p,x-y\rangle}(a(\tfrac{1}{2}(x+y),p)\theta(\varepsilon p)\psi(y)d^nyd^np. \quad (6.34)$$

One then shows using partial integrations:

(i) that the limit $\lim_{\varepsilon\to 0}\widehat{A}_{\theta,\varepsilon}\psi$ exists and is independent of the choice of θ; the convergence is moreover uniform;

(ii) that the limit, denoted by $\widehat{A}\psi(x)$ is given by the integral (6.33) where $N > m+n$. □

Remark 6.18. The right-hand side of formula (6.33) is called an "oscillatory integral", and one writes it as

$$\widehat{A}\psi(x) = \left(\tfrac{1}{2\pi\hbar}\right)^n \widetilde{\iint} e^{\frac{i}{\hbar}\langle p,x-y\rangle} a(\tfrac{1}{2}(x+y),p)\psi(y)d^nyd^np.$$

One can work with oscillatory integrals very much as with ordinary integrals provided that some care is taken in their evaluation (for example, one should not in general attempt to apply Fubini's rule to them!).

To limit ourselves to symbols belonging to the standard pseudo-differential classes $S^m_{\rho,\delta}(\mathbb{R}^{2n}_z)$ is far too restrictive to be useful in quantum mechanics where one is particularly interested in operators $\widehat{A}$ acting on $L^2(\mathbb{R}^n_x)$, or at least, with dense domain $D_{\widehat{A}} \subset L^2(\mathbb{R}^n_x)$. For instance, Hilbert–Schmidt operators play a central role in the study of quantum-mechanical states; we will see in Chapter 9 (Theorem 9.21) that these operators are Weyl operators with symbols belonging to $L^2(\mathbb{R}^n_x)$, and such symbols do not generally belong to any of the classes $S^m_{\rho,\delta}(\mathbb{R}^{2n}_z)$.

Here is one first result which shows that $\widehat{A}$ is a bounded operator on $L^2(\mathbb{R}^n_x)$ provided that the Fourier transform Fa of the symbol a is integrable (in Exercise 6.20 below the Reader is invited to prove that this condition can be replaced by the condition $a \in L^1(\mathbb{R}^n_x)$).

We denote here by Fa the Fourier transform of a with respect to the variables $z = (x, p)$:

$$Fa(\varsigma) = \left(\tfrac{1}{2\pi\hbar}\right)^n \int e^{-\frac{i}{\hbar}\langle z,\zeta\rangle} a(z) d^{2n}z. \tag{6.35}$$

Proposition 6.19. *Assume that $Fa \in L^1(\mathbb{R}^{2n}_z)$. For every $\psi \in \mathcal{S}(\mathbb{R}^n_x)$ we have*

$$||\widehat{A}\psi||_{L^2(\mathbb{R}^n_x)} \leq \left(\tfrac{1}{2\pi\hbar}\right)^{2n} ||Fa||_{L^1(\mathbb{R}^n_x)} ||\psi||_{L^2(\mathbb{R}^n_x)}, \tag{6.36}$$

hence $\widehat{A}$ is a continuous operator in $\mathcal{S}(\mathbb{R}^{2n}_z)$ for the induced L^2-norm, and can thus be extended into a bounded operator $L^2(\mathbb{R}^n_x) \longrightarrow L^2(\mathbb{R}^n_x)$.

Proof. The kernel of $\widehat{A}$ is given by

$$K_{\widehat{A}}(x, y) = \left(\tfrac{1}{2\pi\hbar}\right)^{n/2} F_2(\tfrac{1}{2}(x + y), y - x)$$

where F_2 is the Fourier transform in the p variables. By the Fourier inversion formula we have

$$F_2 a(\tfrac{1}{2}(x + y), y - x) = \left(\tfrac{1}{2\pi\hbar}\right)^{n/2} \int e^{\frac{i}{2\hbar}\langle x+y,\xi\rangle} F(\xi, y - x) d^n\xi$$

and hence

$$K_{\widehat{A}}(x, y) = \left(\tfrac{1}{2\pi\hbar}\right)^n \int e^{\frac{i}{2\hbar}\langle x+y,\xi\rangle} Fa(\xi, y - x) d^n\xi.$$

It follows that

$$\int |K_{\widehat{A}}(x, y)| d^n x \leq \left(\tfrac{1}{2\pi\hbar}\right)^n \int |Fa(\xi, y - x)| d^n\xi d^n x,$$

$$\int |K_{\widehat{A}}(x, y)| d^n y \leq \left(\tfrac{1}{2\pi\hbar}\right)^n \int |Fa(\xi, y - x)| d^n\xi d^n y.$$

Setting $\eta = y - x$ we have

$$\int |Fa(\xi, y - x)| d^n\xi d^n x = \int |Fa(\xi, \eta)| d^n\xi d^n\eta,$$

hence the two inequalities above can be rewritten

$$\int |K_{\widehat{A}}(x,y)|d^n x \le \left(\tfrac{1}{2\pi\hbar}\right)^n ||Fa||_{L^1},$$

$$\int |K_{\widehat{A}}(x,y)|d^n y \le \left(\tfrac{1}{2\pi\hbar}\right)^n ||Fa||_{L^1}.$$

The rest of the proof of the inequality (6.36) is now similar to that of the usual Schur lemma: setting $C = (2\pi\hbar)^{-n}||Fa||_{L^1}$ we have, using Cauchy–Schwarz's inequality,

$$\begin{aligned}|\widehat{A}\psi(x)|^2 &\le \int |K_{\widehat{A}}(x,y)|d^n y \int |K_{\widehat{A}}(x,y)||\psi(y)|^2 d^n y \\ &\le C \int |K_{\widehat{A}}(x,y)||\psi(y)|^2 d^n y\end{aligned}$$

and hence

$$\int |\widehat{A}\psi(x)|^2 dx \le C \int \left(\int |K_{\widehat{A}}(x,y)|d^n x \right) |\psi(y)|^2 d^n x d^n y$$

that is

$$\int |\widehat{A}\psi(x)|^2 dx \le C^2 \int |\psi(y)|^2 d^n y$$

which is precisely (6.36). The last statement of the lemma immediately follows from the continuity of $\widehat{A}$ and the density of $\mathcal{S}(\mathbb{R}^n_x)$ in $L^2(\mathbb{R}^n_x)$. □

Exercise 6.20. By modifying in an adequate way the proof above, show that the conclusions of Proposition 6.19 remain true if one replaces the condition $Fa \in L^1(\mathbb{R}^{2n}_z)$ by $a \in L^1(\mathbb{R}^{2n}_z)$.

Proposition 6.19 implies the following rather general L^2-continuity result:

Corollary 6.21. *If $a \in C^{n+1}(\mathbb{R}^{2n}_z)$ and $\partial^\alpha_z a \in L^1(\mathbb{R}^{2n}_z)$ for every multi-index $\alpha \in \mathbb{N}^{2n}$ such that $|\alpha| \le n+1$, then $\widehat{A} \overset{\text{Weyl}}{\longleftrightarrow} a$ is a bounded operator on $L^2(\mathbb{R}^n_x)$.*

Proof. In view of Proposition 6.19 it suffices to check that the condition on a implies that $Fa \in L^1(\mathbb{R}^{2n}_z)$. Now, for every integer $m > 0$,

$$(1+|z|^2)^m Fa = F((1-\hbar^2\Delta^m_z)a)$$

where Δ^m_z is the mth power of the generalized Laplacian $\Delta_z = \sum_{j=1}^n \partial^2_{x_j} + \partial^2_{p_j}$. It follows that we have the estimate

$$(1+|z|^2)^m |Fa(z)| \le \int |(1-\hbar^2\Delta^m_z)a(z)|d^{2n}z < \infty$$

if $2m \le 2(n+1)$, and we can then find $C > 0$ such that

$$|Fa(z)| \le C(1+|z|^2)^{-n-1}$$

which implies that $Fa \in L^1(\mathbb{R}^{2n}_z)$. □

In most cases of interest one cannot expect L^2-boundedness for a Weyl operator $\widehat{A}$; it is usually only defined on some domain $D_{\widehat{A}} \subset L^2(\mathbb{R}^n_x)$. An excellent substitute for Proposition 6.19 (and its corollary) is then to look for conditions on the symbol that ensure that $\widehat{A} : \mathcal{S}(\mathbb{R}^n_x) \longrightarrow \mathcal{S}(\mathbb{R}^n_x)$. We are going to prove below that it is actually sufficient to assume that the symbol a and its derivatives are growing, for $|z| \to \infty$, slower than some polynomial to ensure $\mathcal{S}(\mathbb{R}^n_x)$ continuity. Let us define this concept precisely:

Definition 6.22. We will call a function $a \in C^\infty(\mathbb{R}^{2n}_z)$ a "polynomially bounded symbol" if there exists real numbers $m \in \mathbb{R}$, $\delta > 0$, such that

$$|\partial_z^\gamma a(z)| \leq C_\gamma (1 + |z|^2)^{\frac{1}{2}(m+\delta|\gamma|)} \tag{6.37}$$

for some $C_\gamma > 0$.

Clearly every function $a \in \mathcal{S}(\mathbb{R}^{2n}_z)$ is polynomially bounded.

Theorem 6.23. *Assume that a satisfies the estimates (6.37) with $0 < \delta < 1$.*

(i) *The associated Weyl operator $\widehat{A}$ is continuous $\mathcal{S}(\mathbb{R}^n_x) \longrightarrow \mathcal{S}(\mathbb{R}^n_x)$ and $\widehat{A}\psi(x)$ is given by the iterated integral*

$$\widehat{A}\psi(x) = \left(\tfrac{1}{2\pi\hbar}\right)^n \int \left(\int e^{\frac{i}{\hbar}\langle p, x-y\rangle} a(\tfrac{1}{2}(x+y), p)\psi(y) d^n y \right) d^n p \tag{6.38}$$

where the integral in y is $O(|p|^{-\infty})$.

(ii) *The operator $\widehat{A}$ extends by transposition into a continuous operator $\mathcal{S}'(\mathbb{R}^n_x) \longrightarrow \mathcal{S}'(\mathbb{R}^n_x)$.*

Proof. The statement (ii) follows from (i) since the Weyl symbol of the transpose of $\widehat{A}$ satisfies the same estimates (6.37) as that of $\widehat{A}$.

Let us prove (i). Setting

$$f(p) = \int e^{\frac{i}{\hbar}\langle p, x-y\rangle} a(\tfrac{1}{2}(x+y), p)\psi(y) d^n y \tag{6.39}$$

we have, for $1 \leq j \leq n$,

$$p_j^2 f(p) = (i\hbar)^2 \int \partial^2_{y_j} (e^{\frac{i}{\hbar}\langle p, x-y\rangle}) a(\tfrac{1}{2}(x+y), p)\psi(y) d^n y$$

and hence

$$|p|^{2N} f(p) = (i\hbar)^{2N} \int \Delta_y^N (e^{\frac{i}{\hbar}\langle p, x-y\rangle}) a(\tfrac{1}{2}(x+y), p)\psi(y) d^n y \tag{6.40}$$

where $\Delta_y = \sum_{j=1}^n \partial^2_{y_j}$. Using Leibniz's rule it is easily seen that the conditions (6.38) imply that the function $y \longmapsto a(\frac{1}{2}(x+y), p)\psi(y)$ and all its derivatives vanish at infinity, hence, performing partial integrations in (6.40):

$$|p|^{2N} f(p) = (i\hbar)^{2N} \int e^{\frac{i}{\hbar}\langle p, x-y\rangle} \Delta_y^N \left[a(\tfrac{1}{2}(x+y), p)\psi(y) \right] d^n y$$

from which follows that we have

$$|p|^{2N}|f(p)| \leq \hbar^{2N} \int |\Delta_y^N \left[a(\tfrac{1}{2}(x+y),p)\psi(y)\right]| d^n y.$$

Using once again Leibniz's rule, $\Delta_y^N \left[a(\frac{1}{2}(x+y),p)\psi(y)\right]$ is a linear combination of terms of the type

$$g_{\alpha\beta}(z) = \partial_y^\alpha a(\tfrac{1}{2}(x+y),p)\partial_y^\beta \psi(y) \quad , \quad |\alpha| + |\beta| = N$$

and each of these terms satisfies, by (6.38), an estimate

$$|g_{\alpha\beta}(z)| \leq C_{\alpha\beta}(1+|z|^2)^{\frac{1}{2}(m+\delta N)}\partial_y^\beta \psi(y).$$

It follows that there exist constants $C_N, C'_N > 0$ such that

$$\begin{aligned}|p|^{2N}|f(p)| &\leq C_N(1+|z|^2)^{\frac{1}{2}(m+\delta N)} \int \partial_y^\beta \psi(y) d^n y \\ &\leq C'_N(1+|z|^2)^{\frac{1}{2}(m+\delta N)}\end{aligned}$$

and hence, in particular,

$$|f(p)| = O(|p|^{m+(\delta-1)N}) \quad , \quad |p| \to \infty$$

for every N. If $\delta < 1$ we have

$$\lim_{N\to\infty} (\delta - 1)N = -\infty$$

and hence $|f(p)| = O(|p|^{-\infty})$ and the integral in p in (6.38) is thus absolutely convergent. Let us now show that $\widehat{A}\psi \in \mathcal{S}(\mathbb{R}_x^n)$ when $\psi \in \mathcal{S}(\mathbb{R}_x^n)$ (the proof of the continuity easily follows: see the exercise below). For any multi-indices α', β' in $\mathbb{N}^n$ the function $x^{\alpha'}\partial_x^{\beta'}\widehat{A}\psi(x)$ is, by Leibniz's rule, a linear combination with complex coefficients of terms of the type

$$F_{\alpha,\beta}(x) = \int p^\beta \left(\int x^\alpha e^{\frac{i}{\hbar}\langle p,x-y\rangle} \partial_x^\beta a(\tfrac{1}{2}(x+y),p)\psi(y)d^n y\right) d^n p \tag{6.41}$$

with $|\alpha|+|\beta| = |\alpha'|+|\beta'|$; since $x^\alpha\partial_x^\beta a$ is polynomially bounded with the same value of δ as a, it follows that the integral in y is again $O(|p|^{-\infty})$ so that $|F_{\alpha,\beta}(x)| < \infty$. Hence $x^{\alpha'}\partial_x^{\beta'}\widehat{A}\psi(x)$ is a bounded function for every pair (α', β') of multi-indices, and $\widehat{A}\psi \in \mathcal{S}(\mathbb{R}_x^n)$ as claimed. □

Exercise 6.24. Finish the proof of Theorem 6.23 by showing the continuity of $\widehat{A}$ on $\mathcal{S}(\mathbb{R}_x^n)$. [Hint: notice that $p^\beta e^{\frac{i}{\hbar}\langle p,x-y\rangle} = (-i\hbar)^{|\beta|}\partial_y^\beta(e^{\frac{i}{\hbar}\langle p,x-y\rangle})$ and use thereafter partial integrations in y in (6.41) to rewrite $F_{\alpha,\beta}(x)$ in a convenient way making terms $\partial_x^\gamma(x^\lambda\psi)$ appear in the integrand. Conclude using Leibniz's rule.]

6.3.2 Composition of Weyl operators

Assume that $\widehat{A}$ and $\widehat{B}$ are Weyl operators corresponding to symbols a and b satisfying conditions of Theorem 6.23:

$$\widehat{A}\psi(x) = \left(\tfrac{1}{2\pi\hbar}\right)^n \widetilde{\iint} e^{\frac{i}{\hbar}\langle p,x-y\rangle} a(\tfrac{1}{2}(x+y),p)\psi(y)d^nyd^np,$$

$$\widehat{B}\psi(x) = \left(\tfrac{1}{2\pi\hbar}\right)^n \widetilde{\iint} e^{\frac{i}{\hbar}\langle p,x-y\rangle} b(\tfrac{1}{2}(x+y),p)\psi(y)d^nyd^np.$$

Since each of these operators sends $\mathcal{S}(\mathbb{R}^n_x)$ into itself we may compose them. The natural question which immediately arises is whether the product $\widehat{A}\widehat{B}$ is itself a Weyl operator, and if it is the case, what is then its Weyl symbol? A first remark is that:

Lemma 6.25. *Assume that A and B are operators with kernels K_A and K_B belonging to $\mathcal{S}(\mathbb{R}^{2n}_{x,y})$. Then the kernel K_C of $C = AB$ is in $\mathcal{S}(\mathbb{R}^{2n}_{x,y})$ and is given by the formula*

$$K_C(x,y) = \int K_A(x,z)K_B(z,y)d^nz. \tag{6.42}$$

Proof. In view of Cauchy–Schwarz's inequality,

$$|K_C(x,y)|^2 \leq \left(\int |K_A(x,x')|^2 d^nx'\right)\left(\int |K_B(x',y)|^2 d^nx'\right)$$

and hence

$$\int |K_C(x,y)|^2 d^nxd^ny \leq \int \left(\int |K_A(x,x')|^2 d^nx'\right)\left(\int |K_B(x',y)|^2 d^nx'\right) d^nxd^ny$$

which yields

$$\int |K_C(x,y)|^2 d^nxd^ny \leq \int \left(\int |K_A(x,x')|^2 d^nx'd^nx\right)\left(\int |K_B(x',y)|^2 d^nx'\right) d^ny,$$

that is

$$\int |K_C(x,y)|^2 d^nxd^ny = ||K_A||_{L^2(\mathbb{R}^{2n}_{x,y})}||K_B||_{L^2(\mathbb{R}^{2n}_{x,y})} < \infty.$$

Let us next show that $K_C(x,y)$ given by (6.42) indeed is the kernel of AB. We have, for $\psi \in \mathcal{S}(\mathbb{R}^n_x)$,

$$\begin{aligned} AB\psi(x) &= \int K_A(x,y)B\psi(y)d^ny \\ &= \int K_A(x,y)\left(\int K_B(y,z)\psi(z)d^nz\right)d^ny; \end{aligned}$$

since the integrands are rapidly decreasing we may apply Fubini's theorem, which yields

$$AB\psi(x) = \int \left(\int K_A(x,y)K_B(y,z)d^n y \right) \psi(z)d^n z,$$

that is (6.42). That $K_C \in \mathcal{S}(\mathbb{R}^{2n}_{x,y})$ is immediate to check by differentiating under the integral sign. □

Remark 6.26. Formula (6.42) remains valid under less stringent conditions than $K_{\widehat{A}}, K_{\widehat{B}} \in \mathcal{S}(\mathbb{R}^{2n}_{x,y})$ (the "integral" can be interpreted as a distributional bracket for fixed x and z). We will study in some detail the case where the kernels $K_{\widehat{A}}$ and $K_{\widehat{B}}$ are in $L^2(\mathbb{R}^{2n}_{x,y})$ in Section 9.2 of Chapter 9.

Proposition 6.27. *Let $\widehat{A} \overset{\text{Weyl}}{\longleftrightarrow} a$ and $\widehat{B} \overset{\text{Weyl}}{\longleftrightarrow} b$ be Weyl operators. Then $\widehat{C} = \widehat{A}\widehat{B}$ has (when defined) Weyl symbol*

$$c(z) = \left(\tfrac{1}{4\pi\hbar}\right)^{2n} \iint e^{\frac{i}{2\hbar}\sigma(z',z'')} a(z + \tfrac{1}{2}z')b(z - \tfrac{1}{2}z'')d^{2n}z'd^{2n}z''. \tag{6.43}$$

Proof. Assume that the Weyl symbols a, b of $\widehat{A}$ and $\widehat{B}$ are in $\mathcal{S}(\mathbb{R}^{2n}_z)$. In view of formula (6.42) for the kernel of $\widehat{A}\widehat{B}$, and formula (6.23) expressing the kernel in terms of the symbol, we have

$$K_{\widehat{A}\widehat{B}}(x,y) = \left(\tfrac{1}{2\pi\hbar}\right)^{2n} \times \iiint e^{\frac{i}{\hbar}(\langle x-\alpha,\zeta\rangle + \langle \alpha - y,\xi\rangle)} a(\tfrac{1}{2}(x+\alpha),\zeta)b(\tfrac{1}{2}(x+y),\xi)d^n\alpha d^n\zeta d^n\xi.$$

In view of formula (6.25), which reads in our case

$$c(x,p) = \int e^{-\frac{i}{\hbar}\langle p,u\rangle} K_{\widehat{A}\widehat{B}}(x + \tfrac{1}{2}u, x - \tfrac{1}{2}u)d^n u,$$

the symbol of $\widehat{A}\widehat{B}$ is thus

$$c(z) = \left(\tfrac{1}{\pi\hbar}\right)^{2n} \times \iiiint e^{\frac{i}{\hbar}Q} a(\tfrac{1}{2}(x + \alpha + \tfrac{1}{2}u),\zeta)b(\tfrac{1}{2}(x+\alpha - \tfrac{1}{2}u),\xi)d^n\alpha d^n\zeta d^n u d^n\xi$$

where the phase Q is given by

$$\begin{aligned} Q &= \langle x - \alpha + \tfrac{1}{2}u, \zeta\rangle + \langle \alpha - x + \tfrac{1}{2}u, \xi\rangle - \langle u, p\rangle \\ &= \langle x - \alpha + \tfrac{1}{2}u, \zeta - p\rangle + \langle \alpha - x + \tfrac{1}{2}u, \xi - p\rangle. \end{aligned}$$

Setting $\zeta' = \zeta - p$, $\xi' = \xi - p$, $\alpha' = \frac{1}{2}(\alpha - x + \frac{1}{2}u)$ and $u' = \frac{1}{2}(\alpha - x - \frac{1}{2}u)$ we have

$$d^n\alpha d^n\zeta d^n u d^n\xi = 2^{2n}(d^n\alpha' d^n\zeta' d^n u' d^n\xi') \quad \text{and} \quad Q = 2\sigma(u',\xi';\alpha',\zeta')$$

and hence

$$c(z) = \left(\tfrac{1}{\pi\hbar}\right)^{2n}$$
$$\times \iiiint e^{\frac{2i}{\hbar}\sigma(u',\xi';\alpha',\zeta')} a(x+\alpha', p+\zeta') b(x+u', p+\xi') d^n\alpha' d^n\zeta' d^n u' d^n\xi';$$

formula (6.43) follows setting $z' = 2(\alpha', \zeta')$ and $z'' = -2(u', \xi')$. □

Remark 6.28. Expanding the integrand in formula (6.43) and using repeated integrations by parts one finds that one can formally write

$$c(z) = a(z) e^{\frac{i\hbar}{2}\overleftrightarrow{L}}(z) \tag{6.44}$$

where $\overleftrightarrow{L} = \overleftarrow{\partial}_z^T J \overrightarrow{\partial}_z$, the arrows indicating the direction in which the partial derivatives act[5]. This notation is commonly used in physics.

Remark 6.29. If we view $\hbar$ as a variable parameter, then the symbol c of the composed operator will generally depend on $\hbar$, even if a and b do not.

We now assume that the Weyl operators

$$\widehat{A} = \left(\tfrac{1}{2\pi\hbar}\right)^n \int a_\sigma(z)\widehat{T}(z) d^{2n}z \quad , \quad \widehat{B} = \left(\tfrac{1}{2\pi\hbar}\right)^n \int b_\sigma(z)\widehat{T}(z) d^{2n}z$$

can be composed (this is the case for instance if a_σ and b_σ are in $\mathcal{S}(\mathbb{R}^{2n}_z)$) and set $\widehat{C} = \widehat{A}\widehat{B}$. Assuming that we can write

$$\widehat{C} = \left(\tfrac{1}{2\pi\hbar}\right)^n \int c_\sigma(z)\widehat{T}(z) d^{2n}z$$

we ask: what is c_σ? The answer is given by the following theorem:

Theorem 6.30. *The twisted symbol of the composed operator $\widehat{C} = \widehat{A}\widehat{B}$ is the function c_σ, defined by*

$$c_\sigma(z) = \left(\tfrac{1}{2\pi\hbar}\right)^n \int e^{\frac{i}{2\hbar}\sigma(z,z')} a_\sigma(z-z') b_\sigma(z') d^{2n}z' \tag{6.45}$$

or, equivalently,

$$c_\sigma(z) = \left(\tfrac{1}{2\pi\hbar}\right)^n \int e^{-\frac{i}{2\hbar}\sigma(z,z')} a_\sigma(z') b_\sigma(z-z') d^{2n}z'. \tag{6.46}$$

Proof. Writing

$$\widehat{A} = \left(\tfrac{1}{2\pi\hbar}\right)^n \int a_\sigma(z_0)\widehat{T}(z_0) d^{2n}z_0 \quad , \quad \widehat{B} = \left(\tfrac{1}{2\pi\hbar}\right)^n \int b_\sigma(z_1)\widehat{T}(z_1) d^{2n}z_1$$

[5] Littlejohn [112] calls $\overleftrightarrow{L}$ the "Janus operator" (it is double-faced!)

we have, using the equality

$$\widehat{T}(z_0+z_1)=e^{-\frac{i}{2\hbar}\sigma(z_0,z_1)}\widehat{T}(z_0)\widehat{T}(z_1)$$

(formula (6.11)),

$$\begin{aligned}\widehat{T}(z_0)\widehat{B}&=\left(\tfrac{1}{2\pi\hbar}\right)^n\int b_\sigma(z_1)\widehat{T}(z_0)\widehat{T}(z_1)d^{2n}z_1\\&=\left(\tfrac{1}{2\pi\hbar}\right)^n\int e^{\frac{i}{2\hbar}\sigma(z_0,z_1)}b_\sigma(z_1)\widehat{T}(z_0+z_1)d^{2n}z_1\end{aligned}$$

and hence

$$\widehat{A}\widehat{B}=\left(\tfrac{1}{2\pi\hbar}\right)^{2n}\iint e^{\frac{i}{2\hbar}\sigma(z_0,z_1)}a_\sigma(z_0)b_\sigma(z_1)\widehat{T}(z_0+z_1)d^{2n}z_0d^{2n}z_1.$$

Setting $z=z_0+z_1$ and $z'=z_1$ this can be written

$$\widehat{A}\widehat{B}=\left(\tfrac{1}{2\pi\hbar}\right)^{2n}\int\left(\int e^{\frac{i}{2\hbar}\sigma(z,z')}a_\sigma(z-z')b_\sigma(z')d^{2n}z'\right)\widehat{T}(z)d^{2n}z,$$

hence (6.45). Formula (6.46) follows by a trivial change of variables and using the antisymmetry of σ. □

Remark 6.31. We urge the reader to note that in the proof above we did not at any moment use the explicit definition of the Weyl–Heisenberg operators $\widehat{T}(z)$, but only the property

$$\widehat{T}(z_0+z_1)=e^{-\frac{i}{2\hbar}\sigma(z_0,z_1)}\widehat{T}(z_0)\widehat{T}(z_1)$$

of these operators. We will keep this remark in mind when we study, later on, phase-space Weyl calculus.

We can express formulae (6.45) and (6.46) in a more compact way using the notion of twisted convolution, essentially due to Grossmann *et al.* [83]:

Definition 6.32. Let λ be a real constant, $\lambda\neq 0$. The "λ-twisted convolution product" of two functions $f,g\in\mathcal{S}(\mathbb{R}^{2n}_z)$ is the function $f*_\lambda g\in\mathcal{S}(\mathbb{R}^{2n}_z)$ defined by

$$(f*_\lambda g)(z)=\int e^{\frac{i}{2\lambda}\sigma(z,z')}f(z-z')g(z')d^nz'. \tag{6.47}$$

Observe that as opposed with ordinary convolution, we have in general $f*_\lambda g\neq g*_\lambda f$. In fact, it immediately follows from the definition, performing a trivial change of variables and using the antisymmetry of the symplectic form, that

$$f*_\lambda g=g*_{-\lambda}f. \tag{6.48}$$

Theorem 6.30 can be restated in terms of the twisted convolution as:

Corollary 6.33. *The twisted Weyl symbol of the composed operator* $\widehat{C} = \widehat{A}\widehat{B}$ *is (when defined) the function*

$$c_\sigma = \left(\tfrac{1}{2\pi\hbar}\right)^{2n} a_\sigma *_{-\hbar} b_\sigma. \tag{6.49}$$

Exercise 6.34. Let $f, g \in \mathcal{S}(\mathbb{R}^{2n}_z)$. Show that:

$$\mathcal{F}_\sigma(a *_{-2\hbar} b) = (\mathcal{F}_\sigma a) *_{-2\hbar} b = a *_{-2\hbar} \mathcal{F}_\sigma b, \tag{6.50}$$

$$\mathcal{F}_\sigma a *_{-2\hbar} \mathcal{F}_\sigma b = a *_{-2\hbar} b \tag{6.51}$$

where $\mathcal{F}_\sigma$ is the symplectic Fourier transform (6.13).

6.3.3 Quantization versus dequantization

As we have pointed out before, "quantization" of classical observables (which are candidates for being pseudo-differential symbols) is not a uniquely well-defined procedure: we have exposed so far one theory of quantization, whose characteristic feature is that it has the property of symplectic covariance: composition of the observable with a linear symplectic transformation corresponds, on the operator level, to conjugation by a metaplectic operator. There are in fact many other ways to quantize an observable, see for instance Nazaikiinskii *et al.* [128] for other schemes which are very efficient in many cases of mathematical and physical interest (Maxwell equations, for instance). From this point of view, the justification of a given quantization scheme is beyond the realm of mathematics: it is the physicist's responsibility to decide, by confronting experiences in the "real world", to decide in each case which is the "good theory". What we mathematicians can do instead is the following: given a quantum observable (= operator) we can try to see whether this quantum observable originates, via some quantization scheme, from a classical observable. In the language of pseudo-differential calculus this amounts to asking if it is possible (within the framework of some given calculus) to associate to an operator a well-defined (and unique) symbol. This is the problem of *dequantization.* We are going to answer (at least partially) that question in the case of Weyl calculus, following Gracia-Bondia [77].

Recall that if $a \in \mathcal{S}(\mathbb{R}^{2n}_z)$, the associated Weyl operator is defined by the Bochner integral

$$\widehat{A} = \left(\tfrac{1}{2\pi\hbar}\right)^n \int a_\sigma(z)\widehat{T}(z)d^{2n}z$$

where $\widehat{T}(z)$ is the Weyl–Heisenberg operator; equivalently

$$\widehat{A} = \left(\tfrac{1}{\pi\hbar}\right)^n \int a(z)\widetilde{T}(z)d^{2n}z \tag{6.52}$$

where $\widetilde{T}(z)$ is the Grossmann–Royer operator:

$$\widetilde{T}(z_0)\psi(x) = e^{\frac{2i}{\hbar}\langle p_0, x-x_0\rangle}\psi(2x_0 - x).$$

Formula (6.52) can be written, for $\psi \in \mathcal{S}(\mathbb{R}^n_x)$, as

$$\widehat{A}\psi = \left(\tfrac{1}{\pi\hbar}\right)^n \left\langle a(\cdot), \widetilde{T}(\cdot)\psi \right\rangle \tag{6.53}$$

where $\langle\cdot,\cdot\rangle$ denotes the usual distributional bracket:

$$\langle\cdot,\cdot\rangle : \mathcal{S}'(\mathbb{R}^{2n}_z) \times \mathcal{S}(\mathbb{R}^{2n}_z) \longrightarrow \mathbb{C}.$$

The problem of dequantization can now, within our framework, be stated as follows: given an operator $\widehat{A} : \mathcal{S}(\mathbb{R}^{2n}_z) \longrightarrow \mathcal{S}'(\mathbb{R}^{2n}_z)$ how can we find a satisfying (6.53)? We are going to answer this question; let us first introduce the following terminology: if $\widehat{A}$ is an operator with kernel $K \in \mathcal{S}(\mathbb{R}^n_x \times \mathbb{R}^n_y)$ we call the real number

$$\operatorname{Tr}\widehat{A} = \int K(x,x)d^nx$$

the trace of $\widehat{A}$. (We will study in detail the notion of trace, and that of trace-class operator, in Chapter 9.)

Theorem 6.35. *Let $a \in \mathcal{S}(\mathbb{R}^{2n}_z)$ and $\widehat{A} \overset{\text{Weyl}}{\longleftrightarrow} a$. The trace of $\widehat{A}\widetilde{T}(z)$ and $\widetilde{T}(z)\widehat{A}$ are defined in the sense above, and:*

(i) *We have*

$$a(z) = 2^n \operatorname{Tr}(\widehat{A}\widetilde{T}(z)) = 2^n \operatorname{Tr}(\widetilde{T}(z)\widehat{A}); \tag{6.54}$$

(ii) *Let $\widehat{A}$ be a bounded operator. The mapping*

$$\mathcal{S}(\mathbb{R}^{2n}_z) \longrightarrow \mathbb{C} \ \ , \ \ b \longmapsto (2\pi\hbar)^n \operatorname{Tr}(\widehat{A}\widehat{B}) \tag{6.55}$$

coincides with a in the distributional sense.

Proof. (i) We have, for $\psi \in \mathcal{S}(\mathbb{R}^n_x)$,

$$\widehat{A}\widetilde{T}(z)\psi(x) = \left(\tfrac{1}{2\pi\hbar}\right)^n \iint e^{\frac{i}{\hbar}\langle p, x-y\rangle} a(\tfrac{1}{2}(x+y), p) e^{\frac{2i}{\hbar}\langle p_0, y-x_0\rangle} \psi(2x_0 - y) d^ny d^np;$$

setting $y' = 2x_0 - y$ the kernel of $\widehat{A}\widetilde{T}(z)$ is thus the function

$$K(x,y') = \left(\tfrac{1}{2\pi\hbar}\right)^n e^{\frac{2i}{\hbar}\langle p_0, x_0 - y'\rangle} \int e^{\frac{i}{\hbar}\langle p, x+y'-2x_0\rangle} a(\tfrac{1}{2}(x-y') + x_0, p) d^np$$

so that

$$K(x,x) = \left(\tfrac{1}{2\pi\hbar}\right)^n e^{\frac{2i}{\hbar}\langle p_0, x_0 - x\rangle} \int e^{\frac{2i}{\hbar}\langle p, x-x_0\rangle} a(x_0, p) d^np.$$

Integrating in x yields

$$\begin{aligned}\int K(x,x)d^nx &= \left(\tfrac{1}{2\pi\hbar}\right)^n \int \left[\int e^{\frac{2i}{\hbar}\langle p-p_0, x\rangle} d^nx\right] e^{\frac{2i}{\hbar}\langle p_0 - p, x_0\rangle} a(x_0, p) d^np \\ &= 2^{-n} \int \delta(p - p_0) e^{\frac{2i}{\hbar}\langle p_0 - p, x_0\rangle} a(x_0, p) d^np \\ &= 2^{-n} a(x_0, p_0)\end{aligned}$$

which proves the first equality (6.54); the second is proven in the same way.

(ii) Assume first $a \in \mathcal{S}(\mathbb{R}^{2n}_z)$. We have, using (6.52),

$$\operatorname{Tr}(\widehat{A}\widehat{B}) = \left(\tfrac{1}{\pi\hbar}\right)^n \operatorname{Tr}\left[\widehat{A}\int b(z)\widetilde{T}(z)d^{2n}z\right],$$

that is, since $\widehat{A}$ is bounded,

$$\begin{aligned}\operatorname{Tr}(\widehat{A}\widehat{B}) &= \left(\tfrac{1}{\pi\hbar}\right)^n \operatorname{Tr}\left[\int b(z)\widehat{A}\widetilde{T}(z)d^{2n}z\right] \\ &= \left(\tfrac{1}{\pi\hbar}\right)^n \int b(z)\operatorname{Tr}(\widehat{A}\widetilde{T}(z))d^{2n}z\end{aligned}$$

(see exercise below for the justification of the second equality). Taking into account (6.54) this is

$$\operatorname{Tr}(\widehat{A}\widehat{B}) = \left(\tfrac{1}{2\pi\hbar}\right)^n \int b(z)a(z)d^{2n}z = \left(\tfrac{1}{2\pi\hbar}\right)^n \langle a, b\rangle .$$

The case of a general bounded $\widehat{A}$ follows by continuity. □

Exercise 6.36. Justify the fact that $\operatorname{Tr}\left[\int b(z)\widehat{A}\widetilde{T}(z)d^{2n}z\right] = \int b(z)\operatorname{Tr}(\widehat{A}\widetilde{T}(z))d^{2n}z$. [Hint: pass to the kernels; alternatively use Proposition 9.19 in Chapter 9.

6.4 The Wigner–Moyal Transform

In this section we study a very important device, the *Wigner–Moyal transform.* It was introduced by Wigner[6] [180] as a device allowing one to express quantum mechanical expectation values in the same form as the averages of classical statistical mechanics (see formula (6.70); in a footnote he however reports that he found the expression for the transform in collaboration with Szilard. Recognition of the fundamental relationship between the Weyl pseudo-differential calculus and the Wigner transform probably goes back to Moyal's paper [127] – many years before mathematicians spoke about "Weyl calculus"! There are of course many good introductory texts for this topic; a nice review can be found in Marchiolli's paper [118]. The subject has experienced an intense revival of interest both in pure and applied mathematics during the last two decades; the Wigner–Moyal transform is also being used as an important tool in signal analysis (time-frequency analysis); see Gröchenig [78] for a review of the state of the art. For applications to quantum optics see for instance Schleich [142] or Simon *et al.* [153]. We will use it in our treatment of phase-space quantum mechanics in Chapter 10 where it will be instrumental in the derivation of the Schrödinger equation in phase space and for the understanding of the meaning of the solutions to that equation.

[6] But he acknowledges previous (unpublished) work of L. Szilard.

6.4.1 Definition and first properties

Consider, for $\psi, \phi \in \mathcal{S}(\mathbb{R}^n_x)$, the tensor product $K_{\psi,\phi} = \psi \otimes \overline{\phi}$, that is

$$K_{\psi,\phi}(x, y) = \psi(x)\overline{\phi}(y)$$

and let us ask what the associated Weyl operator $\widehat{A}_{\psi,\phi}$ looks like. In view of formula (6.25) in Theorem 6.12 the symbol of that operator is

$$\begin{aligned} a_{\psi,\phi}(x,p) &= \int e^{-\frac{i}{\hbar}\langle p,y\rangle} K_{\widehat{A}}(x + \tfrac{1}{2}y, x - \tfrac{1}{2}y) d^n y \\ &= \int e^{-\frac{i}{\hbar}\langle p,y\rangle} \psi(x + \tfrac{1}{2}y)\overline{\phi(x - \tfrac{1}{2}y)} d^n y. \end{aligned}$$

The product of this function by $(2\pi\hbar)^{-n}$ is, by definition, the *Wigner–Moyal transform* of the pair (ψ, ϕ):

Definition 6.37. Let $\psi, \phi \in \mathcal{S}(\mathbb{R}^n_x)$.

(i) The symbol of the Weyl operator with kernel $(2\pi\hbar)^{-n}\psi \otimes \overline{\phi}$ is called the *"Wigner–Moyal transform" $W(\psi, \phi)$ of the pair (ψ, ϕ)*:

$$W(\psi, \phi)(z) = \left(\tfrac{1}{2\pi\hbar}\right)^n \int e^{-\frac{i}{\hbar}\langle p,y\rangle} \psi(x + \tfrac{1}{2}y)\overline{\phi(x - \tfrac{1}{2}y)} d^n y. \tag{6.56}$$

(ii) The function $W\psi = W(\psi, \psi)$ is called the "*Wigner* transform"[7] of ψ:

$$W\psi(z) = \left(\tfrac{1}{2\pi\hbar}\right)^n \int e^{-\frac{i}{\hbar}\langle p,y\rangle} \psi(x + \tfrac{1}{2}y)\overline{\psi(x - \tfrac{1}{2}y)} d^n y. \tag{6.57}$$

Remark 6.38. In some texts one uses different normalizations; for instance the factor $(2\pi\hbar)^{-n/2}$ often appears instead of our $(2\pi\hbar)^{-n}$. Our choice is consistent with the normalization of the symplectic Fourier transform (6.13).

The Wigner transform thus allows us to associate phase space functions to functions defined on "configuration space" $\mathbb{R}^n_x$. It is thus an object of choice for phase space quantization techniques, allowing the passage from representation-dependent quantum mechanics to quantum mechanics in phase space.

The Wigner–Moyal transform can be expressed in a quasi-trivial way using the Grossmann–Royer operators

$$\widetilde{T}(z_0)\psi(x) = e^{\frac{2i}{\hbar}\langle p_0, x - x_0\rangle}\psi(2x_0 - x) \tag{6.58}$$

introduced in Subsection 5.5.3 of Chapter 5:

[7] It is sometimes called the "Wigner–Blokhintsev transform".

Proposition 6.39. *The Wigner–Moyal transform of a pair* (ψ, ϕ) *of functions belonging to* $\mathcal{S}(\mathbb{R}^n_x)$ *is given by*

$$W(\psi,\phi)(z) = \left(\tfrac{1}{\pi\hbar}\right)^n (\widetilde{T}(z)\psi, \phi)_{L^2(\mathbb{R}^n_x)} \tag{6.59}$$

where $\widetilde{T}(z)$ *is the Grossmann-Royer operator associated to* z*; hence in particular:*

$$W\psi(z) = \left(\tfrac{1}{\pi\hbar}\right)^n (\widetilde{T}(z)\psi, \psi)_{L^2(\mathbb{R}^n_x)}. \tag{6.60}$$

Proof. We have

$$(\widetilde{T}(z_0)\psi, \phi)_{L^2(\mathbb{R}^n_x)} = \int e^{\frac{2i}{\hbar}\langle p_0, x - x_0\rangle} \psi(2x_0 - x)\overline{\phi(x)} d^n x.$$

Setting $y = 2(x_0 - x)$ this is

$$\begin{aligned}(\widetilde{T}(z_0)\psi, \phi)_{L^2(\mathbb{R}^n_x)} &= 2^{-n} \int e^{-\frac{i}{\hbar}\langle p_0, y\rangle} \psi(x_0 + \tfrac{1}{2}y)\overline{\phi(x_0 - \tfrac{1}{2}y)} d^n x \\ &= (\pi\hbar)^n W(\psi, \phi)(z_0)\end{aligned}$$

proving (6.59). □

Note that the Wigner–Moyal transform obviously exists for larger classes of functions than those which are rapidly decreasing; in fact $W(\psi, \phi)$ is defined for all $\psi, \phi \in L^2(\mathbb{R}^n_x)$, as follows from the Cauchy–Schwarz inequality

$$|W(\psi,\phi)(z)| \leq \left(\tfrac{1}{\pi\hbar}\right)^n ||\psi||_{L^2(\mathbb{R}^n_x)} ||\phi||_{L^2(\mathbb{R}^n_x)}$$

which immediately follows from formula (6.59). More generally, it can be extended to pairs of tempered distributions; to prove this we will establish an important formula, called the *Moyal identity* in the mathematical and physical literature:

Proposition 6.40. *The Wigner–Moyal transform is a bilinear mapping*

$$W : \mathcal{S}(\mathbb{R}^n_x) \times \mathcal{S}(\mathbb{R}^n_x) \longrightarrow \mathcal{S}(\mathbb{R}^n_x)$$

satisfying the "Moyal identity"

$$(W(\psi,\phi), W(\psi',\phi'))_{L^2(\mathbb{R}^{2n}_z)} = \left(\tfrac{1}{2\pi\hbar}\right)^n (\psi,\psi')_{L^2(\mathbb{R}^n_x)} \overline{(\phi,\phi')_{L^2(\mathbb{R}^n_x)}} \tag{6.61}$$

and hence extends to a bilinear mapping $\mathcal{S}'(\mathbb{R}^n_x) \times \mathcal{S}'(\mathbb{R}^n_x) \longrightarrow \mathcal{S}'(\mathbb{R}^n_x)$*. In particular*

$$(W\psi, W\psi')_{L^2(\mathbb{R}^{2n}_z)} = \left(\tfrac{1}{2\pi\hbar}\right)^n |(\psi,\psi')_{L^2(\mathbb{R}^n_x)}|^2. \tag{6.62}$$

Proof. The first statement is clear, and the last follows from (6.61) in view of the density of $\mathcal{S}(\mathbb{R}^n_x)$ in $\mathcal{S}'(\mathbb{R}^n_x)$. Let us prove (6.61). Setting

$$(2\pi\hbar)^{2n} A = (W(\psi,\phi), W(\psi',\phi'))_{L^2(\mathbb{R}^{2n}_z)}$$

we have

$$A = \iiint e^{-\frac{i}{\hbar}\langle p,y-y'\rangle}\psi(x+\tfrac{1}{2}y)\overline{\psi'}(x+\tfrac{1}{2}y') \\ \times \overline{\phi}(x-\tfrac{1}{2}y)\phi'(x-\tfrac{1}{2}y')d^n y d^n y' d^n x d^n p,$$

the integral in p is $(2\pi\hbar)^n\partial(y-y')$ and hence

$$A = (2\pi\hbar)^n \iint \psi(x+\tfrac{1}{2}y)\overline{\psi'}(x-\tfrac{1}{2}y)\overline{\phi}(x+\tfrac{1}{2}y)\phi'(x-\tfrac{1}{2}y)d^n y d^n y' d^n x.$$

Setting $u = x + \frac{1}{2}y$ and $v = x - \frac{1}{2}y$ we get

$$A = (2\pi\hbar)^n \int \psi(u)\overline{\psi'(u)}d^n u \int \overline{\phi(v)}\phi'(v)d^n v,$$

whence (6.61). □

For further use, let us compute the symplectic Fourier transform

$$W_\sigma(\psi,\phi) = \mathcal{F}_\sigma W(\psi,\phi)$$

of the Wigner–Moyal transform of the pair (ψ,ϕ); we will see that $W_\sigma(\psi,\phi)$ and $W(\psi,\phi)$ are essentially the same, up to scaling factors:

Lemma 6.41. *Let $\psi,\phi \in \mathcal{S}(\mathbb{R}^n_x)$. We have*

$$W_\sigma(\psi,\phi)(z) = \left(\tfrac{1}{2\pi\hbar}\right)^n \int e^{-\frac{i}{\hbar}\langle p,x'\rangle}\psi(x'+\tfrac{1}{2}x)\overline{\phi(x'-\tfrac{1}{2}x)}d^n x', \tag{6.63}$$

that is

$$W_\sigma(\psi,\phi)(z) = 2^{-n}W(\psi,\phi)(\tfrac{1}{2}z). \tag{6.64}$$

Proof. Set $F = \mathcal{F}_\sigma W(\psi,\phi)$; by definition of $\mathcal{F}_\sigma$ and $W(\psi,\phi)$ we have

$$W_\sigma(\psi,\phi)(z) = \left(\tfrac{1}{2\pi\hbar}\right)^{2n} \\ \times \iiint e^{-\frac{i}{\hbar}(\langle p,x'\rangle + \langle p',y-x\rangle)}\psi(x'+\tfrac{1}{2}y)\overline{\phi(x'-\tfrac{1}{2}y)}d^n p' d^n x' d^n y,$$

that is, calculating the integral in p',

$$W_\sigma(\psi,\phi)(z) = \left(\tfrac{1}{2\pi\hbar}\right)^n \iint \delta(x-y)e^{-\frac{i}{\hbar}\langle p,x'\rangle}\psi(x'+\tfrac{1}{2}y)\overline{\phi(x'-\tfrac{1}{2}y)}d^n x' d^n y \\ = \left(\tfrac{1}{2\pi\hbar}\right)^n \iint e^{-\frac{i}{\hbar}\langle p,x'\rangle}\psi(x'+\tfrac{1}{2}x)\overline{\phi(x'-\tfrac{1}{2}x)}d^n x'$$

which is (6.63). Formula (6.64) follows. □

The Heisenberg–Weyl operators behave in a natural way with respect to the Wigner transform:

Proposition 6.42. *For every* $\psi \in L^2(\mathbb{R}^n_x)$ *and* $z_0 \in \mathbb{R}^{2n}_z$ *we have*

$$T(z_0)W\psi = W(\widehat{T}(z_0)\psi). \tag{6.65}$$

Proof. We have, by definition of $\widehat{T}(z_0)$,

$$\widehat{T}(z_0)\psi(x) = e^{\frac{i}{\hbar}(\langle p_0,x\rangle - \frac{1}{2}\langle p_0,x_0\rangle)}\psi(x-x_0)$$

and hence

$$W(\widehat{T}(z_0)\psi) = \left(\tfrac{1}{2\pi\hbar}\right)^n \int e^{-\frac{i}{\hbar}\langle p-p_0,y\rangle}\psi(x-x_0+\tfrac{1}{2}y)\overline{\psi}(x-x_0-\tfrac{1}{2}y)d^ny$$

which is precisely (6.65). □

6.4.2 Wigner transform and probability

One of the best-known – and most widely discussed – properties of the Wigner–Moyal transform is the following. It shows that the Wigner transform has features reminding us of those of a probability distribution:

Proposition 6.43. *Assume that* $\psi, \phi \in L^1(\mathbb{R}^n_x) \cap L^2(\mathbb{R}^n_x)$ *and denote by* $F\psi$ *the* $\hbar$*-Fourier transform of* ψ*. The following properties hold:*

(i) *We have*

$$\int W(\psi,\phi)(z)d^np = \psi(x)\overline{\phi(x)} \quad \textit{and} \quad \int W(\psi,\phi)(z)d^nx = F\psi(p)\overline{F\phi(p)}, \tag{6.66}$$

(ii) *hence, in particular*

$$\int W\psi(z)d^np = |\psi(x)|^2 \quad \textit{and} \quad \int W\psi(z)d^nx = |F\psi(p)|^2. \tag{6.67}$$

Proof. Let us prove the first formula (6.66). Noting that

$$\int e^{-\frac{i}{\hbar}\langle p,y\rangle}d^np = (2\pi\hbar)^n\,\delta(y)$$

we have

$$\begin{aligned}\int W(\psi,\phi)(z)d^np &= \left(\tfrac{1}{2\pi\hbar}\right)^n \int \left(\int e^{-\frac{i}{\hbar}\langle p,y\rangle}d^np\right)\psi(x+\tfrac{1}{2}y)\overline{\phi(x-\tfrac{1}{2}y)}d^ny \\ &= \int \delta(y)\psi(x+\tfrac{1}{2}y)\overline{\phi(x-\tfrac{1}{2}y)}d^ny \\ &= \int \delta(y)\psi(x)\overline{\phi(x)}d^ny \\ &= \psi(x)\overline{\phi(x)}\end{aligned}$$

as claimed. Let us prove the second formula (6.66). Setting $x' = x + \frac{1}{2}y$ and $x'' = x - \frac{1}{2}y$ in the right-hand of the equality

$$\int W\psi(z)d^n x = \left(\tfrac{1}{2\pi\hbar}\right)^n \iint e^{-\frac{i}{\hbar}\langle p,y\rangle}\psi(x+\tfrac{1}{2}y)\overline{\phi(x-\tfrac{1}{2}y)}d^n y d^n x$$

we get

$$\begin{aligned}\int W\psi(z)d^n x &= \left(\tfrac{1}{2\pi\hbar}\right)^n \iint e^{-\frac{i}{\hbar}\langle p,x'\rangle}\psi(x')\overline{e^{-\frac{i}{\hbar}\langle p,x''\rangle}\phi(x'')}d^n x' d^n x'' \\ &= \left(\tfrac{1}{2\pi\hbar}\right)^n \int e^{-\frac{i}{\hbar}\langle p,x'\rangle}\psi(x')d^n x' \overline{\int e^{-\frac{i}{\hbar}\langle p,x''\rangle}\phi(x'')d^n x''} \\ &= F\psi(p)\overline{F\phi(p)}\end{aligned}$$

and we are done. □

It immediately follows from any of the two formulae (6.67) above that

$$\int W\psi(z)d^n z = ||\psi||^2_{L^2(\mathbb{R}^n_x)} = ||F\psi||^2_{L^2(\mathbb{R}^n_p)}. \tag{6.68}$$

If the function ψ is normalized: $||\psi||^2_{L^2(\mathbb{R}^n_x)} = 1$, then so is $W\psi(z)$:

$$\int W\psi(z)d^n z = 1 \text{ if } ||\psi||^2_{L^2(\mathbb{R}^n_x)} = 1.$$

If in addition $W\psi \geq 0$ it would thus be a probability density. For this reason the Wigner transform $W\psi$ of a normalized function is sometimes called a *quasi-probability density* in the literature. In fact, we will prove in Section 8.5 of Chapter 8 (Proposition 8.47) that if ψ is a Gaussian then $W\psi \geq 0$. It is however the only case for which in general $W\psi$ takes negative values:

Exercise 6.44. Assume that $\psi \in \mathcal{S}(\mathbb{R}^n_x)$ is odd: $\psi(-x) = -\psi(x)$. Show that $W\psi$ takes negative values [Hint: calculate $W\psi(0)$.]

The fact that the Wigner transform is pointwise positive only for Gaussians was proved originally by Hudson [94], whose proof was generalized and commented by Janssen [98]. We will however see that when we average $W\psi$ over "sufficiently large" sets (in a sense to be defined), then we will always obtain a non-negative function: this is the manifestation of the fact that the points of the quantum-mechanical phase space are quantum-mechanically admissible sets.

Let us next prove a result due to Moyal and which allows us to express the mathematical expectation of a Weyl operator in terms of the Wigner transform and the Weyl symbol. In fact, formula (6.70) below shows that this expectation is obtained by "averaging" the Weyl symbol $a \overset{\text{Weyl}}{\longleftrightarrow} \widehat{A}$ (viewed as a "classical observable") with respect to $W\psi$.

Proposition 6.45. *Let $a \overset{\text{Weyl}}{\longleftrightarrow} \widehat{A}$; we have, for all $\psi, \phi \in \mathcal{S}(\mathbb{R}^n_x)$,*

$$(\widehat{A}\psi, \phi)_{L^2(\mathbb{R}^n_x)} = \int W(\psi, \phi)(z)a(z)d^{2n}z \tag{6.69}$$

and, in particular

$$(\widehat{A}\psi, \psi)_{L^2(\mathbb{R}^n_x)} = \int W\psi(z)a(z)d^{2n}z. \tag{6.70}$$

Proof. Let us write

$$W = \int W(\psi, \phi)a(z)d^{2n}z.$$

By definition of $W(\psi, \phi)$ we have

$$W = \left(\tfrac{1}{2\pi\hbar}\right)^n \iiint e^{-\frac{i}{\hbar}\langle p,y\rangle} a(z)\psi(x + \tfrac{1}{2}y)\overline{\phi(x - \tfrac{1}{2}y)}d^n y d^n x d^n p,$$

that is, performing the change of variables $u = x + \frac{1}{2}y$, $v = x - \frac{1}{2}y$:

$$W = \left(\tfrac{1}{2\pi\hbar}\right)^n \iiint e^{\frac{i}{\hbar}\langle p,v-u\rangle} a(\tfrac{1}{2}(u + v), p)\psi(u)\overline{\phi(v)}d^n u d^n v d^n p,$$

that is $W = (\widehat{A}\psi, \phi)_{L^2(\mathbb{R}^n_x)}$. □

The following corollary of Proposition 6.45 is due to Moyal:

Corollary 6.46. *Let $a \overset{\text{Weyl}}{\longleftrightarrow} \widehat{A}$ and $\psi \in L^2(\mathbb{R}^n_x)$ be a normed function: $||\psi||_{L^2} = 1$. The mathematical expectation value of $\widehat{A}$ in the state ψ is given by the formula*

$$\langle \widehat{A} \rangle_\psi = \int a(z)W\psi(z)d^{2n}z \quad \text{if} \quad ||\psi||_{L^2} = 1. \tag{6.71}$$

Proof. It is an immediate consequence of formula (6.70) taking into account definition (8.1) of the mathematical expectation value. □

This result is very important, both mathematically, and from the point of view of the statistical interpretation of quantum mechanics. It shows again that the vocation of the Wigner transform is probabilistic, since formula (6.71) expresses the expectation value of the operator $\widehat{A}$ in a given normalized state by averaging its symbol against the Wigner transform of that state. This result, and the fact that the quantities $\int W\psi(z)d^n p$ and $\int W\psi(z)d^n x$ indeed are probability densities have led to metaphysical discussions about "negative probabilities"; see for instance Feynman's contribution in [88].

6.4.3 On the range of the Wigner transform

It is clear that if we pick up an arbitrary function of z it will not in general be the Wigner transform of some function of x. This is because the Wigner transform is a symbol of a very particular type: the Schwartz kernel of the associated Weyl operator is a tensor product. This of course puts strong limits on the range of the Wigner transform. Here is another constraint: it immediately follows from the Cauchy–Schwarz inequality that we have the bound

$$|W(\psi,\phi)(z)| \leq \left(\tfrac{1}{2\pi\hbar}\right)^n ||\psi||_{L^2(\mathbb{R}^n_x)}||\phi||_{L^2(\mathbb{R}^n_x)} \tag{6.72}$$

for ψ, ϕ in $L^2(\mathbb{R}^n_x)$; in particular

$$|W\psi(z)| \leq \left(\tfrac{1}{2\pi\hbar}\right)^n ||\psi||^2_{L^2(\mathbb{R}^n_x)}. \tag{6.73}$$

These inequalities show that a function cannot be the Wigner–Moyal or Wigner transform of functions $\psi,\phi \in L^2(\mathbb{R}^n_x)$ if it is too large in some points.

Let us discuss a little bit more in detail the question of the invertibility of the Wigner transform. We begin by noting that both ψ and $e^{i\alpha}\psi$ ($\alpha \in \mathbb{R}$) have the same Wigner transform, as immediately follows from the definition of $W\psi$. Conversely: if $\psi,\psi' \in \mathcal{S}(\mathbb{R}^n_x)$ then

$$W\psi = W\psi' \Longleftrightarrow \psi = \alpha\psi' \;\;,\;\; |\alpha| = 1. \tag{6.74}$$

Indeed, for fixed x set

$$f(y) = \psi(x + \tfrac{1}{2}y)\overline{\psi}(x - \tfrac{1}{2}y),$$
$$f'(y) = \psi'(x + \tfrac{1}{2}y)\overline{\psi'}(x - \tfrac{1}{2}y);$$

the equality $W\psi = W\psi'$ is then equivalent to the equality of the Fourier transforms of f and f' and hence

$$\psi(x + \tfrac{1}{2}y)\overline{\psi}(x - \tfrac{1}{2}y) = \psi'(x + \tfrac{1}{2}y)\overline{\psi'}(x - \tfrac{1}{2}y)$$

for all x, y; taking $y = 0$ we get $|\psi|^2 = |\psi'|^2$ which proves (6.74).

Let us now show how the Wigner transform can be inverted; this will give us some valuable information on its range.

Proposition 6.47. *Let $\psi \in L^2(\mathbb{R}^n_x)$ and $a \in \mathbb{R}^n_x$ be such that $\psi(a) \neq 0$.*

(i) *Then*

$$\psi(x) = \frac{1}{\overline{\psi(a)}} \int e^{\frac{i}{\hbar}\langle p, x-a\rangle} W\psi(\tfrac{1}{2}(x + a), p)d^np. \tag{6.75}$$

(ii) *Conversely, a function $W \in L^2(\mathbb{R}^n_z)$ is the Wigner transform of some $\psi \in L^2(\mathbb{R}^n_x)$ if and only if there exist functions A and X such that*

$$\int e^{\frac{2i}{\hbar}\langle p, x-a\rangle} W(\tfrac{1}{2}(x + a), p)d^np = A(a)X(x). \tag{6.76}$$

(iii) *When this is the case we have*

$$\psi(x) = \frac{\int e^{\frac{2i}{\hbar}\langle p,x-a\rangle} W(\frac{1}{2}(x+a),p)d^n p}{\left(\int W(a,p)d^n p\right)^{1/2}} \tag{6.77}$$

for every $a \in \mathbb{R}^n_x$ *such that* $\int W(a,p)d^n p \neq 0$.

Proof. (i) For fixed x the formula

$$W\psi(x,p) = \left(\tfrac{1}{2\pi\hbar}\right)^n \int e^{-\frac{i}{\hbar}\langle p,y\rangle} \psi(x+\tfrac{1}{2}y)\overline{\psi(x-\tfrac{1}{2}y)}d^n y$$

shows that the function $p \longmapsto (2\pi\hbar)^{n/2} W\psi(x,p)$ is the Fourier transform of the function $y \longmapsto \psi(x+\frac{1}{2}y)\overline{\psi(x-\frac{1}{2}y)}$; it follows that

$$\psi(x+\tfrac{1}{2}y)\overline{\psi(x-\tfrac{1}{2}y)} = \int e^{-\frac{i}{\hbar}\langle p,y\rangle} W\psi(x,p)d^n p.$$

Setting $y = 2(x-a)$ in the formula above we get

$$\psi(2x-a)\overline{\psi(a)} = \int e^{-\frac{2i}{\hbar}\langle p,x-a\rangle} W\psi(x,p)d^n p$$

and (6.75) follows replacing $2x-a$ by x.

(ii) Assume that there exists ψ such that

$$\psi(x) = \frac{1}{\overline{\psi(a)}} \int e^{\frac{i}{\hbar}\langle p,x-a\rangle} W(\tfrac{1}{2}(x+a),p)d^n p;$$

then the right-hand side must be independent of the choice of a and hence

$$\int e^{\frac{2i}{\hbar}\langle p,x-a\rangle} W(\tfrac{1}{2}(x+a),p)d^n p = \overline{\psi(a)}\psi(x)$$

so that (6.76) holds with $A = \overline{\psi}$ and $B = \psi$. Suppose that, conversely, we can find A and B such that (6.76) holds. Setting $x = a$ we get

$$\int W(a,p)d^n p = A(a)X(a)$$

whence (6.77). □

Chapter 7

The Metaplectic Group

As we have seen, the symplectic group has connected covering groups $\mathrm{Sp}_q(n)$ of all orders $q = 1, 2, \ldots, \infty$. It turns out that the two-fold covering group $\mathrm{Sp}_2(n)$ is particularly interesting in quantum mechanics, because it can be faithfully represented by a group of unitary operators acting on $L^2(\mathbb{R}^n_x)$ (or on $L^2(\mathbb{R}^{2n}_z)$): this group is the *metaplectic group* $\mathrm{Mp}(n)$. It is thus characterized, up to an isomorphism, by the exactness of the sequence

$$0 \longrightarrow \{\pm I\} \longrightarrow \mathrm{Mp}(n) \longrightarrow \mathrm{Sp}(n) \longrightarrow 0.$$

The importance of $\mathrm{Mp}(n)$ in quantum mechanics not only comes from the fact that it intervenes in various symplectic covariance properties, but also because every continuously differentiable path in $\mathrm{Mp}(n)$ passing through the identity is the propagator for the Schrödinger equation associated by the Weyl correspondence to a quadratic Hamiltonian (which is usually time-dependent). In that sense the metaplectic representation of the symplectic group is the shortest and easiest bridge between classical and quantum mechanics.

The metaplectic group has a rather long history, and is a subject of interest both for mathematicians and physicists. The germ of the idea of the metaplectic representation is found in van Hove [169]. It then appears in the work of Segal [147], Shale [149] who observed that the metaplectic representation should be looked upon as an analogue of the spin representation of the orthogonal group. Weil [176] (in a more general setting) elaborated on Siegel's work in number theory; also see the work of Igusa [95] on theta functions. The theory has been subsequently developed by many authors, for instance Buslaev [19], Maslov [119], Leray [107], and the author [56, 58, 61]. The denomination "quadratic Fourier transform" we are using in this book appears in Gaveau *et al.* [45]. For different presentations of the metaplectic group see Folland [42] and Wallach [175]. The first to consider the Maslov index on $\mathrm{Mp}(n)$ was apparently Maslov himself [119]; the general definition we are giving here is due to the author [56, 58, 61]; we will see that it is closely related to the *ALM* index studied in Chapter 3.

7.1 Definition and Properties of Mp(n)

We begin by defining the structure of Mp(n) as a group; that it actually is a connected two-fold covering of Sp(n) will be established in Subsection 7.1.2.

7.1.1 Quadratic Fourier transforms

We have seen in Chapter 2, Subsection 2.2.3, that the symplectic group Sp(n) is generated by the free symplectic matrices

$$S = \begin{bmatrix} A & B \\ C & D \end{bmatrix} \in \mathrm{Sp}(n) \quad , \quad \det B \neq 0.$$

To each such matrix we associated the generating function

$$W(x, x') = \tfrac{1}{2}\left\langle DB^{-1}x, x\right\rangle - \left\langle B^{-1}x, x'\right\rangle + \tfrac{1}{2}\left\langle B^{-1}Ax', x'\right\rangle$$

and we showed that

$$(x, p) = S(x', p') \Longleftrightarrow p = \partial_x W(x, x') \ , \ p' = -\partial_{x'} W(x, x').$$

Conversely, to every polynomial of the type

$$\begin{aligned} &W(x, x') = \tfrac{1}{2}\langle Px, x\rangle - \langle Lx, x'\rangle + \tfrac{1}{2}\langle Qx', x'\rangle \\ &\textit{with } \ P = P^T \ , \ Q = Q^T \ , \textit{ and } \ \det L \neq 0 \end{aligned} \tag{7.1}$$

we can associate a free symplectic matrix, namely

$$S_W = \begin{bmatrix} L^{-1}Q & L^{-1} \\ PL^{-1}Q - L^T & L^{-1}P \end{bmatrix}. \tag{7.2}$$

We now associate an operator $\widehat{S}_{W,m}$ to every S_W by setting, for $f \in \mathcal{S}(\mathbb{R}^n_x)$,

$$\widehat{S}_{W,m} f(x) = \left(\tfrac{1}{2\pi i}\right)^{n/2} \Delta(W) \int e^{iW(x,x')} f(x') d^n x'; \tag{7.3}$$

here $\arg i = \pi/2$ and the factor $\Delta(W)$ is defined by

$$\Delta(W) = i^m \sqrt{|\det L|}; \tag{7.4}$$

the integer m corresponds to a choice of $\arg \det L$:

$$m\pi \equiv \arg \det L \mod 2\pi. \tag{7.5}$$

Notice that we can rewrite definition (7.3) in the form

$$\widehat{S}_{W,m} f(x) = \left(\tfrac{1}{2\pi}\right)^{n/2} \left(e^{-i\frac{\pi}{4}}\right)^{\mu} \Delta(W) \int e^{iW(x,x')} f(x') d^n x' \tag{7.6}$$

where

$$\mu = 2m - n. \tag{7.7}$$

Definition 7.1.

(i) The operator $\widehat{S}_{W,m}$ is called a "quadratic Fourier transform" associated to the free symplectic matrix S_W.

(ii) The class modulo 4 of the integer m is called the "Maslov index" of $\widehat{S}_{W,m}$. The quadratic Fourier transform corresponding to the choices $S_W = J$ and $m = 0$ is denoted by $\widehat{J}$.

The generating function of J being simply $W(x, x') = -\langle x, x'\rangle$, it follows that

$$\widehat{J}f(x) = \left(\tfrac{1}{2\pi i}\right)^{n/2} \int e^{-i\langle x,x'\rangle} f(x')d^n x' = i^{-n/2}Ff(x) \tag{7.8}$$

for $f \in \mathcal{S}(\mathbb{R}^n_x)$; F is the usual Fourier transform. It follows from the Fourier inversion formula that the inverse $\widehat{J}^{-1}$ of $\widehat{J}$ is given by the formula

$$\widehat{J}^{-1}f(x) = \left(\tfrac{i}{2\pi}\right)^{n/2} \int e^{i\langle x,x'\rangle} f(x')dx' = i^{n/2}F^{-1}f(x).$$

Note that the identity operator cannot be represented by an operator $\widehat{S}_{W,m}$ since it is not a free symplectic matrix.

Of course, if m is one choice of Maslov index, then $m+2$ is another equally good choice: to each function W formula (7.3) associates not one but *two* operators $\widehat{S}_{W,m}$ and $\widehat{S}_{W,m+2} = -\widehat{S}_{W,m}$ (this reflects the fact that the operators $\widehat{S}_{W,m}$ are elements of the two-fold covering group of Sp(n)).

Let us define operators $\widehat{V}_{-P}$ and $\widehat{M}_{L,m}$ by

$$\widehat{V}_{-P}f(x) = e^{\frac{i}{2}\langle Px,x\rangle} f(x) \quad , \quad \widehat{M}_{L,m}f(x) = i^m\sqrt{|\det L|}f(Lx). \tag{7.9}$$

We have the following useful factorization result:

Proposition 7.2. *Let W be the quadratic form (7.1).*

(i) *We have the factorization*

$$\widehat{S}_{W,m} = \widehat{V}_{-P}\widehat{M}_{L,m}\widehat{J}\widehat{V}_{-Q}; \tag{7.10}$$

(ii) *The operators $\widehat{S}_{W,m}$ extend to unitary operators $L^2(\mathbb{R}^n_x) \longrightarrow L^2(\mathbb{R}^n_x)$ and the inverse of $\widehat{S}_{W,m}$ is*

$$\widehat{S}_{W,m}^{-1} = \widehat{S}_{W^*,m^*} \text{ with } W^*(x,x') = -W(x',x) \ , \ m^* = n - m. \tag{7.11}$$

Proof. (i) By definition of $\widehat{J}$ we have

$$\widehat{J}f(x) = i^{-n/2}Ff(x) = \left(\tfrac{1}{2\pi i}\right)^{n/2} \int e^{-i\langle x,x'\rangle} f(x')d^n x';$$

the factorization (7.10) immediately follows, noting that

$$\widehat{M}_{L,m}\widehat{J}f(x) = \left(\tfrac{1}{2\pi i}\right)^{n/2} i^m\sqrt{|\det L|} \int e^{-i\langle Lx,x'\rangle} f(x')d^n x'.$$

(ii) The operators $\widehat{V}_{-P}$ and $\widehat{M}_{L,m}$ are trivially unitary, and so is the modified Fourier transform $\widehat{J}$; (ii) We obviously have

$$(\widehat{V}_{-P})^{-1} = \widehat{V}_P \text{ and } (\widehat{M}_{L,m})^{-1} = \widehat{M}_{L^{-1},-m}$$

and $\widehat{J}^{-1}$ is given by

$$\widehat{J}^{-1}f(x) = \left(\tfrac{i}{2\pi\hbar}\right)^{n/2} \int e^{\frac{i}{\hbar}\langle x,x'\rangle} f(x')d^n x'.$$

Writing

$$\widehat{S}_{W,m}^{-1} = \widehat{V}_Q \widehat{J}^{-1} \widehat{M}_{L^{-1},-m} \widehat{V}_P$$

and noting that

$$\begin{aligned}\widehat{J}^{-1}\widehat{M}_{L^{-1},-m}f(x) &= \left(\tfrac{i}{2\pi}\right)^{n/2} i^{-m} \sqrt{|\det L^{-1}|} \int e^{i\langle x,x'\rangle} f(L^{-1}x')d^n x' \\ &= \left(\tfrac{1}{2\pi i}\right)^{n/2} i^{-m+n} \sqrt{|\det L|} \int e^{i\langle L^T x,x'\rangle} f(x')d^n x' \\ &= \widehat{M}_{-L^T,n-m} \widehat{J} f(x),\end{aligned}$$

the inversion formulas (7.11) follow. □

It follows from the proposition above that the operators $\widehat{S}_{W,m}$ form a subset, closed under the operation of inversion, of the group $\mathcal{U}(L^2(\mathbb{R}^n_x))$ of unitary operators acting on $L^2(\mathbb{R}^n_x)$. They thus generate a subgroup of that group.

Definition 7.3. The subgroup of $\mathcal{U}(L^2(\mathbb{R}^n_x))$ generated by the quadratic Fourier transforms $\widehat{S}_{W,m}$ is called the "metaplectic group $\mathrm{Mp}(n)$". The elements of $\mathrm{Mp}(n)$ are called "metaplectic operators".

Every $\widehat{S} \in \mathrm{Mp}(n)$ is thus, by definition, a product $\widehat{S}_{W_1,m_1} \cdots \widehat{S}_{W_k,m_k}$ of metaplectic operators associated to free symplectic matrices.

We will use the following result which considerably simplifies many arguments (*cf.* Proposition 2.36, Subsection 2.2.3, Chapter 2):

Lemma 7.4. *Every $\widehat{S} \in \mathrm{Mp}(n)$ can be written as a product of exactly two quadratic Fourier transforms: $\widehat{S} = \widehat{S}_{W,m}\widehat{S}_{W',m'}$. (Such a factorization is, however, never unique: for instance $I = \widehat{S}_{W,m}\widehat{S}_{W^*,m^*}$ for every generating function W.)*

We will not prove this here, and refer to Leray [107], Ch. 1, or de Gosson [61].

Exercise 7.5. Show that quadratic Fourier transform $\widehat{S}_{W,m}$ cannot be a local operator (a local operator on $\mathcal{S}'(\mathbb{R}^n_x)$ is an operator $\widehat{S}$ such that $\mathrm{Supp}(\widehat{S}f) \subset \mathrm{Supp}(f)$ for $f \in \mathcal{S}'(\mathbb{R}^n_x)$).

7.1.2 The projection $\pi^{\mathrm{Mp}} : \mathrm{Mp}(n) \longrightarrow \mathrm{Sp}(n)$

We are going to show that Mp(n) is a double covering of the symplectic group Sp(n) and hence a faithful representation of $\mathrm{Sp}_2(n)$.

We will denote the elements of the dual $(\mathbb{R}_z^{2n})^*$ of $\mathbb{R}_z^{2n}$ by a, b, etc. Thus $a(z) = a(x,p)$ is the value of the linear form a° at the point $z = (x,p)$.

To every a we associate a first order linear partial differential operator A obtained by formally replacing p in $a(x,p)$ by D_x:

$$A = a(x, D_x) \ , \ D_x = -i\partial_x;$$

thus, if

$$a(x,p) = \langle \alpha, x \rangle + \langle \beta, p \rangle$$

for $\alpha = (\alpha_1, \dots, \alpha_n)$, $\beta = (\beta_1, \dots, \beta_n)$ in $\mathbb{R}^n$, then

$$A = \langle \alpha, x \rangle + \langle \beta, D_x \rangle = \langle \alpha, x \rangle - i \langle \beta, \partial_x \rangle . \tag{7.12}$$

Obviously the sum of two operators of the type above is an operator of the same type, and so is the product of such an operator by a scalar. It follows that these operators form a $2n$-dimensional vector space, which we denote by $\mathrm{Diff}^{(1)}(n)$.

The vector spaces $\mathbb{R}_z^{2n}$, $(\mathbb{R}_z^{2n})^*$ and $\mathrm{Diff}^{(1)}(n)$ are isomorphic since they all have the same dimension $2n$. The following result explicitly describes three canonical isomorphisms between these spaces:

Lemma 7.6.

(i) *The linear mappings*

$$\varphi_1 : \mathbb{R}_z^{2n} \longrightarrow (\mathbb{R}_z^{2n})^* \quad , \quad \varphi_1 : z_0 \longmapsto a,$$
$$\varphi_2 : (\mathbb{R}_z^{2n})^* \longrightarrow \mathrm{Diff}^{(1)}(n) \quad , \quad \varphi_2 : a \longmapsto A$$

where a is the unique linear form on $\mathbb{R}_z^{2n}$ such that $a(z) = \sigma(z, z_0)$ are isomorphisms, hence so is their compose φ :

$$\varphi = \varphi_2 \circ \varphi_1 : \mathbb{R}_z^{2n} \longrightarrow \mathrm{Diff}^{(1)}(n);$$

the latter associates to $z_0 = (x_0, p_0)$ the operator

$$A = \varphi(z_0) = \langle p_0, x \rangle - \langle x_0, D_x \rangle .$$

(ii) *Let $[A, B] = AB - BA$ be the commutator of $A, B \in \mathrm{Diff}^{(1)}(n)$; we have*

$$[\varphi(z_1), \varphi(z_2)] = -i\sigma(z_1, z_2) \tag{7.13}$$

for all $z_1, z_2 \in \mathbb{R}_z^{2n}$.

Proof. (i) The vector spaces $\mathbb{R}_z^{2n}$, $(\mathbb{R}_z^{2n})^*$, and $\mathrm{Diff}^{(1)}(n)$ having the same dimension, it suffices to show that $\ker(\varphi_1)$ and $\ker(\varphi_2)$ are zero. Now, $\varphi_1(z_0) = 0$ is equivalent to the condition $\sigma(z, z_0) = 0$ for all z, and hence to $z_0 = 0$ since a symplectic form non-degenerate. If $\varphi_2(a) = 0$ then

$$Af = \varphi_2(a)f = 0 \quad \textit{for all} \quad f \in \mathcal{S}(\mathbb{R}_x^n)$$

which implies $A = 0$ and thus $a = 0$.

(ii) Let $z_1 = (x_1, p_1)$, $z_2 = (x_2, p_2)$. We have

$$\varphi(z_1) = \langle p_1, x\rangle - \langle x_1, D_x\rangle \quad , \quad \varphi(z_2) = \langle p_2, x\rangle - \langle x_2, D_x\rangle$$

and hence

$$[\varphi(z_1), \varphi(z_2)] = i(\langle x_1, p_2\rangle - \langle x_2, p_1\rangle)$$

which is precisely (7.13). □

We are next going to show that the metaplectic group $\mathrm{Mp}(n)$ acts by conjugation on $\mathrm{Diff}^{(1)}(n)$. This will allow us to explicitly construct a covering mapping $\mathrm{Mp}(n) \longrightarrow \mathrm{Sp}(n)$.

Lemma 7.7. *For $z_0 = (x_0, p_0) \in \mathbb{R}_z^{2n}$ define $A \in \mathrm{Diff}^{(1)}(n)$ by*

$$A = \varphi(z_0) = \langle p_0, x\rangle - \langle x_0, D_x\rangle \,.$$

(i) *Let $\{\widehat{J}, \widehat{M}_{L,m}, \widehat{V}_P\}$ be the set of generators of $\mathrm{Mp}(n)$ defined in Proposition 7.2. We have:*

$$\widehat{J}A\widehat{J}^{-1} = \langle -x_0, x\rangle - \langle p_0, D_x\rangle = \varphi(Jz_0), \tag{7.14}$$

$$\widehat{M}_{L,m}A(\widehat{M}_{L,m})^{-1} = \langle L^T p_0, x\rangle - \langle L^{-1}x_0, D_x\rangle = \varphi(M_L z_0), \tag{7.15}$$

$$\widehat{V}_P A(\widehat{V}_P)^{-1} = \langle p_0 + Px_0, x\rangle - \langle x_0, D_x\rangle = \varphi(V_P z_0). \tag{7.16}$$

(ii) *If $A \in \mathrm{Diff}^{(1)}(n)$ and $\widehat{S} \in \mathrm{Mp}(n)$, then $\widehat{S}A\widehat{S}^{-1} \in \mathrm{Diff}^{(1)}(n)$.*

(iii) *For every $\widehat{S} \in \mathrm{Mp}(n)$ the mapping*

$$\Phi_{\widehat{S}} : \mathrm{Diff}^{(1)}(n) \longrightarrow \mathrm{Diff}^{(1)}(n) \quad , \quad A \longmapsto \widehat{S}A\widehat{S}^{-1}$$

is a vector space automorphism.

Proof. (i) Using the properties of the Fourier transform, it is immediate to verify that:

$$\langle x_0 D_x, f\rangle = \widehat{J}^{-1}\langle x_0, x\rangle \widehat{J}f \quad , \quad \langle p_0, x\rangle f = -\widehat{J}^{-1}\left\langle p_0, D_x\widehat{J}\right\rangle f$$

for $f \in \mathcal{S}(\mathbb{R}^n)$, hence (7.14). To prove (7.15) it suffices to remark that

$$\widehat{M}_{L,m}\langle p_0, x\rangle (\widehat{M}_{L,m})^{-1} f(x) = \langle p_0, Lx\rangle f(x)$$

and

$$\widehat{M}_{L,m} \langle x_0, D_x\rangle (\widehat{M}_{L,m})^{-1} f(x) = x_0 (L^{-1})^T D_x f(x).$$

Let us prove formula (7.16). Recalling that by definition

$$\widehat{V}_{-P} f(x) = e^{\frac{i}{2}\langle Px,x\rangle} f(x)$$

we have, since P is symmetric,

$$\langle x_0, D_x\rangle \widehat{V}_{-P} f(x) = \widehat{V}_{-P} \left(\langle Px_0, x\rangle f(x) + \langle p_0, D_x f(x)\rangle\right),$$

hence

$$\widehat{V}_P A(\widehat{V}_{-P} f)(x) = \langle (p_0 + Px_0), x\rangle f(x) - \langle x_0, D_x\rangle f(x)$$

which is (7.16).

Property (ii) immediately follows since $\widehat{S}$ is a product of operators $\widehat{J}$, $\widehat{M}_{L,m}$, $\widehat{V}_P$.

(iii) The mapping $\Phi_{\widehat{S}}$ is trivially a linear mapping $\mathrm{Diff}^{(1)}(n) \longrightarrow \mathrm{Diff}^{(1)}(n)$. If $B = \widehat{S}A\widehat{S}^{-1} \in \mathrm{Diff}^{(1)}(n)$, then we have also $A = \widehat{S}^{-1}B\widehat{S} \in \mathrm{Diff}^{(1)}(n)$ since $A = \widehat{S}^{-1}B(\widehat{S}^{-1})^{-1}$. It follows that $\Phi_{\widehat{S}}$ is surjective and hence bijective. □

Since the operators $\widehat{J}$, $\widehat{M}_{L,m}$, $\widehat{V}_P$ generate Mp(n) the lemma above shows that for every $\widehat{S} \in \mathrm{Mp}(n)$ there exists a linear automorphism S of $\mathbb{R}^{2n}_z$ such that $\Phi_{\widehat{S}}(A) = \widehat{a \circ S}$, that is

$$\Phi_{\widehat{S}}(\varphi(z_0)) = \varphi(Sz_0). \tag{7.17}$$

Let us show that the automorphism S preserves the symplectic form. For $z, z' \in \mathbb{R}^{2n}_z$ we have, in view of (7.13),

$$\begin{aligned}\sigma(Sz, Sz') &= i\left[\varphi(Sz), \varphi(Sz')\right] \\ &= i\left[\Phi_{\widehat{S}}\varphi(z), \Phi_{\widehat{S}}\varphi(z')\right] \\ &= i\left[\widehat{S}\varphi(z)\widehat{S}^{-1}, \widehat{S}\varphi(z')\widehat{S}^{-1}\right] \\ &= i\widehat{S}\left[\varphi(z), \varphi(z')\right]\widehat{S}^{-1} \\ &= \sigma(z, z'),\end{aligned}$$

hence $S \in \mathrm{Sp}(n)$ as claimed.

This result allows us to define a natural projection

$$\pi^{\mathrm{Mp}} : \mathrm{Mp}(n) \longrightarrow \mathrm{Sp}(n) \quad , \quad \pi^{\mathrm{Mp}} : \widehat{S} \longmapsto S;$$

it is the mapping which to $\widehat{S} \in \mathrm{Mp}(n)$ associates the element $S \in \mathrm{Sp}(n)$ defined by (7.17), that is

$$S = \varphi^{-1}\Phi_{\widehat{S}}\varphi. \tag{7.18}$$

That this mapping is a covering mapping follows from:

Proposition 7.8.

(i) *The mapping* π^{Mp} *is a continuous group epimorphism of* $\mathrm{Mp}(n)$ *onto* $\mathrm{Sp}(n)$ *such that:*

$$\pi^{\mathrm{Mp}}(\widehat{J}) = J \ , \ \pi^{\mathrm{Mp}}(\widehat{M}_{L,m}) = M_L \ , \ \pi^{\mathrm{Mp}}(\widehat{V}_P) = V_P \tag{7.19}$$

and hence

$$\pi^{\mathrm{Mp}}(\widehat{S}_{W,m}) = S_W. \tag{7.20}$$

(ii) *We have* $\ker(\pi^{\mathrm{Mp}}) = \{-I, +I\}$; *hence* $\pi^{\mathrm{Mp}} : \mathrm{Mp}(n) \longrightarrow \mathrm{Sp}(n)$ *is a twofold covering of the symplectic group.*

Proof. (i) Let us first show that π^{Mp} is a group homomorphism. In view of the obvious identity $\Phi_{\widehat{S}}\Phi_{\widehat{S}'} = \Phi_{\widehat{S}\widehat{S}'}$ we have

$$\begin{aligned}\pi^{\mathrm{Mp}}(\widehat{S}\widehat{S}') &= \varphi^{-1}\Phi_{\widehat{S}\widehat{S}'}\varphi \\ &= (\varphi^{-1}\Phi_{\widehat{S}}\varphi)(\varphi^{-1}\Phi_{\widehat{S}'}\varphi) \\ &= \pi^{\mathrm{Mp}}(\widehat{S})\pi^{\mathrm{Mp}}(\widehat{S}').\end{aligned}$$

Let us next prove that π^{Mp} is surjective. We have seen in Chapter 2 (Corollary 2.40 of Proposition 2.39) that the matrices J, M_L, and V_P generate $\mathrm{Sp}(n)$ when L and P range over, respectively, the invertible and symmetric real matrices of order n. It is thus sufficient to show that formulae (7.19) hold. Now, using (7.14), (7.15), and (7.16) we have

$$\varphi\Phi_{\widehat{J}}\varphi^{-1} = J \ , \varphi\Phi_{\widehat{M}_{L,m}}\varphi^{-1} = M_L \ , \varphi\Phi_{\widehat{V}_P}\varphi^{-1} = V_P,$$

hence (7.19). Formula (7.20) follows since every quadratic Fourier transform $\widehat{S}_{W,m}$ can be factorized as

$$\widehat{S}_{W,m} = \widehat{V}_{-P}\widehat{M}_{L,m}\widehat{J}\widehat{V}_{-Q}$$

in view of Proposition 7.2 above. To establish the continuity of the mapping π^{Mp} we first remark that the isomorphism $\varphi : \mathbb{R}^{2n}_z \longrightarrow \mathrm{Diff}^{(1)}(n)$ defined in Lemma 7.6 is trivially continuous, and so is its inverse. Since $\Phi_{\widehat{S}\widehat{S}'} = \Phi_{\widehat{S}}\Phi_{\widehat{S}'}$ it suffices to show that for every $A \in \mathrm{Diff}^{(1)}(n)$, $f_S(A)$ has A as limit when $\widehat{S} \to I$ in $\mathrm{Mp}(n)$. Now, $\mathrm{Mp}(n)$ is a group of continuous automorphisms of $\mathcal{S}(\mathbb{R}^n)$ hence, when $\widehat{S} \to I$, then $\widehat{S}^{-1}f \to f$ for every $f \in \mathcal{S}(\mathbb{R}^n)$, that is $A\widehat{S}^{-1}f \to Af$ and also $\widehat{S}A\widehat{S}^{-1}f \to f$.

(ii) Suppose that $\varphi^{-1}\Phi_{\widehat{S}}\varphi = I$. Then $\widehat{S}A\widehat{S}^{-1} = A$ for every $A \in \mathrm{Diff}^{(1)}(n)$ and this is only possible if $\widehat{S}$ is multiplication by a constant c with $|c| = 1$ (see Exercise below); thus $\ker(\pi^{\mathrm{Mp}}) \subset S^1$. In view of Lemma 7.4 we have $\widehat{S} = \widehat{S}_{W,m}\widehat{S}_{W',m'}$ for some choice of (W, m) and (W', m') hence the condition $\widehat{S} \in \ker(\pi^{\mathrm{Mp}})$ is equivalent to

$$\widehat{S}_{W',m'} = c(\widehat{S}_{W,m})^{-1} = c\widehat{S}_{W^*,m^*}$$

which is only possible if $c = \pm 1$, hence $\widehat{S} = \pm I$ as claimed. □

Exercise 7.9. Let $\widehat{S} \in \mathrm{Mp}(n)$. Prove that $\widehat{S}A\widehat{S}^{-1} = A$ for every $A \in \mathrm{Diff}^{(1)}(n)$ if and only if there exists $c \in S^1$ such that $\widehat{S}f = cf$ for all $f \in \mathcal{S}(\mathbb{R}^n_x)$. [Hint: consider the special cases $A = \langle x_0, D_x \rangle$, $A = \langle p_0, x \rangle$.]

It is useful to have a parameter-dependent version of Mp(n); in the applications to quantum mechanics that parameter is $\hbar$, Planck's constant h divided by 2π.

The main observation is that a covering group can be "realized" in many different ways. Instead of choosing π^{Mp} as a projection, we could as well have chosen any other mapping $\mathrm{Mp}(n) \longrightarrow \mathrm{Sp}(n)$ obtained from π^{Mp} by composing it on the left with an inner automorphism of Mp(n), or on the right with an inner automorphism of Sp(n), or both. The point is here that the diagram

$$\begin{array}{ccc} \mathrm{Mp}(n) & \xrightarrow{F} & \mathrm{Mp}(n) \\ \pi^{\mathrm{Mp}} \downarrow & & \downarrow \pi^{\mathrm{Mp}'} \\ \mathrm{Sp}(n) & \xrightarrow[G]{} & \mathrm{Sp}(n) \end{array}$$

is commutative: $\pi'^{\,\mathrm{Mp}} \circ F = G \circ \pi^{\mathrm{Mp}}$, because for all such $\pi'^{\,\mathrm{Mp}}$ we will have $\mathrm{Ker}(\pi'^{\,\mathrm{Mp}}) = \{\pm I\}$ and

$$\pi'^{\,\mathrm{Mp}} : \mathrm{Mp}(n) \longrightarrow \mathrm{Sp}(n)$$

will then be also be a covering mapping. We find it particularly convenient to define a new projection as follows. Set $\widehat{M_\lambda} = \widehat{M_{\lambda I,0}}$ for $\lambda > 0$, that is

$$\widehat{M_\lambda} f(x) = \lambda^{n/2} f(\lambda x) \quad , \quad f \in L^2(\mathbb{R}^n_x)$$

and denote by M_λ the projection of $\widehat{M_\lambda}$ on Sp(n):

$$M_\lambda(x,p) = (\lambda^{-1}x, \lambda p).$$

We have $\widehat{M_\lambda} \in \mathrm{Mp}(n)$ and $M_\lambda \in \mathrm{Sp}(n)$. For $\widehat{S} \in \mathrm{Mp}(n)$ we define $\widehat{S}^\hbar \in \mathrm{Mp}(n)$ by

$$\widehat{S}^\hbar = \widehat{M}_{1/\sqrt{\hbar}} \widehat{S} \widehat{M}_{\sqrt{\hbar}}. \tag{7.21}$$

The projection of $S^\hbar$ on Sp(n) is then given by:

$$\pi^{\mathrm{Mp}}(\widehat{S}^\hbar) = S^\hbar = M_{1/\sqrt{\hbar}} S M_{\sqrt{\hbar}}.$$

We now define the new projection

$$\pi^{\mathrm{Mp}\,\hbar} : \mathrm{Mp}(n) \longrightarrow \mathrm{Sp}(n)$$

by the formula

$$\pi^{\mathrm{Mp}\,\hbar}(\widehat{S}^\hbar) = M_{\sqrt{\hbar}}(\pi^{\mathrm{Mp}}(\widehat{S}^\hbar)) M_{1/\sqrt{\hbar}}$$

which is of course equivalent to

$$\pi^{\mathrm{Mp}^\hbar}(\widehat{S}^\hbar) = \pi^{\mathrm{Mp}}(\widehat{S}).$$

Suppose for instance that $\widehat{S} = \widehat{S}_{W,m}$; it is easily checked, using the fact that W is homogeneous of degree 2 in (x, x'), that

$$\widehat{S}^\hbar_{W,m} f(x) = \left(\tfrac{1}{2\pi i\hbar}\right)^{n/2} \Delta(W) \int e^{\frac{i}{\hbar}W(x,x')} f(x')\, d^n x'$$

or, equivalently,

$$\widehat{S}^\hbar_{W,m} = \hbar^{-n/2} \widehat{S}_{W/\hbar,m}.$$

Also,

$$(\widehat{S}^\hbar_{W,m})^{-1} f(x) = \left(\tfrac{i}{2\pi\hbar}\right)^{n/2} \Delta(W^*) \int e^{\frac{i}{\hbar}W^*(x,x')} f(x')\, d^n x'.$$

The projection of $\widehat{S}^\hbar_{W,m}$ on $\mathrm{Sp}(n)$ is the free matrix S_W:

$$\pi^{\mathrm{Mp}^\hbar}(\widehat{S}^\hbar_{W,m}) = S_W. \tag{7.22}$$

Exercise 7.10. Show that if $\hbar$ and $\hbar'$ are two positive numbers, then we have

$$\widehat{S}^\hbar_{W,m} = M_{\sqrt{\hbar'/\hbar}} \widehat{S}^{\hbar'}_{W,m} M_{\sqrt{\hbar/\hbar'}}$$

(*i.e.*, $\mathrm{Mp}^\hbar(n)$ and $\mathrm{Mp}^{\hbar'}(n)$ are equivalent representations of the metaplectic group).

Remark 7.11. The metaplectic group $\mathrm{Mp}(n)$ is not an irreducible representation of the double covering group $\mathrm{Sp}_2(n)$ (see Folland [42], p. 194, for a proof).

In what follows we will use the following convention, notation, and terminology which is consistent with quantum mechanics:

Notation 7.12. The projection $\mathrm{Mp}(n) \longrightarrow \mathrm{Sp}(n)$ will always assumed to be the homomorphism

$$\pi^{\mathrm{Mp}^\hbar} : \mathrm{Mp}(n) \longrightarrow \mathrm{Sp}(n)$$

and we will drop all the superscripts referring to $\hbar$: we will write π^{Mp} for $\pi^{\mathrm{Mp}^\hbar}$, $\widehat{S}_{W,m}$ for $\widehat{S}^\hbar_{W,m}$ and $\widehat{S}$ for $\widehat{S}^\hbar$. The functions on which these $\hbar$-dependent metaplectic operators are applied will be denoted by Greek letters ψ, ψ', etc.

7.1.3 Metaplectic covariance of Weyl calculus

The phase space translation operators $T(z_0)$ satisfy the intertwining formula $ST(z_0)S^{-1} = T(Sz_0)$ for every $S \in \mathrm{Sp}(n)$. It is perhaps not so surprising that we have a similar formula at the operator level. Let us show that this is indeed the case. This property, the "metaplectic covariance of Weyl pseudo-differential operators" is one of the hallmarks of the version of quantum mechanics we are studying in this book. Recall that $\widehat{T}(z_0)$ and $\widetilde{T}(z_0)$ are, respectively, the Heisenberg–Weyl and Royer–Grossmann operators.

Theorem 7.13. *Let $\widehat{S}\in \mathrm{Mp}(n)$ and $S=\pi^{\mathrm{Mp}}(S)$.*

(i) *We have*

$$\widehat{S}\widehat{T}(z_0)\widehat{S}^{-1}=\widehat{T}(Sz_0)\quad,\quad \widehat{S}\widetilde{T}(z_0)\widehat{S}^{-1}=\widetilde{T}(Sz_0) \tag{7.23}$$

for every $z_0\in\mathbb{R}^{2n}_z$.

(ii) *For every Weyl operator $\widehat{A}\overset{\mathrm{Weyl}}{\longleftrightarrow} a$ we have the correspondence*

$$a\circ S\overset{\mathrm{Weyl}}{\longleftrightarrow}\widehat{S}^{-1}\widehat{A}\widehat{S}. \tag{7.24}$$

Proof. (i) To prove the first formula (7.23) it is sufficient to assume that $\widehat{S}$ is a quadratic Fourier transform $\widehat{S}_{W,m}$ since these generate Mp(n). Suppose indeed we have shown that

$$\widehat{T}(S_W z_0)=\widehat{S}_{W,m}\widehat{T}(z_0)\widehat{S}^{-1}_{W,m}. \tag{7.25}$$

Writing an arbitrary element S of Mp(n) as a product $S_{W,m}S_{W',m'}$ we have

$$\begin{aligned}\widehat{T}(Sz_0)&=\widehat{S}_{W,m}(\widehat{S}_{W',m'}\widehat{T}(z_0)\widehat{S}^{-1}_{W',m'})\widehat{S}^{-1}_{W,m}\\&=\widehat{S}_{W,m}\widehat{T}(S_{W'}z_0)\widehat{S}^{-1}_{W,m}\\&=\widehat{T}(S_WS_{W'}z_0)\\&=\widehat{S}_{W,m}\widehat{S}_{W',m'}\widehat{T}(z_0)(\widehat{S}_{W,m}\widehat{S}_{W',m'})^{-1}\\&=\widehat{S}\widehat{T}(z_0)\widehat{S}^{-1}\end{aligned}$$

and the case of a general $S\in\mathrm{Mp}(n)$ follows by induction on the number of terms (the argument above actually already proves (7.23) in view of (7.4)). Let us thus prove (7.25); equivalently:

$$\widehat{T}(z_0)\widehat{S}_{W,m}=\widehat{S}_{W,m}\widehat{T}(S_W^{-1}z_0). \tag{7.26}$$

For $\psi\in\mathcal{S}(\mathbb{R}^n_x)$ set

$$g(x)=\widehat{T}(z_0)\widehat{S}_{W,m}\psi(x).$$

By definition of a $\widehat{S}_{W,m}$ and $\widehat{T}(z_0)$ we have

$$g(x)=\left(\tfrac{1}{2\pi i\hbar}\right)^{n/2}\Delta(W)e^{-\frac{1}{2\hbar}\langle p_0,x_0\rangle}\int e^{\frac{i}{\hbar}(W(x-x_0,x')+\langle p_0,x\rangle)}\psi(x')d^nx'.$$

In view of formula (2.45) in Proposition 2.37 (Subsection 2.2.3), the function

$$W_0(x,x')=W(x-x_0,x')+\langle p_0,x\rangle \tag{7.27}$$

is a generating function of the free affine symplectomorphism $T(z_0)\circ S$, hence we have just shown that

$$\widehat{T}(z_0)\widehat{S}_{W,m}=e^{\frac{i}{2\hbar}\langle p_0,x_0\rangle}\widehat{S}_{W_0,m} \tag{7.28}$$

where $\widehat{S}_{W_0,m}$ is one of the metaplectic operators associated to W_0. Let us now set

$$h(x) = \widehat{S}_{W,m}\widehat{T}(S_W^{-1}z_0)\psi(x) \quad\text{and}\quad (x_0', p_0') = \widehat{S}_{W,m}^{-1}(x_0, p_0);$$

we have

$$h(x) = \left(\tfrac{1}{2\pi i\hbar}\right)^{n/2}\Delta(W)\int e^{\frac{i}{\hbar}W(x,x')}e^{-\frac{i}{2\hbar}\langle p_0', x_0'\rangle}e^{\frac{i}{\hbar}\langle p_0', x'\rangle}\psi(x'-x_0')\,d^nx'$$

that is, performing the change of variables $x' \longmapsto x' + x_0'$:

$$h(x) = \left(\tfrac{1}{2\pi i\hbar}\right)^{n/2}\Delta(W)\int e^{\frac{i}{\hbar}W(x,x'+x_0')}e^{\frac{i}{2\hbar}\langle p_0', x_0'\rangle}e^{\frac{i}{\hbar}\langle p_0', x'\rangle}\psi(x')\,d^nx'.$$

We will thus have $h(x) = g(x)$ as claimed, if we show that

$$W(x, x' + x_0') + \tfrac{1}{2}\langle p_0', x_0'\rangle + \langle p_0', x'\rangle = W_0(x, x') - \tfrac{1}{2}\langle p_0, x_0\rangle,$$

that is

$$W(x, x' + x_0') + \tfrac{1}{2}\langle p_0', x_0'\rangle + \langle p_0', x'\rangle = W(x - x_0, x') + \langle p_0, x\rangle - \tfrac{1}{2}\langle p_0, x_0\rangle .$$

Replacing x by $x + x_0$ this amounts to proving that

$$W(x + x_0, x' + x_0') + \tfrac{1}{2}\langle p_0', x_0'\rangle + \langle p_0', x'\rangle = W(x, x') + \tfrac{1}{2}\langle p_0, x_0\rangle + \langle p_0, x\rangle .$$

But this equality immediately follows from Proposition 2.37 and its Corollary 2.38. To prove the second formula (7.23) recall (Proposition 5.52, Subsection 5.5.3 of Chapter 5) that we have

$$\widetilde{T}(z_0) = \widehat{T}(z_0)\widetilde{T}(0)\widehat{T}(z_0)^{-1}.$$

It follows that

$$\widetilde{T}(Sz_0) = \widehat{S}\widehat{T}(z_0)(\widehat{S}^{-1}\widetilde{T}(0)\widehat{S})\widehat{T}(z_0)^{-1}\widehat{S}^{-1}.$$

We claim that $\widehat{S}^{-1}\widetilde{T}(0)\widehat{S} = \widetilde{T}(0)$. It suffices of course to prove this when $\widehat{S} = \widehat{S}_{W,m}$. For $\psi \in \mathcal{S}(\mathbb{R}_x^n)$ we have

$$\begin{aligned}
\widetilde{T}(0)\widehat{S}_{W,m}\psi(x) &= \left(\tfrac{1}{2\pi i\hbar}\right)^{n/2}\Delta(W)\int e^{\frac{i}{\hbar}W(-x,x')}\psi(x')\,d^nx' \\
&= \left(\tfrac{1}{2\pi i\hbar}\right)^{n/2}\Delta(W)\int e^{\frac{i}{\hbar}W(-x,-x'')}\psi(-x'')\,d^nx' \\
&= \widehat{S}_{W,m}\widetilde{T}(0)\psi(x)
\end{aligned}$$

and hence

$$\widehat{S}_{W,m}^{-1}\widetilde{T}(0)\widehat{S}_{W,m} = \widetilde{T}(0);$$

the second formula (7.23).

Let us now prove part (ii) of the theorem. In view of formula (6.23) in Theorem 6.12 we have

$$(a \circ S)^w = \int a_\sigma(Sz)\widehat{T}(z)d^{2n}z$$

where $(a \circ S)^w$ is the Weyl operator associated to the symbol $a \circ S$, that is, performing the change of variables $Sz \longmapsto z$ and taking into account the fact that $\det S = 1$,

$$(a \circ S)^w = \int a_\sigma(z)\widehat{T}(S^{-1}z)d^{2n}z.$$

By formula (7.23) in Theorem 7.13 we have $\widehat{S}^{-1}\widehat{T}(z)\widehat{S} = \widehat{T}(S^{-1}z)$ and hence

$$(a \circ S)^w = \int a_\sigma(z)\widehat{S}^{-1}\widehat{T}(z)\widehat{S}d^{2n}z = \widehat{S}^{-1}\left(\int a_\sigma(z)\widehat{T}(z)d^{2n}z\right)\widehat{S}$$

which is (7.24). □

As a straightforward consequence of Theorem 7.13 we obtain the so-called metaplectic covariance formula for the Wigner–Moyal (and Wigner) transform:

Proposition 7.14. *Let* $\psi, \phi \in \mathcal{S}(\mathbb{R}^n_x)$ *and* $\widehat{S} \in \mathrm{Mp}(n)$*; we denote by* S *the projection of* $\widehat{S}$ *on* $\mathrm{Sp}(n)$*. We have*

$$W(\widehat{S}\psi, \widehat{S}\phi)(z) = W(\psi, \phi)(S^{-1}z) \tag{7.29}$$

and hence in particular

$$W(\widehat{S}\psi)(z) = W\psi(S^{-1}z). \tag{7.30}$$

Proof. In view of formula (6.69) in Proposition 6.45 we have, since $\widehat{S}$ is unitary,

$$\int W(\widehat{S}\psi, \widehat{S}\phi)a(z)d^{2n}z = (\widehat{A}\widehat{S}\psi, \widehat{S}\phi)_{L^2(\mathbb{R}^n_x)} = (\widehat{S}^{-1}\widehat{A}\widehat{S}\psi, \phi)_{L^2(\mathbb{R}^n_x)}.$$

In view of (7.24) we have

$$\begin{aligned}(\widehat{S}^{-1}\widehat{A}\widehat{S}\psi, \phi)_{L^2(\mathbb{R}^n_x)} &= \int W(\psi, \phi)(z)(a \circ S)(z)d^{2n}z \\ &= \int W(\psi, \phi)(S^{-1}z)a(z)d^{2n}z,\end{aligned}$$

which establishes (7.29) since ψ and ϕ are arbitrary; (7.30) trivially follows taking $\psi = \phi$. □

7.2 The Metaplectic Algebra

Let H be some Hamiltonian function and $\widehat{H} \overset{\text{Weyl}}{\longleftrightarrow} H$ the corresponding Weyl operator. The associated Schrödinger equation is by definition

$$i\hbar\partial_t\psi = \widehat{H}\psi$$

where ψ is a function (or distribution) in the x, t variables. For arbitrary H this equation is difficult to solve explicitly; it turns out, however, that when H is a quadratic form in z, then the solutions can be expressed using metaplectic operators. To prove this remarkable fact we will have to identify the Lie algebra of the metaplectic group.

7.2.1 Quadratic Hamiltonians

Let us begin by shortly discussing the properties of quadratic Hamiltonians. Quadratic Hamiltonians intervene in many parts of classical (and quantum) mechanics, for instance in the study of motion near equilibrium (see Cushman and Bates [27] for a thorough study of some specific examples), or for the calculation of the energy spectrum of an electron in a uniform magnetic field.

Let H be a homogeneous polynomial in $z \in \mathbb{R}^{2n}_z$ and with coefficients depending on $t \in \mathbb{R}$:

$$H(z,t) = \tfrac{1}{2}\left\langle H''(t)z, z\right\rangle \tag{7.31}$$

($H''(t) = D_z^2 H(z,t)$ is the Hessian of H); the associated Hamilton equations are

$$\dot{z}(t) = JH''(t)(z(t)). \tag{7.32}$$

Recall the following notation: $(S^H_{t,t'})$ is the time-dependent flow determined by H, that is if $t \longmapsto z(t)$ is the solution of (7.32) with $z(t') = z'$ then

$$z(t) = S^H_{t,t'}(z'). \tag{7.33}$$

We will set $S^H_{t,0} = S^H_t$. Assume that H does not depend on t; then (S^H_t) is the one-parameter subgroup of $\mathrm{Sp}(n)$ given by

$$S^H_t = e^{tJH''}.$$

Conversely, if (S_t) is an arbitrary one-parameter subgroup of $\mathrm{Sp}(n)$, then $S_t = e^{tX}$ for some $X \in \mathfrak{sp}(n)$ and we have $(S_t) = (S^H_t)$ where H is the quadratic Hamiltonian

$$H(z) = -\tfrac{1}{2}\left\langle JXz, z\right\rangle. \tag{7.34}$$

The following elementary result is useful:

Lemma 7.15. *Let H and K be two quadratic Hamiltonians associated by* (7.34) *to $X, Y \in \mathfrak{sp}(n)$. The Poisson bracket $\{H, K\}$ is the quadratic Hamiltonian given by*

$$\{H, K\}(z) = -\tfrac{1}{2}\langle J[X, Y]z, z\rangle$$

where $[X, Y] = XY - YX$.

Proof. We have

$$\begin{aligned}\{H, K\}(z) &= -\sigma(X_H(z), X_K(z)) \\ &= -\sigma(Xz, Yz) \\ &= -\langle JXz, Yz\rangle \\ &= -\left\langle Y^T JXz, z\right\rangle .\end{aligned}$$

Now

$$\left\langle Y^T JXz, z\right\rangle = \tfrac{1}{2}\left\langle (Y^T JX - X^T JY)z, z\right\rangle ,$$

that is, since $Y^T J$ and $X^T J$ are symmetric,

$$\left\langle Y^T JXz, z\right\rangle = -\tfrac{1}{2}\langle J(YX - XY)z, z\rangle$$

whence

$$\{H, K\}(z) = \tfrac{1}{2}\langle J(YX - XY)z, z\rangle$$

which we set out to prove. □

7.2.2 The Schrödinger equation

The metaplectic group $\mathrm{Mp}(n)$ is a covering group of $\mathrm{Sp}(n)$; it follows from the general theory of Lie groups that the Lie algebra $\mathfrak{mp}(n)$ of $\mathrm{Mp}(n)$ is isomorphic to $\mathfrak{sp}(n)$ (the Lie algebra of $\mathrm{Sp}(n)$). We are going to construct explicitly an isomorphism $F : \mathfrak{sp}(n) \longrightarrow \mathfrak{mp}(n)$ making the diagram

$$\begin{array}{ccc} \mathfrak{mp}(n) & \overset{F^{-1}}{\longrightarrow} & \mathfrak{sp}(n) \\ \exp\downarrow & & \downarrow\exp \\ \mathrm{Mp}(n) & \underset{\pi^{\mathrm{Mp}}}{\longrightarrow} & \mathrm{Sp}(n) \end{array} \tag{7.35}$$

commutative.

In Chapter 2, Subsection 2.1.3, we called $\mathfrak{sp}(n)$ the "symplectic algebra"; we will call $\mathfrak{mp}(n)$ the "metaplectic algebra". For an approach using the beautiful and important notion of symplectic Clifford algebra see Crumeyrolle [26] and the monograph by Habermann and Habermann [87].

Theorem 7.16.

(i) *The linear mapping F which to $X \in \mathfrak{sp}(n)$ associates the anti-Hermitian operator $F(X) = -\frac{i}{\hbar}\widehat{H}$ where $\widehat{H} \overset{\text{Weyl}}{\longleftrightarrow} H$ with H given by (7.34) is injective, and we have*

$$[F(X), F(X')] = F([X, X']) \tag{7.36}$$

for all X, $X' \in \mathfrak{sp}(n)$;

(ii) *The image $F(\mathfrak{sp}(n))$ of F is the metaplectic algebra $\mathfrak{mp}(n)$.*

Proof. It is clear that the mapping F is linear and injective. Consider the matrices

$$X_{jk} = \begin{bmatrix} \Delta_{jk} & 0 \\ 0 & -\Delta_{jk} \end{bmatrix}, \; Y_{jk} = \frac{1}{2}\begin{bmatrix} 0 & \Delta_{jk} + \Delta_{kj} \\ 0 & 0 \end{bmatrix},$$

$$Z_{jk} = \frac{1}{2}\begin{bmatrix} 0 & 0 \\ \Delta_{jk} + \Delta_{kj} & 0 \end{bmatrix} \quad (1 \leq j \leq k \leq n)$$

with 1 the only non-vanishing entry at the jth row and kth column; these matrices form a basis of $\mathfrak{sp}(n)$ (see Exercise 2.17). For notational simplicity we will assume that $n = 1$ and set $X = X_{11}$, $Y = Y_{11}$, $Z = Z_{11}$:

$$X = \begin{bmatrix} 1 & 0 \\ 0 & -1 \end{bmatrix}, \; Y = \begin{bmatrix} 0 & 1 \\ 0 & 0 \end{bmatrix}, \; Z = \begin{bmatrix} 0 & 0 \\ 1 & 0 \end{bmatrix};$$

the case of general n is studied in an exactly similar way. To the matrices X, Y, Z correspond via formula (7.34) the Hamiltonians

$$H_X = px \, , \; H_Y = \tfrac{1}{2}p^2, \; H_Z = -\tfrac{1}{2}x^2.$$

The operators $\widehat{H}_X = F(X)$, $\widehat{H}_Y = F(Y)$, $\widehat{H}_Z = F(Z)$ form a basis of the vector space $F(\mathfrak{sp}(1))$; they are given by

$$\widehat{H}_X = -i\hbar x\partial_x - \tfrac{1}{2}i\hbar \, , \; \widehat{H}_Y = -\tfrac{1}{2}\hbar^2\partial_x^2 \, , \; \widehat{H}_Z = -\tfrac{1}{2}x^2.$$

Let us show that formula (7.36) holds. In view of the linearity of F it is sufficient to check that

$$[\widehat{H}_X, \widehat{H}_X] = \widehat{H}_{[X,Y]},$$
$$[\widehat{H}_X, \widehat{H}_Z] = \widehat{H}_{[X,Z]},$$
$$[\widehat{H}_Y, \widehat{H}_Z] = \widehat{H}_{[Y,Z]}.$$

We have thus proved that F is a Lie algebra isomorphism. To show that $F(\mathfrak{sp}(1)) = \mathfrak{mp}(1)$ it is thus sufficient to check that the one-parameter groups

$$t \longmapsto U_t = e^{-\frac{i}{\hbar}\widehat{H}_X t},$$
$$t \longmapsto V_t = e^{-\frac{i}{\hbar}\widehat{H}_Y t},$$
$$t \longmapsto W_t = e^{-\frac{i}{\hbar}\widehat{H}_Z t}$$

are subgroups of $\mathrm{Mp}(1)$. Let $\psi_0 \in \mathcal{S}(\mathbb{R})$ and set $\psi(x,t) = U_t\psi_0(x)$. The function ψ is the unique solution of the Cauchy problem

$$i\hbar\partial_t\psi = -(i\hbar x\partial_x + \tfrac{1}{2}i\hbar)\psi \quad , \quad \psi(\cdot,0) = \psi_0.$$

A straightforward calculation (using for instance the method of characteristics) yields

$$\psi(x,t) = e^{-t/2}\psi_0(e^{-t}x)$$

hence the group (U_t) is given by $U_t = \widehat{M}_{L(t),0}$ where $L(t) = e^{-t}$ and we thus have $U_t \in \mathrm{Mp}(1)$ for all t. Leaving the detailed calculations to the reader one similarly verifies that

$$V_t\psi_0(x) = \left(\frac{1}{2\pi i\hbar t}\right)^{1/2} \int \exp\left[\frac{i}{2\hbar t}(x-x')^2\right]\psi_0(x')dx',$$

$$W_t\psi_0(x) = \exp\left(-\frac{1}{2\hbar}x^2\right)\psi_0(x),$$

so that V_t is a quadratic Fourier transform corresponding to the generating function $W = (x-x')^2/2t$ and W_t is the operator $\widehat{V}_{-tI}$; in both cases we have operators belonging to $\mathrm{Mp}(1)$. □

We leave it to the reader to check that the diagram (7.35) is commutative.

Exercise 7.17. Show that $\exp \circ F^{-1} = \pi^{\mathrm{Mp}} \circ \exp$ where exp is a collective notation for the exponential mappings $\mathfrak{mp}(n) \longrightarrow \mathrm{Mp}(n)$ and $\mathfrak{sp}(n) \longrightarrow \mathrm{Sp}(n)$ [Use the generators of $\mathfrak{mp}(n)$ and $\mathfrak{sp}(n)$.]

Let us apply the result above to the Schrödinger equation associated to a quadratic Hamiltonian (7.31). Since $\mathrm{Mp}(n)$ covers $\mathrm{Sp}(n)$ it follows from the unique path lifting theorem from the theory of covering manifolds that we can lift the path $t \longmapsto S^H_{t,0} = S^H_t$ in a unique way into a path $t \longmapsto \widehat{S}^H_t$ in $\mathrm{Mp}(n)$ such that $\widehat{S}^H_0 = I$. Let $\psi_0 \in \mathcal{S}(\mathbb{R}^n_x)$ and set

$$\psi(x,t) = \widehat{S}_t\psi_0(x).$$

It turns out that ψ satisfies Schrödinger's equation

$$i\hbar\partial_t\psi = \widehat{H}\psi$$

where $\widehat{H}$ is the Weyl operator $\widehat{H} \overset{\mathrm{Weyl}}{\longleftrightarrow} H$. The following two exercises will be used in the proof of this property:

Exercise 7.18. Verify that the operator $\widehat{H}$ is given by

$$\widehat{H} = -\frac{\hbar^2}{2}\langle H_{pp}\partial_x, \partial_x\rangle - i\hbar\langle H_{px}x, \partial_x\rangle + \frac{1}{2}\langle H_{xx}x, x\rangle - \frac{i}{2}\operatorname{Tr}(H_{px}) \tag{7.37}$$

where $\operatorname{Tr}(H_{px})$ is the trace of the matrix H_{px}.

Exercise 7.19. Let $\widehat{H} \overset{\text{Weyl}}{\longleftrightarrow} H$, $\widehat{K} \overset{\text{Weyl}}{\longleftrightarrow} K$ where H and K are quadratic Hamiltonians (7.31). Show that

$$[\widehat{H}, \widehat{K}] = i\hbar\widehat{\{H,K\}} = -i\hbar\sigma(X_H, X_K)$$

where $\{\cdot,\cdot\}$ is the Poisson bracket (5.38). (You might want to use the previous Exercise.)

Let us now show that $\psi = \widehat{S}_t\psi_0$ is a solution of Schrödinger's equation:

Corollary 7.20. *Let $t \longmapsto \widehat{S}_t$ be the lift to* $\mathrm{Mp}(n)$ *of the flow $t \longmapsto S_t^H$. For every $\psi_0 \in \mathcal{S}(\mathbb{R}_x^n)$ the function ψ defined by $\psi(x,t) = \widehat{S}_t\psi_0(x)$ is a solution of the partial differential equation*

$$i\hbar\partial_t\psi = \widehat{H}\psi \quad , \quad \psi(\cdot,0) = \psi_0$$

where $\widehat{H} \overset{\text{Weyl}}{\longleftrightarrow} H$.

Proof. We have

$$i\hbar\partial_t\psi = i\hbar \lim_{\Delta t\to 0}\left[\frac{1}{\Delta t}(\widehat{S}_{\Delta t} - I)\right]\widehat{S}_t\psi_0,$$

hence it suffices to show that

$$\lim_{\Delta t\to 0}\left[\frac{1}{\Delta t}(\widehat{S}_{\Delta t} - I)\right] f = \widehat{H}f$$

for every function $f \in \mathcal{S}(\mathbb{R}_x^n)$. But this equality is an immediate consequence of Theorem 7.16. □

For applications see M. Brown's thesis [16] where the relation of $\mathrm{Mp}(n)$ with the Bohmian approach to quantum mechanics is studied in detail, and applied to various physical problems.

7.2.3 The action of $\mathrm{Mp}(n)$ on Gaussians: dynamical approach

Let us study the action of metaplectic operators on general Gaussian functions centered at some point $z_0 = (x_0, p_0)$. For the calculations that will be involved we find it convenient to write such a Gaussian in the form

$$\psi_0(x) = e^{\frac{i}{\hbar}\langle p_0, x - x_0\rangle} f_0(\tfrac{1}{\sqrt{\hbar}}(x - x_0)) \tag{7.38}$$

where f_0 is a complex exponential:

$$f_0(x) = e^{\frac{i}{2}\langle M_0 x, x\rangle} \quad , \quad M_0 = M_0^T \quad , \quad \operatorname{Im} M_0 > 0. \tag{7.39}$$

Since every $\widehat{S} \in \mathrm{Mp}(n)$ is the product of operators (7.9) and of Fourier transforms, one can in principle calculate $\widehat{S}\psi_0$ using formula (8.30) giving the

Fourier transform of a complex Gaussian. This "frontal attack" approach (which we encourage the reader to try!) however leads to some cumbersome calculations, and has a tendency to hide the dynamical interpretation of the final result. We therefore prefer to use what we have learnt about the relationship between the metaplectic group and Schrödinger's equation. The idea is the following: every $\widehat{S} \in \mathrm{Mp}(n)$ can be joined by a path

$$t \longmapsto \widehat{S}_t \in \mathrm{Mp}(n) \ , \ \widehat{S}_0 = I \ , \ \widehat{S}_1 = \widehat{S}$$

and the function $(x,t) \longmapsto \widehat{S}_t \psi_{z_0,M_0}(x)$ is the solution of the Cauchy problem

$$i\hbar \frac{\partial \psi}{\partial t} = \widehat{H}\psi \ , \ \psi(\cdot,0) = \psi_{z_0,M_0} \tag{7.40}$$

where $\widehat{H}$ is the quadratic Hamiltonian whose flow (S_t^H) is the projection on $\mathrm{Sp}(n)$ of the path $t \longmapsto \widehat{S}_t$. (This follows from the one-to-one correspondence between one-parameter families in $\mathrm{Sp}(n)$ and $\mathrm{Mp}(n)$ established in the previous subsections.)

We will denote by H_{xx}, H_{xp}, etc. the matrices of second derivatives of H with respect to the variables appearing in subscript; the Hessian matrix of H is thus

$$H'' = \begin{bmatrix} H_{xx} & H_{xp} \\ H_{px} & H_{pp} \end{bmatrix} \ , \ H_{xp}^T = H_{px}.$$

Theorem 7.21. *Let ψ_0 be the Gaussian (7.38); the function $\psi_t = \widehat{S}_t\psi_0$ is given by the formula*

$$\psi_t(x) = e^{\frac{i}{\hbar}(\Phi(z_0,t)+\langle p_t, x-x_t\rangle)} f(\tfrac{1}{\sqrt{\hbar}}(x-x_t),t) \tag{7.41}$$

where $t \longmapsto z_t = (x_t,p_t)$ is the solution of Hamilton's equations for H with initial condition $z_0 = (x_0,p_0)$,

$$\Phi(x_0,p_0,t) = \int_{z_0,0}^{z_t,t} pdx - Hdt' \tag{7.42}$$

is the action integral, and

$$f(x,t) = a(t)e^{\frac{i}{2}\langle M(t)x,x\rangle}$$

with

$$a(t) = \exp\left[-\frac{1}{2}\int_0^t Tr(H_{pp}M)dt'\right] \tag{7.43}$$

and $M(t) = X(t)Y(t)^{-1}$ where the matrices $X(t)$ and $Y(t)$ are calculated as follows: find X_0 and Y_0 such that $M_0 = X_0Y_0^{-1}$; then $X = X(t)$ and $Y = Y(t)$ are given by

$$\begin{bmatrix} X \\ Y \end{bmatrix} = S_t^H \begin{bmatrix} X \\ Y \end{bmatrix} \ , \ X(0) = X_0 \ , \ Y(0) = Y_0 \tag{7.44}$$

where (S_t^H) is the linear symplectic (time-dependent) flow determined by the quadratic Hamiltonian H.

Proof. (*Cf.* [128], §2.1, for computational details). Making the Ansatz

$$\psi(x,t) = e^{\frac{i}{\hbar}(\Phi(z_0,t)+\langle p_t, x-x_t\rangle)} f(\tfrac{1}{\sqrt{\hbar}}(x-x_t),t),$$

one finds after insertion in equation (7.40) that $\Phi(z_0,t)$ is indeed the action (7.42) calculated along the Hamiltonian trajectory $t \longmapsto (x_t, p_t)$ starting from (x_0, p_0) at time $t=0$, and that $f(\cdot,t)$ is given by

$$f(x,t) = a(t)e^{\frac{i}{2}\langle M(t)x,x\rangle} \ , \ M(t)^T = M(t) \ , \ \operatorname{Im} M(t) > 0$$

where the functions $t \longmapsto a(t)$ and $t \longmapsto M(t)$ are solutions of the system of coupled first-order differential equations

$$\dot{a} + \tfrac{1}{2}Tr(H_{pp}M)a = 0 \tag{7.45}$$

and

$$\dot{M} + H_{xx} + MH_{px} + H_{xp}M + M^T H_{pp} M = 0 \tag{7.46}$$

with initial conditions $M(0) = M_0$, $a(0) = 1$. The equation (7.46) is recognized to be a matrix Riccati equation; setting $M = XY^{-1}$ we have

$$\dot{M} = XY^{-1} - XY^{-1}\dot{Y}X^{-1};$$

since $M = M^T$, insertion shows that the matrices X and Y satisfy the equations

$$\begin{aligned} &\dot{X} + H_{xx}Y + H_{xp}X = 0, \\ &\dot{Y} - H_{px}Y - H_{pp}X = 0 \end{aligned}$$

or, equivalently,

$$\frac{d}{dt}\begin{bmatrix} X \\ Y \end{bmatrix} = JH'' \begin{bmatrix} X \\ Y \end{bmatrix},$$

hence (7.44) (we leave it to the reader to check that $M = XY^{-1}$ only depends on M_0). One verifies by a straightforward computation that the matrix $M = XY^{-1}$ thus obtained indeed is symmetric, and one finally solves the equation (7.46) explicitly, which yields (7.43). □

Note that the "center" of the Gaussian ψ_t follows the Hamiltonian trajectory determined by the quadratic function H; this typical behavior, well-known in quantum mechanics, is related to Ehrenfest's theorem on average values (see for instance Messiah [123], Chapter VI, for a proof and comments).

7.3 Maslov Indices on $\mathrm{Mp}(n)$

This section is a little bit technical. Recall that we use the term "metaplectic Maslov index" to mean the integer m (uniquely defined modulo 4) appearing in the metaplectic operators $\widehat{S}_{W,m}$. Now, an arbitrary $S \in \mathrm{Mp}(n)$ is generally not of the type $\widehat{S}_{W,m}$, but as we have seen it is always the product $\widehat{S}_{W,m}\widehat{S}_{W',m'}$ of two such operators. While the factorization $S = \widehat{S}_{W,m}\widehat{S}_{W',m'}$ is never unique, it turns out that there is a conserved quantity, namely an integer modulo 4 only depending on $m + m'$ and not on the way we factor the operator $\widehat{S}$. This integer modulo 4 is the "metaplectic Maslov index" of S and is denoted by $m(\widehat{S})$. It turns out that calculations are easier if one works with a variant of $m(\widehat{S})$ which we will call simply "Maslov index on Mp(n)", and which is defined in terms of the Maslov index on $\mathrm{Sp}_{\infty}(n)$ studied in Chapter 3, Section 3.3.2.

Let us begin by establishing the existence of invariants associated with products of quadratic Fourier transforms. We will use the following shorthand notation: we will write $W = (P, L, Q)$ for every quadratic form

$$W(x, x') = \tfrac{1}{2}\langle Px, x\rangle - \langle Lx, x'\rangle + \tfrac{1}{2}\langle Qx', x'\rangle \tag{7.47}$$

with $P = P^T$, $Q = Q^T$, *and* $\det L \neq 0$.

7.3.1 The Maslov index $\widehat{\mu}(\widehat{S})$

In this section we define and study in detail the Maslov index of an arbitrary metaplectic operator $\widehat{S} \in \mathrm{Mp}(n)$. We will see that it can be very simply related to the Maslov index on the universal covering group and that it is in fact identical to the Maslov index on $\mathrm{Sp}_q(n)$ studied in Chapter 3 if we take $q = 2$.

Theorem 7.22. *Let* $W = (P, L, Q)$ *be a quadratic form* (7.47).

(i) *Let* $W' = (P', L', Q')$ *and* $W'' = (P'', L'', Q'')$ *be such that* $\widehat{S}_{W'',m''} = \widehat{S}_{W,m}\widehat{S}_{W',m}$ *(cf. Lemma 7.4), then the metaplectic Maslov indices* m, m', *and* m'' *satisfy the relation*

$$m'' \equiv m + m' - \mathrm{Inert}(P' + Q) \mod 4 \tag{7.48}$$

where $\mathrm{Inert}(P' + Q)$ *is the number of negative eigenvalues of the symmetric matrix* $P' + Q$*;*

(ii) *If* $W''' = (P''', L''', Q''')$ *is such that*

$$\widehat{S}_{W,m}\widehat{S}_{W',m'} = \widehat{S}_{W'',m''}\widehat{S}_{W''',m'''},$$

then we have

$$m + m' + \frac{1}{2}\,\mathrm{sign}(P' + Q) \equiv m'' + m''' + \frac{1}{2}\,\mathrm{sign}(P''' + Q'') \mod 4 \tag{7.49}$$

and also

$$\mathrm{rank}(P' + Q) \equiv \mathrm{rank}(P''' + Q'') \mod 4. \tag{7.50}$$

Proof. We only sketch the proof, since it is rather long and technical. Part (i) (formula (7.48)) is due to Leray [107], Chapter I, §1,2. Part (ii) is due to the author [56, 58] (also see [61], Chapter 3). The idea is the following: let $\psi_k(x) = \psi(x\sqrt{k})$ $(k > 0)$ where $\psi(x) = \exp(-|x|^2/2\hbar)$ (any other radially symmetric element of $\mathcal{S}(\mathbb{R}^n_x)$ would do as well). Using carefully the method of stationary phase one finds that if $\widehat{S} = \widehat{S}_{W,m}\widehat{S}_{W',m'}$, then

$$\widehat{S}\psi_k(0) \sim C' i^{m+m'-n/2}(e^{i\pi/4})^{\operatorname{sign}(P'+Q)}k^{-\operatorname{rank}(P'+Q)/2} \quad \textit{for} \quad k \to \infty$$

where $C' > 0$ is a constant depending on W and W'. If $\widehat{S} = \widehat{S}_{W'',m''}\widehat{S}_{W''',m'''}$ we will thus have

$$\widehat{S}\psi_k(0) \sim C'' i^{m'+m''-n/2}(e^{i\pi/4})^{\operatorname{sign}(P'''+Q'')}k^{-\operatorname{rank}(P'''+Q'')/2} \quad \textit{for} \quad k \to \infty$$

for some new constant $C'' > 0$ depending on W'' and W'''. This is only possible if (7.49) and (7.50) hold. □

Definition 7.23. (i) Let $\widehat{S}_{W,m} \in \mathrm{Mp}(n)$; we call the integer

$$\widehat{\mu}(\widehat{S}_{W,m}) = 2m - n \tag{7.51}$$

the "Maslov index" of the quadratic Fourier transform $\widehat{S}_{W,m}$.

(ii) Let $\widehat{S} = \widehat{S}_{W,m}\widehat{S}_{W',m'}$ be an arbitrary element of $\mathrm{Mp}(n)$: the integer

$$\widehat{\mu}(\widehat{S}) = \widehat{\mu}(\widehat{S}_{W,m}) + \widehat{\mu}(\widehat{S}_{W',m'}) + \widehat{\operatorname{sign}}(P' + Q) \tag{7.52}$$

(which is uniquely defined modulo 4) is called the "Maslov index of $\widehat{S}$". [The hats "$\widehat{\ }$" should be understood as "class modulo 4".]

Notice that the Maslov index is well defined in view of (ii) in Theorem 7.22 which shows that $\widehat{\mu}(\widehat{S})$ is indeed independent of the factorization of $\widehat{S}$ in a product of two quadratic Fourier transforms.

It turns out (not so unexpectedly!) that we can rewrite definition (7.52) in terms of the signature of a triple of Lagrangian plane. Let us begin by proving a technical result:

Lemma 7.24. *Let S_W and $S_{W'}$ be two free symplectic matrices and write*

$$S_W = \begin{bmatrix} A & B \\ C & D \end{bmatrix}, \quad S_{W'} = \begin{bmatrix} A' & B' \\ C' & D' \end{bmatrix}, \quad S_W S_{W'} = \begin{bmatrix} A'' & B'' \\ C'' & D'' \end{bmatrix}.$$

Let $\ell_P = 0 \times \mathbb{R}^n_p$. We have

$$\tau(\ell_P, S_W\ell_P, S_W S_{W'}\ell_P) = \operatorname{sign}(B^{-1}B''(B')^{-1}). \tag{7.53}$$

Proof. Let us first prove (7.53) in the particular case where $S_{W'}\ell_P = \ell_X$, that is when $S_{W'}$ is of the type

$$S_{W'} = \begin{bmatrix} A' & B' \\ C' & 0 \end{bmatrix}$$

in which case $B'' = AB'$. Using the Sp(n)-invariance and the antisymmetry of the signature, we have, since $S_{W'}\ell_P = \ell_X$:

$$\tau(\ell_P, S_W\ell_P, S_W S_{W'}\ell_P) = \tau(\ell_X, S_W^{-1}\ell_P, \ell_P). \tag{7.54}$$

The inverse of S_W being the symplectic matrix

$$S_W^{-1} = \begin{bmatrix} D^T & -B^T \\ -C^T & A^T \end{bmatrix}$$

we thus have

$$S_W^{-1}\begin{bmatrix} 0 \\ p \end{bmatrix} = \begin{bmatrix} -B^T p \\ A^T p \end{bmatrix}$$

so the Lagrangian plane $S_W^{-1}\ell_P$ has for equation $Ax + Bp = 0$; since B is invertible (because S_W is free) this equation can be rewritten $p = -B^{-1}Ax$. Using the second formula (1.24) (Chapter 1, Section 1.4) together with the identity (7.54) we have

$$\begin{aligned}\tau(\ell_P, S_W\ell_P, S_W S_{W'}\ell_P) &= -\operatorname{sign}(-B^{-1}A) \\ &= \operatorname{sign}(B^{-1}(AB')(B')^{-1}) \\ &= \operatorname{sign}(B^{-1}B''(B')^{-1})\end{aligned}$$

which proves (7.53) in the case $S_{W'}\ell_P = \ell_X$. The general case reduces to the former, using the fact that the symplectic group acts transitively on all pairs of transverse Lagrangian planes. In fact, since

$$S_{W'}\ell_P \cap \ell_P = \ell_X \cap \ell_P = 0$$

we can find $S_0 \in \mathrm{Sp}(n)$ such that $(\ell_P, S_{W'}\ell_P) = S_0(\ell_P, \ell_X)$, that is $S_{W'}\ell_P = S_0\ell_X$ and $S_0\ell_P = \ell_P$. It follows, using again the antisymmetry and Sp(n)-invariance of σ that:

$$\tau(\ell_P, S_W\ell_P, S_W S_{W'}\ell_P) = \tau(\ell_X, (S_W S_0)^{-1}\ell_P, \ell_P)$$

which is (7.54) with S_W replaced by $S_W S_0$. Changing $S_{W'}$ into $S_0^{-1}S_{W'}$ (and hence leaving $S_W S_{W'}$ unchanged) we are led back to the first case. Since $S_0\ell_P = \ell_P$, S_0 must be of the type

$$S_0 = \begin{bmatrix} L & 0 \\ P & (L^{-1})^T \end{bmatrix} \quad , \quad \det(L) \neq 0,\ P = P^T;$$

writing again S_W in block-matrix form, the products SS_0 and $S_0^{-1}S_{W'}$ are thus of the type

$$SS_0 = \begin{bmatrix} * & B(L^T)^{-1} \\ * & * \end{bmatrix} \quad , \quad S_0^{-1}S_{W'} = \begin{bmatrix} * & L^{-1}B' \\ * & * \end{bmatrix}$$

(the stars "*" are block-entries that are easily calculated, but that we do not need to write down), and hence

$$\tau(\ell_P, S_W\ell_P, S_W S_{W'}\ell_P) = \operatorname{sign}(L^T B^{-1}B''B'^{-1}L) = \operatorname{sign}(B^{-1}B''B'^{-1})$$

proving (7.53) in the general case. □

Remark 7.25. Formula (3.46) identifies $\tau(\ell_P, S\ell_P, SS'\ell_P)$ with the signature of the "composition form" $Q(S, S')$ defined by Robbin and Salamon in [134]; the reader will however notice that $\tau(\ell_P, S\ell_P, SS'\ell_P)$ is defined for all $S, S' \in \mathrm{Sp}(n)$ while $Q(S, S')$ is only defined for S, S' satisfying transversality conditions.

We now have the machinery needed to express $\widehat{\mu}(\widehat{S})$ in terms of the signature:

Theorem 7.26. *The Maslov index $\widehat{\mu}$ on $\mathrm{Mp}(n)$ has the following properties:*

(i) *Let $\widehat{S} = \widehat{S}_{W,m}\widehat{S}_{W',m'}$ be an arbitrary element of $\mathrm{Mp}(n)$. We have*

$$\widehat{\mu}(\widehat{S}) = \widehat{\mu}(\widehat{S}_{W,m}) + \widehat{\mu}(\widehat{S}_{W',m'}) + \tau(\ell_P, S_W\ell_P, S_W S_{W'}\ell_P). \tag{7.55}$$

(ii) *For all $\widehat{S}$, $\widehat{S}'$ in $\mathrm{Mp}(n)$ we have*

$$\widehat{\mu}(\widehat{S}\widehat{S}') = \widehat{\mu}(\widehat{S}) + \widehat{\mu}(\widehat{S}') + \tau(\ell_P, S\ell_P, SS'\ell_P) \tag{7.56}$$

where S, S' are the projections of $\widehat{S}$, $\widehat{S}'$ on $\mathrm{Sp}(n)$.

Proof. (i) If $W = (P, L, Q)$ and $W' = (P', L', Q')$ we have

$$S_W = \begin{bmatrix} L^{-1}Q & L^{-1} \\ PL^{-1}Q - L^T & L^{-1}P \end{bmatrix},$$

$$S_{W'} = \begin{bmatrix} L'^{-1}Q' & L'^{-1} \\ P'L'^{-1}Q' - L'^T & L'^{-1}P' \end{bmatrix}$$

(formula (2.43) in Chapter 1, Subsection 2.2.3) and thus

$$S_{W''} = S_W S_{W'} = \begin{bmatrix} * & L^{-1}(P'+Q)L'^{-1} \\ * & * \end{bmatrix}.$$

Writing

$$B'' = L^{-1}(P'+Q)L'^{-1} = B(P'+Q)B'$$

we have $P' + Q = B^{-1}B''(B')^{-1}$ and (7.53) implies that

$$\tau(\ell_P, S_W\ell_P, S_W S_{W'}\ell_P) = \operatorname{sign}(P'+Q),$$

hence formula (7.55).

(ii) (We are following de Gosson [56, 61].) Assume first that $\widehat{S}' = \widehat{S}_{W,m}$ and let us show that

$$\widehat{\mu}(\widehat{S}\widehat{S}_{W,m}) = \widehat{\mu}(\widehat{S}) + \widehat{\mu}(\widehat{S}_{W,m}) + \tau(\ell_P, S\ell_P, SS_W\ell_P)$$

for every $\widehat{S} \in \mathrm{Mp}(n)$. Let $\ell \in \mathrm{Lag}(n)$ and define functions $f_W : \mathrm{Mp}(n) \longrightarrow \mathbb{Z}$ and by $g_{W,\ell} : \mathrm{Mp}(n) \longrightarrow \mathbb{Z}$ by

$$f_W(\widehat{S}) = \widehat{\mu}(\widehat{S}\widehat{S}_{W,m}) - \widehat{\mu}(\widehat{S}) - \tau(\ell_P, S\ell_P, SS_W\ell_P),$$
$$g_{W,\ell}(\widehat{S}) = \widehat{\mu}(\widehat{S}) - \tau(S\ell_P, \ell_P, \ell).$$

In view of the cocycle property of τ we have

$$\tau(\ell_P, S\ell_P, SS_W\ell_P)$$
$$= \tau(\ell_P, S\ell_P, \ell) - \tau(\ell_P, SS_W\ell_P, \ell) + \tau(S\ell_P, SS_W\ell_P, \ell);$$

from which follows that

$$f_W(\widehat{S}) = g_{W,\ell}(\widehat{S}\widehat{S}_{W,m}) - g_{W,\ell}(\widehat{S}) - \tau(S\ell_P, SS_W\ell_P, \ell_P).$$

Choose now ℓ such that

$$S\ell_P \cap \ell = SS_W\ell_P \cap \ell = \ell_P \cap \ell = 0;$$

$g_{W,\ell}$ is constant on a neighborhood of $\widehat{S}$ in $\mathrm{Mp}(n)$; since on the other hand

$$S\ell_P \cap SS_W\ell_P = S_W\ell_P \cap \ell_P = 0,$$

the function $S \longmapsto \tau(S\ell_P, SS_W\ell_P, \ell)$ is constant in a neighborhood of S in $\mathrm{Sp}(n)$ hence $f_W(\widehat{S})$ is locally constant on $\mathrm{Mp}(n)$ and hence constant since $\mathrm{Mp}(n)$ is connected. That constant value is

$$f_W(\widehat{S}_{W',m'}) = \widehat{\mu}(\widehat{S}_{W',m'}\widehat{S}_{W,m}) - \widehat{\mu}(\widehat{S}_{W',m'})$$
$$- \tau(\ell_P, S_{W'}\ell_P, S_{W'}S_W\ell_P),$$

that is

$$f_W(\widehat{S}_{W',m'}) = \widehat{\mu}(\widehat{S}_{W,m})$$

proving formula (7.56) in the case $\widehat{S}' = \widehat{S}_{W,m}$. To prove it in the general case we proceed as follows: writing $\widehat{S}' = \widehat{S}_{W,m}\widehat{S}_{W',m'}$ we have

$$\widehat{\mu}(\widehat{S}\widehat{S}') = \widehat{\mu}(\widehat{S}\widehat{S}_{W,m}) + \widehat{\mu}(\widehat{S}_{W',m'}) + \tau(\ell_P, SS_W\ell_P, SS'\ell_P)$$
$$= \widehat{\mu}(\widehat{S}) + \widehat{\mu}(\widehat{S}_{W,m}) + \tau(\ell_P, S\ell_P, SS_W\ell_P) + \widehat{\mu}(\widehat{S}_{W',m'})$$
$$+ \tau(\ell_P, SS_W\ell_P, SS'\ell_P).$$

Set

$$\Sigma = \tau(\ell_P, S\ell_P, SS_W\ell_P) + \tau(\ell_P, SS_W\ell_P, SS'\ell_P);$$

using successively the antisymmetry, the cocycle property, and the $\mathrm{Sp}(n)$-invariance, and of τ we have

$$\begin{aligned}\Sigma &= \tau(\ell_P, SS_W\ell_P, SS'\ell_P) - \tau(\ell_P, SS_W\ell_P, SS'\ell_P)\\ &= \tau(SS_W\ell_P, SS'\ell_P, S\ell_P) - \tau(\ell_P, SS'\ell_P, S\ell_P)\\ &= \tau(S_W\ell_P, S'\ell_P, \ell_P) + \tau(\ell_P, S\ell_P, SS'\ell_P),\end{aligned}$$

hence, noting that $S' = S_W S_{W'}$:

$$\widehat{\mu}(\widehat{S}\widehat{S}') = \widehat{\mu}(\widehat{S}) + \widehat{\mu}(\widehat{S}') + \tau(\ell_P, S\ell_P, SS'\ell_P) \mod 4$$

which is precisely (7.56). □

As an application let us calculate the indices of the identity and of the inverse of a metaplectic operator:

Corollary 7.27. *We have*

$$\widehat{\mu}(\widehat{I}) = 0 \ , \ \widehat{\mu}(\widehat{S}^{-1}) = -\widehat{\mu}(\widehat{S}). \tag{7.57}$$

Proof. Let $\widehat{S}_{W,m}$ be an arbitrary quadratic Fourier transform; in view of (7.55) we have

$$\begin{aligned}\widehat{\mu}(\widehat{I}) &= \widehat{\mu}(\widehat{S}_{W,m}\widehat{S}_{W,m}^{-1})\\ &= \widehat{\mu}(\widehat{S}_{W,m}) + \widehat{\mu}(\widehat{S}_{W,m}^{-1}) + \tau(\ell_P, S_W\ell_P, \ell_P)\\ &= \widehat{\mu}(\widehat{S}_{W,m}) + \widehat{\mu}(\widehat{S}_{W,m}^{-1})\end{aligned}$$

(the last equality because $\tau(\ell_P, S_W\ell_P, \ell_P) = 0$ by antisymmetry of the signature) and hence, using formula (7.11),

$$\widehat{\mu}(\widehat{I}) = 2m - n + 2(n-m) - n = 0.$$

The second formula (7.57) follows from the first in view of (7.56) since

$$\widehat{\mu}(\widehat{I}) = \widehat{\mu}(\widehat{S}\widehat{S}^{-1}) = \widehat{\mu}(\widehat{S}) + \widehat{\mu}(\widehat{S}^{-1}) + \tau(\ell_P, S\ell_P, \ell_P)$$

using the fact that $\tau(\ell_P, S\ell_P, \ell_P) = 0$. □

7.3.2 The Maslov indices $\widehat{\mu}_\ell(\widehat{S})$

The reader will certainly have noted that in all the formulae above a special role is played by the Lagrangian plane $\ell_P = 0 \times \mathbb{R}^n_p$. It turns out that it is possible (and useful!) to define a Maslov index associated to an arbitrary Lagrangian plane $\ell \in \mathrm{Lag}(n)$, whose properties are quite similar to those of the standard Maslov index.

Definition 7.28. Let $\ell \in \operatorname{Lag}(n)$ and $S_0 \in \operatorname{Sp}(n)$ be such that $\ell = S_0\ell_P$. Let $\widehat{S}_0$ be one of the two operators in $\operatorname{Mp}(n)$ such that $\pi^{\mathrm{Mp}}(\widehat{S}_0) = S_0$. The integer modulo 4 defined by

$$\widehat{\mu}_\ell(\widehat{S}) = \widehat{\mu}(\widehat{S}_0^{-1}\widehat{S}\widehat{S}_0) \mod 4$$

is called the "Maslov index of $\widehat{S}$ relatively to ℓ".

That the integer $\widehat{\mu}(\widehat{S}_0^{-1}\widehat{S}\widehat{S}_0)$ is independent of the choice of $\widehat{S}_0$ is clear, since for a given $S_0 \in \operatorname{Sp}(n)$ there are only two operators $\pm\widehat{S}_0$ with projection $S_0 \in \operatorname{Sp}(n)$. Slightly more delicate is the proof that $\widehat{\mu}(\widehat{S}_0^{-1}\widehat{S}\widehat{S}_0)$ does not depend on the choice of $S_0 \in \operatorname{Sp}(n)$ such that $\ell = S_0\ell_P$. To show this we proceed as in de Gosson [56]. Assume that $\ell = S_1\ell_P$. Then $S_0S_1^{-1}\ell_P = \ell_P$ so that $R = S_0S_1^{-1}$ is in $St(\ell_P)$, the stabilizer of ℓ_P in $\operatorname{Sp}(n)$. Now choose $\widehat{R} \in \operatorname{Mp}(n)$ with projection R. We have, using (7.56) together with the cocycle property and $\operatorname{Sp}(n)$-invariance of τ:

$$\begin{aligned}\widehat{\mu}(\widehat{R}^{-1}\widehat{S}_0^{-1}\widehat{S}\widehat{S}_0\widehat{R}) &= \widehat{\mu}(\widehat{R}^{-1}) + \widehat{\mu}(\widehat{S}_0^{-1}\widehat{S}\widehat{S}_0) + \widehat{\mu}(\widehat{R}) \\ &\quad + \tau(\ell_P, R^{-1}\ell_P, R^{-1}S_0^{-1}SS_0\ell_P) + \tau(\ell_P, S_0^{-1}SS_0\ell_P, S_0^{-1}SS_0R\ell_P),\end{aligned}$$

that is, since $\widehat{\mu}(\widehat{R}^{-1}) = -\widehat{\mu}(\widehat{R})$,

$$\begin{aligned}\widehat{\mu}(\widehat{R}^{-1}\widehat{S}_0^{-1}\widehat{S}\widehat{S}_0\widehat{R}) &= -\widehat{\mu}(\widehat{R}) + \widehat{\mu}(\widehat{S}_0^{-1}\widehat{S}\widehat{S}_0) + \widehat{\mu}(\widehat{R}) \\ &\quad + \tau(R\ell_P, \ell_P, S_0^{-1}SS_0\ell_P) + \tau(\ell_P, S_0^{-1}SS_0\ell_P, S_0^{-1}SS_0R\ell_P) \\ &= \widehat{\mu}(\widehat{S}_0^{-1}\widehat{S}\widehat{S}_0) + \tau(\ell_P, \ell_P, S_0^{-1}SS_0\ell_P) + \tau(\ell_P, S_0^{-1}SS_0\ell_P, S_0^{-1}SS_0\ell_P)\end{aligned}$$

and hence, using the antisymmetry of τ,

$$\widehat{\mu}(\widehat{R}^{-1}\widehat{S}_0^{-1}\widehat{S}\widehat{S}_0\widehat{R}) = \widehat{\mu}(\widehat{S}_0^{-1}\widehat{S}\widehat{S}_0).$$

This proves that $\widehat{\mu}(\widehat{S}_0^{-1}\widehat{S}\widehat{S}_0)$ only depends on $\widehat{S}$ and ℓ, as claimed, which completely justifies Definition 7.28.

The properties of $\widehat{\mu}_\ell$ are immediately deduced from those of $\widehat{\mu}$ detailed in Theorem 7.26:

Corollary 7.29. *The Maslov index $\widehat{\mu}_\ell$ on* Mp(n) *has the following properties:*

(i) *For all $\widehat{S}$, $\widehat{S}'$ in* Mp(n) *we have*

$$\widehat{\mu}_\ell(\widehat{S}\widehat{S}') = \widehat{\mu}_\ell(\widehat{S}) + \widehat{\mu}_\ell(\widehat{S}') + \tau(\ell, S\ell, SS'\ell) \mod 4 \tag{7.58}$$

where S, S' are the projections of $\widehat{S}$, $\widehat{S}'$ on Sp(n).

(ii) *We have*

$$\widehat{\mu}_\ell(\widehat{I}) = 0 \textit{ and } \widehat{\mu}_\ell(\widehat{S}^{-1}) = -\widehat{\mu}_\ell(\widehat{S}).$$

Proof. This is an immediate consequence of the definition of $\widehat{\mu}_\ell$ and Theorem 7.26 and Corollary 7.27. □

The relation (7.58) clearly indicates a relationship between the Maslov index $\widehat{\mu}_\ell$ and the Maslov index on $\mathrm{Sp}_\infty(n)$ studied in Chapter 3, Subsection 3.3.2 (such a relation was of course already apparent in formula (7.56) of Theorem 7.26).

Corollary 7.30. *Let $S_\infty \in \mathrm{Sp}_\infty(n)$ have projection $\widehat{S} \in \mathrm{Mp}(n)$. For every $\ell \in \mathrm{Lag}(n)$ we have*

$$\widehat{\mu}_\ell(\widehat{S}) = \mu_\ell(S_\infty) \mod 4,$$

that is

$$\widehat{\mu}_\ell(\widehat{S}) = [\mu_\ell]_2([S]_2)$$

where $[\mu_\ell]_2$ is the Maslov index on $\mathrm{Sp}_2(n)$.

Proof. This is an immediate consequence of Proposition 3.31 on the uniqueness of a function $\mu_\ell : \mathrm{Sp}_\infty(n) \longrightarrow \mathbb{Z}$ satisfying

$$\mu_\ell(S_\infty S'_\infty) = \mu_\ell(S_\infty) + \mu_\ell(S'_\infty) + \tau(\ell, S\ell, SS'\ell)$$

($S = \pi^{\mathrm{Sp}}(S_\infty)$, $S' = \pi^{\mathrm{Sp}}(S'_\infty)$) and locally constant on the set $\{S_\infty : S\ell \cap \ell = 0\}$. □

Remark 7.31. In [58] we have shown that it is possible to reconstruct the ALM index modulo 4 using only the properties of the Maslov indices $\widehat{\mu}_\ell$ on $\mathrm{Mp}(n)$.

7.4 The Weyl Symbol of a Metaplectic Operator

Metaplectic operators are perfect candidates for being treated as Weyl operators; in this section we set out to find the symbol of $\widehat{S} \in \mathrm{Mp}(n)$. We will see that this innocent program leads to quite substantial calculations where the Conley–Zehnder index studied in Section 4.3 of Chapter 3 plays an essential role. For convenience recall here that the Heisenberg–Weyl operators satisfy the relations

$$\widehat{T}(z_0)\widehat{T}(z_1) = e^{-\frac{i}{\hbar}\sigma(z_0,z_1)}\widehat{T}(z_1)\widehat{T}(z_0), \tag{7.59}$$

$$\widehat{T}(z_0 + z_1) = e^{-\frac{i}{2\hbar}\sigma(z_0,z_1)}\widehat{T}(z_0)\widehat{T}(z_1). \tag{7.60}$$

We will moreover use the following Fresnel-type formula: let M be a real symmetric $m \times m$ matrix. If $\det M \neq 0$, then the Fourier transform of $u \longmapsto \exp(i\langle Mu, u\rangle/2\hbar)$ is given by

$$\left(\tfrac{1}{2\pi\hbar}\right)^{m/2} \int e^{-\frac{i}{\hbar}\langle v,u\rangle} e^{\frac{i}{2\hbar}\langle Mu,u\rangle} d^m u = |\det M|^{-1/2} e^{\frac{i\pi}{4}\operatorname{sgn} M} e^{-\frac{i}{2\hbar}\langle M^{-1}v,v\rangle} \tag{7.61}$$

where $\operatorname{sgn} M$, the "signature" of M, is the number of > 0 eigenvalues of M minus the number of < 0 eigenvalues.

7.4.1 The operators $\widehat{R}_\nu(S)$

Let M_S be the symplectic Cayley transform of $S \in \mathrm{Sp}(n)$ such that $\det(S-I) \neq 0$ (Definition 4.13 in Chapter 4, Section 4.3): it is the symmetric matrix

$$M_S = \frac{1}{2}J(S+I)(S-I)^{-1}.$$

To S we associate the operator

$$\widehat{R}_\nu(S) = \left(\frac{1}{2\pi\hbar}\right)^n \frac{i^\nu}{\sqrt{|\det(S-I)|}} \int e^{\frac{i}{2\hbar}\langle M_S z, z\rangle} \widehat{T}(z) d^{2n}z \tag{7.62}$$

interpreted as a Bochner integral. We will show that provided we choose the integer ν correctly this is the Weyl representation of $\pm\widehat{S} \in \mathrm{Mp}(n)$ such that $\pi^{\mathrm{Mp}}(\widehat{S}) = S$ (the definition of the operators (7.62) is due to Mehlig and Wilkinson [121], who however do not make precise the integer ν). Anyhow, formula (7.62) defines a Weyl operator with complex Gaussian twisted symbol

$$a_\sigma^{(S),\nu}(z) = \frac{i^\nu}{\sqrt{|\det(S-I)|}} e^{\frac{i}{2\hbar}\langle M_S z, z\rangle}. \tag{7.63}$$

Assume in addition that $\det(S+I) \neq 0$. Since the symbol and twisted symbol of a Weyl operator are symplectic Fourier transforms of each other, the symbol of $\widehat{R}_\nu(S)$ is $a^{(S),\nu} = F_\sigma a_\sigma^{(S),\nu}$, that is

$$a^{(S),\nu}(z) = \left(\frac{1}{2\pi\hbar}\right)^n \frac{i^\nu}{\sqrt{|\det(S-I)|}} \int e^{-\frac{i}{\hbar}\sigma(z,z')} e^{\frac{i}{2\hbar}\langle M_S z', z'\rangle} d^{2n}z';$$

applying the Fresnel formula (7.61) with $m = 2n$ we then get

$$a^{(S),\nu}(z) = \frac{i^{\nu + \frac{1}{2}\operatorname{sgn} M_S}}{\sqrt{|\det(S-I)|}} |\det M_S|^{-1/2} e^{\frac{i}{2\hbar}\langle JM_S^{-1}Jz, z\rangle}.$$

Since, by definition of M_S,

$$\det M_S = 2^{-n}\det(S+I)\det(S-I),$$

we can rewrite the formula above as

$$a^{(S),\nu}(z) = 2^{n/2} \frac{i^{\nu + \frac{1}{2}\operatorname{sgn} M_S}}{\sqrt{|\det(S+I)|}} e^{\frac{i}{2\hbar}\langle JM_S^{-1}Jz, z\rangle}; \tag{7.64}$$

beware that this formula is only valid when S has no eigenvalue ± 1.

The following alternative forms of the operators $\widehat{R}_\nu(S)$ are interesting:

Lemma 7.32. *Let $S \in \mathrm{Sp}(n)$ be such that $\det(S-I) \neq 0$. The operator $\widehat{R}_\nu(S)$ can be written as*

$$\widehat{R}_\nu(S) = \left(\frac{1}{2\pi\hbar}\right)^n i^\nu \sqrt{|\det(S-I)|} \int e^{-\frac{i}{2\hbar}\sigma(Sz,z)} \widehat{T}((S-I)z) d^{2n}z, \tag{7.65}$$

that is, as

$$\widehat{R}_\nu(S) = \left(\frac{1}{2\pi\hbar}\right)^n i^\nu \sqrt{|\det(S-I)|} \int \widehat{T}(Sz)\widehat{T}(-z) d^{2n}z. \tag{7.66}$$

Proof. We have

$$\frac{1}{2}J(S+I)(S-I)^{-1} = \frac{1}{2}J + J(S-I)^{-1},$$

hence, in view of the antisymmetry of J,

$$\langle M_S z, z\rangle = \langle J(S-I)^{-1}z, z\rangle = \sigma((S-I)^{-1}z, z).$$

Performing the change of variables $z \longmapsto (S-I)^{-1}z$ we can rewrite the integral in the right-hand side of (7.62) as

$$\begin{aligned}\int e^{\frac{i}{2\hbar}\langle M_S z,z\rangle} \widehat{T}(z) d^{2n}z &= \int e^{\frac{i}{2\hbar}\sigma(z,(S-I)z)} \widehat{T}((S-I)z) d^{2n}z \\ &= \int e^{-\frac{i}{2\hbar}\sigma(Sz,z)} \widehat{T}((S-I)z) d^{2n},\end{aligned}$$

hence (7.66). Taking into account the relation (7.60) we have

$$\widehat{T}((S-I)z) = e^{-\frac{i}{2\hbar}\sigma(Sz,z)} \widehat{T}(Sz)\widehat{T}(-z)$$

and formula (7.65) follows. □

Let us begin by studying composition and inversion for the operators $\widehat{R}_\nu(S)$:

Theorem 7.33. *Let S and S' in $\mathrm{Sp}(n)$ be such that $\det(S-I) \neq 0$, $\det(S'-I) \neq 0$.*

(i) *If $\det(SS'-I) \neq 0$, then*

$$\widehat{R}_\nu(S)\widehat{R}_\nu(S') = \widehat{R}_{\nu+\nu'+\frac{1}{2}\operatorname{sgn} M}(SS'). \tag{7.67}$$

(ii) *The operator $\widehat{R}_\nu(S)$ is invertible and its inverse is*

$$\widehat{R}_\nu(S)^{-1} = \widehat{R}_{-\nu}(S^{-1}). \tag{7.68}$$

Proof. (i) The twisted symbols of $\widehat{R}_\nu(S)$ and $\widehat{R}_\nu(S')$ are, respectively,

$$a_\sigma(z) = \frac{i^\nu}{\sqrt{|\det(S-I)|}} e^{\frac{i}{2\hbar}\langle M_S z, z\rangle},$$
$$b_\sigma(z) = \frac{i^\nu}{\sqrt{|\det(S'-I)|}} e^{\frac{i}{2\hbar}\langle M_{S'} z, z\rangle}.$$

The twisted symbol c_σ of the compose $\widehat{R}_\nu(S)\widehat{R}_\nu(S')$ is given by

$$c_\sigma(z) = \left(\tfrac{1}{2\pi\hbar}\right)^n \int e^{\frac{i}{2\hbar}\sigma(z,z')} a_\sigma(z-z') b_\sigma(z') d^{2n}z'$$

(formula (6.45), Theorem 6.30, Subsection 6.3.1 of Chapter 6), that is

$$c_\sigma(z) = K \int e^{\frac{i}{2\hbar}\sigma(z,z')} e^{\frac{i}{2\hbar}\Phi(z,z')} d^{2n}z'$$

where the constant in front of the integral is

$$K = \left(\frac{1}{2\pi\hbar}\right)^n \frac{i^{\nu+\nu'}}{\sqrt{|\det(S-I)(S'-I)|}}$$

and the phase $\Phi(z,z')$ is

$$\Phi(z,z') = \langle M_S(z-z'), z-z'\rangle + \langle M_{S'}z', z'\rangle$$

that is

$$\Phi(z,z') = \langle M_S z, z\rangle - 2\langle M_S z, z'\rangle + \langle (M_S + M_{S'})z', z'\rangle.$$

Observing that

$$\begin{aligned} \sigma(z,z') - 2\langle M_S z, z'\rangle &= \langle (J - 2M_S)z, z'\rangle \\ &= -2\left\langle J(S-I)^{-1}z, z'\right\rangle, \end{aligned}$$

we have

$$\begin{aligned} &\sigma(z,z') + \Phi(z,z') \\ &\qquad = -2\left\langle J(S-I)^{-1}z, z'\right\rangle + \langle M_S z, z\rangle + \langle (M_S + M_{S'})z', z'\rangle \end{aligned}$$

and hence

$$c_\sigma(z) = K e^{\frac{i}{2\hbar}\langle M_S z, z\rangle)} \int e^{-\frac{i}{\hbar}\left\langle J(S-I)^{-1}z, z'\right\rangle} e^{\frac{i}{2\hbar}\left\langle (M_S+M_{S'})z', z'\right\rangle} d^{2n}z'. \tag{7.69}$$

Applying the Fresnel formula (7.61) with $m = 2n$ and replacing K with its value we get

$$c_\sigma(z) = \left(\tfrac{1}{2\pi\hbar}\right)^n |\det[(M_S + M_{S'})(S-I)(S'-I)]|^{-1/2} e^{\frac{i\pi}{4}\operatorname{sgn} M} e^{\frac{i}{\hbar}\Theta(z)} \tag{7.70}$$

where the phase Θ is given by

$$\begin{aligned}\Theta(z) &= \langle M_S z, z\rangle - \langle (M_S + M_{S'})^{-1} J(S-I)^{-1}z, J(S-I)^{-1}z\rangle \\ &= \left\langle M_S + (S^T - I)^{-1} J(M_S + M_{S'})^{-1} J(S-I)^{-1}z, z\right\rangle ,\end{aligned}$$

that is $\Theta(z) = M_{SS'}$ in view of part (ii) of Lemma 4.14 in Chapter 4, Subsection 4.3. Noting that by definition of the symplectic Cayley transform we have

$$M_S + M_{S'} = J(I + (S-I)^{-1} + (S'-I)^{-1}),$$

it follows that

$$\begin{aligned}\det[(M_S + M_{S'})(S-I)(S'-I)] &= \det[(S-I)(M_S + M_{S'})(S'-I)] \\ &= \det[(S-I)(M_S + M_{S'})(S'-I)] \\ &= |\det(SS'-I)|\end{aligned}$$

which concludes the proof of the first part of the proposition.

(ii) Since $\det(S-I) \neq 0$ we also have $\det(S^{-1}-I) \neq 0$. Formula (7.69) in the proof of part (i) shows that the symbol of $\widehat{R}_\nu(S)\widehat{R}_{-\nu}(S^{-1})$ is

$$c_\sigma(z) = Ke^{\frac{i}{2\hbar}\langle M_S z,z\rangle)} \int e^{-\frac{i}{\hbar}\left\langle J(S-I)^{-1}z,z'\right\rangle} e^{\frac{i}{2\hbar}\left\langle (M_S + M_{S^{-1}})z',z'\right\rangle} d^{2n}z'$$

where the constant K is this time

$$K = \left(\frac{1}{2\pi\hbar}\right)^n \frac{1}{\sqrt{|\det(S-I)(S^{-1}-I)|}} = \left(\frac{1}{2\pi\hbar}\right)^n \frac{1}{|\det(S-I)}$$

since $\det(S^{-1}-I) = \det(I-S)$. Using again Lemma 4.14 we have $M_S + M_{S^{-1}} = 0$, hence, setting $z'' = (S^T - I)^{-1}Jz$,

$$\begin{aligned}c_\sigma(z) &= \left(\frac{1}{2\pi\hbar}\right)^n \frac{1}{|\det(S-I)|} e^{\frac{i}{2\hbar}\langle M_S z,z\rangle} \int e^{-\frac{i}{\hbar}\left\langle J(S-I)^{-1}z,z'\right\rangle} d^{2n}z' \\ &= \left(\frac{1}{2\pi\hbar}\right)^n e^{\frac{i}{2\hbar}\langle M_S z,z\rangle} \int e^{-\frac{i}{\hbar}\langle z,z''\rangle} d^{2n}z''.\end{aligned}$$

Since by the Fourier inversion formula

$$\int e^{-\frac{i}{\hbar}\langle z,z''\rangle} d^{2n}z'' = (2\pi\hbar)^{2n}\delta(z),$$

we thus have $c_\sigma(z) = (2\pi\hbar)^n\delta(z)$, and this is precisely the Weyl symbol of the identity operator (Proposition 6.10 in Chapter 6). □

The composition formula above allows us to prove that the operators $\widehat{R}_\nu(S)$ are unitary:

Corollary 7.34. *Let $S \in \mathrm{Sp}(n)$ be such that* $\det(S-I) \neq 0$. *The operators $\widehat{R}_\nu(S)$ are unitary:* $\widehat{R}_\nu(S)^* = \widehat{R}_\nu(S)^{-1}$.

Proof. In view of Proposition 6.16 (Subsection 6.2.2, Chapter 6) giving the symbol of the adjoint of a Weyl operator is the complex conjugate of the symbol of that operator. Since the twisted and Weyl symbol are symplectic Fourier transforms of each other, the symbol a of $\widehat{R}_\nu(S)$ is thus given by

$$(2\pi\hbar)^n\, a(z) = \frac{i^\nu}{\sqrt{|\det(S-I)|}} \int e^{-\frac{i}{\hbar}\sigma(z,z')} e^{\frac{i}{2\hbar}\langle M_S z',z'\rangle} d^{2n}z'.$$

We have

$$(2\pi\hbar)^n\, \overline{a(z)} = \frac{i^{-\nu}}{\sqrt{|\det(S-I)|}} \int e^{\frac{i}{\hbar}\sigma(z,z')} e^{-\frac{i}{2\hbar}\langle M_S z',z'\rangle} d^{2n}z'.$$

Since $M_{S^{-1}} = -M_S$ and $|\det(S-I)| = |\det(S^{-1}-I)|$ we have

$$\begin{aligned}(2\pi\hbar)^n\, \overline{a(z)} &= \frac{i^{-\nu}}{\sqrt{|\det(S^{-1}-I)|}} \int e^{-\frac{i}{\hbar}\sigma(z,z')} e^{\frac{i}{2\hbar}\langle M_{S^{-1}} z',z'\rangle} d^{2n}z' \\ &= \frac{i^{-\nu}}{\sqrt{|\det(S^{-1}-I)|}} \int e^{\frac{i}{\hbar}\sigma(z,z')} e^{\frac{i}{2\hbar}\langle M_{S^{-1}} z',z'\rangle} d^{2n}z',\end{aligned}$$

hence $\overline{a(z)}$ is the symbol of $\widehat{R}_\nu(S)^{-1}$ and this concludes the proof. □

7.4.2 Relation with the Conley–Zehnder index

Theorem 7.33 and its corollary will allow us to prove that if we identify the integer ν with the Conley–Zehnder index studied in Chapter 4 (Section 4.3), then the operators $\widehat{R}_\nu(S)$ are metaplectic operators. This will however require some work. Let us begin by giving a definition:

Definition 7.35. Let $\widehat{S} \in \mathrm{Mp}(n)$ have projection $S \in \mathrm{Sp}(n)$ such that $\det(S-I) \neq 0$ and choose $S_\infty \in \mathrm{Sp}_\infty(n)$ covering $\widehat{S}$. The integer modulo 4 defined by

$$\widehat{i}_{\mathrm{CZ}}(\widehat{S}) \equiv i_{\mathrm{CZ}}(S_\infty) \mod 4 \tag{7.71}$$

is called the "Conley–Zehnder index" on $\mathrm{Mp}(n)$.

The Conley–Zehnder index on $\mathrm{Mp}(n)$ is well defined: assume in fact that S'_∞ is a second element of $\mathrm{Sp}_\infty(n)$ covering $\widehat{S}$; we have $S'_\infty = \alpha^r S_\infty$ for some $r \in \mathbb{Z}$ (α the generator of $\pi_1[\mathrm{Sp}(n)]$); since $\mathrm{Mp}(n)$ is a double covering of $\mathrm{Sp}(n)$ the integer r must be even. Recalling that

$$i_{\mathrm{CZ}}(\alpha^r S_\infty) = i_{\mathrm{CZ}}(S_\infty) + 2r$$

(property 4.13 of the Conley–Zehnder index in Subsection 4.3.1, Chapter 4) the left-hand side of (7.71) only depends on $\widehat{S}$ and not on the element of $\mathrm{Sp}_\infty(n)$ covering it.

Let S and S' in $\mathrm{Sp}(n)$ be such that $\det(S-I)\neq 0$. Let $\widehat{S}$ and $\widehat{S}'$ in $\mathrm{Mp}(n)$ have projections S and S': $\pi^{\mathrm{Mp}}(\widehat{S})=S$ and $\pi^{\mathrm{Mp}}(\widehat{S}')=S'$ (there are two possible choices in each case). Recall now that we have shown in Chapter 4 (Proposition 4.17, Section 4.3) that

$$i_{\mathrm{CZ}}(S_\infty S'_\infty)=i_{\mathrm{CZ}}(S_\infty)+i_{\mathrm{CZ}}(S'_\infty)+\frac{1}{2}\operatorname{sign}M_S$$

hence, taking classes modulo 4,

$$i_{\mathrm{CZ}}(\widehat{S}\widehat{S}')=i_{\mathrm{CZ}}(\widehat{S})+i_{\mathrm{CZ}}(\widehat{S})+\frac{1}{2}\operatorname{sign}M_S.$$

Choosing $\nu=i_{\mathrm{CZ}}(\widehat{S})$, $\nu'=i_{\mathrm{CZ}}(\widehat{S}')$ formula (7.67) becomes

$$\widehat{R}_{i_{\mathrm{CZ}}(\widehat{S})}(S)\widehat{R}_{i_{\mathrm{CZ}}(\widehat{S}')}(S')=\widehat{R}_{i_{\mathrm{CZ}}(\widehat{S}\widehat{S}')}(SS') \tag{7.72}$$

which suggests that the operators $\widehat{R}_{i_{\mathrm{CZ}}(\widehat{S})}(S)$ generate a true (two-sheeted) unitary representation of the symplectic group, that is the metaplectic group. Formula (7.72) is however not sufficient for this claim, because the $\widehat{R}_{i_{\mathrm{CZ}}(\widehat{S})}(S)$ have only been defined for $\det(S-I)\neq 0$. What we must do is to show that these operators generate a group, and that this group is indeed the metaplectic group $\mathrm{Mp}(n)$.

Let us recall the following notation, used in Section 4.3 of Chapter 4: if W is a quadratic form

$$W(x,x')=\tfrac{1}{2}\langle Px,x\rangle-\langle Lx,x'\rangle+\tfrac{1}{2}\langle Qx',x'\rangle$$

($P=P^T$, $Q=Q^T$, $\det L\neq 0$), we denote by W_S the Hessian matrix of the function $x\longmapsto W(x,x)$:

$$W_S=P+Q-L-L^T, \tag{7.73}$$

that is

$$W_S=DB^{-1}+B^{-1}A-B^{-1}-(B^T)^{-1} \tag{7.74}$$

when $S=\begin{bmatrix}A & B\\ C & D\end{bmatrix}$ is the free symplectic matrix generated by W. Also recall that

$$\begin{aligned}\det(S-I)&=(-1)^n\det B\det(B^{-1}A+DB^{-1}-B^{-1}-(B^T)^{-1})\\&=(-1)^n\det L^{-1}\det(P+Q-L-L^T)\end{aligned} \tag{7.75}$$

(Lemma 4.19 of Subsection 4.3.4 in Chapter 4).

Proposition 7.36. *Let $\widehat{S}_{W,m} \in \mathrm{Mp}(n)$ be a quadratic Fourier transform with projection $S = S_W$.*

(i) *We have $\widehat{R}_\nu(S_W) = \widehat{S}_{W,m}$ provided that*

$$\nu \equiv i_{CZ}(\widehat{S}) \mod 4. \tag{7.76}$$

(ii) *When this is the case we have*

$$\arg \det(S - I) \equiv (i_{CZ}(\widehat{S}) - n)\pi \mod 2\pi. \tag{7.77}$$

Proof. (i) Let $\delta \in \mathcal{S}'(\mathbb{R}^n)$ be the Dirac distribution centered at $x = 0$; setting

$$C_{W,\nu} = \left(\frac{1}{2\pi\hbar}\right)^n \frac{i^\nu}{\sqrt{|\det(S-I)|}}$$

we have, by definition of $\widehat{R}_\nu(S)$,

$$\begin{aligned}\widehat{R}_\nu(S)\delta(x) &= C_{W,\nu} \int e^{\frac{i}{2\hbar}\langle M_S z_0, z_0\rangle} e^{\frac{i}{\hbar}(\langle p_0, x\rangle - \frac{1}{2}\langle p_0, x_0\rangle)} \delta(x - x_0) d^{2n} z_0 \\ &= C_{W,\nu} \int e^{\frac{i}{2\hbar}\langle M_S(x,p_0),(x,p_0)\rangle} e^{\frac{i}{2\hbar}\langle p,x\rangle} \delta(x - x_0) d^{2n} z_0,\end{aligned}$$

hence, setting $x = 0$,

$$\widehat{R}_\nu(S)\delta(0) = C_{W,\nu} \int e^{\frac{i}{2\hbar}\langle M_S(0,p_0),(0,p_0)\rangle} \delta(-x_0) d^{2n} z_0,$$

that is, since $\int \delta(-x_0) d^n x_0 = 1$,

$$\widehat{R}_\nu(S)\delta(0) = \left(\frac{1}{2\pi\hbar}\right)^n \frac{i^\nu}{\sqrt{|\det(S-I)|}} \int e^{\frac{i}{2\hbar}\langle M_S(0,p_0),(0,p_0)\rangle} d^n p_0. \tag{7.78}$$

Let us next calculate the scalar product

$$\langle M_S(0,p_0), (0,p_0)\rangle = \sigma((S-I)^{-1}0, p_0), (0,p_0)).$$

The relation $(x,p) = (S-I)^{-1}(0,p_0)$ is equivalent to $S(x,p) = (x, p+p_0)$, that is to

$$p + p_0 = \partial_x W(x,x) \text{ and } p = -\partial_{x'} W(x,x).$$

These relations yield, after a few calculations,

$$x = (P + Q - L - L^T)^{-1} p_0 \ ; \ p = (L - Q)(P + Q - L - L^T)^{-1} p_0$$

and hence

$$\langle M_S(0,p_0), (0,p_0)\rangle = -\left\langle W_S^{-1} p_0, p_0\right\rangle \tag{7.79}$$

where W_S is the symmetric matrix (7.73). Applying Fresnel's formula (7.61) to (7.78) we get

$$\left(\tfrac{1}{2\pi\hbar}\right)^n \int e^{\frac{i}{2\hbar}\langle M_S(0,p_0),(0,p_0)\rangle} d^n p_0 = e^{-\frac{i\pi}{4}\operatorname{sgn} W_S} |\det W_S|^{1/2};$$

observing that in view of formula (7.75) we have

$$\frac{1}{\sqrt{|\det(S_W - I)|}} = |\det L|^{1/2} |\det W_S|^{-1/2},$$

we obtain

$$\widehat{R}_\nu(S_W)\delta(0) = \left(\tfrac{1}{2\pi\hbar}\right)^n i^\nu e^{-\frac{i\pi}{4}\operatorname{sgn} W_S} |\det L|^{1/2}.$$

Now, by definition of $\widehat{S}_{W,m}$,

$$\begin{aligned}\widehat{S}_{W,m}\delta(0) &= \left(\tfrac{1}{2\pi i\hbar}\right)^n i^m \sqrt{|\det L|} \int e^{\frac{i}{\hbar}W(0,x')}\delta(x')d^n x' \\ &= \left(\tfrac{1}{2\pi\hbar}\right)^n i^{m-n/2} \sqrt{|\det L|}\end{aligned}$$

and hence

$$i^\nu e^{-\frac{i\pi}{4}\operatorname{sgn} W_S} = i^{m-n/2}.$$

It follows that we have

$$\nu - \tfrac{1}{2}\operatorname{sgn} W_S \equiv m - \tfrac{1}{2}n \mod 4$$

which is equivalent to formula (7.76) since W_S has rank n.

(ii) In view of formula (7.75) we have

$$\arg\det(S - I) = n\pi + \arg\det B + \arg\det W_S \mod 2\pi.$$

Taking into account the obvious relations

$$\begin{aligned}\arg\det B &\equiv m_{\ell_P}(\widehat{S}) \mod 2\pi, \\ \arg\det W_S &\equiv \pi \operatorname{Inert} W_S \mod 2\pi,\end{aligned}$$

formula (7.77) follows. □

Recall from Chapter 7 that every $\widehat{S} \in \operatorname{Mp}(n)$ can be written (in infinitely many ways) as a product $\widehat{S} = \widehat{S}_{W,m}\widehat{S}_{W',m'}$. We are going to show that $\widehat{S}_{W,m}$ and $\widehat{S}_{W',m'}$ always can be chosen such that $\det(\widehat{S}_{W,m} - I) \neq 0$ and $\det(\widehat{S}_{W',m'} - I) \neq 0$.

Corollary 7.37. *The operators $\widehat{R}_\nu(S_W)$ generate* Mp(n). *In fact, every $\widehat{S} \in \operatorname{Mp}(n)$ can be written as a product*

$$\widehat{S} = \widehat{R}_\nu(S_W)\widehat{R}_{\nu'}(S_{W'}) \tag{7.80}$$

where $\det(S_W - I) \neq 0$, $\det(S_{W'} - I) \neq 0$.

Proof. Recall that the quadratic Fourier transforms $\widehat{S}_{W,m}$ generate $\mathrm{Mp}(n)$; in fact every $\widehat{S} \in \mathrm{Mp}(n)$ can be written in the form $\widehat{S} = \widehat{S}_{W,m}\widehat{S}_{W',m'}$. In view of Proposition 7.36 it thus suffices to show that W and W' can be chosen so that $S_W = \pi^{\mathrm{Mp}}(\widehat{S}_{W,m})$ and $S_{W'} = \pi^{\mathrm{Mp}}(\widehat{S}_{W',m})$ satisfy $\det(S_W - I) \neq 0$, $\det(S_{W'} - I) \neq 0$. That the $\widehat{R}_\nu(S_W)$ generate $\mathrm{Mp}(n)$ follows from formula (7.80) since Let us write $\widehat{S} = \widehat{S}_{W,m}\widehat{S}_{W',m'}$ and apply the factorization (2.51) to each of the factors; writing $W = (P, L, Q)$, $W' = (P', L', Q')$ we have

$$\widehat{S} = \widehat{V}_{-P}\widehat{M}_{L,m}\widehat{J}\widehat{V}_{-(P'+Q)}\widehat{M}_{L',m'}\widehat{J}\widehat{V}_{-Q'}. \tag{7.81}$$

We claim that $\widehat{S}_{W,m}$ and $\widehat{S}_{W',m'}$ can be chosen in such a way that $\det(S_W - I) \neq 0$ and $\det(S_{W'} - I) \neq 0$, that is,

$$\det(P + Q - L - L^T) \neq 0 \quad \text{and} \quad \det(P' + Q' - L' - L'^T) \neq 0;$$

this will prove the assertion. We first remark that the right-hand side of (7.81) obviously does not change if we replace P' by $P' + \lambda I$ and Q by $Q - \lambda I$ where $\lambda \in \mathbb{R}$. Choose now λ such that it is not an eigenvalue of $P + Q - L - L^T$ and $-\lambda$ is not an eigenvalue of $P' + Q' - L' - L'^T$; then

$$\det(P + Q - \lambda I - L - L^T) \neq 0,$$
$$\det(P' + \lambda I + Q' - L - L^T) \neq 0$$

and we have $\widehat{S} = \widehat{S}_{W_1,m_1}\widehat{S}_{W_1',m_1'}$ with

$$W_1(x, x') = \frac{1}{2}\langle Px, x\rangle - \langle Lx, x'\rangle + \frac{1}{2}\langle (Q - \lambda I)x', x'\rangle,$$
$$W_1'(x, x') = \frac{1}{2}\langle (P' + \lambda I)x, x\rangle - \langle L'x, x'\rangle + \frac{1}{2}\langle Q'x', x'\rangle;$$

this concludes the proof. □

So far, so good. But we haven't told the whole story yet: there remains to prove that every $\widehat{S} \in \mathrm{Mp}(n)$ such that $\det(S - I) \neq 0$ can be written in the form $\widehat{R}_\nu(S)$.

Proposition 7.38. *For every* $\widehat{S} \in \mathrm{Mp}(n)$ *such that* $\det(S - I) \neq 0$, *we have* $\widehat{S} = \widehat{R}_{\nu(S)}(S)$ *with*

$$\nu(S) = \nu + \nu' + \tfrac{1}{2}\operatorname{sgn}(M + M'). \tag{7.82}$$

Proof. Let us write $\widehat{S} = \widehat{R}_\nu(S_W)\widehat{R}_{\nu'}(S_{W'})$. A straightforward calculation using the composition formula for Weyl operators (6.45) in Theorem 6.30 of Chapter 6, together with the Fresnel integral (7.61), shows that we have

$$\widehat{S} = \left(\tfrac{1}{2\pi\hbar}\right)^n \frac{i^{\nu+\nu'+\frac{1}{2}sgn(M+M')}}{\sqrt{|\det(S_W - I)(S_{W'} - I)(M + M')|}} \int e^{\frac{i}{2\hbar}\langle Nz, z\rangle}\widehat{T}(z)d^{2n}z \tag{7.83}$$

where M and M' correspond to S_W and $S_{W'}$ by

$$M = \frac{1}{2}J(S_W + I)(S_W - I)^{-1},$$
$$M' = \frac{1}{2}J(S_{W'} + I)(S_{W'} - I)^{-1}$$

and N is given by

$$N = M - (M + \tfrac{1}{2}J)(M + M')^{-1}(M - \tfrac{1}{2}J).$$

We claim that

$$\det(S_W - I)(S_{W'} - I)(M + M') = \det(S - I) \tag{7.84}$$

(hence $M + M'$ is indeed invertible), and

$$N = \tfrac{1}{2}J(S + I)(S - I)^{-1} = M_S. \tag{7.85}$$

The first of these identities is easy to check by a direct calculation: by definition of M and M' we have, since $\det J = 1$,

$$\det(S_W - I)(S_{W'} - I)(M + M') = \det(S_W - I)(I + (S_W - I)^{-1} + (S_W - I)^{-1})(S_{W'} - I)$$

that is

$$\det(S_W - I)(S_{W'} - I)(M + M') = \det(S_W S_{W'} - I)$$

which is precisely (7.84). Formula (7.85) is at first sight more cumbersome; there is however an easy way out: assume that $\widehat{S} = \widehat{S}_{W'',m''}$; we *know* that we *must* have in this case

$$N = \tfrac{1}{2}J(S_W S_{W'} + I)(S_W S_{W'} - I)^{-1}$$

and this algebraic identity then holds for all $S = S_W S_{W'}$ since the free symplectic matrices are dense in $\operatorname{Sp}(n)$. Thus,

$$\widehat{S} = \left(\frac{1}{2\pi\hbar}\right)^n \frac{i^{\nu+\nu'+\frac{1}{2}sgn(M+M')}}{\sqrt{|\det(S - I)|}} \int e^{\frac{i}{2\hbar}\langle M_S z, z\rangle}\widehat{T}(z)d^{2n}z$$

and to conclude the proof there remains to prove that

$$\nu(S)\pi = (\nu + \nu' + \tfrac{1}{2}\operatorname{sgn}(M + M'))\pi$$

is effectively one of the two possible choices for $\arg\det(S - I)$. We have

$$(\nu + \nu' + \tfrac{1}{2}\operatorname{sgn}(M + M'))\pi = -\arg\det(S_W - I) - \arg\det(S_{W'} - I) + \tfrac{1}{2}\pi\operatorname{sgn}(M + M');$$

we next note that if R is any real invertible $2n \times 2n$ symmetric matrix with q negative eigenvalues, we have $\arg\det R = q\pi \mod 2\pi$ and $\frac{1}{2}\operatorname{sgn} R = 2n - q$ and hence

$$\arg\det R = \tfrac{1}{2}\pi \operatorname{sgn} R \mod 2\pi.$$

It follows, taking (7.84) into account, that

$$(\nu + \nu' + \tfrac{1}{2}\operatorname{sgn}(M + M'))\pi = \arg\det(S - I) \mod 2\pi$$

which concludes the proof. □

Exercise 7.39. Prove in detail the equality (7.83) used in the proof of Proposition 7.38.

Part III

Quantum Mechanics in Phase Space

Chapter 8

The Uncertainty Principle

This chapter is devoted to one of the most important, basic, and at the same time characteristic features of quantum mechanics, the *uncertainty principle.* That principle can be stated in its crudest (and perhaps most well-known) form as

$$\Delta X_j \Delta P_j \geq \tfrac{1}{2}\hbar \ , \ j = 1, 2, \ldots, n,$$

where ΔX_j and ΔP_j are the standard deviations associated to the position and momentum random variables $X = (X_1, \ldots, X_n)$ and $P = (P_1, \ldots, P_n)$. The idea goes back to Heisenberg's 1927 *Zeitschrift für Physik* paper [89]; but while Heisenberg used a thought experiment to arrive at the formula $\delta q \delta p = h/2\pi$, the modern and more accurate form above was rigorously proven by Robertson [136]. The sharp form

$$(\Delta X_j)^2_\psi (\Delta P_j)^2_\psi \geq \tfrac{1}{4}\operatorname{Cov}(X_j, P_j)_\psi + \tfrac{1}{4}\hbar^2$$

of the uncertainty principle we will use in this chapter (Proposition 8.8) was proven by Schrödinger [143] in 1930. We refer to Jammer's book [97] (especially Chapter 7) for a very detailed discussion of the history of the uncertainty relations, and of their philosophical implications.

We will give a purely geometric interpretation of the sharp uncertainty principle and relate that principle to the notion of linear symplectic capacity introduced in Chapter 2, Subsection 8.3.3. We will then use this interpretation to construct a quantum-mechanical phase space whose elements are no longer points $z = (x, p)$, but rather a class of particular ellipsoids we have called "quantum blobs" in [65, 66, 69], and relate these to Wigner transforms of Gaussians (de Gosson [71, 72]). The fact that topological notions such as symplectic capacities could play a pivotal role in quantization has been realized only recently (see for instance Dragoman [32] for a system of axioms for quantum mechanics where quantum blobs intervene; also Elskens and Escande [38] and Giacchetta *et al.* [48]).

We begin by reviewing the notions of *states* and *observables.* For a very readable detailed analysis of these concepts see Dubin *et al.* [33].

8.1 States and Observables

In this section we introduce some basic concepts and terminology from classical and quantum mechanics; we will return to the discussion in Chapter 9.

Both classical and quantum mechanics are "observable-state" systems; what an observable or a state in quantum mechanics should *really* be is varying from author to author: to paraphrase Quintus Horatius Fallcus (65–8 BC):

Grammatici certant et adhuc sub judice lis est.[1]

The definitions of states and observables we will give in the subsequent subsections are far from being the only possible (nor perhaps the most used); for other points of view see Mackey [115, 116, 117]; Hermann's appendix to Wallach's book [175] also contains a valuable discussion of the topic.

8.1.1 Classical mechanics

A classical physical system $(\mathcal{C})$ consists of a finite number of point-like particles; we assume that the positions and momenta of these particles are exactly known; they can thus be collectively identified (at a given time t) by a point z of $\mathbb{R}_z^{2n}$ (or, more generally, of some symplectic manifold), hereafter called the *phase-space* of $(\mathcal{C})$. Such a point is called a *pure state*; the integer n is called the number of *degrees of freedom* of $(\mathcal{C})$. In its most naive acceptation a *classical observable* is a function on the phase space (or on the extended time-dependent phase space), whose choice is limited by whatever condition of regularity is required by the physics of the problem. If the forces are regular enough, the time evolution of $(\mathcal{C})$ is governed by a privileged observable, the *Hamiltonian function*, whose value at (z,t) is, by definition, the *energy*. In practice it is convenient to extend the notion of state, and that of observable (see Dubin *et al.* [44]):

Definition 8.1.

(i) A (classical) observable is a function $a \in C^\infty(\mathbb{R}_z^{2n})$;

(ii) A (classical) state is a normalized positive distribution $\rho \in \mathcal{E}'(\mathbb{R}_z^{2n})$: $\langle \rho, a\rangle \geq 0$ for every $a \geq 0$ and $\langle \rho, 1\rangle = 1$.

(iii) A pure state is a Dirac distribution $\delta(z - z_0)$; a state which is not pure is called a "mixed state".

The set of all states clearly is a convex set: if ρ and ρ' are two states, then $\alpha\rho + (1-\alpha)\rho'$ is also a state for $0 \leq \alpha \leq 1$.

Standard examples of mixed states are provided by probability densities: assume that ρ is a compactly supported integrable function on $\mathbb{R}_z^{2n}$ such that $\rho \geq 0$ and with integral equal to 1. It is an element of $\mathcal{E}'(\mathbb{R}_z^{2n})$ and we have

$$\langle \rho, a\rangle = \int \rho(z)a(z)d^{2n}z \geq 0 \quad , \quad \langle \rho, 1\rangle = \int \rho(z)d^{2n}z = 1$$

[1] "Scholars dispute, and the case is still before the courts".

so that ρ is indeed a state. More general states which are genuinely compactly supported distributions occur naturally in statistical mechanics; here is one basic example:

Example 8.2. The "microcanonical Gibbs state". Let H be a time-dependent Hamiltonian function and assume that the energy shell $\Sigma_E : H(z) = E$ has compact smooth boundary. The microcanonical state is defined by

$$\rho_{\text{micro}}(z) = \frac{1}{\Sigma(E,V)}\delta(H(z) - E)$$

where the normalization constant $\Sigma(E, V)$ is given by

$$\Sigma(E,V) = \int \delta(H(z) - E)d^{2n}z.$$

Obviously $\operatorname{Supp}\rho_{\text{micro}} \subset \Sigma_E$.

Other examples of classical states are presented from a very elementary point of view in Dubin *et al.* [33].

8.1.2 Quantum mechanics

As we said in the introduction to this chapter, we make ours, with Mackey [117], the point of view that quantum mechanics is a refinement of Hamiltonian mechanics. In other words, for us quantum mechanics is the "better theory" which falsifies (in the Popperian sense) classical mechanics. This implies that instead of focusing on "correspondence rules" associating to a classical observable a quantum operator, we postulate that self-adjoint operators on Hilbert spaces are, by definition, the observables of quantum mechanics.

In traditional quantum mechanics, a *pure state* of a quantum system $(\mathcal{Q})$ is an element of a Hilbert space $\mathcal{H}$. The usual choice in standard quantum mechanics is $\mathcal{H} = L^2(\mathbb{R}^n_x)$, but we will see later on that this choice is absolutely not compelling, and that $L^2(\mathbb{R}^n_x)$ can be replaced by a particular closed subspace of $L^2(\mathbb{R}^{2n}_z)$ using a "wave-packet transform", leading to *quantum mechanics in phase space.* Quantum observables are self-adjoint operators on the Hilbert space $\mathcal{H}$. These operators need not be bounded; for instance multiplication by x_j is not a bounded operator if $\mathcal{H} = L^2(\mathbb{R}^n_x)$; see for instance [139] for a general discussion of operators intervening in quantum mechanics.

8.2 The Quantum Mechanical Covariance Matrix

Quantum mechanics is (at least in its applications) a statistical theory. The notion of covariance matrix is a useful device, both from a theoretical and practical point of view. It turns out, as we will see, that the uncertainty principle can be restated in a very elegant and concise form in terms of this matrix.

8.2.1 Covariance matrices

We refer to Appendix D for the basics of probability theory that are being used in this section.

Definition 8.3. Let $Z = (Z_1, \dots, Z_m)$ be a random variable on $\mathbb{R}^m$. The symmetric $m \times m$ matrix $\Sigma = (\mathrm{Cov}(Z_j, Z_k))_{1\leq j,k\leq m}$ is called the "covariance matrix" of the vector-valued random variable $Z = (Z_1, \dots, Z_m)$.

The correlation coefficients $\rho_{jk} = \rho(Z_j, Z_k)$ are equal to 1 when $j = k$, hence the principal diagonal of the covariance matrix consists of the variances $(\Delta Z_j)^2$.

Example 8.4. For instance, in the case $m = 2$, writing $Z = (X, P)$:

$$\Sigma = \begin{bmatrix} (\Delta X)^2 & \Delta \\ \Delta & (\Delta P)^2 \end{bmatrix} \quad , \quad \Delta = \mathrm{Cov}(X, P).$$

This example is a particular case of the following situation: consider the random vector

$$Z = (X_1, \dots, X_n; P_1, \dots, P_n)$$

where $X_j : \mathbb{R}^n_x \longrightarrow \mathbb{R}$ and $P_j : \mathbb{R}^n_p \longrightarrow \mathbb{R}$; denoting by Σ_{XX} and Σ_{PP} the covariance matrices of $X = (X_1, \dots, X_n)$ and $P = (P_1, \dots, P_n)$, and by

$$\Sigma_{XP} = (\mathrm{Cov}(X_j, P_k))_{1\leq j,k\leq n}$$

the covariance matrix Σ of Z is the symmetric matrix

$$\Sigma = \begin{bmatrix} \Sigma_{XX} & \Sigma_{XP} \\ \Sigma_{PX} & \Sigma_{PP} \end{bmatrix} \quad \text{with} \quad \Sigma_{PX} = \Sigma_{XP}^T.$$

Remark 8.5. The functions X_j and P_k are usually interpreted, as the notation is intended to suggest, as the random variables corresponding to measurements of the position and momentum coordinates x_j and p_k.

8.2.2 The uncertainty principle

Let $\widehat{A}$ be a self-adjoint operator on $L^2(\mathbb{R}^n_x)$ and $\psi \in L^2(\mathbb{R}^n_x)$, $\psi \neq 0$. Since the arguments below are of a quite general character we do not assume here that $\widehat{A}$ is a Weyl operator, nor even that it is a bounded operator on $L^2(\mathbb{R}^n_x)$.

Definition 8.6. The *mathematical expectation* (or: *average value*) of $\widehat{A}$ in the state ψ is by definition the real number

$$\langle \widehat{A} \rangle_\psi = \frac{(\widehat{A}\psi, \psi)_{L^2(\mathbb{R}^n_x)}}{(\psi, \psi)_{L^2(\mathbb{R}^n_x)}}.$$

Since we obviously have

$$\langle\widehat{A}\rangle_{\alpha\psi} = \langle\widehat{A}\rangle_{\psi} \text{ if } \alpha \in \mathbb{C}\backslash\{0\}$$

we may replace ψ by $\psi/||\psi||_{L^2}$ in the definition of $\langle\widehat{A}\rangle_{\psi}$ so that it is no restriction to assume that ψ is normalized to unity:

$$\langle\widehat{A}\rangle_{\psi} = (\widehat{A}\psi,\psi)_{L^2(\mathbb{R}^n_x)} \text{ if } ||\psi||_{L^2(\mathbb{R}^n_x)} = 1. \tag{8.1}$$

This shows that what counts in practice is actually the "ray" $\{\alpha\psi : \alpha \neq 0, \alpha \in \mathbb{C}\}$: the replacement of ψ by $\alpha\psi$ in formula (8.1) has no effect, whatsoever, on the mathematical expectation: the "true" state space of quantum mechanics is actually a *projective* space. The definition of $\langle\widehat{A}\rangle_{\psi}$ is motivated by the following argument: assume that the spectrum of $\widehat{A}$ is discrete, and consists of real numbers λ_j, $j \in \mathbb{N}$ (recall that $\widehat{A}$ is assumed to be self-adjoint); let $(\psi_j)_{j\in\mathbb{N}}$ be the corresponding orthonormal basis of eigenfunctions of $\widehat{A}$. A basic axiom of quantum mechanics is that the probability of obtaining the value λ_j for the observable $\widehat{A}$ when measuring it in the normalized state ψ is

$$\Pr_{\psi}(\lambda = \lambda_j) = |(\psi,\psi_j)_{L^2(\mathbb{R}^n_x)}|^2.$$

The mathematical expectation of the observable $\widehat{A}$ in the state ψ is in this case

$$\langle\widehat{A}\rangle_{\psi} = \sum_{j=1}^{\infty} \lambda_j |(\psi,\psi_j)_{L^2(\mathbb{R}^n_x)}|^2 = (\widehat{A}\psi,\psi)_{L^2(\mathbb{R}^n_x)}$$

which is of course consistent with formula (8.1).

Definition 8.7. Let $\widehat{A}$ be an observable;

(i) If $\widehat{A}^2$ also is an observable and ψ is a quantum state, then

$$(\Delta\widehat{A})^2_{\psi} = \langle\widehat{A}^2\rangle_{\psi} - \langle\widehat{A}\rangle^2_{\psi}$$

is called the "variance of $\widehat{A}$ in the state ψ"; its positive square root $(\Delta\widehat{A})_{\psi}$ is called "standard deviation".

(ii) If $\widehat{B}$ is a second observable with the same property, then

$$\operatorname{Cov}(\widehat{A},B)_{\psi} = \tfrac{1}{2}\langle\widehat{A}\widehat{B} + \widehat{B}\widehat{A}\rangle_{\psi}$$

is the "covariance" of the pair $(\widehat{A},\widehat{B})$ in the state ψ.

The following result relates commutation relations between operators to quantum uncertainty:

Proposition 8.8. *If the variances and covariances of two self-adjoint operators $\widehat{A}$ and $\widehat{B}$ exist and satisfy the commutation relation*

$$[\widehat{A},\widehat{B}]=\widehat{A}\widehat{B}-\widehat{B}\widehat{A}=i\hbar I,$$

then the following inequality holds:

$$(\Delta\widehat{A})^2_\psi(\Delta\widehat{B})^2_\psi \geq \tfrac{1}{4}\operatorname{Cov}(\widehat{A},\widehat{B})_\psi+\tfrac{1}{4}\hbar^2. \tag{8.2}$$

Proof. It is no restriction to take $||\psi||_{L^2}=1$. It is also sufficient to assume that $\langle\widehat{A}\rangle_\psi=\langle\widehat{B}\rangle_\psi=0$ because the general case is reduced to the case $\langle\widehat{A}\rangle_\psi=\langle\widehat{B}\rangle_\psi=0$ applying the result to the operators $\widehat{A}-\langle\widehat{A}\rangle_\psi I$ and $\widehat{B}-\langle\widehat{B}\rangle_\psi I$. The proof of (8.2) is thus reduced to the proof of the inequality

$$\langle\widehat{A}^2\rangle_\psi\langle\widehat{B}^2\rangle_\psi \geq \tfrac{1}{4}\operatorname{Cov}(\widehat{A},\widehat{B})_\psi+\tfrac{1}{4}\hbar^2 \tag{8.3}$$

when $\langle\widehat{A}\rangle_\psi=\langle\widehat{B}\rangle_\psi=0$. We have, since $\widehat{A}$ is self-adjoint,

$$\langle\widehat{A}^2\rangle_\psi=(\widehat{A}^2\psi,\psi)_{L^2(\mathbb{R}^n_x)}=||\widehat{A}\psi||^2_{L^2(\mathbb{R}^n_x)},$$
$$\langle\widehat{B}^2\rangle_\psi=(\widehat{B}^2\psi,\psi)_{L^2(\mathbb{R}^n_x)}=||\widehat{B}\psi||^2_{L^2(\mathbb{R}^n_x)},$$

hence, using Cauchy–Schwarz's inequality:

$$\langle\widehat{A}^2\rangle_\psi\langle\widehat{B}^2\rangle_\psi \geq |(\widehat{A}\psi,\widehat{B}\psi)_{L^2(\mathbb{R}^n_x)}|^2=|(\widehat{A}\widehat{B}\psi,\psi)_{L^2(\mathbb{R}^n_x)}|^2.$$

Noting that

$$\widehat{A}\widehat{B}=\tfrac{1}{2}(\widehat{A}\widehat{B}+\widehat{B}\widehat{A})+\tfrac{1}{2}[\widehat{A},\widehat{B}]=\tfrac{1}{2}(\widehat{A}\widehat{B}+\widehat{B}\widehat{A})+\tfrac{i\hbar}{2}I_d$$

and $||\psi||_{L^2}=1$ this can be rewritten as

$$\langle\widehat{A}^2\rangle_\psi\langle\widehat{B}^2\rangle_\psi \geq |\tfrac{1}{2}(\widehat{A}\widehat{B}+\widehat{B}\widehat{A})\psi,\psi)|_{L^2(\mathbb{R}^n_x)}+\tfrac{i\hbar}{2}|^2;$$

since $(\widehat{A}\widehat{B}+\widehat{B}\widehat{A})\psi,\psi)_{L^2(\mathbb{R}^n_x)}$ is real, (8.3) follows. □

The following consequence of this result is immediate:

Corollary 8.9. *Let $\widehat{X}_j$ be the operator of multiplication by x_j and $\widehat{P}_j=-i\hbar\partial/\partial x_j$. We have*

$$(\Delta X_j)^2_\psi(\Delta P_j)^2_\psi \geq \tfrac{1}{4}\operatorname{Cov}(X_j,P_j)_\psi+\tfrac{1}{4}\hbar^2 \tag{8.4}$$

and, in particular, $(\Delta X_j)_\psi(\Delta P_j)_\psi \geq \frac{1}{2}\hbar$.

A *caveat*: in the corollary above none of the operators $\widehat{X}_j,\widehat{P_j}$ are bounded on $L^2(\mathbb{R}^n_x)$. It turns out that this is a quite general feature, as is shown by the following argument of Winter and Wielandt (see Theorem 4.11 in the book [33] by Dubin *et al.*): let $\mathcal{H}$ be an arbitrary Hilbert space and assume that $\widehat{X}$ and $\widehat{P}$

are such that $[\widehat{X}, \widehat{P}] = i\hbar$. Then $\widehat{X}$ or $\widehat{P}$ is unbounded. Assume in fact that $\widehat{P}$ is bounded. A straightforward induction argument shows that for every integer k we have $[\widehat{X}, \widehat{P}^{k+1}] = i\hbar(k+1)\widehat{P}^k$ and hence

$$\hbar(k+1)||\widehat{P}^k||_{\mathcal{H}} \leq 2||\widehat{X}||_{\mathcal{H}}||\widehat{P}^k||_{\mathcal{H}}$$

so that

$$\hbar(k+1) \leq 2||\widehat{X}||_{\mathcal{H}}||\widehat{P}||_{\mathcal{H}};$$

since this inequality holds for every k the operator $\widehat{X}$ can thus not be bounded.

From now on we will always assume that $\widehat{A}$ is the Weyl operator with symbol a, defined for $\psi \in \mathcal{S}(\mathbb{R}^n_x)$ by

$$\widehat{A}\psi(x) = \left(\tfrac{1}{2\pi\hbar}\right)^n \iint e^{\frac{i}{\hbar}\langle p, x-y\rangle} a(\tfrac{1}{2}(x+y), p)\psi(y) d^n y d^n p.$$

We claim that the "uncertainty principle" (8.2) is invariant under the action of the symplectic group $\mathrm{Sp}(n)$. Let us glorify this important statement by giving it the status of a theorem:

Theorem 8.10. *Assume that the Weyl operators $\widehat{A} \overset{\text{Weyl}}{\longleftrightarrow} a$ and $\widehat{B} \overset{\text{Weyl}}{\longleftrightarrow} b$ satisfy the uncertainty relations (8.2). Let $S \in \mathrm{Sp}(n)$ and let $\widehat{A}_S \overset{\text{Weyl}}{\longleftrightarrow} a \circ S^{-1}$ and $\widehat{B}_S \overset{\text{Weyl}}{\longleftrightarrow} b \circ S^{-1}$.*

(i) *We have*

$$(\Delta\widehat{A}_S)^2_{\widehat{S}\psi}(\Delta\widehat{B}_S)^2_{\widehat{S}\psi} \geq \tfrac{1}{4}\operatorname{Cov}(\widehat{A}, \widehat{B})_\psi - \tfrac{1}{4}[\widehat{A}, \widehat{B}]^2_\psi \tag{8.5}$$

for every $\widehat{S} \in \mathrm{Mp}(n)$ with $\pi(\widehat{S}) = S$.

(ii) *In particular*

$$(\Delta\widehat{X}_j)^2_{\widehat{S}\psi}(\Delta\widehat{P}_j)^2_{\widehat{S}\psi} \geq \tfrac{1}{4}\operatorname{Cov}(\widehat{X}_j, \widehat{P}_j)_\psi + \tfrac{1}{4}\hbar^2. \tag{8.6}$$

Proof. Using the metaplectic covariance property (7.24) in Theorem 7.13 we have

$$\langle\widehat{A}_S\rangle^2_{\widehat{S}\psi} = \int |\widehat{S}\widehat{A}\psi(x)|^2 d^n x = ||\widehat{S}\widehat{A}\psi||^2_{L^2},$$

hence, since $\widehat{S}$ is a unitary operator:

$$\langle\widehat{A}_S\rangle_{\widehat{S}\psi} = ||\widehat{A}\psi||^2_{L^2} = \langle\widehat{A}\rangle^2_\psi.$$

Writing similar relations for $\langle B_S\rangle_\psi$, $\langle A_S^2\rangle_\psi$ and $\langle B_S^2\rangle_\psi$ formulae (8.5) and (8.6) follow. □

We will use this result in the forthcoming subsections to restate the uncertainty principle in a geometric form. Let us first explain what we mean by a "quantum mechanically admissible covariance matrix".

8.3 Symplectic Spectrum and Williamson's Theorem

The message of Williamson's theorem is that one can diagonalize any positive definite symmetric matrix M using a symplectic matrix, and that the diagonal matrix has the very simple form

$$D = \begin{bmatrix} \Lambda_\sigma & 0 \\ 0 & \Lambda_\sigma \end{bmatrix}$$

where the diagonal elements of Λ_σ are the moduli of the eigenvalues of JM. This is a truly remarkable result which will allow us to construct a precise phase space quantum mechanics in the ensuing chapters. One can without exaggeration say that this theorem carries the germs of the recent developments of symplectic topology; it leads immediately to a proof of Gromov's famous non-squeezing theorem in the linear case and has many applications both in mathematics and physics. Williamson proved this result in 1963 and it has been rediscovered several times since that – with different proofs.

8.3.1 Williamson normal form

Let M be a real $m \times m$ symmetric matrix: $M = M^T$. Elementary linear algebra tells us that all the eigenvalues $\lambda_1, \lambda_2, \dots, \lambda_m$ of M are real, and that M can be diagonalized using an orthogonal transformation: $M = R^T D R$ with $R \in O(m)$ and $D = \operatorname{diag}[\lambda_1, \lambda_2, \dots, \lambda_{nm}]$. Williamson's theorem provides us with the symplectic variant of this result. It says that every symmetric and positive definite matrix M can be diagonalized using *symplectic* matrices, and this in a very particular way. Because of its importance in everything that will follow, let us describe Williamson's diagonalization procedure in detail.

Theorem 8.11. *Let M be a positive-definite symmetric real $2n \times 2n$ matrix.*

(i) *There exists $S \in \mathrm{Sp}(n)$ such that*

$$S^T M S = \begin{bmatrix} \Lambda & 0 \\ 0 & \Lambda \end{bmatrix} \quad , \ \Lambda \text{ diagonal,} \tag{8.7}$$

the diagonal entries λ_j of Λ being defined by the condition

$$\pm i\lambda_j \text{ is an eigenvalue of } JM^{-1}. \tag{8.8}$$

(ii) *The sequence $\lambda_1, \dots, \lambda_n$ does not depend, up to a reordering of its terms, on the choice of S diagonalizing M.*

Proof. (*Cf.* Folland [42], Ch.4.) (i) A quick examination of the simple case $M = I$ shows that the eigenvalues are $\pm i$, so that it is a good idea to work in the space $\mathbb{C}^{2n}$ and to look for complex eigenvalues and vectors for JM. Let us denote by $\langle \cdot, \cdot \rangle_M$ the scalar product associated with M, that is $\langle z, z' \rangle_M = \langle Mz, z \rangle$. Since both $\langle \cdot, \cdot \rangle_M$ and the symplectic form are non-degenerate we can find a unique invertible matrix K of order $2n$ such that

$$\langle z, Kz' \rangle_M = \sigma(z, z')$$

for all z, z'; that matrix satisfies

$$K^T M = J = -MK.$$

Since the skew-product is antisymmetric we must have $K = -K^M$ where $K^M = -M^{-1}K^T M$ is the transpose of K with respect to $\langle \cdot, \cdot \rangle_M$; it follows that the eigenvalues of $K = -M^{-1}J$ are of the type $\pm i\lambda_j$, $\lambda_j > 0$, and so are those of JM^{-1}. The corresponding eigenvectors occurring in conjugate pairs $e'_j \pm i f'_j$, we thus obtain a $\langle \cdot, \cdot \rangle_M$-orthonormal basis $\{e'_i, f'_j\}_{1 \leq i,j \leq n}$ of $\mathbb{R}^{2n}_z$ such that $Ke'_i = \lambda_i f'_i$ and $Kf'_j = -\lambda_j e'_j$. Notice that it follows from these relations that

$$K^2 e'_i = -\lambda_i^2 e'_i \quad , \quad K^2 f'_j = -\lambda_j^2 f'_j$$

and that the vectors of the basis $\{e'_i, f'_j\}_{1 \leq i,j \leq n}$ satisfy the relations

$$\begin{aligned}
\sigma(e'_i, e'_j) &= \langle e'_i, Ke'_j \rangle_M = \lambda_j \langle e'_i, f'_j \rangle_M = 0,\\
\sigma(f'_i, f'_j) &= \langle f'_i, Kf'_j \rangle_M = -\lambda_j \langle f'_i, e'_j \rangle_M = 0,\\
\sigma(f'_i, e'_j) &= \langle f'_i, Ke'_j \rangle_M = \lambda_i \langle e'_i, f'_j \rangle_M = -\lambda_i \delta_{ij}.
\end{aligned}$$

Setting $e_i = \lambda_i^{-1/2} e'_i$ and $f_j = \lambda_j^{-1/2} f'_j$, the basis $\{e_i, f_j\}_{1 \leq i,j \leq n}$ is symplectic. Let S be the element of $\mathrm{Sp}(n)$ mapping the canonical symplectic basis to $\{e_i, f_j\}_{1 \leq i,j \leq n}$. The $\langle \cdot, \cdot \rangle_M$-orthogonality of $\{e_i, f_j\}_{1 \leq i,j \leq n}$ implies (8.7) with $\Lambda = \operatorname{diag}[\lambda_1, \ldots, \lambda_n]$. To prove the uniqueness statement (ii) it suffices to show that if there exists $S \in \mathrm{Sp}(n)$ such that $S^T L S = L'$ with $L = \operatorname{diag}[\Lambda, \Lambda]$, $L' = \operatorname{diag}[\Lambda', \Lambda']$, then $\Lambda = \Lambda'$. Since S is symplectic we have $S^T J S = J$ and hence $S^T L S = L'$ is equivalent to $S^{-1} J L S = J L'$ from which follows that JL and JL' have the same eigenvalues. These eigenvalues are precisely the complex numbers $\pm i/\lambda_j$. □

The diagonalizing matrix S in the theorem above has no reason to be unique. However:

Proposition 8.12. *Assume that S and S' are two elements of $\mathrm{Sp}(n)$ such that*

$$M = (S')^T D S' = S^T D S$$

where D is the Williamson diagonal form of M. Then $S(S')^{-1} \in \mathrm{U}(n)$.

Proof. Set $U = S(S')^{-1}$; we have $U^T D U = D$. We are going to show that $UJ = JU$; the lemma will follow. Setting $R = D^{1/2} U D^{-1/2}$ we have

$$R^T R = D^{-1/2}(U^T D U) D^{-1/2} = D^{-1/2} D D^{-1/2} = I$$

hence $R \in O(2n)$. Since J commutes with each power of D we have, since $JU = (U^T)^{-1}J$,

$$\begin{aligned} JR &= D^{1/2}JUD^{-1/2} = D^{1/2}(U^T)^{-1}JD^{-1/2} \\ &= D^{1/2}(U^T)^{-1}D^{-1/2}J = (R^T)^{-1}J, \end{aligned}$$

hence $R \in \mathrm{Sp}(n) \cap O(2n)$ so that $JR = RJ$. Now $U = D^{-1/2}RD^{1/2}$ and therefore

$$\begin{aligned} JU &= JD^{-1/2}RD^{1/2} = D^{-1/2}JRD^{1/2} \\ &= D^{-1/2}RJD^{1/2} = D^{-1/2}RD^{1/2}J \\ &= UJ \end{aligned}$$

which was to be proven. □

8.3.2 The symplectic spectrum

Let M be a positive-definite and symmetric real matrix: $M > 0$. We have seen above that the eigenvalues of JM are of the type $\pm i\lambda_{\sigma,j}$ with $\lambda_{\sigma,j} > 0$. We will always order the positive numbers $\lambda_{\sigma,j}$ as a decreasing sequence:

$$\lambda_{\sigma,1} \geq \lambda_{\sigma,2} \geq \cdots \geq \lambda_{\sigma,n} > 0. \tag{8.9}$$

Definition 8.13. With the ordering convention above $(\lambda_{\sigma,1}, \ldots, \lambda_{\sigma,n})$ is called the "symplectic spectrum of M and is denoted by $\mathrm{Spec}_\sigma(M)$:

$$\mathrm{Spec}_\sigma(M) = (\lambda_{\sigma,1}, \ldots, \lambda_{\sigma,n})$$

Here are two important properties of the symplectic spectrum:

Proposition 8.14. *Let $\mathrm{Spec}_\sigma(M) = (\lambda_{\sigma,1}, \ldots, \lambda_{\sigma,n})$ be the symplectic spectrum of M.*

(i) *$\mathrm{Spec}_\sigma(M)$ is a symplectic invariant:*

$$\mathrm{Spec}_\sigma(S^TMS) = \mathrm{Spec}_\sigma(M) \quad \text{for every } S \in \mathrm{Sp}(n); \tag{8.10}$$

(ii) *the sequence $(\lambda_{\sigma,n}^{-1}, \ldots, \lambda_{\sigma,1}^{-1})$ is the symplectic spectrum of M^{-1}:*

$$\mathrm{Spec}_\sigma(M^{-1}) = (\mathrm{Spec}_\sigma(M))^{-1}. \tag{8.11}$$

Proof. (i) is an immediate consequence of the definition of $\mathrm{Spec}_\sigma(M)$.
(ii) The eigenvalues of JM are the same as those of $M^{1/2}JM^{1/2}$; the eigenvalues of JM^{-1} are those of $M^{-1/2}JM^{-1/2}$. Now

$$M^{-1/2}JM^{-1/2} = -(M^{1/2}JM^{1/2})^{-1},$$

hence the eigenvalues of JM and JM^{-1} are obtained from each other by the transformation $t \longmapsto -1/t$. The result follows since the symplectic spectra are obtained by taking the moduli of these eigenvalues. □

Here is a result allowing us to compare the symplectic spectra of two positive definite symmetric matrices. It is important, because it is an algebraic version of Gromov's non-squeezing theorem [81] in the linear case. We are following the lines of Giedke *et al.* [49]; for a proof using a variational argument see Hofer and Zehnder [91]. (Hörmander [92], §21.5, gives a detailed classification of general quadratic forms on symplectic space; also see the listing due to Galin in Arnol'd [3], Appendix 6.)

Theorem 8.15. *Let M and M' be two symmetric positive definite matrices of same dimension. We have*

$$M \leq M' \Longrightarrow \operatorname{Spec}_\sigma(M) \leq \operatorname{Spec}_\sigma(M'). \tag{8.12}$$

Proof. When two matrices A and B have the same eigenvalues we will write $A \simeq B$. When those of A are smaller than or equal to those of B (for a common ordering) we will write $A \leq B$. Notice that when A or B is invertible we have $AB \simeq BA$. With this notation, the statement is equivalent to

$$M \leq M' \Longrightarrow (JM')^2 \leq (JM)^2$$

since the eigenvalues of JM and JM' occur in pairs $\pm i\lambda$, $\pm i\lambda'$ with λ and λ' real. The relation $M \leq M'$ is equivalent to $z^T Mz \leq z^T M'z$ for every $z \in \mathbb{R}^{2n}_z$. Replacing z by successively $(JM^{1/2})z$ and $(JM'^{1/2})z$ in $z^T Mz \leq z^T M'z$ we thus have, taking into account the fact that $J^T = -J$, that is, since $J^T = -J$,

$$M^{1/2}JM'JM^{1/2} \leq M^{1/2}JMJM^{1/2}, \tag{8.13}$$

$$M'^{1/2}JM'JM'^{1/2} \leq M'^{1/2}JMJM'^{1/2}. \tag{8.14}$$

Noting that we have

$$M^{1/2}JM'JM^{1/2} \simeq MJM'J,$$
$$M'^{1/2}JMJM'^{1/2} \simeq M'JMJ \simeq MJM'J,$$

we can rewrite the relations (8.13) and (8.14) as

$$MJM' \leq JM^{1/2}JM'JM^{1/2},$$
$$M'^{1/2}JM'JM'^{1/2} \leq MJM'J$$

and hence, by transitivity

$$M'^{1/2}JM'JM'^{1/2} \leq M^{1/2}JMJM^{1/2}. \tag{8.15}$$

Since we have

$$M^{1/2}JMJM^{1/2} \simeq (MJ)^2 \quad , \quad M'^{1/2}JM'JM'^{1/2} \simeq (M'J)^2$$

the relation (8.15) is equivalent to $(M'J)^2 \leq (MJ)^2$, which was to be proven. □

Let M be a positive-definite and symmetric real matrix $2n \times 2n$; we denote by $\mathbb{M}$ the ellipsoid in $\mathbb{R}_z^{2n}$ defined by the condition $\langle Mz, z\rangle \leq 1$:

$$\mathbb{M} : \langle Mz, z\rangle \leq 1.$$

In view of Williamson's theorem there exist $S \in \mathrm{Sp}(n)$ such that $S^T MS = D$ with $D = \mathrm{diag}[\Lambda, \Lambda]$ and that $\Lambda = \mathrm{diag}[\lambda_{1,\sigma}, \ldots, \lambda_{n,\sigma}]$ where $(\lambda_{1,\sigma}, \ldots, \lambda_{n,\sigma})$ is the symplectic spectrum of M. It follows that

$$S^{-1}(\mathbb{M}) : \sum_{j=1}^{n} \lambda_{j,\sigma}(x_j^2 + p_j^2) \leq 1.$$

Definition 8.16. The number $R_\sigma(\mathbb{M}) = 1/\sqrt{\lambda_{1,\sigma}}$ is called the *symplectic radius* of the phase-space ellipsoid $\mathbb{M}$; $c_\sigma(\mathbb{M}) = \pi R_\sigma^2 = \pi/\lambda_{1,\sigma}$ is its symplectic *area*.

The properties of the symplectic area are summarized in the following result, whose "hard" part follows from Theorem 8.15:

Corollary 8.17. *Let* $\mathbb{M}$ *and* $\mathbb{M}'$ *be two ellipsoids in* $(\mathbb{R}_z^{2n}, \sigma)$.

(i) *If* $\mathbb{M} \subset \mathbb{M}'$ *then* $c_\sigma(\mathbb{M}) \leq c_\sigma(\mathbb{M}')$;
(ii) *For every* $S \in \mathrm{Sp}(n)$ *we have* $c_\sigma(S(\mathbb{M})) = c_\sigma(\mathbb{M})$;
(iii) *For every* $\lambda > 0$ *we have* $c_\sigma(\lambda\mathbb{M}) = \lambda^2 c_\sigma(\mathbb{M})$.

Proof. (i) Assume that $\mathbb{M} : \langle Mz, z\rangle \leq 1$ and $\mathbb{M}' : \langle M'z, z\rangle \leq 1$. If $\mathbb{M} \subset \mathbb{M}'$ then $M \geq M'$ and hence $\mathrm{Spec}_\sigma(M) \geq \mathrm{Spec}_\sigma(M')$ in view of the implication (8.12) in Theorem 8.15; in particular $\lambda_{1,\sigma} \leq \lambda'_{1,\sigma}$.

Let us prove (ii). We have $S(\mathbb{M}) : \langle M'z, z\rangle \leq 1$ with $S' = (S^{-1})^T MS^{-1}$ and M' thus have the same symplectic spectrum as M in view of Proposition 8.14, (i).

Property (iii) is obvious. □

In the next subsection we generalize the notion of symplectic radius and area to arbitrary subsets of phase space.

8.3.3 The notion of symplectic capacity

Let us now denote $B(R)$ the phase-space ball $|z| \leq R$ and by $Z_j(R)$ the phase-space cylinder with radius R based on the conjugate coordinate plane x_j, p_j:

$$Z_j(R) : x_j^2 + p_j^2 \leq R^2.$$

Since we have

$$Z_j(R) : \langle Mz, z\rangle \leq 1$$

where the matrix M is diagonal and only has two entries different from zero, we can view $Z_j(R)$ as a degenerate ellipsoid with symplectic radius R. This observation motivates the following definition; recall that $\mathrm{Symp}(n)$ is the group of all symplectomorphisms of the standard symplectic space $(\mathbb{R}_z^{2n}, \sigma)$.

Definition 8.18. A "symplectic capacity" on $(\mathbb{R}_z^{2n}, \sigma)$ is a mapping c which to every subset Ω of $\mathbb{R}_z^{2n}$ associates a number $c_{\text{lin}}(\Omega) \geq 0$, or ∞, and having the following properties:

(i) $c(\Omega) \leq c(\Omega')$ if $\Omega \subset \Omega'$;

(ii) $c(f(\Omega)) = c(\Omega)$ for every $f \in \operatorname{Symp}(n)$;

(iii) $c(\lambda\Omega) = \lambda^2 c(\Omega)$ for every $\lambda \in \mathbb{R}$;

(iv) $c(B(R)) = c(Z_j(R)) = \pi R^2$.

When the properties (i)–(iv) above only hold for affine symplectomorphisms $f \in \operatorname{ISp}(n)$, we say that c is a "linear symplectic capacity" and we write $c = c_{\text{lin}}$.

While the construction of general symplectic capacities is very difficult (the existence of any symplectic capacity is equivalent to Gromov's non-squeezing theorem [81] as we will see below), it is reasonably easy to exhibit linear symplectic capacities; we encourage the reader to work out in detail the two following exercises (if the need is urgent, a proof can be found in [114] or [91]):

Exercise 8.19. For $\Omega \subset \mathbb{R}_z^{2n}$ set

$$\underline{c}_{\text{lin}}(\Omega) = \sup_{f \in \operatorname{ISp}(n)} \{\pi R^2 : f(B^{2n}(R)) \subset \Omega\}, \tag{8.16}$$

$$\overline{c}_{\text{lin}}(\Omega) = \inf_{f \in \operatorname{ISp}(n)} \{\pi R^2 : f(\Omega) \subset Z_j(R)\}. \tag{8.17}$$

Show that $\underline{c}_{\text{lin}}$ and $\overline{c}_{\text{lin}}$ are linear symplectic capacities.

The linear symplectic capacity $\underline{c}_{\text{lin}}$ defined in the exercise above can be interpreted as follows: for every $\Omega \subset \mathbb{R}_z^{2n}$ the number $\underline{c}_{\text{lin}}(\Omega)$ (which can be $+\infty$) is the supremum of all the πR^2 of phase space balls $B(R)$ that can be "stuffed" inside Ω using elements of $\operatorname{Sp}(n)$ and translations; similarly $\overline{c}_{\text{lin}}(\Omega)$ is the infimum of all πR^2 such that a cylinder $Z_j(R)$ can contain the deformation of Ω by elements of $\operatorname{Sp}(n)$ and translations. It turns out that $\underline{c}_{\text{lin}}$ and $\overline{c}_{\text{lin}}$ are respectively the smallest and largest linear symplectic capacities:

Exercise 8.20. Let $\underline{c}_{\text{lin}}(\Omega)$ and $\overline{c}_{\text{lin}}(\Omega)$ be defined by (8.16) and (8.17). Show that every linear symplectic capacity c_{lin} on $(\mathbb{R}_z^{2n}, \sigma)$ is such that

$$\underline{c}_{\text{lin}}(\Omega) \leq c_{\text{lin}}(\Omega) \leq \overline{c}_{\text{lin}}(\Omega)$$

for every $\Omega \subset \mathbb{R}_z^{2n}$.

We have several times mentioned Gromov's non-squeezing theorem in this chapter. It is time now to state it. Let us first define

Definition 8.21. Let Ω be an arbitary subset of $\mathbb{R}_z^{2n}$. Let R_σ be the supremum of the set

$$\{R : \exists f \in \operatorname{Symp}(n) \text{ such that } f(B(R)) \subset \Omega\}.$$

The number $c_G(\Omega) = \pi R_\sigma^2$ is called "symplectic area" (or "Gromov width") of Ω.

For $r > 0$ let

$$Z_j(r) = \{z = (x,p) : x_j^2 + p_j^2 \leq r^2\}$$

be a cylinder with radius R based on the x_j, p_j plane.

Theorem 8.22 (Gromov [81]). *We have $c_G(\Omega) = \pi R_\sigma^2$; equivalently: there exists a symplectomorphism f of $\mathbb{R}_z^{2n}$ such that $f(B^{2n}(z_0, R)) \subset Z_j(r)$ if and only if $R \leq r$.*

(The sufficiency of the condition $R \leq r$ is trivial since if $R \leq r$, then the translation $z \longmapsto z - z_0$ sends $B^{2n}(z_0, R)$ to any cylinder $Z_j(r)$.)

All known proofs of this theorem are notoriously difficult; Gromov used pseudo-holomorphic tools to establish it; in addition he showed in his paper [81] that many results from complex Kähler geometry remain true in symplectic geometry; his work was continued by several authors and is today a very active branch of topology.

As pointed out above, Gromov's theorem and the existence of one single symplectic capacity are equivalent. Let us prove that Gromov's theorem implies that the symplectic area c_G indeed is a symplectic capacity:

Corollary 8.23. *Let $\Omega \subset \mathbb{R}_z^{2n}$ and let R_σ be the supremum of the set*

$$\{R : \exists f \in \mathrm{Symp}(n) \text{ such that } f(B(R)) \subset \Omega\}.$$

The formulae $c_G(\Omega) = \pi R_\sigma^2$ if $R_\sigma < \infty$, $c_G(\Omega) = \infty$ if $R_\sigma = \infty$, define a symplectic capacity on $(\mathbb{R}_z^{2n}, \sigma)$.

Proof. Let us show that the axioms (i)–(iv) of Definition 8.18 are verified by c_G.

Axiom (i) (that is $c_G(\Omega) \leq c_G(\Omega')$ if $\Omega \subset \Omega'$) is trivially verified since a symplectomorphism sending $B(R)$ to Ω' also sends $B(R)$ to any set Ω' containing Ω.

Axiom (ii) requires that $c_G(f(\Omega)) = c_G(\Omega)$ for every symplectomorphism f; to prove that this is true, let $g \in \mathrm{Symp}(n)$ be such that $g(B(R)) \subset \Omega$; then $(f \circ g)(B(R)) \subset f(\Omega)$ for every $f \in \mathrm{Symp}(n)$ hence $c_G(f(\Omega)) \geq c_G(\Omega)$. To prove the opposite inequality we note that replacing Ω by $f^{-1}(\Omega)$ leads to $c_G(\Omega)) \geq c_G(f^{-1}(\Omega))$; since f is arbitrary we have in fact $c_G(\Omega)) \geq c_G(f(\Omega))$ for every $f \in \mathrm{Symp}(n)$.

Axiom (iii), which says that one must have $c_G(\lambda\Omega) = \lambda^2 c_G(\Omega)$ for all $\lambda \in \mathbb{R}$, is trivially satisfied if f is linear (cf. Exercise 8.19). To prove it holds true in the general case as well, first note that it is no restriction to assume $\lambda \neq 0$ and define, for $f : \mathbb{R}_z^{2n} \longrightarrow \mathbb{R}_z^{2n}$, a mapping f_λ by $f_\lambda(z) = \lambda f(\lambda^{-1}z)$. It is clear that f_λ is a symplectomorphism if and only f is. The condition $f(B(R)) \subset \Omega$ being equivalent to $\lambda^{-1} f_\lambda(\lambda B(R)) \subset \Omega$, that is to $f_\lambda(B(\lambda R)) \subset \lambda\Omega$, it follows that $c_G(\lambda\Omega) = \pi(\lambda R_\sigma)^2 = \lambda^2 c_G(\Omega)$.

Let us finally prove that Axiom (iv) is verified by $c_G(\Omega)$. The equality $c_G(B(R)) = \pi R^2$ is obvious: every ball $B(r)$ with $R' \leq R$ is sent into $B(R)$ by the identity and if $R' \geq R$ there exists no $f \in \mathrm{Symp}(n)$ such that $f(B(R')) \subset B(R)$ because symplectomorphisms are volume-preserving. There remains to show that $c_G(Z_j(R)) = \pi R^2$; it is at this point – and only at this point! – we will use Gromov's theorem. If $R' \leq R$ then the identity sends $B(R')$ in $Z_j(R)$ hence $c_G(Z_j(R)) \leq \pi R^2$. Assume that $c_G(Z_j(R)) > \pi R^2$; then there exists a ball $B(R')$ with $R' > R$ and a symplectomorphism f such that $f(B(R')) \subset Z_j(R)$ and this would violate Gromov's theorem. □

The reader is invited to show that, conversely, the existence of a symplectic capacity implies Gromov's theorem:

Exercise 8.24. Assume that you have constructed a symplectic capacity c on $(\mathbb{R}_z^{2n}, \sigma)$. Use the properties of c to prove Gromov's Theorem 8.22.

The number R_σ defined by $c_G(\Omega) = \pi R_\sigma^2$ is called the *symplectic radius* of Ω; that this terminology is consistent with that introduced in Definition 8.16 above follows from the fact that all symplectic capacities (linear or not) agree on ellipsoids. Let us prove this important property:

Proposition 8.25. *Let $\mathbb{M} : \langle Mz, z\rangle \leq 1$ be an ellipsoid in $\mathbb{R}_z^{2n}$ and c an arbitrary linear symplectic capacity on $(\mathbb{R}_z^{2n}, \sigma)$. Let $\lambda_{1,\sigma} \geq \lambda_{2,\sigma} \geq \cdots \geq \lambda_{n,\sigma}$ be the symplectic spectrum of the symmetric matrix M. We have*

$$c(\mathbb{M}) = \frac{\pi}{\lambda_{n,\sigma}} = c_{lin}(\mathbb{M}) \tag{8.18}$$

where c_{lin} is any linear symplectic capacity.

Proof. Let us choose $S \in \mathrm{Sp}(n)$ such that the matrix $S^T M S = D$ is in Williamson normal form; $S^{-1}(\mathbb{M})$ is thus the ellipsoid

$$\sum_{j=1}^{n} \lambda_{j,\sigma}(x_j^2 + p_j^2) \leq 1. \tag{8.19}$$

Since $c(S^{-1}(\mathbb{M})) = c(\mathbb{M})$ it is sufficient to assume that the ellipsoid $\mathbb{M}$ is represented by (8.19). In view of the double inequality

$$\lambda_{n,\sigma}(x_n^2 + p_n^2) \leq \sum_{j=1}^{n} \lambda_{j,\sigma}(x_j^2 + p_j^2) \leq \lambda_{n,\sigma} \sum_{j=1}^{n} (x_j^2 + p_j^2) \tag{8.20}$$

we have

$$B(\lambda_{n,\sigma}^{-1/2}) \subset \mathbb{M} \subset Z(\lambda_{n,\sigma}^{-1/2}),$$

hence, using the monotonicity axiom (i) for symplectic capacities,

$$c(B(\lambda_{n,\sigma}^{-1/2})) \subset c(\mathbb{M}) \subset (Z(\lambda_{n,\sigma}^{-1/2})).$$

The first equality in formula (8.18) now follows from Gromov's Theorem 8.22; the second equality is obvious since we have put $\mathbb{M}$ in normal form using a linear symplectomorphism. □

8.3.4 Admissible covariance matrices

From now on we assume that

$$\Sigma = \begin{bmatrix} \Sigma_{XX} & \Sigma_{XP} \\ \Sigma_{PX} & \Sigma_{PP} \end{bmatrix}, \quad \Sigma_{PX} = \Sigma_{XP}^T \tag{8.21}$$

is a real, symmetric, and positive definite $2n \times 2n$ matrix. Observe that since $J^T = -J$ the matrix $\Sigma + i\frac{\hbar}{2}J$ is Hermitian:

$$(\Sigma + i\tfrac{\hbar}{2}J)^* = \Sigma - i\tfrac{\hbar}{2}J^T = \Sigma + i\tfrac{\hbar}{2}J.$$

It follows, in particular, that all the eigenvalues of $\Sigma + i\frac{\hbar}{2}J$ are *real.*

The following Definition 8.26 of quantum mechanically admissible covariance matrices is due to Simon *et al.* [152]:

Definition 8.26. We will say that a real symmetric $2n \times 2n$ matrix Σ is "quantum mechanically *admissible*" (or, for short: "admissible") if it satisfies the condition

$$\Sigma + i\tfrac{\hbar}{2}J \text{ is positive semi-definite.} \tag{8.22}$$

The property for a covariance matrix to be admissible is invariant under linear symplectic transformations: the equality $S^TJS = J$ implies that

$$S\Sigma S^T + i\tfrac{\hbar}{2}J = S(\Sigma + i\tfrac{\hbar}{2}J)S^T$$

and the Hermitian matrix $S\Sigma S^T + i\frac{\hbar}{2}J$ is thus semi-definite positive if and only $\Sigma + i\frac{\hbar}{2}J$ is.

We are going to restate the quantum admissibility condition (8.22) in terms of the symplectic spectrum of Σ, *i.e.*, in terms of the eigenvalues $\pm i\lambda_{\sigma,j}$, $\lambda_{\sigma,j} > 0$ of $J\Sigma$. The result is the starting point of our phase space quantization scheme.

Proposition 8.27. *The matrix $\Sigma > 0$ is quantum mechanically admissible if and only if the moduli $\lambda_{\sigma,j}$ of the eigenvalues of $J\Sigma$ are $\geq \frac{1}{2}\hbar$, that is, if and only if the symplectic spectrum of Σ satisfies*

$$\operatorname{Spec}_\sigma(\Sigma) \geq (\tfrac{1}{2}\hbar, \tfrac{1}{2}\hbar, \ldots, \tfrac{1}{2}\hbar).$$

Proof. Let us choose $S \in \operatorname{Sp}(n)$ such that $S^T\Sigma S = D$ is in Williamson diagonal form. In view of the discussion above the condition $\Sigma + i\frac{\hbar}{2}J \geq 0$ is equivalent to $D + i\frac{\hbar}{2}J \geq 0$. The characteristic polynomial $\mathcal{P}(\lambda)$ of $D + i\frac{\hbar}{2}J \geq 0$ is $\mathcal{P}(\lambda) = \mathcal{P}_1(\lambda)\cdots\mathcal{P}_n(\lambda)$ where

$$\mathcal{P}_j(\lambda) = \lambda^2 - 2\lambda_{\sigma,j}\lambda + \lambda_{\sigma,j}^2 - \tfrac{1}{4}\hbar^2$$

and $(\lambda_{\sigma,1}, \ldots, \lambda_{\sigma,n})$ is the symplectic spectrum of Σ. The eigenvalues of $D + i\frac{\hbar}{2}J$ are thus the numbers $\lambda_j = \lambda_{\sigma,j} \pm \frac{\hbar}{2} \geq 0$; the condition $\Sigma + i\frac{\hbar}{2}J \geq 0$ means that $\lambda_j \geq 0$ for $j = 1, \ldots, n$ and is thus equivalent to $\lambda_{\sigma,j} \geq \frac{1}{2}\hbar$ for $j = 1, \ldots, n$. □

Let us illustrate this characterization of admissibility in the case $n = 1$:

Example 8.28. Consider the 2×2 matrix from Example 8.4. The characteristic polynomial of $J\Sigma$ is

$$\mathcal{P}(\lambda) = \lambda^2 - (\Delta^2 - (\Delta X)^2(\Delta P)^2)).$$

Since $\Sigma > 0$ its roots $\lambda_{\sigma,1}, \lambda_{\sigma,2}$ must be purely imaginary, hence the condition that $|\lambda_{\sigma,1}| \geq \frac{\hbar}{2}$ and $|\lambda_{\sigma,2}| \geq \frac{\hbar}{2}$ is equivalent to the inequality

$$(\Delta X)^2(\Delta P)^2 \geq \Delta^2 + \tfrac{1}{4}\hbar^2. \tag{8.23}$$

This example shows that when $n = 1$ the condition of quantum admissibility defined above is equivalent to the inequality (8.23); the latter *implies* the "household" version $\Delta X \Delta P \geq \frac{1}{2}\hbar$ of the uncertainty principle, but *is not equivalent to* it. It turns out that it is actually (8.23) which is the *true* uncertainty principle of quantum mechanics, and not the weaker textbook inequality $\Delta X \Delta P \geq \frac{1}{2}\hbar$.

This observation is actually true in arbitrary phase space dimension $2n$. More precisely, if the covariance matrix Σ is quantum mechanically admissible, then

$$(\Delta X_j)^2(\Delta P_j)^2 \geq \Delta_j^2 + \tfrac{1}{4}\hbar^2 \tag{8.24}$$

for $j = 1, 2, \ldots, n$. We leave the proof of this property as an exercise:

Exercise 8.29. Prove that if Σ is a quantum-mechanically admissible covariance matrix, then the inequalities (8.24) hold. [Hint: For every $\varepsilon > 0$ the Hermitian matrix $\Sigma_\varepsilon = \Sigma + \varepsilon I + \frac{1}{2}i\hbar J$ is definite positive, hence every principal minor of Σ_ε is positive.]

8.4 Wigner Ellipsoids

In this section we take a crucial step by rewriting the uncertainty principle in a simple geometric form, using the notion of symplectic capacity.

8.4.1 Phase space ellipsoids

We begin by noting that the datum of a real symmetric positive definite matrix is equivalent to that of a definite quadratic form, that is, ultimately, to that of an ellipsoid. This remark justifies the following definition:

Definition 8.30. Let Σ be a quantum-mechanically admissible covariance matrix associated to a random variable Z such that $\langle Z \rangle = \bar{z}$. We will call the subset

$$\mathbb{W}_\Sigma : \tfrac{1}{2}\langle \Sigma^{-1}(z - \bar{z}), z - \bar{z} \rangle \leq 1 \tag{8.25}$$

of $\mathbb{R}_z^{2n}$ the "Wigner ellipsoid" associated to Z. (Beware the factor 1/2 in the left-hand side!) We will say that $\mathbb{W}_\Sigma$ is a "centered Wigner ellipsoid" if $\bar{z} = 0$.

Here is a trivial example, which is nevertheless useful to keep in mind:

Example 8.31. Consider the bivariate normal probability density

$$\rho(x,p) = \frac{1}{2\pi\sigma_x\sigma_p} \exp\left[-\tfrac{1}{2}\left(\tfrac{x^2}{\sigma_x^2} + \tfrac{p^2}{\sigma_p^2}\right)\right].$$

To ρ we associate the Wigner ellipsoid

$$\tfrac{1}{2}\left(\tfrac{x^2}{\sigma_x^2} + \tfrac{p^2}{\sigma_p^2}\right) \leq 1.$$

The following criterion easily follows from Definition 8.26 of an admissible ellipsoid:

Lemma 8.32. *Let* $\mathbb{M} : \langle M(z-\bar{z}), z-\bar{z}\rangle \leq 1$ *be an ellipsoid in* $(\mathbb{R}_z^{2n}, \sigma)$. $\mathbb{M}$ *is a Wigner ellipsoid if and only if the symplectic spectrum* $(\mu_{\sigma,1}, \dots, \mu_{\sigma,n})$ *of* M *is such that* $\mu_{\sigma,j} \leq \hbar$ *for* $j = 1, \dots, n$*:*

$$\operatorname{Spec}_\sigma(\mathbb{M}) \leq (\hbar, \hbar, \dots, \hbar).$$

Proof. Define a symmetric positive-definite matrix Σ by $M = \frac{1}{2}\Sigma^{-1}$. Let $(\lambda_{1,\sigma}, \dots, \lambda_{1,\sigma})$ be the symplectic spectrum of Σ; then $(\lambda_{1,\sigma}^{-1}, \dots, \lambda_{1,\sigma}^{-1})$ is that of Σ^{-1} (see Proposition 8.14) hence the symplectic spectrum of M is

$$(\mu_{\sigma,1}, \dots, \mu_{\sigma,n}) = (\tfrac{1}{2}\lambda_{1,\sigma}^{-1}, \dots, \tfrac{1}{2}\lambda_{1,\sigma}^{-1}).$$

The matrix Σ is admissible if and only if $\lambda_{j,\sigma} \geq \frac{1}{2}\hbar$ for every j and this is equivalent to the conditions $\mu_{\sigma,j} \leq \hbar$ for every j. □

We are going to see that Wigner ellipsoids have quite remarkable properties; in particular they will allow us to give a purely geometrical statement of the uncertainty principle. We will for instance prove that:

> *An ellipsoid in* $\mathbb{R}_z^{2n}$ *is a Wigner ellipsoid if and only if its intersection with any affine symplectic plane passing through its center has area at least* $\frac{1}{2}h$.

Let us begin by reviewing some very basic facts about matrices and ellipsoids. Recall that a Hermitian matrix M is positive-definite ($M > 0$) if the two following equivalent conditions are satisfied:

- $\langle Mz, z\rangle > 0$ for every $z \neq 0$;
- Every eigenvalue of M is > 0.

Similarly, M is semi-definite positive ($M \geq 0$) if these two conditions hold with $>$ replaced by $\geq$. When $M - M' > 0$ (resp. $M - M' \geq 0$) we will write $M > M'$ or $M' < M$ (resp. $M \geq M'$ or $M' \leq M$).

Let $\mathbb{M} : \langle Mz, z \rangle \leq 1$ and $\mathbb{M}' : \langle M'z, z \rangle \leq 1$ be two concentric ellipsoids. We obviously have

$$\mathbb{M} \subset \mathbb{M}' \Longleftrightarrow M \geq M'.$$

If $\mathbb{M}$ and $\mathbb{M}'$ no longer are concentric it is trivially false that $M \geq M'$ implies $\mathbb{M} \subset \mathbb{M}'$. The converse however remains true:

Lemma 8.33. *Let* $\mathbb{M} : \langle M(z - \bar{z}), z - \bar{z} \rangle \leq 1$ *and* $\mathbb{M}' : \langle M'(z - \bar{z}'), z - \bar{z}' \rangle \leq 1$ *be two ellipsoids. If* $\mathbb{M} \subset \mathbb{M}'$*, then* $M \geq M'$ *and we can translate* $\mathbb{M}$ *so that it becomes an ellipsoid* $\mathbb{M}''$ *contained in and concentric with* $\mathbb{M}'$.

Proof. The statement being invariant under a translation of both ellipsoids by a same vector it is no restriction to assume that $\bar{z}' = 0$. The inclusion $\mathbb{M} \subset \mathbb{M}'$ is equivalent to the inequality

$$\langle M(z - \bar{z}), z - \bar{z} \rangle \geq \langle M'z, z \rangle$$

for all z, that is to

$$\langle (M - M')z, z \rangle \geq 2 \langle Mz, \bar{z} \rangle - \langle M\bar{z}, \bar{z} \rangle \tag{8.26}$$

for all z. Suppose that we do not have $M \geq M'$. Then there exists $z_0 \neq 0$ such that $\langle (M - M')z_0, z_0 \rangle < 0$. The inequality (8.26) implies that we have

$$\lambda^2 \langle (M - M')z_0, z_0 \rangle \geq 2\lambda \langle Mz_0, \bar{z} \rangle - \langle M\bar{z}, \bar{z} \rangle,$$

for every $\lambda \in \mathbb{R}$; dividing both sides of this inequality by λ^2 and letting thereafter $\lambda \to \infty$ we get $\langle (M - M')z_0, z_0 \rangle \geq 0$ which is only possible if $z_0 = 0$. This contradiction shows that $M \geq M'$, as claimed. The second statement in the lemma immediately follows. □

8.4.2 Wigner ellipsoids and quantum blobs

The image of an ellipsoid by an invertible linear transformation is still an ellipsoid. Particularly interesting are the ellipsoids obtained by deforming a ball in $(\mathbb{R}^{2n}_z, \sigma)$ using elements of $\mathrm{Sp}(n)$.

We will denote by $B^{2n}(z_0, R)$ the closed ball in $\mathbb{R}^{2n}_z$ with center z_0 and radius R; when the ball is centered at the origin, *i.e.*, when $z_0 = 0$, we will simply write $B^{2n}(R)$.

Definition 8.34.

(i) A *"symplectic ball"* $\mathbb{B}^{2n}$ *in* $(\mathbb{R}^{2n}_z, \sigma)$ *is the image of a ball* $B^{2n}(z_0, R)$ by some $S \in \mathrm{Sp}(n)$; we will say that R is the *radius* of $\mathbb{B}^{2n}$ and Sz_0 its *center*.

(ii) A "quantum blob" $\mathbb{Q}^{2n}$ is any symplectic ball with radius

$$\sqrt{\hbar} : \mathbb{Q}^{2n} = S(B^{2n}(z_0, \sqrt{\hbar})).$$

We will drop any reference to the dimension when no confusion can arise, and drop the superscript $2n$ and write $B(z_0, R)$, $\mathbb{B}$, $\mathbb{Q}$ instead of $B(z_0, R)$, $\mathbb{B}^{2n}$, $\mathbb{Q}^{2n}$.

The definition of a symplectic ball can evidently be written as

$$\mathbb{B}^{2n} = S(B^{2n}(z_0, R)) = T(Sz_0)S(B^{2n}(R))$$

for some $S \in \mathrm{Sp}(n)$ and $z_0 \in \mathbb{R}^{2n}_z$; $T(Sz_0)$ is the translation $z \longmapsto z + Sz_0$. That is:

A symplectic ball with radius R in (R^{2n}, σ) is the image of $B^{2n}(R)$ by an element of the affine symplectic group $\mathrm{ISp}(n)$.

For instance, since the elements of $\mathrm{Sp}(1)$ are just the area preserving linear automorphisms of $\mathbb{R}^2$, a symplectic ball (respectively, a quantum blob) in the plane is just any phase plane ellipse with area πR^2 (respectively $\frac{1}{2}h$). For arbitrary n we note that since symplectomorphisms are volume-preserving (they have determinant equal to 1) the volume of a quantum blob is just the volume of the ball $B^{2n}(\sqrt{\hbar})$. Since

$$\operatorname{Vol} B^{2n}(R) = \frac{\pi^n}{n!} R^{2n},$$

the volume of a quantum blob is thus

$$\operatorname{Vol} \mathbb{Q}^{2n} = \frac{(\pi\hbar)^n}{n!} = \frac{h^{2n}}{2^n n!}. \tag{8.27}$$

(It is thus smaller than that of the cubic "quantum cells" used in statistical mechanics by a factor of $n!2^n$.)

Lemma 8.35. *An ellipsoid $\mathbb{M} : \langle Mz, z\rangle \leq 1$ in $\mathbb{R}^{2n}_z$ is a symplectic ball with radius one if and only $M \in \mathrm{Sp}(n)$ and we then have $\mathbb{M} = S(B(1))$ with $M = (SS^T)^{-1}$.*

Proof. Assume that $\mathbb{M} = S(B(1)$. Then $\mathbb{M}$ is the set of all $z \in \mathbb{R}^{2n}_z$ such that $\langle S^{-1}z, S^{-1}z\rangle \leq 1$ hence $M = (S^{-1})^T S^{-1}$ is a symmetric definite-positive symplectic matrix. Assume conversely that $M \in \mathrm{Sp}(n)$. Since $M > 0$ we also have $M^{-1} > 0$ and there exists $S \in \mathrm{Sp}(n)$ such that $M^{-1} = SS^T$. Hence $\mathbb{M} : \langle S^{-1}z, S^{-1}z\rangle \leq 1$ is just $S(B(1)$. □

Another very useful observation is that we do not need all symplectic matrices to produce all symplectic balls:

Lemma 8.36. *For every centered symplectic ball $\mathbb{B}^{2n} = S(B^{2n}(R))$ there exist unique real symmetric $n \times n$ matrices L ($\det L \neq 0$) and Q such that $\mathbb{B}^{2n} = S_0(B^{2n}(R))$ and*

$$S_0 = \begin{bmatrix} L & 0 \\ Q & L^{-1} \end{bmatrix} \in \mathrm{Sp}(n). \tag{8.28}$$

Proof. In view of Corollary 2.30 in Chapter 2, Subsection 2.2.2 we can factorize $S \in \mathrm{Sp}(n)$ as $S = S_0 U$ where $U \in \mathrm{U}(n)$ and S_0 is of the type (8.28). The claim follows since $U(B^{2n}(R)) = B^{2n}(R)$. (That L and Q are uniquely defined is clear for if $S_0(B^{2n}(R)) = S_0'(B^{2n}(R))$ then $S_0(S_0')^{-1} \in \mathrm{U}(n)$ and S_0' can only be of the type (8.28) if it is identical to S_0.) □

The proof above shows that *every* symplectic ball (and hence every quantum blob) can be obtained from a ball with the same radius by first performing a symplectic rotation which takes it into another ball, and by thereafter applying two successive symplectic transformations of the simple types

$$M_{L^{-1}} = \begin{bmatrix} L & 0 \\ 0 & L^{-1} \end{bmatrix} \ , \ V_P = \begin{bmatrix} I & 0 \\ -P & 0 \end{bmatrix} \ (L = L^T,\ P = P^T);$$

the first of these transformations is essentially a symplectic rescaling of the coordinates, and the second a "symplectic shear".

One of the main interests of quantum blobs comes from the following property which shows that, if we cut a quantum blob by an affine symplectic plane, we will always obtain an ellipse with area exactly $\frac{1}{2}h$. Assuredly, this property is not really intuitive: if we cut an arbitrary ellipsoid in $\mathbb{R}^m$ by 2-planes, we will always get elliptic sections, but these do not usually have the same areas! Here is a direct elementary proof. It is of course sufficient to assume that the ball is centered at 0.

Proposition 8.37. *The intersection of* $\mathbb{B} = S(B(R))$ *with a symplectic plane* $\mathbb{P}$ *is an ellipse with area* πR^2. *Hence the intersection of a quantum blob with any affine symplectic plane has area at most* $\frac{1}{2}h$ *(and equal to* $\frac{1}{2}h$ *if that plane passes through the center of that blob).*

Proof. We have $\mathbb{B} \cap \mathbb{P} = S_{|\mathbb{P}'}(B(R) \cap \mathbb{P}')$ where $S_{|\mathbb{P}'}$ is the restriction of S to the symplectic plane $\mathbb{P} = S^{-1}(\mathbb{P})$. The intersection $B(R) \cap \mathbb{P}'$ is a circle with area πR^2 and $S_{|\mathbb{P}'}$ is a symplectic isomorphism $\mathbb{P}' \longrightarrow \mathbb{P}$, and is hence area preserving. The area of the ellipse $\mathbb{B} \cap \mathbb{P}$ is thus πR^2 as claimed. □

This property is actually a particular case of a more general result, which shows that the intersection of a symplectic ball with any symplectic subspace is a symplectic ball of this subspace. We will prove this in detail below (Theorem 8.41).

We urge the reader to notice that the assumption that we are cutting $S(B^{2n}(R))$ with *symplectic* planes is essential. The following exercise provides a counterexample which shows that the conclusion of Proposition 8.37 is falsified if we intersect $S(B^{2n}(R))$ with a plane that is not symplectic.

Exercise 8.38. Assume $n = 2$ and take $S = \operatorname{diag}[\lambda_1, \lambda_2, 1/\lambda_1, 1/\lambda_2]$ with $\lambda_1 > 0$, $\lambda_2 > 0$, and $\lambda_1 \neq \lambda_2$. Show that S is symplectic, but that the intersection of $S(B^4(R))$ with the x_2, p_1 plane (which is not conjugate) does not have area πR^2.

The assumption that S is symplectic is also essential in Proposition 8.37:

Exercise 8.39. Assume that we swap the two last diagonal entries of the matrix S in the exercise above so that it becomes $S' = \operatorname{diag}[\lambda_1, \lambda_2, 1/\lambda_2, 1/\lambda_1]$.

(i) Show that S' is not symplectic;

(ii) show that the section $S'(B^{2n}(R))$ by the symplectic x_2, p_2 plane does not have area πR^2.

From Proposition 8.37 we deduce the following important characterizations of Wigner ellipsoids:

Theorem 8.40. *Let* $\mathbb{M} : \langle M(z-\bar{z}), z-\bar{z}\rangle \leq 1$ *be an ellipsoid in* $\mathbb{R}^{2n}_z$. *The three following statements are equivalent:*

(i) $\mathbb{M}$ *is a Wigner ellipsoid* $\mathbb{W}_\Sigma$;

(ii) $\mathbb{M}$ *contains a quantum blob* $\mathbb{Q}$;

(iii) *There exists* $S \in \mathrm{Sp}(n)$ *such that* $M \leq \hbar S^T S$.

Proof. It is no restriction to assume that the ellipsoid $\mathbb{M}$ is centered at $z_0 = 0$. Let us prove the implication (i) $\Longrightarrow$ (ii). In view of Lemma 8.32 the symplectic spectrum $(\mu_{\sigma,1},\ldots,\mu_{\sigma,n})$ of M is such that $\mu_{\sigma,j} \leq \hbar$ for $1 \leq j \leq n$; setting $R_j^2 = 1/\mu_{\sigma,j}$ this means that there exists $S \in \mathrm{Sp}(n)$ such that

$$S^{-1}(\mathbb{M}) : \sum_{j=1}^{n} \mu_{\sigma,j}(x_j^2 + p_j^2) \leq 1,$$

hence $S^{-1}(\mathbb{M})$ contains the ball $B(\mu_{\sigma,n}^{-1/2})$; since $\mu_{\sigma,n}^{-1/2} \geq \sqrt{\hbar}$ it contains a fortiori $B(\sqrt{\hbar})$ and hence $\mathbb{M} \supset S(B(\sqrt{\hbar}))$.

Let us next prove that (ii) $\Longrightarrow$ (iii). Assume that $\mathbb{M}$ contains a quantum blob $\mathbb{Q}$; In view of Lemma 8.33 we may assume that this quantum blob has the same center as $\mathbb{M}$, that is $\mathbb{Q} = S(B(\sqrt{\hbar}))$. By definition, we can find $S' \in \mathrm{Sp}(n)$ such that $\mathbb{Q} = S'(B(\sqrt{\hbar}))$; we have $z \in \mathbb{Q}$ if and only if

$$\langle (S'^T)^{-1}(S')^{-1}z, z\rangle \leq \hbar$$

so the inclusion $S'(B(\sqrt{\hbar})) \subset \mathbb{M}$ is equivalent to the inequality

$$\tfrac{1}{\hbar}\langle (S'^T)^{-1}(S')^{-1}z, z\rangle \leq \langle Mz, z\rangle \quad \text{for every } z$$

and hence to $(S'^T)^{-1}(S')^{-1} \leq M$; set now $S = (S')^{-1}$.

There remains to prove that (iii) $\Longrightarrow$(i). Recall from Lemma 8.32 that $\mathbb{M}$ is a Wigner ellipsoid if and only if the eigenvalues $\pm i\mu_{\sigma,j}$ ($\mu_{\sigma,j} > 0$) of JM are such that $\mu_{\sigma,j} \leq \hbar$. Suppose that $M \leq \hbar S^T S$, then by Theorem 8.15 of Chapter we have $\mu_{\sigma,j} \leq \lambda_{\sigma,j}$ where the $\pm i\lambda_{\sigma,j}$ ($\lambda_{\sigma,j} > 0$) are the eigenvalues of $J(\hbar S^T S) = \hbar J$. The latter being $\pm i\hbar$ we have $\mu_{\sigma,j} \leq \hbar$. □

8.4.3 Wigner ellipsoids of subsystems

Physically the phase space $\mathbb{R}^{2n}_z$ corresponds to a system with "n degrees of freedom". For instance, if we are dealing with N particles moving without constraints in three-dimensional physical space we would take $n = 3N$. Often one is not interested in the total system, but only in a part thereof: we are then working with a *subsystem* having a smaller number of freedoms; we postulate that the phase space of that subsystem is $\mathbb{R}^{2k}_z$ for some $k < n$.

We begin by proving the following extension of Proposition 8.37:

Theorem 8.41. *Let $\mathbb{B}^{2n} = S(B^{2n}(R))$ be a symplectic ball and $(\mathbb{F}, \sigma_{|\mathbb{F}})$ a $2k$-dimensional symplectic subspace of $(\mathbb{R}^{2n}_z, \sigma)$.*

(i) *There exists $S_{\mathbb{F}} \in \operatorname{Sp}(\mathbb{F}, \sigma_{|\mathbb{F}})$ such that $\mathbb{B}^{2n} \cap \mathbb{F} = S_{\mathbb{F}}(B^{2k}(R))$ where $B^{2k}(R)$ is the ball with radius R in $\mathbb{F}$ centered at 0.*

(ii) *Conversely if $\mathbb{B}^{2k} = S_{\mathbb{F}}(B^{2k}(R))$ for some $S_{\mathbb{F}} \in \operatorname{Sp}(\mathbb{F}, \sigma_{|\mathbb{F}})$, then there exists $S \in \operatorname{Sp}(n)$ such that $\mathbb{B}^{2k} = \mathbb{F} \cap S(B^{2n}(R))$.*

Proof. (i) Let $\mathcal{B}$ and $\mathcal{B}'$ be orthosymplectic bases of $(\mathbb{R}^{2n}_z, \sigma)$ such that the subsets

$$\mathcal{B}_{\mathbb{F}} = \{e_1, \dots, e_k; f_1, \dots, f_k\} \quad , \quad \mathcal{B}_{\mathbb{F}'} = \{e'_1, \dots, e'_k; f'_1, .., f'_k\}$$

are bases of $\mathbb{F}$ and $\mathbb{F}' = S^{-1}(\mathbb{F})$, respectively and define $U \in \operatorname{U}(n)$ by the conditions $U(e'_i) = e_i$, $U(f'_j) = f_j$ for $1 \leq i, j \leq n$; clearly $U_{|\mathbb{F}'} : (\mathbb{F}', \sigma_{|\mathbb{F}'}) \longrightarrow (\mathbb{F}, \sigma_{|\mathbb{F}})$. Let $S_{|\mathbb{F}'}$ be the restriction of S to $\mathbb{F}'$. We claim that

$$S_{\mathbb{F}} = S_{|\mathbb{F}'} \circ U \in \operatorname{Sp}(\mathbb{F}, \sigma_{|\mathbb{F}})$$

is such that $S_{\mathbb{F}}(B^{2k}(R) = \mathbb{B}^{2n} \cap \mathbb{F}$. We have, since $U^{-1}(B^{2n}(R) = B^{2n}(R)$,

$$\begin{aligned}\mathbb{B}^{2n} \cap \mathbb{F} &= S_{|\mathbb{F}'}(B^{2n}(R) \cap \mathbb{F}') \\ &= (S_{|\mathbb{F}'} \circ U)(U^{-1}(B^{2n}(R)) \\ &= S_{\mathbb{F}}(B^{2n}(R))\end{aligned}$$

as claimed.

(ii) Assume conversely that there exists $S_{\mathbb{F}} \in \operatorname{Sp}(\mathbb{F}, \sigma_{|\mathbb{F}})$ such that

$$\mathbb{B}^{2k} = S_{\mathbb{F}}(B^{2k}(R)).$$

Define $S \in \operatorname{Sp}(n)$ by the conditions $S = S_{|\mathbb{F}}$ and $S(z) = z$ for $z \notin \mathbb{F}$. We have

$$S_{\mathbb{F}}(B^{2k}(R)) = S_{\mathbb{F}}(B^{2k}(R) \cap \mathbb{F}) = S_{\mathbb{F}}(B^{2n}(R) \cap \mathbb{F})$$

that is, since $S_{\mathbb{F}}(\mathbb{F}) = \mathbb{F}$,

$$S_{\mathbb{F}}(B^{2k}(R)) = S(B^{2n}(R)) \cap \mathbb{F}. \qquad \square$$

An ellipsoid $\mathbb{M}$ of $\mathbb{R}^{2n}_z$ containing a Wigner ellipsoid $\mathbb{W}_\Sigma$ is itself a Wigner ellipsoid: this is an obvious consequence of the definition, for if $S(B^{2n}(z_0, \sqrt{\hbar})) \subset \mathbb{W}_\Sigma$ then $S(B^{2n}(z_0, \sqrt{\hbar})) \subset \mathbb{M}$. (Notice that this property does not even require that $\mathbb{M}$ and $\mathbb{W}_\Sigma$ be concentric in view of Lemma 8.33).

More generally:

Proposition 8.42. *Let* $\mathbb{M} : \langle Mz, z\rangle \leq 1$ *be an ellipsoid in* $\mathbb{R}^{2n}_z$.

(i) *If* $\mathbb{M}$ *is a Wigner ellipsoid* $\mathbb{W}_\Sigma$, *then* $\mathbb{M} \cap \mathbb{F}$ *for every symplectic subspace* $(\mathbb{F}, \sigma_{|\mathbb{F}})$ *of* $(\mathbb{R}^{2n}_z, \sigma)$.

(ii) *Let* $(\mathbb{F}_1, \sigma_1), \ldots, (\mathbb{F}_m, \sigma_m)$ $(\sigma_j = \sigma_{|F_j}$ *for* $j = 1, \ldots, m)$ *be symplectic subspaces of* $(\mathbb{R}^{2n}_z, \sigma)$ *such that*

$$(\mathbb{F}_1, \sigma_1) \oplus \cdots \oplus (\mathbb{F}_m, \sigma_m) = (\mathbb{R}^{2n}_z, \sigma).$$

$\mathbb{M}$ *is a Wigner ellipsoid if and only if each* $\mathbb{M} \cap \mathbb{F}_j$ *is a Wigner ellipsoid in* $(\mathbb{F}_j, \sigma_j)$ *for* $1 \leq j \leq m$.

Proof. (i) In view of Theorem 8.40 above $\mathbb{W}_\Sigma$ contains a quantum blob $\mathbb{Q}$ hence $\mathbb{W}_\Sigma \cap \mathbb{F}$ contains the set $\mathbb{Q} \cap \mathbb{F}$ which is a quantum blob in $\mathbb{W}_\Sigma \cap \mathbb{F}$ by Theorem 8.41. A new use of Theorem 8.40 implies that $\mathbb{W}_\Sigma \cap \mathbb{F}$ is a Wigner ellipsoid in $\mathbb{F}$.

(ii) The case $m = 1$ is trivial and it is sufficient to give the proof for $m = 2$, the general case following by an immediate induction on the number of terms in the direct sum. Writing

$$(\mathbb{R}^{2n}_z, \sigma) = (\mathbb{F}_1, \sigma_1) \oplus (\mathbb{F}_2, \sigma_2)$$

with $\dim \mathbb{F}_1 = 2n_1$ and $\dim \mathbb{F}_2 = 2n_2$, we thus assume that $\mathbb{W}_\Sigma \cap \mathbb{F}_1$ and $\mathbb{W}_\Sigma \cap \mathbb{F}_2$ are Wigner ellipsoids in $\mathbb{F}_1$ and $\mathbb{F}_2$ respectively. Let us prove that $\mathbb{W}_\Sigma$ contains a quantum blob $\mathbb{Q}^{2n}$, and is thus a Wigner ellipsoid. In view of Theorem 8.40 we can find $S_1 \in \mathrm{Sp}(\mathbb{F}_1, \sigma_1)$ and $S_2 \in \mathrm{Sp}(\mathbb{F}_2, \sigma_2)$ such that

$$\mathbb{Q}^{2n_1} = S_1(B^{2n_1}(\sqrt{\hbar})) \subset \mathbb{W}_\Sigma \cap \mathbb{F}_1$$

and

$$\mathbb{Q}^{2n_2} = S_2(B^{2n_2}(\sqrt{\hbar})) \subset \mathbb{W}_\Sigma \cap \mathbb{F}_2.$$

Set now $S = S_1 \oplus S_2$ and $\mathbb{Q}^{2n} = S(B^{2n}(\sqrt{\hbar}))$. We claim that $\mathbb{Q}^{2n} \subset \mathbb{W}_\Sigma$; this will prove (ii). Let $z \in \mathbb{Q}^{2n}$, that is $|Sz|^2 \leq \hbar$; writing $z = z_1 + z_2$ with $z_1 \in \mathbb{F}_1$ and $z_2 \in \mathbb{F}_2$ we have $|S_1 z_1|^2 + |S_2 z_2|^2 \leq \hbar$, hence

$$z_1 \in \mathbb{Q}^{2n_1} \subset \mathbb{W}_\Sigma \cap \mathbb{F}_1 \quad , \quad z_2 \in \mathbb{Q}^{2n_2} \subset \mathbb{W}_\Sigma \cap \mathbb{F}_2$$

so that

$$z \in (\mathbb{W}_\Sigma \cap \mathbb{F}_1) + (\mathbb{W}_\Sigma \cap \mathbb{F}_2) \subset \mathbb{W}_\Sigma$$

and $\mathbb{Q}^{2n} \subset \mathbb{W}_\Sigma$ as claimed. □

The following consequence of Proposition 8.42 is immediate:

Corollary 8.43.

(i) *An ellipsoid* $\mathbb{M} : \langle Mz, z\rangle \leq 1$ *in* $(\mathbb{R}^{2n}_z, \sigma)$ *is a Wigner ellipsoid if and only if there exists symplectic planes* $\mathbb{P}_1, \ldots, \mathbb{P}_n$ *with* $\mathbb{P}_i \cap \mathbb{P}_j = 0$ *for* $i \neq j$ *such that the area of* $\mathbb{M} \cap \mathbb{P}_j$ *is at least* $\frac{1}{2}h$ *for each* j.

(ii) *If this condition is satisfied, then the area of* $\mathbb{M} \cap \mathbb{P}$ *is at least* $\frac{1}{2}h$ *for every symplectic plane in* $(\mathbb{R}^{2n}_z, \sigma)$.

Proof. The conditions $\mathbb{P}_i \cap \mathbb{P}_j = 0$ for $i \neq j$ are equivalent to $\mathbb{R}_z^{2n} = \mathbb{P}_1 \oplus \cdots \oplus \mathbb{P}_n$; it now suffices to apply Proposition 8.42 and to remark that an ellipse in $\mathbb{P}_j$ with area $\geq \frac{1}{2}h$ is trivially a Wigner ellipsoid. □

This corollary shows that if one wants to check whether an ellipsoid $\mathbb{M}$: $\langle Mz, z \rangle \leq 1$ is a Wigner ellipsoid, it suffices to cut it with the n conjugate coordinate planes x_j, p_j. The property obviously extends to the case of general ellipsoids

$$\mathbb{M} : \langle M(z - \bar{z}), (z - \bar{z}) \rangle \leq 1$$

replacing the symplectic planes $\mathbb{P}_1, \ldots, \mathbb{P}_n$ by affine symplectic planes through $\bar{z}$.

8.4.4 Uncertainty and symplectic capacity

The reader has certainly been very pleased to learn that one can express the uncertainty inequalities

$$(\Delta X_j)^2 (\Delta P_j)^2 \geq \operatorname{Cov}(X_j, P_j)^2 + \tfrac{1}{4}\hbar^2$$

very concisely by a statement on the covariance matrix: these relations are equivalent to $\Sigma + i\frac{\hbar}{2}J \geq 0$. We are going to do even better, and express these inequalities in a *very* concise form using the notion of symplectic capacity introduced in Subsection 8.3.3.

Theorem 8.44. *An ellipsoid $\mathbb{M}$ in $\mathbb{R}_z^{2n}$ is a Wigner ellipsoid if and only if*

$$c(\mathbb{M}) \geq \tfrac{1}{2}h \tag{8.29}$$

for every symplectic capacity c on $\mathbb{R}_z^{2n}$.

Proof. In view of Proposition 8.25, condition (8.29) is equivalent to the inequality $\bar{c}(\mathbb{M}_\Sigma) \geq \frac{1}{2}h$. But this is, in turn, equivalent to saying that there exists a quantum blob $\mathbb{Q}^{2n} = S(B^{2n}(z_0, \sqrt{\hbar}))$ contained in $\mathbb{M}$. □

The case where we have equality in (8.29) deserves some very special attention. When $n = 1$ the equality $c(\mathbb{M}_\Sigma) = \frac{1}{2}h$ means that the area of the Wigner ellipse associated to the covariance matrix is exactly $\frac{1}{2}h$ and we thus have equality in the Heisenberg uncertainty relation:

$$(\Delta x)^2 (\Delta p)^2 = \Delta^2 + \tfrac{1}{4}\hbar^2.$$

We are going to see that the limiting case $c(\mathbb{M}_\Sigma) = \frac{1}{2}h$ has a quite interesting property.

Proposition 8.45. *A Wigner ellipsoid $\mathbb{W}_\Sigma$ for which $c(\mathbb{W}_\Sigma) = \frac{1}{2}h$ contains a unique quantum blob $\mathbb{Q}^{2n}$.*

Proof. It is no restriction to assume that $\mathbb{W}_\Sigma$ is a centered ellipsoid in normal form; relabeling if necessary the symplectic coordinates, the condition $c(\mathbb{W}_\Sigma) = \frac{1}{2}$ means that $\mathbb{W}_\Sigma$ is the set of all points (x, p) such that

$$\frac{1}{\hbar}(x_1^2 + p_1^2) + \sum_{2 \leq j \leq n} \frac{1}{R_j^2}(x_j^2 + p_j^2) \leq 1$$

with $R_j^2 \geq \hbar$. Assume that there exists a quantum blob $\mathbb{Q}^{2n} = S(B^{2n}(z_0, \sqrt{\hbar})$ contained in $\mathbb{M}$. It follows from Corollary 8.43 that we must then have $z_0 = 0$.

Let $S, S' \in \mathrm{Sp}(n)$ and $z_0 \in \mathbb{R}_z^{2n}$ be such that $S(B^{2n}(\sqrt{\hbar})) \subset \mathbb{W}_\Sigma$ and $S'(B^{2n}(\sqrt{\hbar})) \subset \mathbb{W}_\Sigma$. In view of Proposition 8.12 we then have $S(S')^{-1} \in \mathrm{U}(n)$ hence $S(B^{2n}(\sqrt{\hbar})) = S'(B^{2n}(\sqrt{\hbar})$. □

Let us end this subsection by shortly discussing the notion of joint quantum probability. Some quantum physicists contend that there is no true collective ("joint") probability density for pairs of conjugate observables, say $\widehat{X}$ and $\widehat{P}$ when $n = 1$. It is in fact possible to construct infinitely many density probabilities ρ having arbitrary densities ρ_X and ρ_P as marginal probability densities, as the following exercise shows:

Exercise 8.46. Suppose $n = 1$ and let ρ_X and ρ_P be defined as above and set

$$\rho(z) = \rho_X(x)\rho_P(p)(1 - f(u(x), v(p)))$$

where f is an arbitrary function and

$$u(x) = \int_{-\infty}^{x} \rho_X(x')dx' \quad , \quad v(p) = \int_{-\infty}^{p} \rho_P(p')dp'.$$

Show that ρ is a probability density on $\mathbb{R}_z^2$ whose marginal probability densities are ρ_X and ρ_P. Extend this result to any n.

We will see in Chapter 10, when we study the phase-space Schrödinger equation, that one actually can associate to a normalized state ψ a quantum phase space probability density whose marginal densities are precisely $|\psi|^2$ and $|F\psi|^2$ in the limit $\hbar \to 0$.

8.5 Gaussian States

Among all states of a quantum system, Gaussian states are certainly among the most interesting, and this not only because they are easy to calculate with: in addition to the fact that Gaussian states play a pivotal role in many parts of quantum mechanics (*e.g.* quantum optics) they will allow us to connect the notions of quantum blobs and Wigner ellipsoid to the density operator machinery.

Schrödinger introduced the notion of coherent states (but the name was coined by Glauber) in 1926 as the states of the quantized harmonic oscillator that minimized the uncertainty relations. Lisiecki [105] gives a very interesting survey of coherent state representations.

We will use the following classical result about the Fourier transform of a complex Gaussian:

Let M be a symmetric complex $m \times m$ matrix such that $\operatorname{Re} M > 0$. We have

$$\left(\tfrac{1}{2\pi\hbar}\right)^{m/2} \int e^{-\frac{i}{\hbar}\langle u,v\rangle} e^{-\frac{1}{2\hbar}\langle Mv,v\rangle} d^m v = (\det M)^{-1/2} e^{-\frac{1}{2\hbar}\langle M^{-1}u,u\rangle} \tag{8.30}$$

where $(\det M)^{-1/2}$ is given by the formula

$$(\det M)^{-1/2} = \lambda_1^{-1/2} \cdots \lambda_m^{-1/2},$$

the numbers $\lambda_1^{-1/2}, \ldots, \lambda_m^{-1/2}$ being the square roots with positive real parts of the eigenvalues $\lambda_1^{-1}, \ldots, \lambda_m^{-1}$ of M^{-1}.

Formula (8.30) can be easily proven using the standard Gauss integral

$$\int_{-\infty}^{\infty} e^{-x^2/2} dx = \sqrt{2\pi}$$

and diagonalizing of M.

8.5.1 The Wigner transform of a Gaussian

Let us in fact explicitly calculate the Wigner transform

$$W\psi(z) = \left(\tfrac{1}{2\pi\hbar}\right)^n \int e^{-\frac{i}{\hbar}\langle p,y\rangle} \psi(x + \tfrac{1}{2}y)\overline{\psi(x - \tfrac{1}{2}y)} d^n y$$

of a Gaussian of the type

$$\psi(x) = e^{-\frac{1}{2\hbar}\langle (X+iY)x,x\rangle} \tag{8.31}$$

where X and Y are real symmetric $n \times n$ matrices, $X > 0$.

Proposition 8.47. *Let ψ be the Gaussian (8.31). Then:*

(i) *The Wigner transform $W\psi$ is the phase space Gaussian*

$$W\psi(z) = \left(\tfrac{1}{\pi\hbar}\right)^{n/2} (\det X)^{-1/2} e^{-\frac{1}{\hbar}\langle Gz,z\rangle} \tag{8.32}$$

where G is the symmetric matrix

$$G = \begin{bmatrix} X + YX^{-1}Y & YX^{-1} \\ X^{-1}Y & X^{-1} \end{bmatrix}. \tag{8.33}$$

(ii) *The matrix G is in addition positive definite and symplectic; in fact $G = S^T S$ where*

$$S = \begin{bmatrix} X^{1/2} & 0 \\ X^{-1/2}Y & X^{-1/2} \end{bmatrix} \in \operatorname{Sp}(n). \tag{8.34}$$

Proof. Set $M = X + iY$. By definition of the Wigner transform we have

$$W\psi(z) = \left(\tfrac{1}{2\pi\hbar}\right)^n \int e^{-\frac{i}{\hbar}\langle p,y\rangle} e^{-\frac{1}{2\hbar}F(x,y)} d^n y$$

where

$$\begin{aligned} F(x,y) &= \langle M(x+\tfrac{1}{2}y), x+\tfrac{1}{2}y\rangle + \langle \overline{M}(x-\tfrac{1}{2}y), x-\tfrac{1}{2}y\rangle \\ &= 2\langle Xx,x\rangle + 2i\langle Yx,y\rangle + \tfrac{1}{2}\langle Xy,y\rangle \end{aligned}$$

and hence

$$W\psi(z) = \left(\tfrac{1}{2\pi\hbar}\right)^n e^{-\frac{1}{\hbar}\langle Xx,x\rangle} \int e^{-\frac{i}{\hbar}\langle p+Yx,y\rangle} e^{-\frac{1}{4\hbar}\langle Xy,y\rangle} d^n y.$$

In view of formula (8.30) above we have

$$\begin{aligned} \int e^{-\frac{i}{\hbar}\langle p+Yx,y\rangle} e^{-\frac{1}{4\hbar}\langle Xy,y\rangle} d^n y = (2\pi\hbar)^{n/2} &\left[\det(\tfrac{1}{2}X)\right]^{-1/2} \\ &\times \exp\left[-\tfrac{1}{\hbar}\langle X^{-1}(p+Yx), p+Yx\rangle\right] \end{aligned}$$

and hence

$$W\psi(z) = \left(\tfrac{1}{\pi\hbar}\right)^{n/2} (\det X)^{-1/2} e^{-\frac{1}{\hbar}\langle Gz,z\rangle}$$

where

$$\langle Gz,z\rangle = \langle (X+YX^{-1})x,x\rangle + 2\langle X^{-1}Yx,p\rangle + \langle X^{-1}p,p\rangle;$$

this proves part (i) of the proposition. The symmetry of G is of course obvious. Property (ii) immediately follows since

$$\begin{bmatrix} X+YX^{-1}Y & YX^{-1} \\ X^{-1}Y & X^{-1} \end{bmatrix} = \begin{bmatrix} X^{1/2} & YX^{-1/2} \\ 0 & X^{-1/2} \end{bmatrix} \begin{bmatrix} X^{1/2} & 0 \\ X^{-1/2}Y & X^{-1/2} \end{bmatrix}$$

and

$$S = \begin{bmatrix} X^{1/2} & 0 \\ X^{-1/2}Y & X^{-1/2} \end{bmatrix}$$

obviously is symplectic. □

The function (8.31) is not normalized; in fact a straightforward calculation (using for instance formula (8.30)) shows that

$$||\psi||_{L^2(\mathbb{R}^n_x)} = \int |\psi(x)|^2 d^n x = \left(\tfrac{\det X}{(\pi\hbar)^n}\right)^{-1/2}. \tag{8.35}$$

Corollary 8.48. *The Wigner transform of the normalized Gaussian*

$$\psi_0(x) = \left(\tfrac{\det X}{(\pi\hbar)^n}\right)^{1/4} \exp\left[-\tfrac{1}{2\hbar}\langle (X+iY)x,x\rangle\right] \tag{8.36}$$

is given by the formula

$$W\psi_0(z) = \left(\tfrac{1}{\pi\hbar}\right)^n e^{-\frac{1}{\hbar}\langle Gz,z\rangle} \tag{8.37}$$

where G is the symplectic matrix (8.33).

Proof. This immediately follows from Proposition 8.47 and formula (8.35). □

It follows, as announced in Chapter 6, that the Wigner transform of a Gaussian always is a positive function. It is noticeable that, conversely, if ψ is a function such that $W\psi \geq 0$, then ψ must be a Gaussian, of the type $\psi(x) = C\exp\left[-Q(x)\right]$ where C is a complex number and Q a positive definite quadratic form (this is a well-known result going back to Hudson [94]).

Exercise 8.49. Verify that we have $\int W\psi_0(z)d^{2n}z = 1$ when ψ_0 is the normalized Gaussian (8.36).

8.5.2 Gaussians and quantum blobs

There is a fundamental relationship between Gaussians and quantum blobs. Let us introduce some notation.

We will denote by $\mathrm{Gauss}_0(n)$ the set of all centered and normalized Gaussians (8.31) and by $\mathrm{Blob}_0(n)$ the set of all quantum blobs centered at $z_0 = 0$.

Recall from Chapter 2, Subsection 2.2.2, that every symplectic matrix can be factored as $S = S_0 R$ where

$$S_0 = \begin{bmatrix} L & 0 \\ Q & L^{-1} \end{bmatrix} \text{ and } R = \begin{bmatrix} U & V \\ -V & U \end{bmatrix} \tag{8.38}$$

the blocks L, Q, U and V being given by

$$L = (AA^T + BB^T)^{1/2}, \tag{8.39}$$

$$U + iV = (AA^T + BB^T)^{-1/2}(A + iB), \tag{8.40}$$

$$Q = (CA^T + DB^T)(AA^T + BB^T)^{-1/2}. \tag{8.41}$$

Theorem 8.50. *The mapping $\mathbb{Q}(\cdot)$ which to every centered Gaussian ψ associates the quantum blob $\mathbb{Q}(\psi) = S(B^{2n}(\sqrt{\hbar}))$ where S is defined by (8.34) is a bijection.*

Proof. In view of Lemma 8.36 every quantum blob $\mathbb{Q} = S(B(\sqrt{\hbar}))$ can be written, in a unique way as $\mathbb{Q} = S_0(B^{2n}(\sqrt{\hbar}))$ where S_0 is given by formula (8.38). Set now $X = L^2$ and $Y = LQ$. Since S_0 is symplectic we must have $LQ = QL$ hence both X and Y are symmetric. This shows that $\mathbb{Q} = \mathbb{Q}(\psi)$ where ψ is the Gaussian (8.31). The mapping $\mathbb{Q}(\cdot)$ is thus surjective. It is also injective because S_0 is uniquely determined by $\mathbb{Q}$ by the already quoted Lemma 8.36. □

The extension of Theorem 8.50 to the case of Gaussians with an arbitrary center is straightforward.

A remarkable consequence of these results is that we can associate in a canonical manner a "companion Gaussian state" to every quantum mechanically admissible ellipsoid with minimum capacity $\frac{1}{2}h$:

Corollary 8.51. *Let $\mathbb{B}$ be an admissible ellipsoid with $\underline{c}(\mathbb{B}) = \frac{1}{2}h$. Then there exists a unique $\psi_{\mathbb{B}} \in \mathrm{Gauss}_0(n)$ such that*

$$W\psi_{\mathbb{B}}(z) = \left(\tfrac{1}{\pi\hbar}\right)^n e^{-\frac{1}{\hbar}\langle Gz,z\rangle} \quad , \; G = S^T S \tag{8.42}$$

where $S \in \mathrm{Sp}(n)$ is any symplectic matrix such that $S(B(\sqrt{\hbar})) \subset \mathbb{B}$.

Proof. We must check that the right hand side of (8.42) is independent of the choice of the symplectic matrix S putting M in the Williamson diagonal form. Let S and S' be two such choices; in view of Proposition 8.12 there exists $U \in \mathrm{U}(n)$ such that $S = US'$ and hence

$$\langle S^T Sz, z\rangle = \langle S'^T U^T US'z, z\rangle = \langle S'^T S'z, z\rangle$$

whence the result. □

8.5.3 Averaging over quantum blobs

An important question in any pseudo-differential calculus is that of the positivity of operators: an operator $\widehat{A} : \mathcal{S}(\mathbb{R}^n_x) \longrightarrow \mathcal{S}'(\mathbb{R}^n_x)$ is positive: $\widehat{A} \geq 0$ if $(\widehat{A}\psi, \psi)_{L^2(\mathbb{R}^n_x)} \geq 0$ for all $\psi \in \mathcal{S}(\mathbb{R}^n_x)$ (the L^2-scalar product is interpreted in the sense of distributions and should be viewed as the distributional bracket $\left\langle \widehat{A}\psi, \overline{\psi}\right\rangle$). It turns out that the Weyl correspondence $a \overset{\text{Weyl}}{\longleftrightarrow} \widehat{A}$ does not preserve positivity in the sense that the condition $a \geq 0$ does not ensure $\widehat{A} \geq 0$, and conversely. Here is a simple example in the case $n = 1$:

Example 8.52. Let $H = \frac{1}{2}(p^2 + x^2)$ and $\widehat{H} = \frac{1}{2}(-\hbar^2\partial_x^2 + x^2)$ be the associated Weyl operator. The functions $\psi_N(x) = h_N(x/\sqrt{\hbar})e^{-x^2/2\sqrt{\hbar}}$ (h_N the Nth Hermite function) form a complete orthonormal system and the corresponding eigenvalues are $\lambda_N = (N + \frac{1}{2})\hbar$. Set now $a = H - \frac{1}{2}\hbar$; then $\widehat{A}\psi_N = N\hbar\psi_N$ hence $\widehat{A} \geq 0$; but a is not a positive function.

We will see that this fact is not a weakness of Weyl calculus but rather a manifestation of the uncertainty principle, which can be alleviated by averaging the symbol a over a quantum blob in a sense to be defined. Let us however mention that if $a \geq 0$ and belongs to some of the standard pseudo-differential classes $S^m_{\rho,\delta}(\mathbb{R}^{2n}_z)$ with $0 \leq \delta < \rho \leq 1$, then one proves that $\widehat{A}$ is the sum of a positive operator with positive symbol in $S^m_{\rho,\delta}(\mathbb{R}^{2n}_z)$ and of an operator with symbol in $S^{m-(\rho-\delta)}_{\rho,\delta}(\mathbb{R}^{2n}_z)$

(see Folland [42], Chapter 2, §6; for related results see Fefferman and Phong [41] and the references therein).

The possible non-positivity of the Weyl operator associated to a positive symbol is of course related to the generic non-positivity of the Wigner transform: recall (formula (6.70) in Proposition 6.45, Chapter 6) that the mathematical expectation of $\widehat{A}$ in a (normalized) state ψ is, when defined, given by the formula

$$\langle\widehat{A}\rangle_\psi = \int W\psi(z)a(z)d^{2n}z.$$

It has been known since de Bruijn [17] that the "average" $W\psi * \Phi_R$ over a Gaussian $\Phi_R(z) = e^{-|z|^2/R^2}$ satisfies

$$W\psi * \Phi_R \geq 0 \text{ if } R^2 = \hbar \text{ , } W\psi * \Phi_R > 0 \text{ if } R^2 > \hbar; \tag{8.43}$$

this actually follows from a positivity result for convolutions of Wigner transforms:

Proposition 8.53. *Let $\psi, \phi \in \mathcal{S}(\mathbb{R}^n_x)$. We have $W\psi * W\phi \geq 0$; more precisely*

$$W\psi * W\phi = (2\pi\hbar)^n \left|W_\sigma(\psi, \phi^\vee)\right|^2 \geq 0 \tag{8.44}$$

where $\phi^\vee(x) = \phi(-x)$ and $W_\sigma(\psi, \phi^\vee)$ is the symplectic Fourier transform of $W_\sigma(\psi, \phi^\vee)$.

Proof. Observing that $W\phi^\vee = (W\phi)^\vee$ we have

$$\begin{aligned} W\psi * W\phi(z) &= \int W\psi(z - z')W\phi(z')d^{2n}z' \\ &= \int W\psi(z + z')W\phi(-z')d^{2n}z' \\ &= \int T(-z)W\psi(z')W\phi^\vee(z')d^{2n}z' \end{aligned}$$

that is, since $T(-z)W\psi = W(\widehat{T}(-z)\psi)$,

$$W\psi * W\phi(z) = \int W(\widehat{T}(-z)\psi)(z')W\phi^\vee(z')d^{2n}z'.$$

In view of the Moyal identity we thus have

$$W\psi * W\phi(z) = \left(\tfrac{1}{2\pi\hbar}\right)^n |(\widehat{T}(-z)\psi, \phi^\vee)_{L^2(\mathbb{R}^n_x)}|^2 \geq 0.$$

Let us prove the identity (8.44). By definition of $\widehat{T}(-z)\psi$ we have

$$(\widehat{T}(-z)\psi, \phi^\vee)_{L^2(\mathbb{R}^n_x)} = \int e^{\frac{i}{\hbar}(-\langle p,x'\rangle - \frac{1}{2}\langle p,x\rangle)}\psi(x' + x)\overline{\phi^\vee}(x')d^n x'.$$

Setting $x' = y - \frac{1}{2}x$ this is

$$(\widehat{T}(-z)\psi, \phi^{\vee})_{L^2(\mathbb{R}^n_x)} = \int e^{-\frac{i}{\hbar}\langle p,y\rangle}\psi(y+\tfrac{1}{2}x)\overline{\phi^{\vee}}(y-\tfrac{1}{2}x)d^n x'$$

which is precisely $y - \frac{1}{2}x$ in view of Lemma 6.41 in Section 6.4, Chapter 6. □

Let us extend de Bruijn's result (8.43) to arbitrary Gaussians. To any symmetric $2n \times 2n$ matrix $\Sigma > 0$ we associate the normalized Gaussian ρ_Σ defined by

$$\rho_\Sigma(z) = \left(\tfrac{1}{2\pi\hbar}\right)^n (\det\Sigma)^{-1/2} e^{-\frac{1}{2\hbar}\langle\Sigma^{-1}z,z\rangle}.$$

A straightforward calculation shows that ρ_Σ has integral 1 (and is hence a probability density); computing the Fourier transform of $\rho_\Sigma * \rho_{\Sigma'}$ using formula (8.30) one moreover easily verifies that

$$\rho_\Sigma * \rho_{\Sigma'} = \rho_{\Sigma+\Sigma'} \tag{8.45}$$

(see Appendix D).

Proposition 8.54. *Let Σ be a covariance matrix and $\mathbb{M}_\Sigma$ the associated ellipsoid.*

(i) *If there exists $S \in \mathrm{Sp}(n)$ such that $\mathbb{M}_\Sigma = S(B(\sqrt{\hbar}))$ (and hence $c(\mathbb{M}_\Sigma) = \frac{1}{2}h$) then $W\psi * \rho_\Sigma \geq 0$.*

(ii) *If $c(\mathbb{M}_\Sigma) > \frac{1}{2}h$, then $W\psi * \rho_\Sigma > 0$.*

Proof. (i) The condition $\mathbb{M}_\Sigma = S(B(\sqrt{\hbar}))$ is equivalent to $\Sigma = \frac{1}{\hbar}S^T S$ and hence there exists a Gaussian $\psi_{X,Y} \in L^2(\mathbb{R}^n_x)$ such that $\rho_\Sigma = W\psi_{X,Y}$. It follows from (8.44) that we have

$$W\psi * \rho_\Sigma = W\psi * W\psi_{X,Y} \geq 0$$

which proves the assertion.

(ii) If $c(\mathbb{M}_\Sigma) > \frac{1}{2}h$, then there exists $S \in \mathrm{Sp}(n)$ such that $\mathbb{M}_\Sigma$ contains $\mathbb{M}_{\Sigma_0} = S(B(\sqrt{\hbar}))$ as a proper subset; it follows that $\Sigma - \Sigma_0 > 0$ and hence

$$W\psi * \rho_\Sigma = (W\psi * \rho_{\Sigma_0}) * \rho_{\Sigma-\Sigma_0} > 0$$

which was to be proven. □

Using Proposition 8.54 we can prove a positivity result for the average of Weyl operators with positive symbol. This result is actually no more than a predictable generalization of calculations that can be found elsewhere (a good summary being [42]); it however very clearly shows that phase space "coarse graining" by ellipsoids with symplectic capacity $\geq \frac{1}{2}\hbar$ (and in particular by quantum blobs) eliminates positivity difficulties appearing in Weyl calculus: we are going to see that the Gaussian average of a symbol $a \geq 0$ over an ellipsoid with symplectic capacity $\geq \frac{1}{2}\hbar$ *always leads to a positive operator*:

Corollary 8.55. *Assume that* $a \geq 0$. *If* $c(\mathrm{M}_\Sigma) \geq \frac{1}{2}h$, *then the operator* $\widehat{A}_\Sigma \overset{\text{Weyl}}{\longleftrightarrow} a * \rho_\Sigma$ *satisfies*

$$(\widehat{A}_\Sigma\psi, \psi)_{L^2(\mathbb{R}^n_x)} \geq 0$$

for all $\psi \in \mathcal{S}(\mathbb{R}^n_x)$.

Proof. In view of formula (6.71) (Subsection 6.4.2, Chapter 6) we have

$$(\widehat{A}_\Sigma\psi, \psi)_{L^2(\mathbb{R}^n_x)} = \int (a * \rho_\Sigma)(z) W\psi(z) d^{2n}z,$$

hence, taking into account the fact that ρ_Σ is an even function,

$$\begin{aligned}(\widehat{A}_\Sigma\psi, \psi)_{L^2(\mathbb{R}^n_x)} &= \iint a(u)\rho_\Sigma(z-u)W\psi(z)d^n u d^n z \\ &= \int a(u)\left(\int \rho_\Sigma(u-z)W\psi(z)d^n z\right)d^n u \\ &= \int a(u)(\rho_\Sigma * W\psi)(u)d^n u.\end{aligned}$$

In view of Proposition 8.54 we have $W\psi * \rho_\Sigma \geq 0$, hence the result. □

Let us illustrate this result on a simple and (hopefully) suggestive example. Consider the harmonic oscillator Hamiltonian

$$H(z) = \frac{1}{2m}(p^2 + m^2\omega^2 x^2)\ ,\ m > 0,\ \omega > 0$$

(with $n = 1$) and choose for ρ_Σ the Gaussian

$$\rho_\sigma(z) = \frac{1}{2\pi\sigma^2} e^{-\frac{1}{2\sigma^2}(x^2+p^2)}.$$

Set $H_\sigma = H * \rho_\sigma$; a straightforward calculation yields the sharp lower bound

$$H_\sigma(z) = H(z) + \frac{\sigma^2}{2m}(1 + m\omega^2) \geq H(z) + \sigma^2\omega.$$

Assume that the ellipsoid $(x^2+p^2)/2\sigma^2 \leq 1$ is admissible; this is equivalent to the condition $\sigma^2 \geq \frac{1}{2}\hbar$ and hence $H_\sigma(z) \geq \frac{1}{2}\hbar\omega$: we have thus recovered the ground state energy of the one-dimensional harmonic oscillator. We leave it to the reader to check that we would still get the same result if we had used a more general Gaussian ρ_Σ instead of ρ_σ. We leave it to the reader to generalize this result to a quadratic Hamiltonian by using Williamson's symplectic diagonalization theorem.

Chapter 9

The Density Operator

One of the traditional approaches to quantum mechanics, common to most introductory textbooks, starts with a discussion of de Broglie's matter waves, and then proceeds to exhibit the wave equation governing their evolution, which is Schrödinger's equation. We prefer an algebraic approach, based on the Heisenberg group and its representations. Here is why. Schrödinger's equation

$$i\hbar\frac{\partial\psi}{\partial t} = \widehat{H}\psi$$

describes the evolution of the matter wave ψ in terms of a self-adjoint operator $\widehat{H}$ associated to a Hamiltonian function by some "quantization rule". For instance, when H is of the physical type "kinetic energy + potential" one replaces the momentum coordinates p_j by the differential operator $-i\hbar\partial/\partial x_j$ and to let the position coordinates x_j stand as they are. Now, if we believe – and we have every reason to do so – that quantum mechanics supersedes classical mechanics as being the "better theory", we are in trouble from a logical point of view, and risk inconsistencies. This is because if we view quantum mechanics as the quest for an algorithm for attaching a self-adjoint operator to a function (or "symbol", to use the terminology of the theory of pseudo-differential operators) we are at risk of overlooking several facts (see Mackey's review article [117])). For instance:

- It has long been known by scientists working in foundational questions that the modern formulation of quantum mechanics[1] rules out a large number of classical Hamiltonians;
- There are features of quantum mechanics (*e.g.*, spin) which are not preserved in the classical limit. We therefore postulate, with Mackey [116, 117], that classical mechanics is only what quantum mechanics looks like under certain limiting conditions, that is, that quantum mechanics is a *refinement* of

[1]I am referring here to the so-called "von Neumann formulation".

Hamiltonian mechanics (this is, by the way, the point of view of the "deformation quantization" of Bayen *et al.* [7], which is an autonomous theory originating in Moyal's trailblazer [127]).

For all these reasons it is worth trying to construct quantum mechanics *ex nihilo* without reference to classical mechanics. This does not mean, of course, that we will not "quantize" Hamiltonians (or other "observables") when possible, or desirable, but one should understand that *dequantization* is perhaps, after all, more important than *quantization*, which may well be a chimera (as exemplified by well-known "no-go" in geometric quantization theories (see for instance Gotay [75] and Tuynman [166]).

The concept of density operator (or "density matrix"), as it is commonly called in physics) comes from statistical physics in defining Gibbs quantum states. For detailed discussions see Ruelle [138], Khinchin [103], Landau and Lifschitz [106], Messiah [123]. See Nazaikinskii *et al.* [128] for an introduction to the topic from the point of view of asymptotics.

We begin by giving a self-contained account of two essential tools from functional analysis, the Hilbert–Schmidt and the trace-class operators.

The notion of quantum state represented (up to a complex factor) by a function ψ is not always convenient, because in practice one is not always certain what state the system is in. This ignorance adds a new uncertainty of a non-quantum nature to the quantum-mechanical one. This motivates the introduction of the notion of *density operator*, which is a powerful tool for describing general quantum systems. To fully understand and exploit this notion we have to make a detour and spend some time to study the basic notions of *trace-class* and *Hilbert–Schmidt operators*, which besides their interest in quantum mechanics play an important role in functional analysis.

9.1 Trace-Class and Hilbert–Schmidt Operators

Classical references for this section are Simon [150], Reed and Simon [134], Kato [100]; also see Bleecker and Booss-Bavnbek [11] and the Appendix I to Chapter 7 in Wallach [175] for a study of trace-class operators. For general results on Hilbert spaces see Dieudonné [30] or Friedman [44]; in the latter the notion of orthogonal projection on closed subspaces of Hilbert spaces is studied in much detail. A classical reference for operator theory is Gohberg and Goldberg [52].

9.1.1 Trace-class operators

Trace-class operators are compact operators for which a trace may be defined, such that the trace is finite and independent of the choice of basis. As *Wikipedia* notes, trace-class operators are essentially the same as nuclear operators, though many authors reserve the term "trace-class operator" for the special case of nuclear operators on Hilbert spaces. The notion is essential for the definition of the density operators of quantum mechanics.

Let $\mathcal{H}$ be a complex separable Hilbert space and (e_j) an orthonormal system of vectors in $\mathcal{H}$ (not necessarily a basis); we denote by F the closed subspace of $\mathcal{H}$ spanned by the e_j. Then the following results hold:

- For every $u \in \mathcal{H}$ the series $\sum_j |(u, e_j)_{\mathcal{H}}|^2$ is convergent and we have *Bessel's inequality*
$$\sum_j |(u, e_j)_{\mathcal{H}}|^2 = ||P_F u||^2_{\mathcal{H}} \leq ||u||^2_{\mathcal{H}} \tag{9.1}$$
where P_F is the orthogonal projection $\mathcal{H} \longrightarrow F$ (we have equality in (9.1) if the e_j span $\mathcal{H}$, that is if (e_j) is an orthonormal basis);
- More generally, for all $u, v \in \mathcal{H}$,
$$\sum_j (u, e_j)_{\mathcal{H}} \overline{(v, e_j)_{\mathcal{H}}} = (P_F u | P_F v)_{\mathcal{H}}; \tag{9.2}$$
- The Fourier series $\sum_j (u, e_j)_{\mathcal{H}} e_j$ is convergent in $\mathcal{H}$ and we have
$$\sum_j (u, e_j)_{\mathcal{H}} e_j = P_F u; \tag{9.3}$$
in particular the series converges to u if (e_j) is an orthonormal basis.

We will denote by $\mathcal{L}(\mathcal{H})$ the normed algebra of all continuous linear operators $\mathcal{H} \longrightarrow \mathcal{H}$; the operator norm is given by
$$||A|| = \sup_{||u||_{\mathcal{H}} \leq 1} ||Au||_{\mathcal{H}} = \sup_{||u||_{\mathcal{H}} = 1} ||Au||_{\mathcal{H}}$$
and we have $||A^*|| = ||A||$.

Definition 9.1. An operator $A \in \mathcal{L}(\mathcal{H})$ is said to be of *"trace class"* if there exist two orthonormal bases $(e_i)_i$ and $(f_j)_j$ of $\mathcal{H}$ such that
$$\sum_{i,j} |(Ae_i, f_j)_{\mathcal{H}}| < \infty \tag{9.4}$$
(we are assuming that both bases are indexed by the same sets). The vector space of all trace-class operators $\mathcal{H} \longrightarrow \mathcal{H}$ is denoted by $\mathcal{L}_{\mathrm{Tr}}(\mathcal{H})$.

Obviously A is of trace class if and only if its adjoint A^* is.

We are going to prove that if the condition (9.4) characterizing trace-class operators holds for *one* pair of orthonormal basis, then it holds for *all*. This property will allow us to prove that $\mathcal{L}_{\mathrm{Tr}}(\mathcal{H})$ is indeed a vector space, and to define the trace of an element of $\mathcal{L}_{\mathrm{Tr}}(\mathcal{H})$ by the formula
$$\mathrm{Tr}\, A = \sum_i (Ae_i, e_i)_{\mathcal{H}}|$$
justifying *a posteriori* the terminology.

Proposition 9.2. *Suppose that If A is of trace class. Then:*

(i) *We have*

$$\sum_{i,j} |(Ae_i, f_j)_{\mathcal{H}}| < \infty \tag{9.5}$$

for all orthonormal bases $(e_i)_i$, $(f_j)_j$ of $\mathcal{H}$ with same index set;

(ii) *If $(e_i)_i$ and $(f_i)_i$ are two orthonormal bases, then the following equality holds:*

$$\sum_i (Ae_i, e_i)_{\mathcal{H}} = \sum_i (Af_i, f_i)_{\mathcal{H}} \tag{9.6}$$

(absolutely convergent series).

(iii) *The set $\mathcal{L}_{\mathrm{Tr}}(\mathcal{H})$ of all trace-class operators is a vector space.*

Proof. (i) Writing Fourier expansions

$$e_i' = \sum_j (e_i', e_j)_{\mathcal{H}} e_j \ \ , \ \ f_\ell' = \sum_k (f_\ell', f_k)_{\mathcal{H}} f_k$$

we have

$$(Ae_i', f_\ell')_{\mathcal{H}} = \sum_{j,k} (e_i', e_j)_{\mathcal{H}} \overline{(f_\ell', f_k)_{\mathcal{H}}} (Ae_j, f_k)_{\mathcal{H}} \tag{9.7}$$

and hence, by the triangle inequality,

$$\sum_{i,\ell} |(Ae_i', f_\ell')_{\mathcal{H}}| \leq \sum_{j,k} \left(\sum_{i,\ell} |(e_i', e_j)_{\mathcal{H}}||(f_\ell', f_k)_{\mathcal{H}}| \right) |(Ae_j, f_k)_{\mathcal{H}}|). \tag{9.8}$$

Using successively the trivial inequality $ab \leq \frac{1}{2}(a^2 + b^2)$ and the Bessel inequality (9.1) we get

$$\begin{aligned}\sum_{i,\ell} |(e_i', e_j)_{\mathcal{H}}||(f_\ell', f_k)_{\mathcal{H}}| &\leq \tfrac{1}{2} \sum_i |(e_i', e_j)_{\mathcal{H}}|^2 + \tfrac{1}{2} \sum_\ell |(f_\ell', f_k)_{\mathcal{H}}|^2 \\ &= \tfrac{1}{2}(||e_j||^2_{\mathcal{H}} + ||f_k||^2_{\mathcal{H}}) = \tfrac{1}{2},\end{aligned}$$

and hence

$$\sum_{i,\ell} |(Ae_i', f_\ell')_{\mathcal{H}}| \leq \tfrac{1}{2} \sum_{j,k} |(Ae_j, f_k)_{\mathcal{H}}|) < \infty$$

which proves (9.5).

(ii) Assume now $e_i = f_i$ and $e_i' = f_i'$. In view of (9.7) we have

$$(Ae_i', e_i')_{\mathcal{H}} = \sum_{j,k} (e_i', e_j)_{\mathcal{H}} \overline{(e_i', e_k)_{\mathcal{H}}} (Ae_j, e_k)_{\mathcal{H}}$$

and hence

$$\sum_i (Ae'_i, e'_i)_{\mathcal{H}} = \sum_{j,k} \left(\sum_i (e'_i, e_j)_{\mathcal{H}} \overline{(e'_i, e_k)_{\mathcal{H}}} \right) (Ae_j, e_k)_{\mathcal{H}}).$$

In view of (9.2)

$$\sum_i (e'_i, e_j)_{\mathcal{H}} \overline{(e'_i, e_k)_{\mathcal{H}}} = (e_j, e_k)_{\mathcal{H}} = \delta_{jk}$$

which establishes (9.6); that the series is absolutely convergent follows from (9.5) with the choice $e_i = f_i$.

(iii) That $\lambda A \in \mathcal{L}_{\mathrm{Tr}}(\mathcal{H})$ if $\lambda \in \mathbb{C}$ and $A \in \mathcal{L}_{\mathrm{Tr}}(\mathcal{H})$ is obvious. Let $A, B \in \mathcal{L}_{\mathrm{Tr}}(\mathcal{H})$; then

$$\sum_i ((A+B)e_i, e_i)_{\mathcal{H}} = \sum_i (Ae_i, e_i)_{\mathcal{H}} + \sum_i (Be_i, e_i)_{\mathcal{H}}$$

and each sum on the right-hand side is absolutely convergent, implying that $A + B \in \mathcal{L}_{\mathrm{Tr}}(\mathcal{H})$. □

Formula (9.6) motivates the following definition:

Definition 9.3. Let A be a trace-class operator in $\mathcal{H}$ and $(e_i)_i$ an orthonormal basis of $\mathcal{H}$. The sum of the absolutely convergent series

$$\mathrm{Tr}(A) = \sum_i (Ae_i, e_i)_{\mathcal{H}} \tag{9.9}$$

(which does not depend on the choice of $(e_i)_i$) is called the "trace" of the operator A.

The following properties of the trace are obvious consequences of (9.9):

$$\mathrm{Tr}(A+B) = \mathrm{Tr}(A) + \mathrm{Tr}(B) \quad , \quad \mathrm{Tr}(A^*) = \overline{\mathrm{Tr}(A)}. \tag{9.10}$$

In particular, self-adjoint trace-class operators have real trace.

Here is an important example of a trace-class operator; we will use it in our study of the density operator to characterize the "purity" of a quantum state:

Example 9.4. Let F be a finite-dimensional subspace of the Hilbert space $\mathcal{H}$. The orthogonal projection $P_F : \mathcal{H} \longrightarrow F$ is an operator of trace class with trace $\mathrm{Tr}(P_F) = \dim F$. Let in fact $(e_i)_{1 \leq i \leq k}$ be an orthonormal basis of F and $(e_i)_i$ a full orthonormal basis of $\mathcal{H}$ containing $(e_i)_{1 \leq i \leq k}$. We have $(P_F e_i, e_i) = 1$ if $1 \leq i \leq n$ and 0 otherwise; the claim follows by (9.9).

Notice that an orthogonal projection on an infinite-dimensional subspace is never an operator of trace class.

Trace-class operators do not only form a vector space, they also form a normed *algebra*:

Proposition 9.5. *Let A and B be trace-class operators A on the Hilbert space $\mathcal{H}$.*

(i) *The product AB is a trace-class operator and we have $\operatorname{Tr}(AB) = \operatorname{Tr}(BA)$.*

(ii) *The formula $||A||_{Tr} = (\operatorname{Tr}(A^*A))^{1/2}$ defines a norm on the algebra $\mathcal{L}_{Tr}(\mathcal{H})$; that norm is associated to the scalar product $(A, B)_{Tr} = \operatorname{Tr}(A^*B)$.*

Proof. (i) Let $(e_i)_i$ be an orthonormal basis of $\mathcal{H}$. Writing

$$(ABe_i, e_i)_{\mathcal{H}} = (Be_i, A^*e_i)_{\mathcal{H}}$$

formula (9.2) yields

$$\sum_i (ABe_i, e_i)_{\mathcal{H}} = \sum_{i,j} (Be_i, e_j)_{\mathcal{H}} (Ae_i, e_j)_{\mathcal{H}} \tag{9.11}$$

and hence

$$\left| \sum_i (ABe_i, e_i)_{\mathcal{H}} \right| \leq \left(\sum_{i,j} |(Be_i, e_j)_{\mathcal{H}}| \right) \left(\sum_{i,j} |(Ae_i, e_j)_{\mathcal{H}}| \right) < \infty$$

so that AB is indeed of trace class in view of Lemma 9.2 (i). Formula (9.11) implies that

$$\sum_i (ABe_i, e_i)_{\mathcal{H}} = \sum_i (BAe_i, e_i)_{\mathcal{H}} \tag{9.12}$$

that is $\operatorname{Tr}(AB) = \operatorname{Tr}(BA)$.

(ii) Since A^*A is self-adjoint its trace is real so $(\operatorname{Tr}(A^*A))^{1/2}$ is well defined. If $||A||_{\mathrm{Tr}} = 0$ then $(Ae_i, e_j)_{\mathcal{H}} = 0$ for all i hence $A = 0$. The relation $||\lambda A||_{\mathrm{Tr}} = |\lambda| ||A||_{\mathrm{Tr}}$ being obvious there only remains to show that the triangle inequality holds. If $A, B \in \mathcal{L}_{\mathrm{Tr}}(\mathcal{H})$ then

$$\begin{aligned} ||A + B||_{\mathrm{Tr}}^2 &= \operatorname{Tr}((A^* + B^*)(A + B)) \\ &= \operatorname{Tr}(A^*A) + \operatorname{Tr}(B^*B) + \operatorname{Tr}(A^*B) + \operatorname{Tr}(B^*A). \end{aligned}$$

In view of the second formula (9.10) we have

$$\operatorname{Tr}(A^*B) + \operatorname{Tr}(B^*A) = 2 \operatorname{Re} \operatorname{Tr}(A^*B)$$

and hence

$$||A + B||_{\mathrm{Tr}}^2 = ||A||_{\mathrm{Tr}}^2 + ||B||_{\mathrm{Tr}}^2 + 2 \operatorname{Re} \operatorname{Tr}(A^*B). \tag{9.13}$$

We have

$$\operatorname{Tr}(A^*B) = \sum_i (Be_i, Ae_i)_{\mathcal{H}},$$

hence, noting that $\operatorname{Re}\operatorname{Tr}(A^*B) \leq |\operatorname{Tr}(A^*B)|$ and using the Cauchy–Schwarz inequality,

$$\operatorname{Re}\operatorname{Tr}(A^*B) \leq \sum_i (Be_i, Be_i)_{\mathcal{H}}^{1/2} (Ae_i, Ae_i)_{\mathcal{H}}^{1/2},$$

that is

$$\operatorname{Re}\operatorname{Tr}(A^*B) \leq \sum_i (B^*Be_i, e_i)_{\mathcal{H}}^{1/2} (A^*Ae_i, e_i)_{\mathcal{H}}^{1/2} \leq ||B||_{\mathrm{Tr}} ||A||_{\mathrm{Tr}}$$

which proves the triangle inequality in view of (9.13). □

Exercise 9.6. Show that if A, B, C are of trace class then

$$\operatorname{Tr}(ABC) = \operatorname{Tr}(CAB) = \operatorname{Tr}(BCA).$$

An essential feature of trace-class operators is that they are compact. (An operator A on $\mathcal{H}$ is compact if the image of the unit ball in $\mathcal{H}$ by A is relatively compact.) The sum and the product of two compact operators is again a compact operator; in fact compact operators form a two-sided ideal in $\mathcal{L}(\mathcal{H})$. A compact operator A on $\mathcal{H}$ is never invertible if $\dim \mathcal{H} = \infty$: suppose that A has an inverse A^{-1}, then $AA^{-1} = I$ would also be compact, and hence any ball $B(R)$ in $\mathcal{H}$ would be a compact set. This contradicts the theorem of F. Riesz that says that closed balls are not compact in an infinite-dimensional Hilbert (or Banach) space.

Proposition 9.7. *A trace-class operator A on a Hilbert space $\mathcal{H}$ is a compact operator.*

Proof. Let (u_j) be a sequence in $\mathcal{H}$ such that $||u_j||_{\mathcal{H}} \leq 1$ for every j. We are going to show that (Au_j) contain a convergent subsequence; this will prove the claim. Let (e_i) be an orthonormal basis of $\mathcal{H}$, writing $u_j = \sum_i (u_j, e_i)_{\mathcal{H}} e_i$ we have

$$\begin{aligned} ||Au_j||_{\mathcal{H}}^2 &= (A^*Au_j, u_j)_{\mathcal{H}} \\ &= \sum_{i,k} (u_j, e_i)_{\mathcal{H}} \overline{(u_j, e_k)_{\mathcal{H}}} (A^*Ae_k, e_i)_{\mathcal{H}}. \end{aligned}$$

Since we have, by Cauchy–Schwarz's inequality,

$$|(u_j, e_i)_{\mathcal{H}}| \leq ||u_j||_{\mathcal{H}} ||e_i||_{\mathcal{H}} \leq 1$$

it follows that

$$||Au_j||_{\mathcal{H}}^2 \leq \sum_{i,k} (A^*Ae_k, e_i)_{\mathcal{H}};$$

A^*A being of trace class the sequence (Au_j) is contained in a compact subset of $\mathcal{H}$ and the result follows. □

We are next going to prove a spectral decomposition result for self-adjoint trace-class operators. Recall that if A is a self-adjoint compact operator on a Hilbert space $\mathcal{H}$, then its spectrum consists of a sequence

$$\cdots \leq \lambda_{-2} \leq \lambda_{-1} < 0 = \lambda_0 \leq \lambda_1 \leq \cdots$$

of real numbers. That 0 belongs to the spectrum follows from the fact that a compact operator is not invertible, as pointed out above. Each $\lambda_j \neq 0$ is moreover an eigenvalue of A. We denote by $\mathcal{H}_j$ the eigenspace corresponding to the eigenvalue $\lambda_j \neq 0$; $\mathcal{H}_j$ is *finite-dimensional* and for $j \neq k$ the spaces $\mathcal{H}_j$ and $\mathcal{H}_k$ are orthogonal and $\mathcal{H}$ splits into the Hilbert sum

$$\mathcal{H} = \operatorname{Ker} A \oplus (\mathcal{H}_{-1} \oplus \mathcal{H}_{-2} \oplus \cdots) \oplus (\mathcal{H}_1 \oplus \mathcal{H}_2 \oplus \cdots).$$

A consequence of these properties is that the trace of an operator of trace class coincides with its *spectral trace* (*i.e.*, the sum of its eigenvalues, counting their multiplicities) and that we have a spectral decomposition result for these operators in terms of projections. Let us prove this in the case that is of interest to us:

Proposition 9.8. *Let A be a positive self-adjoint operator A of trace class on a Hilbert space $\mathcal{H}$; let $\lambda_1 \leq \lambda_2 \cdots$ be its eigenvalues of A and $\mathcal{H}_1, \mathcal{H}_2, \ldots$ the corresponding eigenspaces.*

(i) *We have*

$$A = \sum_j \lambda_j P_j \tag{9.14}$$

where P_j is the orthogonal projection $\mathcal{H} \longrightarrow \mathcal{H}_j$;

(ii) *The trace of A is given by the formula*

$$\operatorname{Tr}(A) = \sum_j \lambda_j \dim \mathcal{H}_j \tag{9.15}$$

and is hence identical with the spectral trace of A;

(iii) *Conversely, every operator of the type (9.14) with $\lambda_j > 0$ and P_j being a projection on a finite-dimensional space is of trace-class if the condition*

$$\sum_j \lambda_j \dim \mathcal{H}_j < \infty$$

holds.

Proof. (i) A is compact, so it verifies the properties listed before the statement of the proposition. Choose an orthonormal basis $(e_{ij})_i$ in each eigenspace $\mathcal{H}_j$ and complete the union $\cup_i (e_{ij})_i$ of these bases into a full orthonormal basis of $\mathcal{H}$ by selecting orthonormal vectors $(f_i)_i$ in $\operatorname{Ker} A$ such that $(f_i, e_{jk})_{\mathcal{H}} = 0$ for all j, k

(this can always be done using a Gram–Schmidt orthonormalization process). Let u be an arbitrary element of $\mathcal{H}$ and expand Au in a Fourier series

$$Au = \sum_i (Au, f_i)_{\mathcal{H}} f_i + \sum_{i,j} (Au, e_{ij})_{\mathcal{H}} e_{ij}.$$

Since A is self-adjoint we have

$$(Au, f_i)_{\mathcal{H}} = (u, Af_i)_{\mathcal{H}} = 0$$

and

$$(Au, e_{ij}) = (u, Ae_{ij}) = \lambda_j (u, e_{ij})$$

so that

$$Au = \sum_j \lambda_j \left(\sum_i (u, e_{ij})_{\mathcal{H}} e_{ij} \right).$$

The operator P_j defined by

$$P_j u = \sum_i (u, e_{ij})_{\mathcal{H}} e_{ij}$$

is the orthogonal projection on $\mathcal{H}_j$; this is (9.14).

(ii) By definition (9.9) of the trace we have

$$\operatorname{Tr}(A) = \sum_i (Af_i, f_i)_{\mathcal{H}} + \sum_{i,j} (Ae_{ij}, e_{ij})_{\mathcal{H}};$$

since $(Af_i, f_i)_{\mathcal{H}} = 0$ and $(Ae_{ij}, e_{ij})_{\mathcal{H}} = \lambda_j$ for every i this reduces to

$$\operatorname{Tr}(A) = \sum_{i,j} \lambda_j (e_{ij}, e_{ij})_{\mathcal{H}} = \sum_j \lambda_j \left(\sum_i (e_{ij}, e_{ij})_{\mathcal{H}} \right), \tag{9.16}$$

hence (9.15) since the sum indexed by i is equal to the dimension of the eigenspace $\mathcal{H}_j$.

(iii) Any operator A that can be written in the form (9.14) is self-adjoint because each projection is self-adjoint (notice that it is also compact because the spaces $\mathcal{H}_j$ are finite-dimensional); moreover the operator A is positive and the condition $\sum_j \lambda_j \dim \mathcal{H}_j < \infty$ is precisely equivalent to A being of trace class in view of the second equality (9.16). □

9.1.2 Hilbert–Schmidt operators

The notion of Hilbert–Schmidt operator is intimately connected to that of trace operator, and gives insight in the properties of these. In fact, as we will see, every self-adjoint trace-class operator is a Hilbert–Schmidt operator (more generally, every trace-class operator can be written as a product of two Hilbert–Schmidt operators). We will see, among other properties, that an operator is a Hilbert–Schmidt operator if and only if it is the limit of a sequence of operators of finite rank.

Definition 9.9. An operator A in $\mathcal{H}$ is called a "Hilbert–Schmidt operator" if there exists an orthonormal basis (e_j) of $\mathcal{H}$ such that

$$\sum_j ||Ae_j||^2_{\mathcal{H}} < \infty.$$

The vector space of all Hilbert–Schmidt operators on $\mathcal{H}$ is denoted by $\mathcal{L}_{\mathrm{HS}}(\mathcal{H})$.

The following result shows that it does not matter which orthonormal basis we use to check that an operator is of the Hilbert–Schmidt class, and that $\mathcal{L}_{\mathrm{HS}}(\mathcal{H})$ is thus indeed a vector space, as claimed in the definition:

Proposition 9.10. *Let A be an operator on the Hilbert space $\mathcal{H}$.*

(i) *A is a Hilbert–Schmidt operator if and only if*

$$\sum_j ||Ae_j||^2_{\mathcal{H}} = \sum_j ||Af_j||^2_{\mathcal{H}} < \infty \tag{9.17}$$

for all orthonormal bases (e_j) and (f_j) of $\mathcal{H}$;

(ii) *A is a Hilbert–Schmidt operator if and only if its adjoint A^* is;*

(iii) *The set $\mathcal{L}_{HS}(\mathcal{H})$ of all Hilbert–Schmidt operators on $\mathcal{H}$ is a vector subspace of $\mathcal{L}(\mathcal{H})$ and the function $||\cdot||_{HS} \geq 0$ defined by the formula*

$$||A||^2_{HS} = \sum_i ||Ae_i||^2_{\mathcal{H}} \tag{9.18}$$

is a norm on that subspace (the "Hilbert–Schmidt norm").

Proof. (i) The condition is sufficient by definition of a Hilbert–Schmidt operator. Assume conversely that $\sum_j ||Ae_j||^2_{\mathcal{H}} < \infty$ for one orthonormal basis (e_j); if (f_j) is an arbitrary orthonormal basis we have,

$$||Ae_i||^2_{\mathcal{H}} = \sum_j |(Ae_j, f_j)^2_{\mathcal{H}}$$

in view of (9.1) and hence

$$\sum_i ||Ae_i||^2_{\mathcal{H}} = \sum_{i,j} |(Ae_j, f_j)^2_{\mathcal{H}} = \sum_{i,j} |(e_j, A^* f_j)^2_{\mathcal{H}},$$

that is, again by (9.1),

$$\sum_i ||Ae_i||^2_{\mathcal{H}} = \sum_i ||A^* f_i||^2_{\mathcal{H}}.$$

Choosing $(e_j)_j = (f_j)_j$ we have in particular

$$\sum_i ||Af_i||^2_{\mathcal{H}} = \sum_i ||A^* f_i||^2_{\mathcal{H}} \tag{9.19}$$

and hence

$$\sum_i ||Ae_i||^2_{\mathcal{H}} = \sum_i ||Af_i||^2_{\mathcal{H}}$$

as claimed. Property

(ii) Immediately follows from (9.19).

(iii) If A and B are Hilbert–Schmidt operators then λA is trivially a Hilbert–Schmidt operator and $||\lambda A||_{\mathrm{HS}} = |\lambda| ||A||_{\mathrm{HS}}$ for every $\lambda \in \mathbb{C}$; on the other hand, by the triangle inequality

$$\sum_j ||(A+B)e_j||^2_{\mathcal{H}} \leq \sum_j ||Ae_j||^2_{\mathcal{H}} + \sum_j ||Be_j||^2_{\mathcal{H}} < \infty$$

for every orthonormal basis $(e_j)_j$ hence $A+B$ is also a Hilbert–Schmidt operator and we have

$$||A+B||^2_{\mathrm{HS}} \leq ||A||^2_{\mathrm{HS}} + ||B||^2_{\mathrm{HS}},$$

hence also

$$||A+B||_{\mathrm{HS}} \leq ||A||_{\mathrm{HS}} + ||B||_{\mathrm{HS}}.$$

Finally, $||A||_{\mathrm{HS}} = 0$ is equivalent to $Ae_j = 0$ for every index j that is to $A = 0$. □

Remark 9.11. The norm (9.18) on $\mathcal{L}_{\mathrm{HS}}(\mathcal{H})$ is called the "Hilbert–Schmidt norm" in the literature.

Every Hilbert–Schmidt operator is actually the limit (in the topology defined by the norm $||\cdot||_{\mathrm{HS}}$) of a sequence of operators of finite rank, and hence compact:

Proposition 9.12. *Let A be an operator on the Hilbert space $\mathcal{H}$.*

(i) *If A is of finite rank, then it is a Hilbert–Schmidt operator;*

(ii) *If A is a Hilbert–Schmidt operator A, then it is the limit of a sequence of operators of finite rank in $\mathcal{L}_{HS}(\mathcal{H})$;*

(iii) *In particular, every Hilbert–Schmidt operator is compact.*

Proof. (i) Since A is of finite rank its kernel $\operatorname{Ker} A$ has finite codimension in $\mathcal{H}$. Choose now an orthonormal basis $(e_j)_{j\in\mathbb{K}}$, $\mathbb{K} \subset \mathbb{N}$, of $\operatorname{Ker} A$ and complete it to a full orthonormal basis $(e_j)_{j\in\mathbb{J}}$ of $\mathcal{H}$. We have

$$\sum_{j\in J} ||Ae_j||^2_{\mathcal{H}} = \sum_{j\in J\setminus N} ||Ae_j||^2_{\mathcal{H}} < \infty$$

since the index subset $\mathbb{J}\setminus\mathbb{K}$ is finite.

(ii) Assume that the set $\{Ae_j : j \in \mathbb{J}\}$ is finite: then A is of finite rank and there is nothing to prove in view of (i). If that set is infinite, it is no restriction, replacing if necessary J by a smaller set to assume that all the Ae_j are distinct, and that $\mathbb{J} = \mathbb{N}$. Let A_k be the linear operator defined by $A_k e_j = Ae_j$ for $j \leq k$ and $A_k e_j = 0$ for all other indices j. We have $\lim_{k\to\infty} A_k = A$ and each A_k is of finite rank.

Property (iii) follows because operators of finite rank are compact, and the limit of a sequence of compact operators is compact. □

A particularly interesting situation occurs when the trace-class operator is self-adjoint (this will be the case for the density matrices we will study in the next section).

Proposition 9.13. *Every self-adjoint trace class operator in a Hilbert space $\mathcal{H}$ is a Hilbert–Schmidt operator (and is hence compact).*

Proof. Let $(e_i)_i$ be an orthonormal basis of $\mathcal{H}$; we have

$$||Ae_i||^2_{\mathcal{H}} = (Ae_i, Ae_i)_{\mathcal{H}} = (A^2 e_i, e_i)_{\mathcal{H}}$$

and in view of Proposition 9.5 the operator A^2 is of trace class. It follows that

$$\sum_i ||Ae_i||^2_{\mathcal{H}} = \sum_i (A^2 e_i, e_i)_{\mathcal{H}} < \infty$$

hence A is a Hilbert–Schmidt operator as claimed. □

It turns out that, more generally, every trace-class operator can be obtained by composing two Hilbert-Schmidt operators; since we will not use this fact we propose its proof as an exercise:

Exercise 9.14. Show that every trace-class operator is the product of two Hilbert–Schmidt operators, and is hence compact. [Hint: use an adequate polar decomposition $A = U(A^*A)^{1/2}$.]

9.2 Integral Operators

Let us specialize our discussion to the case where $\mathcal{H}$ is the Hilbert space $L^2(\mathbb{R}^n_x)$. We begin by discussing the general theory of integral operators with L^2-kernel.

9.2.1 Operators with L^2 kernels

In what follows we assume that A is an operator defined on the Schwartz space $\mathcal{S}(\mathbb{R}^n_x)$ by

$$A\psi(x) = \int K_A(x,y)\psi(y)d^n y \quad , \quad K_A \in L^2(\mathbb{R}^{2n}_{x,y}). \tag{9.20}$$

Such an operator is bounded on $L^2(\mathbb{R}^n_x)$: we have

$$||A\psi||^2_{L^2(\mathbb{R}^n_x)} = \int \left| \int K(x,y)\psi(y)d^n y \right|^2 d^n x$$

and, by Cauchy–Schwarz's inequality

$$\left| \int K_A(x,y)\psi(y)d^n y \right|^2 \leq \int |K_A(x,y)|^2 d^n y \int |\psi(y)|^2 d^n y$$

hence the estimate

$$||A\psi||_{L^2(\mathbb{R}^n_x)} \leq ||K_A||_{L^2(\mathbb{R}^{2n}_{x,y})} ||\psi||_{L^2(\mathbb{R}^n_x)}. \tag{9.21}$$

It follows that the operator A is indeed bounded.

The integral operators (9.20) form an algebra:

Proposition 9.15. *The sum and the product of two integral operators (9.20) is an operator of the same type. More precisely, if A and B have kernels K_A and K_B in $L^2(\mathbb{R}^{2n}_{x,y})$, then $A + B$ has kernel $K_A + K_B \in L^2(\mathbb{R}^{2n}_{x,y})$ and AB has a kernel $K_{AB} \in L^2(\mathbb{R}^{2n}_{x,y})$ given by*

$$K_{AB}(x,y) = \int K_A(x,x')K_B(x',y)d^n x'. \tag{9.22}$$

Proof. That the kernel of $A+B$ is $K_A + K_B$ is obvious; that $K_A + K_B \in L^2(\mathbb{R}^{2n}_{x,y})$ follows from the fact that $L^2(\mathbb{R}^{2n}_{x,y})$ is a vector space. Let us next prove that $K_{AB} \in L^2(\mathbb{R}^{2n}_{x,y})$. In view of Cauchy–Schwarz's inequality

$$|K_{AB}(x,y)|^2 \leq \left(\int |K_A(x,x')|^2 d^n x' \right) \left(\int |K_B(x',y)|^2 d^n x' \right)$$

and hence

$$\int |K_{AB}(x,y)|^2 d^n x d^n y \leq \int \left(\int |K_A(x,x')|^2 d^n x' \right) \times \left(\int |K_B(x',y)|^2 d^n x' \right) d^n x d^n y$$

which yields

$$\int |K_{AB}(x,y)|^2 d^n x d^n y \leq \int \left(\int |K_A(x,x')|^2 d^n x' d^n x \right) \times \left(\int |K_B(x',y)|^2 d^n x' \right) d^n y,$$

that is

$$\int |K_{AB}(x,y)|^2 d^n x d^n y = ||K_A||_{L^2(\mathbb{R}^{2n}_{x,y})} ||K_B||_{L^2(\mathbb{R}^{2n}_{x,y})} < \infty.$$

Let us next show that $K_{AB}(x,y)$ indeed is the kernel of AB. We have, for $\psi \in L^2(\mathbb{R}^n_x)$,

$$\begin{aligned} AB\psi(x) &= \int K_A(x,x')B\psi(x')d^n x' \\ &= \int K_A(x,x')\left(\int K_B(x',y)\psi(y)d^n y\right) d^n x'. \end{aligned}$$

The result will follow if we prove that we can change the order of integration, that is:

$$\begin{aligned} &\int K_A(x,x')\left(\int K_B(x',y)\psi(y)d^n y\right) d^n x' \\ &\qquad\qquad = \int \left(\int K_A(x,x')K_B(x',y)d^n x'\right) \psi(y)d^n y. \end{aligned}$$

For this it suffices to show that the function

$$F(x) = \iint |K_A(x,x')K_B(x',y)\psi(y)|d^n x' d^n y$$

is bounded for almost every x. Now, using again Cauchy–Schwarz's inequality,

$$\begin{aligned} F(x) &= \int \left(\int |K_A(x,x')K_B(x',y)|d^n x'\right) |\psi(y)|d^n y \\ &\leq \left[\int \left(\int |K_A(x,x')K_B(x',y)|^2 d^n x'\right) d^n y'\right]^{1/2} \left(\int |\psi(y)|^2 d^n y\right)^{1/2} \\ &= \left[\int \left(\int |K_A(x,x')K_B(x',y)|^2 d^n x'\right) d^n y'\right]^{1/2} ||\psi||_{L^2(\mathbb{R}^n_x)}. \end{aligned}$$

The function $x \longmapsto G(x)$ defined by

$$G(x) = \int \left(\int |K_A(x,x')K_B(x',y)|^2 d^n x'\right) d^n y'$$

being integrable, we have $G(x) < \infty$ for almost every x, which concludes the proof. □

The algebra of integral operators (9.20) is closed under the operation of taking adjoints:

Exercise 9.16. Show, using similar precise estimates, that the adjoint A^* of the operator (9.20) is the integral operator with kernel K_{A^*} defined by $K_{A^*}(x,y) = \overline{K_A(y,x)}$.

9.2.2 Integral trace-class operators

In many physics books, even the best, one frequently encounters the following claim: let A be an integral operator

$$A\psi(x)=\int K_A(x,y)\psi(y)d^ny$$

such that

$$\int |K_A(x,x)|d^nx<\infty;$$

then A is of trace class and

$$\operatorname{Tr}(A)=\int K_A(x,x)d^nx.$$

This statement is false without more stringent conditions on the kernel K_A. (See however the interesting discussion in §8.4.3 of Dubin *et al.* [33].) Here is one a rigorous result; for the state of the art see Wong's review paper [182]. Additional information can be found in Gröchenig's book [78] and paper [79]; also see Toft [160]; an interesting precursor is Pool [133]. A very nice treatment of operators of trace class in general spaces $L^2(X,\mu)$ is Duistermaat's contribution in [35].

Recall from standard distribution theory that the Sobolev space $H^s(\mathbb{R}^{2n}_z)$ ($s\in\mathbb{R}$) consists of all $a\in\mathcal{S}'(\mathbb{R}^{2n}_z)$ such that the Fourier transform Fa is a function satisfying

$$\int |Fa(z)|^2(1+|z|^2)^s d^{2n}z<\infty.$$

Theorem 9.17. *Let a be a function $\mathbb{R}^{2n}_z\longmapsto\mathbb{C}$.*

(i) *If $\widehat{A}\overset{\text{Weyl}}{\longleftrightarrow}a$ is of trace class and $a\in L^1(\mathbb{R}^{2n}_z)$, then*

$$\operatorname{Tr}(A)=\left(\tfrac{1}{2\pi\hbar}\right)^n\int a(z)d^{2n}z. \tag{9.23}$$

(ii) *If a is such that a and Fa both are in $H^s(\mathbb{R}^{2n}_z)$ with $s>2n$, then $\widehat{A}\overset{\text{Weyl}}{\longleftrightarrow}a$ is of trace class.*

Notice that since $\mathcal{S}(\mathbb{R}^{2n}_z)\subset H^s(\mathbb{R}^{2n}_z)$ for all $s\in\mathbb{R}$, Weyl operators with symbols $a\in\mathcal{S}(\mathbb{R}^{2n}_z)$ are automatically of trace class; since such a symbol a trivially is in $L^1(\mathbb{R}^{2n}_z)$ the trace of such an operator is indeed given by formula (9.23).

Let us give an independent proof of this property when the kernel of the operator is rapidly decreasing (this is the case for instance for Gaussian states):

Proposition 9.18. *Let $K_A\in\mathcal{S}(\mathbb{R}^{2n}_{x,y})$, $K_A(x,y)=\overline{K_A(y,x)}$. The self-adjoint Weyl operator $\widehat{A}\overset{\text{Weyl}}{\longleftrightarrow}a$ with kernel K_A is of trace class and*

$$\operatorname{Tr}(A)=\int K_A(x,x)d^nx \tag{9.24}$$

with $K(x,x)\geq 0$.

Proof. The inequality $K_A(x,x) \geq 0$ trivially follows from $K_A(x,y) = \overline{K_A(y,x)}$, and this condition is equivalent to $A = A^*$. Let $(\psi_j)_j$ be an orthonormal basis of $L^2(\mathbb{R}^n_x)$; then $(\psi_j \otimes \overline{\psi}_k)_{j,k}$ is an orthonormal basis of $L^2(\mathbb{R}^{2n}_z)$. Expanding K_A in a Fourier series in that basis we have

$$K_A(x,y) = \sum_{j,k} c_{j,k}\psi_j(x)\overline{\psi}_k(y)$$

with

$$c_{j,k} = (K_A, \psi_j \otimes \overline{\psi}_k)_{L^2(\mathbb{R}^{2n}_{x,y})}.$$

It follows that

$$\int K_A(x,x)d^nx = \sum_{j,k} c_{j,k} \int \psi_j(x)\overline{\psi}_k(x)d^nx = \sum_j c_{jj}.$$

On the other hand, by definition of the c_{jk},

$$(A\psi_j, \psi_j)_{L^2(\mathbb{R}^n_x)} = \int K_A(x,y)\psi_j(y)\overline{\psi_j(x)}d^nx = c_{jj},$$

hence

$$\operatorname{Tr}(A) = \sum_j (A\psi_j, \psi_j)_{L^2(\mathbb{R}^n_x)} = \int K_A(x,x)d^nx$$

as claimed. □

Self-adjoint trace-class operators are, as we have seen, Hilbert–Schmidt operators with L^2 kernels; they are thus also Weyl operators with L^2 symbols (Theorem 9.21). The first part of the following result which expresses the trace in terms of the symbol is a consequence of Proposition 9.18.

Proposition 9.19. *Let $\widehat{A} \overset{\text{Weyl}}{\longleftrightarrow} a$ and $\widehat{B} \overset{\text{Weyl}}{\longleftrightarrow} b$ be self-adjoint Weyl operators of trace class.*

(i) *We have*

$$\operatorname{Tr}\widehat{A} = \left(\tfrac{1}{2\pi\hbar}\right)^n \int a(z)d^{2n}z = a_\sigma(0) \tag{9.25}$$

(a_σ the twisted symbol of $\widehat{A}$);

(ii) *The trace of the compose $\widehat{A}\widehat{B}$ is given by the formula*

$$\operatorname{Tr}(\widehat{A}\widehat{B}) = \left(\tfrac{1}{2\pi\hbar}\right)^n \int a(z)b(z)d^{2n}z. \tag{9.26}$$

Proof. (i) The second equality (9.25) is obvious since $a_\sigma = \mathcal{F}_\sigma a$. Writing $\widehat{A}\psi$ in pseudo-differential form

$$\widehat{A}\psi(x) = \left(\tfrac{1}{2\pi\hbar}\right)^n \iint e^{\frac{i}{\hbar}\langle p, x-y\rangle} a(\tfrac{1}{2}(x+y), p)\psi(y)d^nyd^np,$$

the kernel $K_{\widehat{A}}$ of $\widehat{A}$ is

$$K_{\widehat{A}}(x,y) = \left(\tfrac{1}{2\pi\hbar}\right)^n \int e^{\frac{i}{\hbar}\langle p,x-y\rangle} a(\tfrac{1}{2}(x+y),p)d^n p$$

and hence, by (9.24),

$$\operatorname{Tr}\widehat{A} = \left(\tfrac{1}{2\pi\hbar}\right)^n \int a(x,p)d^n p d^n x$$

which is the first equality (9.25); the second follows in view of the definition of the symplectic Fourier transform $\mathcal{F}_\sigma a = a_\sigma$.

(ii) We have

$$\operatorname{Tr}(\widehat{A}\widehat{B}) = \left(\tfrac{1}{2\pi\hbar}\right)^n \int c(z)d^{2n}z$$

where $c(z)$ is the symbol of $\widehat{C} = \widehat{A}\widehat{B}$; in view of formula (6.43) (Subsection 6.3.2 of Chapter 6) we have

$$c(z) = \left(\tfrac{1}{4\pi\hbar}\right)^{2n} \iint e^{\frac{i}{2\hbar}\sigma(z',z'')} a(z+\tfrac{1}{2}z')b(z-\tfrac{1}{2}z'')d^{2n}z'd^{2n}z''.$$

Performing the change of variables $u = z + \frac{1}{2}z'$, $v = z - \frac{1}{2}z''$ we have $d^{2n}z'd^{2n}z'' = 4^{2n}d^{2n}ud^{2n}v$ and hence

$$c(z) = \left(\tfrac{1}{\pi\hbar}\right)^{2n} \iint e^{\frac{i}{2\hbar}\sigma(u-z,v-z)} a(u)b(v)d^{2n}ud^{2n}v.$$

Integrating $c(z)$ yields

$$\int c(z)d^{2n}z = \left(\tfrac{1}{\pi\hbar}\right)^{2n} \iiint e^{\frac{2i}{\hbar}\sigma(u-z,v-z)} a(u)b(v)d^{2n}ud^{2n}vd^{2n}z,$$

that is, expanding $\sigma(u-z, v-z)$,

$$\int c(z)d^{2n}z = \left(\tfrac{1}{\pi\hbar}\right)^{2n} \iint \left(\int e^{\frac{2i}{\hbar}\sigma(z,u-v)}d^{2n}z\right) e^{\frac{2i}{\hbar}\sigma(u,v)} a(u)b(v)d^{2n}ud^{2n}v.$$

Now

$$\int e^{\frac{2i}{\hbar}\sigma(z,u-v)}d^{2n}z = \int e^{\frac{2i}{\hbar}\langle Jz,u-v\rangle}d^{2n}z = (\pi\hbar)^{2n}$$

and hence

$$\begin{aligned}\int c(z)d^{2n}z &= \iint \delta(u-v)e^{\frac{2i}{\hbar}\sigma(u,v)} a(u)b(v)d^{2n}ud^{2n}v \\ &= \iint \delta(u-v)a(u)b(v)d^{2n}ud^{2n}v;\end{aligned}$$

formula (9.26) follows. □

9.2.3 Integral Hilbert–Schmidt operators

Let us now prove that the integral operators (9.20), defined by

$$A\psi(x) = \int K_A(x,y)\psi(y)d^n y$$

with $K_A \in L^2(\mathbb{R}^{2n}_{x,y})$ are Hilbert–Schmidt operators, and conversely:

Theorem 9.20. *Let A be a bounded operator on $L^2(\mathbb{R}^n_x)$.*

(i) *If A is a Hilbert–Schmidt operator, then it has kernel $K_A \in L^2(\mathbb{R}^{2n}_{x,y})$;*

(ii) *Conversely, any integral operator A with kernel $K_A \in L^2(\mathbb{R}^{2n}_{x,y})$ is a Hilbert–Schmidt operator.*

Proof. (i) In view of Schwarz's kernel theorem there exists $K_A \in \mathcal{S}'(\mathbb{R}^{2n}_{x,y})$ such that

$$A\psi(x) = \int K_A(x,y)\psi(y)d^n y$$

(the integral be interpreted in the sense of distributions). Let us show that

$$(K_A, \Psi)_{L^2(\mathbb{R}^{2n}_{x,y})} < \infty \quad \textit{for all} \quad \Psi \in L^2(\mathbb{R}^{2n}_{x,y});$$

the claim will follow. Let $(\psi_j)_j$ be an orthonormal basis of $L^2(\mathbb{R}^n_x)$; then $(\psi_j \otimes \overline{\psi}_k)_{j,k}$ is an orthonormal basis of $L^2(\mathbb{R}^{2n}_z)$. Writing

$$\Psi = \sum_{j,k} \lambda_{jk}\psi_j \otimes \overline{\psi}_k$$

we have

$$\begin{aligned}(K_A, \Psi)_{L^2(\mathbb{R}^{2n}_{x,y})} &= \sum_{j,k} \bar{\lambda}_{jk}(K_A, \psi_j \otimes \overline{\psi}_k)_{L^2(\mathbb{R}^{2n}_{x,y})} \\ &= \sum_{j,k} \bar{\lambda}_{jk} \iint K_A(x,y)\psi_k(y)\overline{\psi}_j(x)d^n y d^n x \\ &= \sum_{j,k} \bar{\lambda}_{jk}(A\psi_k, \psi_j)_{L^2(\mathbb{R}^n_x)}.\end{aligned}$$

Using Cauchy–Schwarz's inequality and recalling that $||\psi_j||_{L^2(\mathbb{R}^n_x)} = 1$ we get

$$\begin{aligned}|(K_A, \Psi)_{L^2(\mathbb{R}^{2n}_{x,y})}|^2 &\leq \sum_{j,k} |\lambda_{jk}|^2 ||A\psi_k||_{L^2(\mathbb{R}^n_x)} \\ &= \sum_k \left(\sum_j |\lambda_{jk}|^2\right) ||A\psi_k||^2_{L^2(\mathbb{R}^n_x)},\end{aligned}$$

that is

$$|(K_A, \Psi)_{L^2(\mathbb{R}^{2n}_{x,y})}|^2 \leq ||\Psi||^2_{L^2(\mathbb{R}^{2n}_{x,y})} \sum_k ||A\psi_k||^2_{L^2(\mathbb{R}^n_x)}$$

noting that

$$\sum_j |\lambda_{jk}|^2 \leq \sum_{j,k} |\lambda_{jk}|^2 = ||\Psi||^2_{L^2(\mathbb{R}^{2n}_{x,y})}.$$

Since A is a Hilbert–Schmidt operator we have $\sum_k ||A\psi_k||^2_{L^2(\mathbb{R}^n_x)} < \infty$ and hence $(K_A, \Psi)_{L^2(\mathbb{R}^{2n}_{x,y})} < \infty$ which we set out to prove.

(ii) Let (ψ_j) be an orthonormal basis of $L^2(\mathbb{R}^n_x)$; we have to show that

$$\sum_j ||A\psi_j||^2_{L^2(\mathbb{R}^n_x)} = \sum_j |(A\psi_j, A\psi_j)_{L^2(\mathbb{R}^n_x)}|^2 < \infty,$$

that is, equivalently,

$$\sum_j |(A^*A\psi_j, \psi_j)_{L^2(\mathbb{R}^n_x)}|^2 < \infty.$$

The operator $B = A^*A$ has kernel K_B in $L^2(\mathbb{R}^{2n}_{x,y})$ (Proposition 9.15 and Exercise 9.16) hence it suffices to show that

$$\sum_j |(B\psi_j, \psi_j)_{L^2(\mathbb{R}^n_x)}|^2 < \infty$$

for every integral operator with $K_B \in L^2(\mathbb{R}^{2n}_{x,y})$. We have

$$\begin{aligned}
\sum_j |(B\psi_j, \psi_j)_{L^2(\mathbb{R}^n_x)}|^2 &= \sum_j \left| \int \left(\int K_B(x,y)\psi_j(y)d^ny \right) \overline{\psi_j(x)} d^nx \right|^2 \\
&= \sum_j \left| \iint K_B(x,y)\psi_j(y)d^ny\overline{\psi_j(x)}d^nx \right|^2 \\
&= \sum_j \left| (K_B, \psi_j \otimes \overline{\psi_j})_{L^2(\mathbb{R}^{2n}_{x,y})} \right|^2 .
\end{aligned}$$

The sequence $(\psi_j \otimes \overline{\psi_j})_j$ is orthonormal in $L^2(\mathbb{R}^{2n}_{x,y})$, hence, in view of Bessel's inequality:

$$\sum_j |(B\psi_j, \psi_j)_{L^2(\mathbb{R}^n_x)}|^2 \leq ||K_B||_{L^2(\mathbb{R}^{2n}_{x,y})} < \infty$$

and we are done. □

We are next going to show that the space of integral Hilbert–Schmidt operators on $L^2(\mathbb{R}^n_x)$ is precisely the set of Weyl operators with symbol $a \in L^2(\mathbb{R}^{2n}_z)$ (Stein [158], Ch. XII, §4; Wong [181, 182]). This is indeed an important result, since it will allow us a precise description of the density matrices of quantum states in Section 9.3 using the methods of Weyl calculus.

Theorem 9.21. *The Weyl correspondence* $a \overset{\text{Weyl}}{\longleftrightarrow} \widehat{A}$ *induces an isomorphism between* $L^2(\mathbb{R}_z^{2n})$ *and the space of Hilbert–Schmidt operators on* $L^2(\mathbb{R}_x^n)$*:*

(i) *If* $a \in L^2(\mathbb{R}_z^{2n})$*, then* $\widehat{A}$ *is a Hilbert–Schmidt operator on* $L^2(\mathbb{R}_x^n)$.

(ii) *If conversely the kernel of a Weyl operator* $\widehat{A} \overset{\text{Weyl}}{\longleftrightarrow} a$ *is in* $L^2(\mathbb{R}_{x,y}^{2n})$*, then* $a \in L^2(\mathbb{R}_z^{2n})$.

(iii) *Let* A *be an arbitrary integral Hilbert–Schmidt operator on* $L^2(\mathbb{R}_x^n)$*. There exists* $a \in L^2(\mathbb{R}_z^{2n})$ *such that* $A = \widehat{A} \overset{\text{Weyl}}{\longleftrightarrow} a$ *and we have*

$$||a||^2_{L^2(\mathbb{R}_z^{2n})} = (2\pi\hbar)^n ||K_A||^2_{L^2(\mathbb{R}_{x,y}^{2n})}. \tag{9.27}$$

Proof. The statement (iii) follows from (i) and (ii) in view of Theorem 9.20.

(i) and (ii). We have seen (formula (6.26) in the proof of Theorem 6.12, Chapter 6) that the kernel $K_{\widehat{A}}$ of the Weyl operator $\widehat{A}$ and its symbol a are related by

$$K_{\widehat{A}}(x,y) = \left(\tfrac{1}{2\pi\hbar}\right)^n \int e^{\frac{i}{\hbar}\langle p, x-y\rangle} a(\tfrac{1}{2}(x+y), p) d^n p$$

and also that (formula 6.25, *ibid.*)

$$a(z) = \int e^{-\frac{i}{\hbar}\langle p,y\rangle} K_A(x + \tfrac{1}{2}y, x - \tfrac{1}{2}y) d^n y.$$

Assume first that $K_A \in \mathcal{S}(\mathbb{R}_z^{2n})$. Then the function

$$(x,y) \longmapsto K_A(x + \tfrac{1}{2}y, x - \tfrac{1}{2}y)$$

is also in $\mathcal{S}(\mathbb{R}_z^{2n})$ and we have

$$|a(z)|^2 = \iint e^{-\frac{i}{\hbar}\langle p, y-y'\rangle} K_A(x + \tfrac{1}{2}y, x - \tfrac{1}{2}y)\overline{K_A(x + \tfrac{1}{2}y, x - \tfrac{1}{2}y)} d^n y d^n y',$$

hence, integrating successively in p and x:

$$\begin{aligned} ||a||^2_{L^2(\mathbb{R}_z^{2n})} &= (2\pi\hbar)^n \iiint \delta(y-y')|K_A(x + \tfrac{1}{2}y, x - \tfrac{1}{2}y)|^2 d^n y d^n y' d^n x \\ &= (2\pi\hbar)^n \iiint |K_A(x + \tfrac{1}{2}y, x - \tfrac{1}{2}y)|^2 d^n y d^n x. \end{aligned}$$

Setting $u = x + \frac{1}{2}y$ and $v = x + \frac{1}{2}y$ we obtain (9.27) hence the result for $K_A \in \mathcal{S}(\mathbb{R}_z^{2n})$. Since $\mathcal{S}(\mathbb{R}_z^{2n})$ is dense in $L^2(\mathbb{R}_z^{2n})$ this equality holds for all a in $L^2(\mathbb{R}_z^{2n})$ or K_A in $L^2(\mathbb{R}_{x,y}^{2n})$; properties (i) and (ii) follow. □

9.3 The Density Operator of a Quantum State

A density operator is a device containing all the information needed to calculate the probabilities of the results of measurements one might perform on one part of a quantum system, where it is assumed that no information is accessible on the remaining part of the system. Admittedly, this explanation is somewhat sibylline, but we will explain in detail what it really is about. While the literature devoted to the density operator is immense, reading Fano's foundational paper [40] is a must.

9.3.1 Pure and mixed quantum states

Let us begin by giving a strict mathematical definition of a density operator in terms of the concepts introduced in the previous sections.

Definition 9.22. A "density operator" on a separable Hilbert space $\mathcal{H}$ is an operator $\widehat{\rho}:\mathcal{H}\longrightarrow\mathcal{H}$ having the following properties:

(i) $\widehat{\rho}$ is self-adjoint and semi-definite positive: $\widehat{\rho}=\widehat{\rho}^*$, $\widehat{\rho}\geq 0$;
(ii) $\widehat{\rho}$ is of trace class and $\operatorname{Tr}(\widehat{\rho})=1$.

It follows from Proposition 9.13 which says that a self-adjoint trace-class operator that is a density operator is in particular a Hilbert–Schmidt operator (and is hence compact).

In quantum mechanics the Hilbert space $\mathcal{H}$ is usually realized as a space of square-integrable functions, typically $L^2(\mathbb{R}^n_x)$ (or a closed subspace of $L^2(\mathbb{R}^{2n}_z)$, see Chapter 10).

Here is a first example of a density operator. Let us assume that we are in presence of a well-defined quantum state, represented by an element $\psi\neq 0$ of $\mathcal{H}$. Such a state is called a *pure state* in quantum mechanics. It is no restriction to assume that ψ is normalized, that is $||\psi||_{\mathcal{H}}=1$, so that the mathematical expectation of $\widehat{A}$ is

$$\langle\widehat{A}\rangle_\psi=(\widehat{A}\psi,\psi)_{\mathcal{H}}. \tag{9.28}$$

Consider now the projection operator

$$\widehat{\rho}_\psi:\mathcal{H}\longrightarrow\{\alpha\psi:\alpha\in\mathbb{C}\} \tag{9.29}$$

of $\mathcal{H}$ on the "ray" generated by ψ. For each $\phi\in\mathcal{H}$ we have

$$\widehat{\rho}_\psi\phi=\alpha\psi\quad,\quad\alpha=(\phi,\psi)_{\mathcal{H}}. \tag{9.30}$$

We will call $\widehat{\rho}_\psi$ the *pure density operator* associated with ψ; it is a trace-class operator with trace equal to 1.

Exercise 9.23. Check this statement in detail.

Observe that when $\mathcal{H} = L^2(\mathbb{R}^n_x)$, formula (9.30) can be written

$$\widehat{\rho}_\psi\phi(x) = \int \psi(x)\overline{\psi(y)}\phi(y)d^n y,$$

hence the kernel of $\widehat{\rho}_\psi$ is just the tensor product

$$K_{\widehat{\rho}_\psi} = \psi \otimes \overline{\psi}. \tag{9.31}$$

We are going to see that the pure density operator $\widehat{\rho}_\psi$ is in this case a Weyl operator. We have in fact already encountered its symbol ρ_ψ in Chapter 6: it is the Wigner transform of the function ψ! Let us prove this essential property:

Proposition 9.24. *Let $\widehat{\rho}_\psi$ be the density operator associated to a pure state ψ by* (9.30).

(i) *The Weyl symbol ρ_ψ of $\widehat{\rho}_\psi$ is the Wigner transform $W\psi$ of ψ and thus*

$$\widehat{\rho}_\psi\phi(x) = \left(\tfrac{1}{2\pi\hbar}\right)^n \iint e^{\frac{i}{\hbar}\langle p, x-y\rangle} W\psi(\tfrac{1}{2}(x+y), p)\phi(y)d^n y d^n p, \tag{9.32}$$

that is

$$\widehat{\rho}_\psi = \left(\tfrac{1}{2\pi\hbar}\right)^n \int W_\sigma\psi(z)\widehat{T}(z)d^{2n}z \tag{9.33}$$

where $W_\sigma\psi = \mathcal{F}_\sigma W\psi$ is the symplectic Fourier transform of $W\psi$.

(ii) *Let $\widehat{A} \overset{\text{Weyl}}{\longleftrightarrow} a$. If $||\psi||_{L^2(\mathbb{R}^n_x)} = 1$, then*

$$\langle\widehat{A}\rangle_\psi = \operatorname{Tr}(\widehat{\rho}_\psi\widehat{A}) \tag{9.34}$$

($\langle\widehat{A}\rangle_\psi = (\widehat{A}\psi, \psi)_{L^2(\mathbb{R}^n_x)}$ the mathematical expectation of $\widehat{A}$ in the state ψ).

Proof. (i) We have

$$\widehat{\rho}_\psi\phi(x) = (\phi, \psi)_{L^2(\mathbb{R}^n_x)}\psi(x) = \int \phi(y)\psi(x)\overline{\psi(y)}d^n y,$$

hence the kernel of $\widehat{\rho}_\psi$ is $\psi\otimes\overline{\psi}$. In view of formula (6.25) (Theorem 6.12, Subsection 6.2.2 of Chapter 6) the Weyl symbol ρ_ψ of $\widehat{\rho}_\psi$ is given by

$$\rho_\psi(z) = \left(\tfrac{1}{2\pi\hbar}\right)^n \int e^{-\frac{i}{\hbar}\langle p, y\rangle}\psi(x + \tfrac{1}{2}y)\overline{\psi(x + \tfrac{1}{2}y)}d^n y,$$

that is $\rho_\psi(z) = W\psi(z)$ as claimed.

(ii) In view of formula (6.71) in Proposition 6.45 (Subsection 6.4.2 of Chapter 6) we have

$$\langle\widehat{A}\rangle_\psi = \int a(z)W\psi(z)d^{2n}z.$$

Formula (9.34) follows from (i) using the expression (9.26) in Proposition 9.19 giving the trace of the compose of two Weyl operators. □

Exercise 9.25. Check directly part (i) of Proposition 9.24 assuming that $\widehat{A}$ has a discrete spectrum $(\lambda_j)_j$ and corresponding orthonormal eigenvector basis $(\psi_j)_{j\in\mathbb{N}}$.

Exercise 9.26. Look after formula (9.34) in any classical book on quantum mechanics and analyze in detail the way it is derived.

We have so far been assuming that the quantum system under consideration was in a well-known state characterized by a function ψ. Unfortunately things are not always that simple in the quantum world. Suppose for instance that we have the choice between a finite or infinite number of states, described by functions $\psi_1, \psi_2, \ldots$, each ψ_j having a probability α_j to be the "true" description. We can then form a weighted "mixture" of the ψ_j by forming the convex sum

$$\psi = \sum_{j=1}^{\infty} \alpha_j \psi_j \ , \ \sum_{j=1}^{\infty} \alpha_j = 1 \ , \ \alpha_j \geq 0. \tag{9.35}$$

We will say that ψ is a *mixed state*.

Definition 9.27. The density operator of the mixed state (9.35) is the self-adjoint operator

$$\widehat{\rho} = \sum_{j=1}^{\infty} \alpha_j \widehat{\rho}_{\psi_j} \tag{9.36}$$

where the real numbers α_j satisfy the conditions (9.35) above.

It is clear that $\widehat{\rho}$ is a density operator in the sense of Definition 9.22: since trace-class operators form a vector space, $\widehat{\rho}$ is indeed of trace class and its trace is 1 since $\operatorname{Tr}(\widehat{\rho}_j) = 1$ and the α_j sum to 1. That $\widehat{\rho} = \widehat{\rho}^*$ is obvious, and the positivity of $\widehat{\rho}$ follows from the fact that $\alpha_j \geq 0$ for each j. We will see below (Corollary 9.29) that any density operator on $L^2(\mathbb{R}^n_x)$ is actually of the type (9.36).

The importance of the distinction between pure and mixed states can be seen from the following argument: in quantum mechanics one wants to calculate the mathematical expectations (or "averages") of observables (position, momentum, the energy, to name a few). Assume that we want to study a system *(S)*; in practice *(S)* is always a part of a larger system $(\tilde{S})$ (for instance the Universe...). The Hilbert space of the states is thus $\tilde{\mathcal{H}} = \mathcal{H} \otimes \mathcal{H}'$ where $\mathcal{H}$ is the Hilbert space of *(S)* and $\mathcal{H}'$ corresponds to degrees of freedom external to *(S)*. Assume that the total system $(\tilde{S})$ is in some pure state $\tilde{\psi} \in \tilde{\mathcal{H}}$ and let $A : \mathcal{H} \longrightarrow \mathcal{H}$ be an observable of its subsystem *(S)*. Define the mathematical expectation of A by

$$\langle A \rangle_{\tilde{\psi}} = ((A \otimes I')\tilde{\psi}, \tilde{\psi})_{\tilde{\mathcal{H}}}$$

where I' is the identity operator in $\mathcal{H}'$. Unless $\tilde{\psi} = \psi \otimes \psi'$ there exists no $\psi \in \mathcal{H}$ such that $\langle A \rangle_{\tilde{\psi}} = \langle A \rangle_{\psi}$ and we can thus not calculate the expectation of the observable A unless we incorporate in one way or another some statistical information about the total system.

Here is a very important result that describes *all* density matrices in a Hilbert space $\mathcal{H}$.

Theorem 9.28. *Let $\widehat{\rho}$ be an operator on a Hilbert space $\mathcal{H}$.*

(i) *$\widehat{\rho}$ is a density operator if and only if there exists a (finite or infinite) sequence (α_j) of positive numbers and finite-dimensional pairwise orthogonal subspaces $\mathcal{H}_j$ of $\mathcal{H}$ such that*

$$\widehat{\rho} = \sum_j \alpha_j \widehat{\rho}_{\mathcal{H}_j} \quad \text{and} \quad \sum_j \alpha_j \dim \mathcal{H}_j = 1 \tag{9.37}$$

where $\widehat{\rho}_{\mathcal{H}_j}$ is the orthogonal projection $\mathcal{H} \longrightarrow \mathcal{H}_j$;

(ii) *We have*

$$0 \leq \operatorname{Tr}(\widehat{\rho}^2) \leq \operatorname{Tr}(\widehat{\rho}) \leq 1 \tag{9.38}$$

for every density operator $\widehat{\rho}$; and $\operatorname{Tr}(\widehat{\rho}^2) = 1$ if and only if $\widehat{\rho}$ is a pure-state density operator.

Proof. The statement (i) is just Proposition 9.8, taking into account that the orthogonal projections $\widehat{\rho}_j$ are self-adjoint.

(ii) Since the spaces $\mathcal{H}_j$ are pairwise orthogonal we have $\widehat{\rho}_{\mathcal{H}_j}\widehat{\rho}_{\mathcal{H}_k} = 0$ if $j \neq k$ and hence $\widehat{\rho}^2 = \sum_j \alpha_j^2 \widehat{\rho}_{\mathcal{H}_j}$. The condition $\sum_j \alpha_j \dim \mathcal{H}_j = 1$ implies that we must have $\alpha_j \leq 1$ for each j so that

$$\operatorname{Tr} \widehat{\rho}^2 = \sum_j \alpha_j^2 \dim \mathcal{H}_j \leq \sum_j \alpha_j \dim \mathcal{H}_j = 1.$$

We have seen that if $\widehat{\rho}$ is a pure-state density operator then it is a projection of rank 1, hence $\widehat{\rho}^2 = \widehat{\rho}$ has trace 1. Suppose conversely that $\operatorname{Tr}(\widehat{\rho}^2) = 1$, that is

$$\sum_j \alpha_j^2 \dim \mathcal{H}_j = \sum_j \alpha_j \dim \mathcal{H}_j = 1.$$

Since $\dim \mathcal{H}_j > 0$ for every j, this equality is only possible if the numbers α_j are either zero or 1; since the case $\alpha_j = 0$ is excluded it follows that the sum $\sum_j \alpha_j \dim \mathcal{H}_j = 1$ reduces to one single term, say $\alpha_{j_0} \dim \mathcal{H}_{j_0} = 1$ so that $\widehat{\rho} = \alpha_{j_0} P_{j_0}$ and $\widehat{\rho}^2 = \alpha_{j_0}^2 \widehat{\rho}_{\mathcal{H}_{j_0}}$. The equality $\operatorname{Tr}(\widehat{\rho}) = \operatorname{Tr}(\widehat{\rho}^2) = 1$ can hold if and only if $\alpha_{j_0} = 1$, hence $\dim \mathcal{H}_{j_0} = 1$ and $\widehat{\rho}$ is a projection of rank 1, and hence a pure-state density operator. □

Specializing to the case where $\mathcal{H} = L^2(\mathbb{R}^n_x)$ we get:

Corollary 9.29. *An operator $\widehat{\rho} : L^2(\mathbb{R}^n_x) \longrightarrow L^2(\mathbb{R}^n_x)$ is a density operator if and only if there exists a family $(\psi_j)_{j\in\mathbb{J}}$ in $L^2(\mathbb{R}^n_x)$, a sequence $(\lambda_j)_{j\in\mathbb{J}}$ of numbers $\lambda_j \geq 0$ with $\sum_{j\in\mathbb{J}} \lambda j = 1$ such that the Weyl symbol ρ of $\widehat{\rho}$ is given by*

$$\rho = \sum_{j\in\mathbb{J}} \lambda_j W\psi_j. \tag{9.39}$$

Proof. Assume that the Weyl symbol of $\widehat{\rho}$ is given by (9.39); in view of Proposition 9.24 and the discussion preceding it we have

$$\widehat{\rho} = \sum_{j \in \mathbb{J}} \alpha_j \widehat{\rho}_j$$

where $\widehat{\rho}_j$ is the orthogonal projection on the ray $\{\alpha\psi_j : \alpha \in \mathbb{C}\}$. It follows that $\widehat{\rho}$ is a density operator. If conversely $\widehat{\rho}$ is a density operator on $L^2(\mathbb{R}^n_x)$, then there exist pairwise orthogonal finite dimensional subspaces $\mathcal{H}_1, \mathcal{H}_2, \ldots$ of $L^2(\mathbb{R}^n_x)$ such that

$$\widehat{\rho} = \sum_j \alpha_j \widehat{\rho}_{\mathcal{H}_j} \ , \ \sum_j m_j \alpha_j = 1$$

with $\widehat{\rho}_{\mathcal{H}_j}$ the orthogonal projection on $\mathcal{H}_j$ and $m_j = \dim \mathcal{H}_j$. Choose now an orthonormal basis $\psi_1, \ldots, \psi_{m_1}$ of $\mathcal{H}_1$, an orthonormal basis $\psi_{m_1+1}, \ldots, \psi_{m_1+m_2+1}$ of $\mathcal{H}_2$, and so on. The Weyl symbol of $\widehat{\rho}$ is

$$\rho = \alpha_1 \sum_{j=1}^{m_1} W\psi_j + \alpha_2 \sum_{j=m_1}^{m_1+m_2+1} W\psi_j + \cdots$$

which is (9.39), setting $\lambda_j = m_j \alpha_j$. □

Part (iii) of Theorem 9.28 above motivates the definition of the notion of *purity* of a quantum state which plays an important role in quantum optics and the theory of squeezed states:

Definition 9.30. Let $\widehat{\rho}$ be a density operator on $\mathcal{H}$; the number $\mu(\widehat{\rho}) = \operatorname{Tr}(\widehat{\rho}^2)$ is called the "purity of the quantum state" $\widehat{\rho}$ represents.

The purity of a state satisfies the double inequality $0 \leq \mu(\widehat{\rho}) \leq 1$. We will come back to it in the next subsection when we study Gaussian states.

Exercise 9.31. Show that a quantum state is pure if and only if its purity equals 1.

Recalling (formula (9.31) that the operator kernel of the density operator of a pure state ψ is just the tensor product $\psi \otimes \overline{\psi}$, we have more generally:

Corollary 9.32. *Let the density operator $\widehat{\rho}$ be given by formula (9.37) and let $(\psi_{jk})_{j,k}$ be a double-indexed family of orthonormal vectors in $L^2(\mathbb{R}^n_x)$ such that the subfamily $(\psi_{jk})_k$ is a basis of $\mathcal{H}_j$ for each j.*

(i) *The kernel $K_{\widehat{\rho}}$ of $\widehat{\rho}$ is given by*

$$K_{\widehat{\rho}}(x,y) = \sum_{j,k} \lambda_j \psi_{jk}(x) \otimes \overline{\psi_{jk}}(y) \quad \textit{with } \textstyle\sum_j \lambda_j = 1. \tag{9.40}$$

(ii) *The Weyl symbol of $\widehat{\rho}$ is given by*

$$a(z) = \sum_{j,k} \lambda_j W\psi_{jk}(z) \tag{9.41}$$

($W\psi_{jk}$ the Wigner transform of ψ_{jk}).

Proof. (i) We have

$$\widehat{\rho}\psi = \sum_j \lambda_j \widehat{\rho}\psi = \sum_{j,k} \lambda_j (\psi, \psi_{jk})_{\mathcal{H}} \psi_{jk}$$

that is, by definition of the scalar product:

$$\widehat{\rho}\psi = \sum_j \lambda_j \widehat{\rho}\psi = \sum_{j,k} \lambda_j \int \psi_{jk}(x)\psi(y)\overline{\psi_{jk}}(y)d^n y$$

which is (9.40).

(ii) Formula (9.41) for the symbol immediately follows from (9.40) in view of Proposition 9.24(i). □

9.3.2 Time-evolution of the density operator

Assume now that we are studying a partial known quantum system which is changing with time; we assume that it is described by a mixed state (9.35):

$$\psi = \sum_{j=1}^{\infty} \alpha_j \psi_j \ , \ \sum_{j=1}^{\infty} \alpha_j = 1 \ , \ \alpha_j \geq 0 \tag{9.42}$$

and that the time evolution of each ψ_j is governed by a collective Schrödinger equation

$$i\hbar \frac{\partial \psi}{\partial t} = \widehat{H}\psi.$$

We denote by $(\widehat{U}_t)$ the evolution operator determined by that equation: $\psi(x,t) = \widehat{U}_t \psi(x,0)$; this operator is unitary and satisfies

$$i\hbar \frac{d}{dt}\widehat{U}_t = \widehat{H}\widehat{U}_t \ , \ i\hbar \frac{d}{dt}\widehat{U}_t^* = -\widehat{U}_t^* \widehat{H}. \tag{9.43}$$

Proposition 9.33. *Let $\widehat{\rho}$ be the density operator at time $t = 0$ of the mixed state* (9.42).

(i) *At time t this density operator is given by the formula*

$$\widehat{\rho}_t = \widehat{U}_t \widehat{\rho} \widehat{U}_t^*;$$

(ii) *The mapping $t \longmapsto \widehat{\rho}_t$ satisfies the operator equation*

$$i\hbar \frac{d}{dt}\widehat{\rho}_t = [\widehat{H}, \widehat{\rho}_t]. \tag{9.44}$$

Proof. To prove (i) it suffices to note that for all $\psi, \phi \in L^2(\mathbb{R}^n_x)$

$$\begin{aligned}\widehat{\rho}_{\widehat{U}_t\psi}\phi &= (\phi, \widehat{U}_t\psi)_{L^2(\mathbb{R}^n_x)}\widehat{U}_t\psi \\ &= (\widehat{U}_t^*\phi, \psi)_{L^2(\mathbb{R}^n_x)}\widehat{U}_t\psi \\ &= \widehat{U}_t\widehat{\rho}_\psi\widehat{U}_t^*\phi.\end{aligned}$$

Differentiating $\widehat{\rho}_t = \widehat{U}_t\widehat{\rho}\widehat{U}_t^*$ in t now yields, using formulae (9.43),

$$\begin{aligned}i\hbar\frac{d}{dt}\widehat{\rho}_t &= i\hbar\frac{d}{dt}(\widehat{U}_t\widehat{\rho}\widehat{U}_t^*) \\ &= \widehat{H}\widehat{U}_t\widehat{\rho}\widehat{U}_t^* + \widehat{U}_t\widehat{\rho}(-\widehat{U}_t^*\widehat{H}) \\ &= \widehat{H}\widehat{\rho}_t - \widehat{\rho}_t\widehat{H}\end{aligned}$$

which proves (ii). □

Remark 9.34. The evolution equation (9.44) should not be confounded with the Heisenberg equation of elementary quantum mechanics[2]; it is actually the quantized form of Liouville's equation

$$\frac{\partial \rho_{\text{cl}}}{\partial t} = \{H, \rho_{\text{cl}}\} \tag{9.45}$$

from classical statistical mechanics.

We emphasize that there are many theoretical (and practical) advantages in using the density operator formalism instead of wave-functions (see Messiah [123], Chapter VIII, for a lucid discussion of the notion of density operator from a leading physicist's viewpoint). First, it is always possible to represent a quantum state by a density operator, whether this state is pure or mixed; one can *a posteriori* determine the purity of this state by calculating the trace of the square of the density operator of this state. Secondly, the wave-function of a pure state is only defined up to a complex factor (or a phase, when the wave-function is normalized), while the corresponding density operator is uniquely defined. For mixed states the situation is even more clear-cut: there is a great arbitrariness in a statistical mixture

$$\psi = \sum_{j=1}^{\infty} \alpha_j\psi_j.$$

Unless the ψ_j are all linearly independent (which is generally not the case!), the decomposition above is never unique in opposition to the associated density operator which is independent of the way the mixture is written.

[2] All the quantities involved are written in the "Schrödinger representation".

9.3.3 Gaussian mixed states

Gaussian functions, which we already have encountered several times, are objects of particular interest. This is not only because they are relatively easy to calculate with: in addition to being intensively studied in quantum optics they will allow us to link the quantum blobs of Chapter 8 to the Wigner transform of a mixed state.

Recall (Corollary 8.48, Section 8.5) that the Wigner transform of a normalized pure state

$$\psi_0(x) = \left(\tfrac{\det X}{(\pi\hbar)^n}\right)^{1/4} \exp\left[-\tfrac{1}{2\hbar}\langle (X+iY)x, x\rangle\right]$$

is given by the formula

$$W\psi_0(z) = \left(\tfrac{1}{\pi\hbar}\right)^n e^{-\frac{1}{\hbar}\langle Gz,z\rangle}$$

where G is the symmetric positive symplectic matrix

$$G = \begin{bmatrix} X+YX^{-1}Y & YX^{-1} \\ X^{-1}Y & X^{-1} \end{bmatrix}.$$

The function $W\psi_0$ satisfies

$$\int W\psi_0(z)d^{2n}z = ||\psi_0||^2_{L^2(\mathbb{R}^n_x)} = 1$$

(formula (6.68) of Chapter 6, Section 6.4). If we want $W\psi_0$ to be the Weyl symbol of a density operator we have to renormalize ψ_0 and replace it by $(2\pi\hbar)^n\psi_0$. More generally, consider a phase-space Gaussian

$$W_F(z) = 2^n\sqrt{\det F}e^{-\frac{1}{\hbar}\langle Fz,z\rangle} \tag{9.46}$$

where $F = F^T > 0$. Setting $F = \frac{\hbar}{2}\Sigma^{-1}$ shows that W_F is the probability distribution

$$W_F(z) = (2\pi\hbar)^n\rho_\Sigma(z) \tag{9.47}$$

where

$$\rho_\Sigma(z) = \left(\tfrac{1}{2\pi}\right)^n (\det\Sigma)^{-1/2}e^{-\frac{1}{2}\langle\Sigma^{-1}z,z\rangle};$$

in particular

$$\left(\tfrac{1}{2\pi\hbar}\right)^n \int W_F(z)d^{2n}z = 1$$

so that W_F is a priori a good candidate for being the symbol of a density operator. Define in fact the Weyl operator $\widehat{\rho}_\Sigma$,

$$\widehat{\rho}_\Sigma\psi(x) = \left(\tfrac{1}{2\pi\hbar}\right)^n \iint e^{\frac{i}{\hbar}\langle p,x-y\rangle} W_F(z)(\tfrac{1}{2}(x+y),p)\psi(y)d^ny d^np, \tag{9.48}$$

that is[3]

$$\widehat{\rho}_\Sigma\psi(x) = \iint e^{\frac{i}{\hbar}\langle p,x-y\rangle}\rho_\Sigma(z)(\tfrac{1}{2}(x+y),p)\psi(y)d^ny d^np; \tag{9.49}$$

this operator will be the density operator of some quantum state if the three following conditions are fulfilled:

- $\widehat{\rho}_\Sigma$ is self-adjoint: $\widehat{\rho}_\Sigma = \widehat{\rho}_\Sigma^*$;
- $\widehat{\rho}_\Sigma$ is of trace class and $\operatorname{Tr}\widehat{\rho}_\Sigma = 1$;
- $\widehat{\rho}_\Sigma$ is non-negative: $\widehat{\rho}_\Sigma \geq 0$.

It is clear that $\widehat{\rho}_\Sigma = \widehat{\rho}_\Sigma^*$ since ρ_Σ is real; the operator $\widehat{\rho}_\Sigma = \widehat{\rho}_\Sigma^*$ is of trace-class because of Proposition 9.18, which tells us that in addition

$$\operatorname{Tr}\widehat{\rho}_\Sigma = \left(\tfrac{1}{2\pi\hbar}\right)^n \int W_F(z)d^{2n}z = 1.$$

So far, so good. How about the positive semi-definiteness property $\widehat{\rho}_\Sigma \geq 0$? As we already have noticed before, in Subsection 8.5.3 of the last chapter, the fact that the symbol of a Weyl operator is positive does not imply that the operator itself is positive. The obstruction, as we hinted at, is related to the uncertainty principle of quantum mechanics: if the Gaussian ρ_Σ is too sharply peaked, then $\widehat{\rho}_\Sigma$ will *not* be the density of a quantum state. It however follows from Corollary 8.55 of Proposition 8.54 that if we "average" ρ_Σ over a quantum blob, then we will always obtain the Wigner transform of a mixed state. In fact, the following precise result holds (*cf.* de Gosson [72]):

Theorem 9.35. *Let $\mathbb{Q} = S(B^{2n}(\sqrt{\hbar}))$, $S \in \operatorname{Sp}(n)$, be a quantum blob and $W_\mathbb{Q}$ the associated normalized Gaussian:*

$$W_\mathbb{Q}(z) = \left(\tfrac{1}{\pi\hbar}\right)^n e^{-\frac{1}{\hbar}\langle (SS^T)^{-1}z,z\rangle}.$$

(i) *The convolution product $W_{\Sigma,\mathbb{Q}} = \rho_\Sigma * W_\mathbb{Q}$ is the Wigner transform of a mixed state $\widehat{\rho}_{\Sigma,\mathbb{Q}}$;*

(ii) *We have*

$$W_{\Sigma,\mathbb{Q}}(z) = W\widehat{\rho}(z) = \left(\tfrac{1}{\pi\hbar}\right)^n \sqrt{\det K} e^{-\frac{1}{\hbar}\langle Kz,z\rangle} \tag{9.50}$$

where the symplectic spectrum of K is the image of that of $F = \frac{\hbar}{2}\Sigma^{-1}$ by the mapping $\lambda \longmapsto \lambda/(1+\lambda)$.

Proof. (i) The Weyl operator $\widehat{\rho}_{\Sigma,\mathbb{Q}}$ with symbol $W_{\Sigma,\mathbb{Q}}(z)$ is self-adjoint, and assuming that (9.50) holds we have

$$\operatorname{Tr}\widehat{\rho}_{\Sigma,\mathbb{Q}} = \int W_{\Sigma,\mathbb{Q}}(z)d^{2n}z = 1$$

[3]Beware that ρ_Σ is not the symbol of $\widehat{\rho}_\Sigma$!

so there remains to check the positivity of $\widehat{\rho}_{\Sigma,\mathbb{Q}}$; the latter follows from Corollary 8.55 of Proposition 8.54.

(ii) Setting $G = (SS^T)^{-1}$ we have

$$W_{\Sigma,\mathbb{Q}}(z) = \left(\tfrac{1}{\pi\hbar}\right)^{2n} \sqrt{\det F} \int e^{-\frac{1}{\hbar}\langle F(z-z'),(z-z')\rangle} e^{-\frac{1}{\hbar}\langle Gz',z'\rangle} d^{2n}z';$$

and the change of variables $z'' = S^{-1}z'$ yields

$$W_{\Sigma,\mathbb{Q}}(Sz) = \left(\tfrac{1}{\pi\hbar}\right)^{2n} \sqrt{\det F} \int e^{-\frac{1}{\hbar}\langle S^TFS(z-z''),(z-z'')\rangle} e^{-\frac{1}{\hbar}|z''|^2} d^{2n}z''.$$

Replacing if necessary S by another symplectic matrix we may assume, in view of Williamson's theorem, that

$$S^TFS = D = \begin{bmatrix} \Lambda_\sigma & 0 \\ 0 & \Lambda_\sigma \end{bmatrix}$$

where $\Lambda_\sigma = \operatorname{diag}[\lambda_1, \dots, \lambda_n]$, $(\lambda_1, \dots, \lambda_n)$ the symplectic spectrum of F, and hence

$$W_{\Sigma,\mathbb{Q}}(Sz) = \left(\tfrac{1}{\pi\hbar}\right)^{2n} \sqrt{\det D} \int e^{-\frac{1}{\hbar}\langle D(z-z''),z-z''\rangle} e^{-\frac{1}{\hbar}|z''|^2} d^{2n}z''.$$

Using the elementary convolution formula

$$\int_{-\infty}^{\infty} e^{-a(u-t)^2} e^{-bt^2} dt = \sqrt{\frac{\pi}{a+b}} \exp\left(-\frac{ab}{a+b}u^2\right)$$

valid for all $a, b > 0$ together with the fact that the matrix D is diagonal, we find

$$W_{\Sigma,\mathbb{Q}}(Sz) = \left(\frac{1}{\pi\hbar}\right)^n (\det D(I+D)^{-1})^{1/2} \exp\left[-\frac{1}{\hbar}\langle D(I+D)^{-1}z, z\rangle\right],$$

that is

$$W_{\Sigma,\mathbb{Q}}(z) = \exp\left[-\frac{1}{\hbar} z^T (S^{-1})^T D(I+D)^{-1} S^{-1} z\right].$$

Setting

$$K = (S^{-1})^T D(I+D)^{-1} S^{-1}$$

we obtain formula (9.50). There remains to show that the moduli of the eigenvalues of JK are not superior to 1. Since $S \in \operatorname{Sp}(n)$ we have $J(S^{-1})^T = SJ$ and hence

$$JK = SJD(I+D)^{-1}S^{-1}$$

has the same eigenvalues as $JD(I+D)^{-1}$. Now,

$$JD(I+D)^{-1} = \begin{bmatrix} 0 & \Lambda_\sigma(I+\Lambda_\sigma)^{-1} \\ -\Lambda_\sigma(I+\Lambda_\sigma)^{-1} & 0 \end{bmatrix}$$

so that the eigenvalues $\mu_{\sigma,j}$ of JF are the numbers

$$\mu_{\sigma,j} = \pm i \frac{\lambda_{\sigma,j}}{1+\lambda_{\sigma,j}} \tag{9.51}$$

which are such that $|\mu_{\sigma,j}| \leq 1$ which was to be proven. □

The result above leads us to ask the following question: which conditions (if any) should one impose on the covariance matrix Σ in the formula

$$\rho_\Sigma(z) = \left(\tfrac{1}{2\pi}\right)^n (\det \Sigma)^{-1/2} e^{-\frac{1}{2}\langle \Sigma^{-1}z,z\rangle}$$

in order that it be the Wigner transform of some quantum state? Writing (9.47) in the form

$$W(z) = \left(\tfrac{1}{\pi\hbar}\right)^n \sqrt{\det F} e^{-\frac{1}{\hbar}\langle Fz,z\rangle}$$

a partial answer is already given by Proposition 8.47 (Section 8.5, Chapter 8) which implies that if F is symplectic, then W is the Wigner transform of the pure Gaussian state

$$\psi(x) = \left(\tfrac{\det X}{(\pi\hbar)^n}\right)^{1/4} \exp\left[-\tfrac{1}{2\hbar}\langle (X+iY)x,x\rangle\right]$$

where the symmetric matrices X, Y are determined from F. Let us discuss the case of more general covariance matrices Σ.

The following result, which is no more than a consequence of Theorem 9.35 above, shows that in a sense the Gaussians which are Wigner transforms of pure Gaussian states are the lower limit for what is acceptable as a candidate for the Wigner transform of a quantum state. Recall that the covariance matrix Σ is said to be quantum mechanically admissible if its symplectic spectrum is such that

$$\operatorname{Spec}_\sigma(\Sigma) \geq \left(\tfrac{1}{2}\hbar, \tfrac{1}{2}\hbar, \ldots, \tfrac{1}{2}\hbar\right) \tag{9.52}$$

(Proposition 8.27 in Subsection 8.3.4 of Chapter 8).

Corollary 9.36. *Let ρ_Σ be the phase-space Gaussian*

$$\rho_\Sigma(z) = \left(\tfrac{1}{2\pi}\right)^n (\det \Sigma)^{-1/2} e^{-\frac{1}{2}\langle \Sigma^{-1}z,z\rangle}.$$

(i) *The operator $\widehat{\rho}_\Sigma$ associated to ρ_Σ by formula (9.49) is the Wigner transform of the density operator of a mixed quantum state, if and only if the covariance matrix Σ is quantum mechanically admissible;*

(ii) *When this is the case, the purity of that state is*

$$\mu(\widehat{\rho}_\Sigma) = \left(\frac{\hbar}{2}\right)^n (\det \Sigma)^{1/2}.$$

Proof. In view of Williamson's symplectic diagonalization theorem there exists $S \in \mathrm{Sp}(n)$ such that

$$S^T \Sigma S = \begin{bmatrix} \Lambda_\sigma & 0 \\ 0 & \Lambda_\sigma \end{bmatrix} \quad , \quad \Lambda_\sigma = \mathrm{diag}[\lambda_1, \dots, \lambda_n]$$

where $\lambda_1, \dots, \lambda_n$ is the symplectic spectrum of Σ. In view of (9.52) we have $\lambda_1 \geq \cdots \geq \lambda_n \geq \frac{1}{2}\hbar$ so that $S^T \Sigma S \geq S^T \Sigma_0 S$ where $\Sigma_0 = \frac{1}{2}\hbar I$ (I the $2n \times 2$ identity matrix), and hence also $\Sigma \geq \Sigma_0$. This allows us to write

$$\rho_\Sigma = \rho_{\Sigma - \Sigma_0} * \rho_{\Sigma_0}$$

and (i) now follows from Theorem 9.35. Let us calculate the purity of $\widehat{\rho}_\Sigma$. In view of formula (9.26) in Proposition 9.19 we have

$$\mu(\widehat{\rho}_\Sigma) = \mathrm{Tr}(\widehat{\rho}_\Sigma^2) = \left(\tfrac{1}{2\pi\hbar}\right)^n \int W_F^2(z) d^{2n}z,$$

that is, since $W_F = (2\pi\hbar)^n \rho_\Sigma$,

$$\mu(\widehat{\rho}_\Sigma) = (2\pi\hbar)^n \int \rho_\Sigma^2(z) d^{2n}z.$$

We have

$$\int \rho_\Sigma^2(z) d^{2n}z = \left(\tfrac{1}{2\pi}\right)^{2n} (\det \Sigma)^{-1} \int e^{-\langle \Sigma^{-1} z, z\rangle} d^{2n}z.$$

Recalling that if M is a positive definite symmetric matrix, then

$$\int e^{-\langle Mz,z\rangle} d^{2n}z = \pi^n (\det M)^{-1/2}$$

we have

$$\int \rho_\Sigma^2(z) d^{2n}z = \left(\tfrac{1}{4\pi}\right)^n (\det \Sigma)^{1/2}$$

and hence

$$\mu(\widehat{\rho}_\Sigma) = \left(\frac{\hbar}{2}\right)^n (\det \Sigma)^{1/2}.$$

Notice that since $\det \Sigma \geq (\frac{1}{2}\hbar)^n$ for an admissible Σ in view of (9.52) we indeed have $\mu(\widehat{\rho}_\Sigma) \leq 1$. □

Chapter 10

A Phase Space Weyl Calculus

Most traditional presentations of quantum mechanics (in its wave-mechanical form) begin with the Schrödinger equation

$$i\hbar\frac{\partial\psi}{\partial t} = H(x, -i\hbar\partial_x)\psi \tag{10.1}$$

where $H(x, -i\hbar\partial_x)$ is an operator acting on functions of the position variables and associated in some convenient (often *ad hoc*) way to the Hamilton function H; one possible choice is the Weyl operator $\widehat{H} \overset{\text{Weyl}}{\longleftrightarrow} H$. Our own presentation in the previous chapters has complied with this tradition, but the reader will probably remember that we have at several times mentioned that it is possible to write a phase space Schrödinger equation. In this chapter we will actually show that simple considerations involving no more than the invariance properties of the Poincaré–Cartan form $pdx - Hdt$ lead to such an equation.

In the first section we begin by discussing the relevance and logical need for Schrödinger equations in phase space, and thereafter define a fundamental transform which will allow us to give equivalent formulations of quantum mechanics in phase space. We thereafter introduce the phase-space Schrödinger equations, with a special emphasis on its most symmetric variant, which we write formally – and hopefully suggestively – in the form

$$i\hbar\frac{\partial\Psi}{\partial t} = H(\tfrac{1}{2}x + i\hbar\partial_p, \tfrac{1}{2}p - i\hbar\partial_x)\Psi \tag{10.2}$$

when $n = 1$; the function Ψ depends on $z = (x,p)$ and t. This equation has a certain aesthetic appeal, because it reinstates in quantum mechanics the symmetry of classical mechanics in its Hamiltonian formulation, where the time-evolution is governed by the equations

$$\dot{x} = \partial_p H(x,p,t) \quad , \quad \dot{p} = -\partial_x H(x,p,t);$$

in both cases the variables x and p are placed, up to a change of sign, on the same footing.

10.1 Introduction and Discussion

In Chapter 6 the Heisenberg–Weyl operators were used as the cornerstones for the definition of Weyl pseudo-differential calculus: to a conveniently chosen symbol a we associated an operator $\widehat{A}$ acting on the Schwartz space $\mathcal{S}(\mathbb{R}^n_x)$ by

$$\widehat{A}\psi(x) = \left(\tfrac{1}{2\pi\hbar}\right)^n \int a_\sigma(z_0)\widehat{T}(z_0)\psi(x)d^{2n}z_0$$

(Definition 6.9 in Chapter 6, Section 6.2); the twisted symbol a_σ is the symplectic Fourier transform $\mathcal{F}_\sigma a$ of a; equivalently, using the Grossmann–Royer operators:

$$\widehat{A}\psi(x) = \left(\tfrac{1}{\pi\hbar}\right)^n \int a(z_0)\tilde{T}(z_0)\psi(x)d^{2n}z_0.$$

The considerations above inexorably lead us to associate to $\widehat{A}$ an operator $\widehat{A}_{\mathrm{ph}}$ acting on $\mathcal{S}(\mathbb{R}^{2n}_z)$ by the formula

$$\widehat{A}_{\mathrm{ph}}\,\Psi(z) = \left(\tfrac{1}{2\pi\hbar}\right)^n \int a_\sigma(z_0)\widehat{T}_{\mathrm{ph}}(z_0)\Psi(z)d^{2n}z_0$$

where $\widehat{T}_{\mathrm{ph}}$ and $\tilde{T}_{\mathrm{ph}}$ are variants of the Heisenberg–Weyl operator acting on phase-space functions defined by

$$\widehat{T}_{\mathrm{ph}}(z_0)\Psi(z) = e^{-\frac{i}{2\hbar}\sigma(z,z_0)}\Psi(z-z_0)$$

where σ is the standard symplectic form. Equivalently, redefining the Grossmann–Royer operators in a suitable way one has the formula

$$\widehat{A}_{\mathrm{ph}}\,\psi(x) = \left(\tfrac{1}{\pi\hbar}\right)^n \int a(z_0)\tilde{T}_{\mathrm{ph}}(z_0)\psi(x)d^{2n}z_0.$$

We are going to study these operators in detail in Section 10.3; let us first motivate our constructions.

10.1.1 Discussion of Schrödinger's argument

Schrödinger's equation (10.1) governing the time-evolution of matter waves can be rigorously justified for quadratic or linear potentials if one uses the theory of the metaplectic group, but it *cannot* be mathematically justified for arbitrary Hamiltonian functions. The fact that Schrödinger's equation governs the evolution of matter waves for arbitrary Hamiltonians is a *physical postulate* which can only be made *plausible* by using formal analogies: this is what is done in all texts on quantum mechanics (see for instance Dirac [31], *p.* 108–111). So how did Schrödinger arrive at his equation? He actually elaborated on Hamilton's optical–mechanical analogy, and took several mathematically illegitimate steps applying the Hamilton–Jacobi theory (see Jammer [97] or Moore [125] for a thorough discussion of Schrödinger's argument). Here is the idea, somewhat oversimplified. Schrödinger's starting point was that a "matter wave" consists – as all waves do – of an amplitude and of a phase. Consider now a system with Hamiltonian H represented by a phase-space point $z(0)$ at time $t = 0$. That system evolves with time and is represented by a point $z(t)$ at time t. Schrödinger postulated that the phase Φ of the associated matter wave satisfied by Hamilton–Jacobi's equation

$$\frac{\partial W}{\partial t} + H(x, \partial_x \Phi, t) = 0. \tag{10.3}$$

That postulate allowed him, using some non-rigorous mathematical manipulations, to arrive at his equation (10.1). Let us now make the following observation: the phase Φ of the matter wave is just the phase of the Lagrangian manifold $p = \partial_x \Phi(x, t)$; in view of our discussion in Chapter 5 the *change* of that phase is the line integral

$$\Delta\Phi = \int_\Gamma pdx - Hdt \tag{10.4}$$

calculated along the arc of extended phase-space trajectory Γ joining $(z(0), 0)$ to $(z(t), t)$. Let us assume that the Hamiltonian flow (f_t^H) is free for small $t \neq 0$ (this is the case if H is of the type "kinetic energy + potential", for instance); then the initial and final *position* vectors $x(0)$ and $x(t)$ uniquely determine $p(0)$ and $p(t)$, so that $\Delta\Phi$ can be identified with the generating function $W(x_0, x, t)$. This property is intimately related to the fact that the Poincaré–Cartan form

$$\alpha_H = pdx - Hdt$$

is a *relative integral invariant*; equivalently the exterior derivative

$$d\alpha_H = dp \wedge dx - dH \wedge dt$$

vanishes on trajectory tubes. As already discussed in Subsection 5.2.1 of Chapter 5, this property is actually shared by any of the differential forms

$$\alpha_H^{(\lambda)} = \lambda pdx + (\lambda - 1)xdp - Hdt$$

where λ is an arbitrary real number. It turns out that a particularly neat choice is $\lambda = 1/2$ because it leads to the form

$$\alpha_H^{(1/2)} = \frac{1}{2}\sigma(z, dz) - Hdt \tag{10.5}$$

where the variables x and p play (up to the sign) similar roles. Let us investigate the quantum-mechanical consequences of this choice. In Section 5.5 of Chapter 5 we considered the Schrödinger equation

$$i\hbar\frac{\partial\psi}{\partial t} = \widehat{\sigma}(\hat{z}, z_0)\psi \quad , \psi(\cdot, 0) = \psi_0 \tag{10.6}$$

where the operator

$$\widehat{\sigma}(\hat{z}, z_0) = \langle -i\hbar\partial_x, x_0\rangle - \langle p_0, x\rangle$$

was obtained using Weyl correspondence from the translation Hamiltonian $H_{z_0}(z) = \sigma(z, z_0)$. We checked that the solution of (10.6) was given by

$$\psi(x,t) = \widehat{T}(tz_0)\psi_0(x)$$

where $\widehat{T}(z_0)$ is the usual Heisenberg–Weyl operator; explicitly

$$\psi(x,t) = e^{\frac{i}{\hbar}\Phi(x,t)}\psi_0(x - tx_0)$$

where the phase

$$\Phi(x,t) = t\langle p_0, x\rangle - \tfrac{1}{2}t^2\langle p_0, x_0\rangle$$

is obtained by integrating the Poincaré–Cartan form $\alpha_{H_{z_0}}$ along the extended phase-space trajectory Γ leading from $(z - tz_0, 0)$ to (z, t). Suppose now that we replace in the procedure above $\alpha_{H_{z_0}}$ by $\alpha^{(1/2)}_{H_{z_0}}$; mimicking the proof of Proposition 5.46 we find that the integral of $\alpha^{(1/2)}_{H_{z_0}}$ is

$$\int_\Gamma \alpha^{(1/2)}_{H_{z_0}} = -\frac{t}{2}\sigma(z, z_0).$$

In analogy with the definition of the Heisenberg–Weyl operators we can thus define an operator $\widehat{T}_{\mathrm{ph}}(z_0)$ acting on functions of $z = (x,p)$ by the formula

$$\widehat{T}_{\mathrm{ph}}(z_0)\Psi_0(z) = e^{-\frac{i}{2\hbar}\sigma(z,z_0)}T(z_0)\Psi_0(z).$$

A straightforward calculation shows that the function

$$\Psi(z,t) = \widehat{T}_{\mathrm{ph}}(tz_0)\Psi_0(z) = e^{-\frac{i}{2\hbar}\sigma(z,z_0)}\Psi_0(z - tz_0)$$

satisfies the first-order partial differential equation

$$i\hbar\frac{\partial\Psi}{\partial t} = \frac{1}{2}\sigma(z, z_0)\Psi - i\hbar\left\langle z_0, \partial_z\right\rangle\Psi;$$

rearranging the terms in the right-hand side of this equation shows that it can be rewritten formally as

$$i\hbar\frac{\partial\Psi}{\partial t} = \sigma\left(\tfrac{1}{2}x + i\hbar\partial_p, \tfrac{1}{2}p - i\hbar\partial_x; x_0, p_0\right)\Psi. \tag{10.7}$$

The replacement of α_H by its symmetrized variant $\alpha_{H_{z_0}}^{(1/2)}$ has thus led us to the replacement of the usual quantum rules $x_j \longrightarrow x_j$ by the *phase-space quantum rules*

$$x_j \longrightarrow \widehat{X}_j = \frac{1}{2}x_j + i\hbar\frac{\partial}{\partial p_j} \quad , \quad p_j \longrightarrow \widehat{P}_j = \frac{1}{2}p_j - i\hbar\frac{\partial}{\partial x_j}. \tag{10.8}$$

Notice that the operators $\widehat{X}_j, \widehat{P}_j$ obey the usual canonical commutation relations

$$[\widehat{X}_j, \widehat{P}_k] = -i\hbar\delta_{jk}$$

so that our discussion of the Heisenberg algebra and group in the first section of Chapter 6 suggests that these quantization rules could be consistent with the existence of an irreducible representation of the Heisenberg group in phase space. We will prove that this is indeed the case.

The equation (10.7) corresponds, as we have seen, to the choice $\lambda = 1/2$ for the integral invariant $\alpha_H^{(\lambda)}$. Of course any other choice is *a priori* equally good from a purely mathematical point of view. For instance the choice $\lambda = 0$ would lead to another phase-space Schrödinger equation, namely

$$i\hbar\frac{\partial\Psi}{\partial t} = \sigma(x + i\hbar\partial_p, -i\hbar\partial_x; x_0, p_0)\Psi \tag{10.9}$$

and thus corresponds to the quantum rules

$$x_j \longrightarrow x_j + i\hbar\frac{\partial}{\partial p_j} \quad , \quad p_j \longrightarrow -i\hbar\frac{\partial}{\partial x_j}.$$

These have been considered and exploited by Torres-Vega and Frederick [162, 163] (see our discussion of their phase-space Schrödinger equation in [73]). It turns out, as we will see, that the "symmetrized" choice (10.8) is the most convenient for the determination of which solutions correspond to true quantum states (pure or mixed), and directly related to the "quantum blobs" introduced and studied in Chapter 8.

10.1.2 The Heisenberg group revisited

Recall from Chapter 6 that the Heisenberg group $\mathbf{H}_n$ is the extended phase space $\mathbb{R}_{z,t}^{2n+1}$ equipped with the composition law

$$(z,t)\maltese(z',t) = (z + z', t + t' + \tfrac{1}{2}\sigma(z,z'));$$

the operators

$$\widehat{T}(z_0, t_0) = e^{\frac{i}{\hbar}(\langle p_0, x\rangle - \frac{1}{2}\langle p_0, x_0\rangle + t_0)}T(z_0)$$

then formed a unitary representation of $\mathbf{H}_n$ in the Hilbert space $L^2(\mathbb{R}_x^n)$.

Let us *define* operators $\widehat{T}_{\text{ph}}(tz_0)$ by letting them act on $L^2(\mathbb{R}_z^{2n})$ by the formula

$$\widehat{T}_{\text{ph}}(tz_0)\Psi_0(z) = e^{\frac{i}{\hbar}\varphi'(z,t)}T(tz_0)\Psi_0(z)$$

(the subscript "ph" stands for "phase space"), where the phase $\varphi'(z,t)$ is obtained by integrating, not the Poincaré–Cartan form

$$\alpha_{H_{z_0}} = pdx - \sigma(z, z_0)$$

corresponding to the translation Hamiltonian, but rather its symmetrized variant

$$\alpha_{H_{z_0}}^{(1/2)} = \tfrac{1}{2}\sigma(z, dz) - H_{z_0}dt.$$

This yields after a trivial calculation

$$\varphi'(z,t) = -\tfrac{1}{2}H_{z_0}(z)t = -\tfrac{1}{2}\sigma(z, z_0)t \tag{10.10}$$

so that

$$\widehat{T}_{\text{ph}}(tz_0)\Psi_0(z) = e^{-\frac{it}{2\hbar}\sigma(z,z_0)t}\Psi_0(z - tz_0). \tag{10.11}$$

Let us denote by $\mathcal{U}(L^2(\mathbb{R}_z^{2n}))$ the set of unitary operators acting on $L^2(\mathbb{R}_z^{2n})$.

Definition 10.1. The "symplectic phase-space representation" of the Heisenberg group $\mathbf{H}_n$ is the mapping $(z_0, t_0) \longmapsto \widehat{T}_{\text{ph}}(z_0, t_0)$ from $\mathbf{H}_n$ to $\mathcal{U}(L^2(\mathbb{R}_z^{2n}))$ defined by

$$\widehat{T}_{\text{ph}}(z_0, t_0)\Psi(z) = e^{\frac{i}{\hbar}t_0}\widehat{T}_{\text{ph}}(z_0)\Psi(z) \tag{10.12}$$

where the operator

$$\widehat{T}_{\text{ph}}(z_0) = e^{-\frac{i}{2\hbar}\sigma(z,z_0)t}T(z_0)$$

is obtained by (10.11) setting $t = 1$.

That the operators $\widehat{T}_{\text{ph}}(z_0, t_0)$ are unitary for each $(z_0, t_0) \in \mathbb{R}_z^{2n} \times \mathbb{R}_t$ is obvious since we have

$$|\widehat{T}_{\text{ph}}(z_0, t_0)\Psi(z)|^2 = |\Psi(z - z_0)|^2$$

and hence

$$\int |\widehat{T}_{\text{ph}}(z_0, t_0)\Psi(z)|^2 d^{2n}z = ||\Psi||^2_{L^2(\mathbb{R}_z^{2n})}\,.$$

A straightforward calculation moreover shows that

$$\widehat{T}_{\text{ph}}(z_0, t_0)\widehat{T}_{\text{ph}}(z_1, t_1) = \widehat{T}_{\text{ph}}(z_0 + z_1, t_0 + t_1 + \tfrac{1}{2}\sigma(z_0, z_1)) \tag{10.13}$$

so that $\widehat{T}_{\text{ph}}$ is indeed a representation of $\mathbf{H}_n$ on some subspace of $L^2(\mathbb{R}_z^{2n})$ (*cf.* formula (6.12 in Chapter 6, Subsection 6.1.2). We will describe this space in Subsection 10.2.2 but let us make the following remark: we could actually have decided to let the Heisenberg–Weyl operators themselves act on phase-space functions by the formula

$$\widehat{T}(z_0, t_0)\Psi(z) = e^{\frac{i}{\hbar}(\langle p_0, x\rangle - \frac{1}{2}\langle p_0, x_0\rangle)}\Psi(z - z_0).$$

The reason for our choice

$$\widehat{T}_{\mathrm{ph}}(z_0)\Psi(z) = e^{-\frac{i}{2\hbar}\sigma(z,z_0)t}\Psi(z - z_0)$$

is that it places, from the very beginning, the variables x and p on the same footing, and is thus consistent with our program, which is to arrive at a Schrödinger equation in phase space having the desired symmetries.

Observe that the operators $\widehat{T}_{\mathrm{ph}}$ satisfy the same commutation relation as the usual Weyl–Heisenberg operators:

$$\widehat{T}_{\mathrm{ph}}(z_1)\widehat{T}_{\mathrm{ph}}(z_0) = e^{-\frac{i}{\hbar}\sigma(z_0,z_1)}\widehat{T}_{\mathrm{ph}}(z_0)\widehat{T}_{\mathrm{ph}}(z_1) \tag{10.14}$$

and that we have

$$\widehat{T}_{\mathrm{ph}}(z_0)\widehat{T}_{\mathrm{ph}}(z_1) = e^{\frac{i}{2\hbar}\sigma(z_0,z_1)}\widehat{T}_{\mathrm{ph}}(z_0 + z_1). \tag{10.15}$$

We will see in Section 10.2 that the operators $\widehat{T}_{\mathrm{ph}}(z_0, t_0)$ lead to a new irreducible unitary representation of the Heisenberg group $\mathbf{H}_n$ on a closed subspace of $L^2(\mathbb{R}^{2n}_z)$ which is unitarily equivalent to the Schrödinger representation. Let us first briefly state and discuss the Stone–von Neumann theorem.

10.1.3 The Stone–von Neumann theorem

As we showed in Chapter 6 (subsection 6.1.2) the Heisenberg–Weyl operators lead, via the formula

$$\widehat{T}(z_0, t_0) = e^{\frac{i}{\hbar}t_0}\widehat{T}(z_0)$$

to an irreducible unitary representation of $\mathbf{H}_n$ in the Hilbert space $L^2(\mathbb{R}^n_x)$: the only closed subspaces of $L^2(\mathbb{R}^n_x)$ which are invariant under all operators $\widehat{T}(z_0, t_0)$ are $\{0\}$ or $L^2(\mathbb{R}^n_x)$ itself. The Stone–von Neumann theorem classifies the irreducible representations of the Heisenberg group:

Theorem 10.2. *Let $(T, \mathcal{H})$ be a unitary irreducible representation of the Heisenberg group $\mathbf{H}_n$ in some separable Hilbert space $\mathcal{H}$. Then there exists $\mu \in \mathbb{R}$ such that $T(0, t) = e^{i\mu t}$, and:*

(i) *If $\mu = 0$, then $\dim \mathcal{H} = 1$ and there exists $z_0 \in \mathbb{R}^{2n}_z$ such that $T(z, t) = e^{i\langle z_0, z\rangle}$;*

(ii) *If $\mu \neq 0$, then the representation $(T, \mathcal{H})$ is equivalent to the Schrödinger representation $(\widehat{T}, L^2(\mathbb{R}^n_x))$.*

We will not prove this result here, and refer to Wallach [175] (who reproduces with minor changes a proof by Barry Simon), or to Folland [42] (who reproduces

Stone's original proof); the reader will find an alternative proof in the first chapter of Lion and Vergne [111].

Stone and von Neumann's theorem is actually a consequence of a much more general theorem, due to Frobenius and Mackey, the theorem on *systems of imprimitivities*; see Jauch [99] (Ch. 12, §3) for a detailed discussion of that deep result and of its consequences for quantum mechanics (Mackey discusses in detail the genesis of the notion in [117]).

It is important to understand that, contrary to what is sometimes claimed in the literature on representation theory, Stone–von Neumann's theorem does *not* say that the Schrödinger representation is the only possible irreducible representation of $\mathbf{H}_n$. What it says is that if we succeed in one way or another in constructing a unitary and irreducible representation T' of $\mathbf{H}_n$ in some Hilbert space $\mathcal{H}$, then there will exist an isomorphism $U : L^2(\mathbb{R}^n_x) \longrightarrow \mathcal{H}$ such that

$$U \circ \widehat{T}(z_0, t_0) = \widehat{T}'(z_0, t_0) \circ U \quad \text{for all } (z_0, t_0) \in \mathbb{R}^{2n}_z.$$

We are going to show in the next section that we can in fact construct, using a particular form of "wave-packet transform", infinitely many unitary and irreducible representations of $\mathbf{H}_n$, each on a closed space of $L^2(\mathbb{R}^{2n}_z)$. This will automatically lead us to a Schrödinger equation in phase space.

10.2 The Wigner Wave-Packet Transform

We are going to show that the symplectic phase-space representation

$$\widehat{T}_{\text{ph}}(z_0)\Psi(z) = e^{-\frac{i}{2\hbar}\sigma(z,z_0)t}\Psi(z - z_0)$$

is unitarily equivalent to the Schrödinger representation, and hence irreducible on a closed subspace of $L^2(\mathbb{R}^{2n}_z)$. For this purpose we introduce a variant of the "wave-packet transform" studied by Nazaikiinskii *et al.* [128], Chapter 2, §2. We want to associate in a both reasonable and useful way to every $\psi \in L^2(\mathbb{R}^n_x)$ a function $\Psi \in L^2(\mathbb{R}^{2n}_z)$. That such a correspondence will not be bijective is intuitively clear since functions on phase space depend on twice as many variables as those on configuration space; we will therefore be confronted with problems of range. For instance, we will see that elements of $L^2(\mathbb{R}^{2n}_z)$ that are too sharply peaked and concentrated around a phase space point cannot correspond to any quantum state, pure or mixed.

10.2.1 Definition of U_ϕ

In what follows ϕ will be be a rapidly decreasing function normalized to unity:

$$\phi \in \mathcal{S}(\mathbb{R}^n_x) \quad , \quad ||\phi||^2_{L^2(\mathbb{R}^n_x)} = 1. \tag{10.16}$$

Recall (Chapter 6, Section 6.4) that the Wigner–Moyal transform of $\psi, \phi \in \mathcal{S}(\mathbb{R}^n_x)$ is defined by

$$W(\psi,\phi)(z) = \left(\tfrac{1}{2\pi\hbar}\right)^n \int e^{-\frac{i}{\hbar}\langle p,y\rangle}\psi(x+\tfrac{1}{2}y)\overline{\phi}(x-\tfrac{1}{2}y)d^ny.$$

Definition 10.3. The integral operator $U_\phi : L^2(\mathbb{R}^n_x) \longrightarrow L^2(\mathbb{R}^{2n}_z)$ defined by

$$U_\phi\psi(z) = \left(\tfrac{\pi\hbar}{2}\right)^{n/2} W(\psi,\phi)(\tfrac{1}{2}z) \tag{10.17}$$

where $W(\psi,\phi)$ is the Wigner–Moyal transform of the pair (ψ,ϕ) will be called the "Wigner wave-packet transform" associated with ϕ; equivalently

$$U_\phi\psi(z) = \left(\tfrac{1}{2\pi\hbar}\right)^{n/2} \left\langle \widetilde{T}(\tfrac{1}{2}z)\psi, \overline{\phi}\right\rangle \tag{10.18}$$

where $\widetilde{T}(z)$ is the Grossmann–Royer operator and $\langle\cdot,\cdot\rangle$ the distributional bracket.

That (10.17) are equivalent for $\phi \in \mathcal{S}(\mathbb{R}^n_x)$ follows from the fact that we have established in Chapter 6, Section 6.4 (formula (6.59) in Proposition 6.39) that

$$W(\psi,\phi)(z) = \left(\tfrac{1}{\pi\hbar}\right)^n (\widetilde{T}(z)\psi,\phi)_{L^2(\mathbb{R}^n_x)}.$$

Notice that formula (10.18) makes sense for $\psi \in \mathcal{S}'(\mathbb{R}^n_x)$, allowing the extension of the Wigner wave-packet transform to tempered distributions.

We will prove below that the range of U_ϕ is a closed subspace of $L^2(\mathbb{R}^{2n}_z)$ (and hence a Hilbert space); let us first give an explicit formula for $U_\phi\psi(z)$ in terms of integrals and exponentials. Since

$$W(\psi,\phi)(x,p) = \left(\tfrac{1}{2\pi\hbar}\right)^n \int_{\mathbb{R}^n} e^{-\frac{i}{\hbar}\langle p,y\rangle}\psi(x+\tfrac{1}{2}y)\overline{\phi}(x-\tfrac{1}{2}y)d^ny$$

we have

$$U_\phi\psi(z) = \left(\tfrac{1}{2\pi\hbar}\right)^{n/2} 2^{-n} \int e^{-\frac{i}{2\hbar}\langle p,y\rangle}\psi(\tfrac{1}{2}(x+y))\overline{\phi}(\tfrac{1}{2}(x-y))d^ny$$

that is, setting $x+\frac{1}{2}y = x'$,

$$U_\phi\psi(z) = \left(\tfrac{1}{2\pi\hbar}\right)^{n/2} e^{\frac{i}{2\hbar}\langle p,x\rangle} \int e^{-\frac{i}{\hbar}\langle p,x'\rangle}\psi(x')\overline{\phi}(x-x')d^nx'. \tag{10.19}$$

Remark 10.4. An interesting (but not at all mandatory!) choice is to take for ϕ the real Gaussian

$$\phi_\hbar(x) = \left(\tfrac{1}{\pi\hbar}\right)^{n/4} e^{-\frac{1}{2\hbar}|x|^2}. \tag{10.20}$$

The corresponding operator $U_{\phi_\hbar}$ is then (up to an exponential factor) the "coherent state representation" familiar to quantum physicists. In [128] Nazaikiinskii *et al.* define an alternative wave-packet transform U by the formula

$$U\psi(z) = \left(\tfrac{1}{2\pi\hbar}\right)^{n/2} \int e^{\frac{i}{\hbar}\langle p,x-x'\rangle}\phi_\hbar(x-x')\psi(x')d^nx'. \tag{10.21}$$

The operators U and $U_{\phi_\hbar}$ are obviously related by the simple formula

$$U_{\phi_\hbar} = e^{-\frac{i}{2\hbar}\langle p,x\rangle}U. \tag{10.22}$$

The Wigner wave-packet transform is "metaplectically covariant":

Proposition 10.5. *For every $\psi \in L^2(\mathbb{R}^n_x)$ and $\widehat{S} \in \mathrm{Mp}(n)$ we have the following "metaplectic covariance formula" for the wave-packet transform:*

$$U_\phi(\widehat{S}\psi)(z) = U_{\widehat{S}^{-1}\phi}\psi(S^{-1}z) \tag{10.23}$$

where $\pi^{\mathrm{Mp}}(\widehat{S}) = S$.

Proof. Recalling the formula

$$W(\widehat{S}\psi, \widehat{S}\phi)(z) = W(\psi,\phi)(S^{-1}z)$$

(Proposition 7.14, Chapter 7) we have, by definition of U_ϕ,

$$\begin{aligned} U_\phi(\widehat{S}\psi)(z) &= \left(\tfrac{\pi\hbar}{2}\right)^{n/2} W(\widehat{S}\psi,\phi)(\tfrac{1}{2}z) \\ &= \left(\tfrac{\pi\hbar}{2}\right)^{n/2} W(\psi,\widehat{S}^{-1}\phi)(\tfrac{1}{2}S^{-1}z) \\ &= U_{\widehat{S}^{-1}\phi}\psi(S^{-1}z) \end{aligned}$$

which was to be proven. □

The interest of the Wigner wave-packet transform U_ϕ comes from the fact that it is an isometry of $L^2(\mathbb{R}^n_x)$ onto a closed subspace $\mathcal{H}_\phi$ of $L^2(\mathbb{R}^{2n}_z)$ and that it takes the operators x and $-i\hbar\partial_x$ into the operators $x/2 + i\hbar\partial_p$ and $p/2 - i\hbar\partial_x$, respectively:

Theorem 10.6. *The Wigner wave-packet transform U_ϕ has the following properties:*

(i) *U_ϕ is an isometry: the Parseval formula*

$$(U_\phi\psi, U_\phi\psi')_{L^2(\mathbb{R}^{2n}_z)} = (\psi,\psi')_{L^2(\mathbb{R}^n_x)} \tag{10.24}$$

holds for all $\psi,\psi' \in \mathcal{S}(\mathbb{R}^n_x)$. In particular $U^*_\phi U_\phi = I$ on $L^2(\mathbb{R}^n_x)$.

(ii) *The range $\mathcal{H}_\phi$ of U_ϕ is closed in $L^2(\mathbb{R}^{2n}_z)$ (and is hence a Hilbert space), and the operator $P_\phi = U_\phi U^*_\phi$ is the orthogonal projection in $L^2(\mathbb{R}^{2n}_z)$ onto $\mathcal{H}_\phi$.*

(iii) *The intertwining relations*

$$\left(\frac{1}{2}x_j + i\hbar\frac{\partial}{\partial p_j}\right)U_\phi\psi = U_\phi(x_j\psi), \tag{10.25a}$$

$$\left(\frac{1}{2}p_j - i\hbar\frac{\partial}{\partial x_j}\right)U_\phi\psi = U_\phi(-i\hbar\frac{\partial}{\partial x_j}\psi) \tag{10.25b}$$

hold for $\psi \in \mathcal{S}(\mathbb{R}^n_x)$.

Proof. (i) Formula (10.24) is an immediate consequence of the property

$$(W(\psi,\phi),W(\psi',\phi'))_{L^2(\mathbb{R}^{2n}_z)} = \left(\tfrac{1}{2\pi\hbar}\right)^n (\psi,\psi')_{L^2(\mathbb{R}^n_x)}\overline{(\phi,\phi')}_{L^2(\mathbb{R}^n_x)} \tag{10.26}$$

of the Wigner–Moyal transform (Proposition 6.40, Section 6.4 of Chapter 6). In fact, taking $\phi = \phi'$ we have

$$\begin{aligned}(U_\phi\psi, U_\phi\psi')_{L^2(\mathbb{R}^{2n}_z)} &= \left(\tfrac{\pi\hbar}{2}\right)^n \int W(\psi,\phi)(\tfrac{1}{2}z)\overline{W(\psi',\phi)}(\tfrac{1}{2}z)d^{2n}z \\ &= (2\pi\hbar)^n (W(\psi,\phi),W(\psi',\phi))_{L^2(\mathbb{R}^{2n}_z)} \\ &= (\psi,\psi')_{L^2(\mathbb{R}^n_x)}(\phi,\phi)_{L^2(\mathbb{R}^n_x)}\end{aligned}$$

which is formula (10.24) since ϕ is normalized to unity.

To prove (ii) we note that, since $P^*_\phi = P_\phi$ and $U^*_\phi U_\phi = I$, we have

$$\begin{aligned}P^2_\phi &= (U_\phi U^*_\phi)(U_\phi U^*_\phi) \\ &= U^*_\phi U_\phi \\ &= P_\phi,\end{aligned}$$

hence P_ϕ is indeed an orthogonal projection. Let us show that its range is $\mathcal{H}_\phi$; the closedness of $\mathcal{H}_\phi$ will follow since the range of a projection in a Hilbert space always is closed. Since $U^*_\phi U_\phi = I$ on $L^2(\mathbb{R}^n_x)$ we have $U^*_\phi U_\phi \psi = \psi$ for every ψ in $L^2(\mathbb{R}^n_x)$ and hence the range of U^*_ϕ is $L^2(\mathbb{R}^n_x)$. It follows that the range of U_ϕ is that of $U_\phi U^*_\phi = P_\phi$ and we are done.

(iii) The verification of the formulae (10.25) is purely computational, using differentiations and partial integrations; it is therefore left to the reader. □

The intertwining formulae (10.25) show that the Wigner wave-packet transform replaces the usual quantization rules

$$x_j \longrightarrow x \quad , \quad p_j \longrightarrow -i\hbar\frac{\partial}{\partial x_j}$$

leading to the standard Schrödinger equation to the phase-space quantization rules

$$x_j \longrightarrow \frac{1}{2}x_j + i\hbar\frac{\partial}{\partial p_j} \quad , \quad x_j \longrightarrow \frac{1}{2}p_j - i\hbar\frac{\partial}{\partial x_j};$$

observe that these rules are independent of the choice of ϕ and are in this sense "intrinsic".

Exercise 10.7. Show that the adjoint U^*_ϕ of the Wigner wave-packet transform U_ϕ is given by the formula

$$U^*_\phi\Psi(x) = \left(\tfrac{1}{2\pi\hbar}\right)^{n/2} \iint e^{\frac{i}{2\hbar}\langle p, x-x''\rangle}\Psi(x+x'',p)\phi(x'')d^npd^nx''$$

or, equivalently,

$$U_\phi^*\Psi(x) = \left(\tfrac{1}{2\pi\hbar}\right)^{n/2} \iint e^{\frac{i}{2\hbar}\langle p, 2x-q\rangle}\Psi(q,p)\phi(q-x)d^npd^nq.$$

[Hint: evaluate $(U_\phi\psi, \Psi)_{L^2(\mathbb{R}_x^n)}$ using formula (10.19).]

10.2.2 The range of U_ϕ

One should be aware of the fact that the range $\mathcal{H}_\phi$ of U_ϕ is smaller than $L^2(\mathbb{R}_z^{2n})$. This is intuitively clear since functions in $L^2(\mathbb{R}_z^{2n})$ depend on twice as many variables as those in $L^2(\mathbb{R}_x^n)$ of which $\mathcal{H}_\phi$ is an isometric copy. Let us discuss this in some detail.

Theorem 10.8. *Let $\phi_\hbar$ be the Gaussian (10.20): $\phi_\hbar(x) = (\pi\hbar)^{-n/4}e^{-\frac{1}{2\hbar}|x|^2}$.*

(i) *The range of the Wigner wave-packet transform $U_{\phi_\hbar}$ consists of the functions $\Psi \in L^2(\mathbb{R}_z^{2n})$ for which the conditions*

$$\left(\frac{\partial}{\partial x_j} - i\frac{\partial}{\partial p_j}\right)\left[e^{\frac{1}{2\hbar}|z|^2}\Psi(z)\right] = 0 \tag{10.27}$$

hold for $1 \leq j \leq n$.

(ii) *For every ϕ the range of the Wigner wave-packet transform U_ϕ is isometric to $\mathcal{H}_{\phi_\hbar}$.*

Proof. (i) We have $U_{\phi_\hbar} = e^{-\frac{i}{2\hbar}\langle p,x\rangle}U$ where U is the operator (10.21). It is shown in Nazaikinskii *et al.* [128] (Chapter 2, §2) that the range of U consists of all $\Psi \in L^2(\mathbb{R}_z^{2n})$ such that

$$\left(\frac{\partial}{\partial x_j} - i\frac{\partial}{\partial p_j}\right)\left[e^{\frac{1}{2\hbar}|p|^2}\Psi(z)\right] = 0 \text{ for } 1 \leq j \leq n$$

(the proof of this property, based on the Weierstrass theorem, is rather long and technical and is therefore omitted here). That the range of $U_{\phi_\hbar}$ is characterized by (10.27) follows by an immediate calculation that is left to the reader.

(ii) If U_{ϕ_1} and U_{ϕ_2} are two Wigner wave-packet transforms corresponding to the choices ϕ_1, ϕ_2 then $U_{\phi_2}U_{\phi_1}^*$ is an isometry of $\mathcal{H}_{\phi_1}$ onto $\mathcal{H}_{\phi_2}$. □

The result above leads us to address the following related question: given $\Psi \in L^2(\mathbb{R}_z^{2n})$, can we find ϕ and ψ in $L^2(\mathbb{R}_x^n)$ such that $\Psi = U_\phi\psi$? We are going to see that the answer is in general *no*. Intuitively speaking the reason is the following: if Ψ is too "concentrated" in phase space, it cannot correspond via the inverse transform $U_\phi^{-1} = U_\phi^*$ to a solution of the standard Schrödinger equation, because the uncertainty principle would be violated. Let us make this precise when the function Ψ is a Gaussian. We first make the following obvious remark: in view of condition (10.27) every Gaussian

$$\Psi_0(z) = Ce^{-\frac{1}{2\hbar}|z|^2} \quad , \quad C \in \mathbb{C}$$

is in the range of $U_{\phi_\hbar}$, $\phi_\hbar$ the "coherent state" (10.20).

Theorem 10.9. *Let G be a real symmetric positive-definite $2n \times 2n$ matrix and denote by Ψ_G the Gaussian defined by*

$$\Psi_G(z) = e^{-\frac{1}{2\hbar}\langle Gz,z\rangle}. \tag{10.28}$$

(i) *There exist functions $\psi, \phi \in \mathcal{S}(\mathbb{R}^n_x)$ such that $U_\phi \psi = \Psi_G$ if and only if $G \in \mathrm{Sp}(n)$. When this is the case we have*

$$\phi = \alpha \widehat{S}^{-1}\phi_\hbar \quad , \quad \psi = 2^{n/2}\overline{\alpha}\,(\pi\hbar)^{n/4}\,\widehat{S}^{-1}\phi_\hbar$$

where $\phi_\hbar$ is the Gaussian (10.20), α an arbitrary complex number with $|\alpha| = 1$, and $\widehat{S} \in \mathrm{Mp}(n)$ has projection $S \in \mathrm{Sp}(n)$ such that $G = S^T DS$ is a Williamson diagonalization of G.

(ii) *Equivalently, $|\Psi_G|^2$ must be the Wigner transform $W\psi$ of a Gaussian state*

$$\psi(x) = c\,(\pi\hbar)^{n/2}\,(\det X)^{1/2} e^{-\frac{1}{2\hbar}\langle Mx,x\rangle} \tag{10.29}$$

with $|c| = 1$, $M = M^T$, $\operatorname{Re} M > 0$.

Proof. In view of the relation (10.17) between U_ϕ and the Wigner–Moyal transform, $U_\phi\psi = \Psi_G$ is equivalent to

$$W(\psi,\phi)(z) = \left(\tfrac{2}{\pi\hbar}\right)^{n/2} e^{-\frac{2}{\hbar}\langle Gz,z\rangle}.$$

By Williamson's symplectic diagonalization theorem (Theorem 8.11, Subsection 8.3.1) there exists $S \in \mathrm{Sp}(n)$ such that $G = S^T DS$ where D is the diagonal matrix

$$D = \begin{bmatrix} \Lambda & 0 \\ 0 & \Lambda \end{bmatrix} \quad , \quad \Lambda = \operatorname{diag}[\lambda_1, \dots, \lambda_n],$$

the numbers $\pm i\lambda_1, \dots, \lambda_n$ ($\lambda_j > 0$) being the eigenvalues of JG^{-1}; since $\langle Gz, z\rangle = \langle DSz, Sz\rangle$,

$$W(\psi,\phi)(S^{-1}z) = \left(\tfrac{2}{\pi\hbar}\right)^{n/2} e^{-\frac{2}{\hbar}\langle Dz,z\rangle}.$$

In view of the metaplectic covariance property of the Wigner–Moyal transform (Proposition 7.14 of Chapter 7, Subsection 7.1.3) we have

$$W(\psi,\phi)(S^{-1}z) = W(\widehat{S}\psi, \widehat{S}\phi)(z)$$

where $\widehat{S} \in \mathrm{Mp}(n)$ has projection S. By definition of the Wigner–Moyal transform this is the same thing as

$$\left(\tfrac{1}{2\pi\hbar}\right)^{n/2} \int e^{-\frac{i}{\hbar}\langle p,y\rangle} \widehat{S}\psi(x + \tfrac{1}{2}y)\overline{\widehat{S}\phi}(x - \tfrac{1}{2}y) d^n y = 2^n e^{-\frac{2}{\hbar}\langle Dz,z\rangle},$$

that is, in view of the Fourier inversion formula,

$$\widehat{S}\psi(x+\tfrac{1}{2}y)\overline{\widehat{S}\phi}(x-\tfrac{1}{2}y) = 2^n\left(\tfrac{1}{2\pi\hbar}\right)^{n/2}\int e^{\frac{i}{\hbar}\langle p,y\rangle}e^{-\frac{2}{\hbar}\langle Dz,z\rangle}d^np$$
$$= \left(\tfrac{2}{\pi\hbar}\right)^{n/2}e^{-\frac{1}{\hbar}\langle\Lambda x,x\rangle}\int e^{\frac{i}{\hbar}\langle p,y\rangle}e^{-\frac{1}{\hbar}\langle\Lambda p,p\rangle}d^np.$$

Setting $M = 2\Lambda$ in the generalized Fresnel formula

$$\left(\tfrac{1}{2\pi\hbar}\right)^{n/2}\int e^{-\frac{i}{\hbar}\langle p,y\rangle}e^{-\frac{1}{2\hbar}\langle Mp,p\rangle}d^np = (\det M)^{-1/2}e^{-\frac{1}{2\hbar}\langle M^{-1}y,y\rangle},$$

valid for all positive definite symmetric matrices M (*cf.* formula (7.61) in Chapter 6, Subsection 7.4), we thus have

$$\widehat{S}\psi(x+\tfrac{1}{2}y)\overline{\widehat{S}\phi}(x-\tfrac{1}{2}y) = 2^{n/2}(\det\Lambda)^{-1/2}e^{-\frac{1}{\hbar}\left(\langle\Lambda x,x\rangle+\frac{1}{4}\langle\Lambda y,y\rangle\right)}.$$

Setting $u = x+\frac{1}{2}y$ and $v = x-\frac{1}{2}y$ this is

$$\widehat{S}\psi(u)\overline{\widehat{S}\phi}(v) = 2^{n/2}(\det\Lambda)^{-1/2}$$
$$\times\exp\left[-\frac{1}{4\hbar}\left((\Lambda+\Lambda^{-1})(|u|^2+|v|^2)+2(\Lambda-\Lambda^{-1})\langle u,v\rangle\right)\right]$$

and this equality can only be true if there are no products $\langle u,v\rangle$ in the right-hand side. This condition requires that $\Lambda = \Lambda^{-1}$ and since Λ is positive definite this implies $\Lambda = I$ and hence $\Delta = I$. It follows that

$$\widehat{S}\psi(u)\overline{\widehat{S}\phi}(v) = 2^{n/2}e^{-\frac{1}{2\hbar}(|u|^2+|v|^2)}$$

so that setting successively $u = x$, $v = 0$ and $u = 0$, $v = x$,

$$\widehat{S}\psi(x)\overline{\widehat{S}\phi}(0) = \widehat{S}\psi(0)\overline{\widehat{S}\phi}(x) = 2^{n/2}e^{-\frac{1}{2\hbar}x^2}.$$

It follows that both $\widehat{S}\psi$ and $\widehat{S}\phi$ are Gaussians of the type

$$\widehat{S}\psi(x) = \psi(0)e^{-\frac{1}{2\hbar}|x|^2}\quad,\quad \widehat{S}\phi(x) = \widehat{S}\phi(0)e^{-\frac{1}{2\hbar}|x|^2};$$

ϕ being normalized to unity so is $\widehat{S}\phi$ (because $\widehat{S}$ is unitary); this requires that $\widehat{S}\phi = \alpha\phi_\hbar$ with $|\alpha| = 1$ and hence $\phi(0) = \alpha(\pi\hbar)^{-n/4}$. Since $\psi(0)\phi(0) = 2^{n/2}$ we must thus have

$$\widehat{S}\psi(0) = 2^{n/2}\overline{\alpha}\,(\pi\hbar)^{n/4}$$

which concludes the proof of part (i) of the theorem.

To prove (ii) we note that it follows from formula (8.32) in Proposition 8.47 (Chapter 8) that the Wigner transform of a Gaussian

$$\psi(x) = (\pi\hbar)^{n/2}(\det X)^{1/2}e^{-\frac{1}{2\hbar}\langle(X+iY)x,x\rangle}$$

is given by

$$W\psi(z) = e^{-\frac{1}{\hbar}\langle Gz,z\rangle}$$

where G is the positive definite symplectic matrix

$$G = \begin{bmatrix} X + YX^{-1}Y & YX^{-1} \\ X^{-1}Y & X^{-1} \end{bmatrix}.$$

Conversely, every such $S \in \mathrm{Sp}(n)$ can be put in the form above, which ends the proof of (ii) since the datum of $W\psi$ determines ψ up to a complex factor with modulus one. □

10.3 Phase-Space Weyl Operators

We now have the technical tools that are needed to define and study the phase-space Weyl calculus which will lead us to the Schrödinger equation in phase space.

10.3.1 Useful intertwining formulae

The relation between the Wigner wave-packet transform and the operators $\widehat{T}_{\mathrm{ph}}(z_0, t_0)$ is not immediately obvious; we will see that U_ϕ actually acts, for *every* ϕ, as an intertwining operator for $\widehat{T}_{\mathrm{ph}}(z_0, t_0)$ and $\widehat{T}(z_0, t_0)$. Because of the importance of this result we give it the status of a theorem:

Theorem 10.10. *Let U_ϕ, $\phi \in \mathcal{S}(\mathbb{R}^n_x)$, be an arbitrary Wigner wave-packet transform.*

(i) *We have, for all $(z_0, t_0) \in \mathbb{R}^{2n+1}_{z,t}$,*

$$\widehat{T}_{ph}(z_0, t_0)U_\phi = U_\phi \widehat{T}(z_0, t_0); \tag{10.30}$$

(ii) *The following intertwining formula holds for every operator $\widehat{A}_{ph}$:*

$$\widehat{A}_{ph}U_\phi = U_\phi \widehat{A}. \tag{10.31}$$

Proof. (i) It suffices to prove formula (10.30) for $t_0 = 0$, that is

$$\widehat{T}_{\mathrm{ph}}(z_0)U_\phi = U_\phi \widehat{T}(z_0). \tag{10.32}$$

Let us write the operator U_ϕ in the form $U_\phi = e^{\frac{i}{2\hbar}p\cdot x}W_\phi$ where the operator W_ϕ is thus defined by

$$W_\phi\psi(z) = \left(\tfrac{1}{2\pi\hbar}\right)^{n/2} \int e^{-\frac{i}{\hbar}\langle p,x'\rangle}\phi(x-x')\psi(x')d^nx'. \tag{10.33}$$

We have, by definition of $\widehat{T}_{\mathrm{ph}}(z_0)$,

$$\widehat{T}_{\mathrm{ph}}(z_0)U_\phi\psi(z) = \exp\left[-\frac{i}{2\hbar}(\sigma(z,z_0) + \langle p-p_0, x-x_0\rangle)\right] W_\phi\psi(z-z_0)$$
$$= \exp\left[\frac{i}{2\hbar}(-2\langle p,x_0\rangle + \langle p_0,x_0\rangle + \langle p,x\rangle)\right] W_\phi\psi(z-z_0)$$

and, by definition of $W_\phi\psi$,

$$W_\phi\psi(z-z_0) = \left(\tfrac{1}{2\pi\hbar}\right)^{n/2} \int e^{-\frac{i}{\hbar}\langle p-p_0,x'\rangle}\phi(x-x'-x_0)\psi(x')d^nx'$$
$$= \left(\tfrac{1}{2\pi\hbar}\right)^{n/2} e^{\frac{i}{\hbar}\langle p-p_0,x_0\rangle} \int e^{-\frac{i}{\hbar}\langle p-p_0,x''\rangle}\phi(x-x'')\psi(x'')d^nx''$$

where we have set $x'' = x' + x_0$. The overall exponential in $\widehat{T}_{\mathrm{ph}}(z_0)U_\phi\psi(z)$ is thus

$$u_1 = \exp\left[\frac{i}{2\hbar}(-\langle p_0,x_0\rangle + \langle p,x\rangle - 2\langle p,x''\rangle + 2\langle p_0,x''\rangle)\right].$$

Similarly,

$$U_\phi(\widehat{T}(z_0)\psi)(z) = \left(\tfrac{1}{2\pi\hbar}\right)^{n/2} e^{\frac{i}{2\hbar}\langle p,x\rangle}$$
$$\times \int e^{-\frac{i}{\hbar}\langle p,x''\rangle}\phi(x-x'')e^{\frac{i}{\hbar}(\langle p_0,x''\rangle - \frac{1}{2}\langle p_0,x_0\rangle)}\psi(x''-x_0)d^nx''$$

yielding the overall exponential

$$u_2 = \exp\left[\frac{i}{\hbar}\left(\frac{1}{2}\langle p,x\rangle - \langle p,x''\rangle + \langle p_0,x''\rangle - \frac{1}{2}\langle p_0,x_0\rangle\right)\right] = u_1.$$

Let us prove formula (10.31). In view of formula (10.32) we have

$$\widehat{A}_{\mathrm{ph}}U_\phi\psi = \left(\tfrac{1}{2\pi\hbar}\right)^n \int (\mathcal{F}_\sigma A)(z_0)\widehat{T}_{\mathrm{ph}}(z_0)(U_\phi\psi)(z)d^{2n}z_0$$
$$= \left(\tfrac{1}{2\pi\hbar}\right)^n \int (\mathcal{F}_\sigma A)(z_0)U_\phi(\widehat{T}(z_0)\psi)(z)d^{2n}z_0$$
$$= \left(\tfrac{1}{2\pi\hbar}\right)^n U_\phi\left(\int (\mathcal{F}_\sigma A)(z_0)\widehat{T}(z_0)\psi)(z)d^{2n}z_0\right)$$
$$= U_\phi(\widehat{A}\psi)(z).$$

(The passage from the second equality to the third is legitimated by the fact that U_ϕ is both linear and continuous.) □

An immediate consequence of Theorem 10.10 is:

Corollary 10.11. *The representation $\widehat{T}_{ph}$ of the Heisenberg group $\mathbf{H}_n$ is unitarily equivalent to the Schrödinger representation and is thus an irreducible representation of $\mathbf{H}_n$ on each of the Hilbert spaces $\mathcal{H}_\phi$.*

Proof. The intertwining formula (10.30) implies that $\widehat{T}_{\mathrm{ph}}$ and $\widehat{T}$ are unitarily equivalent representations of $\mathbf{H}_n$; the irreducibility of the representation $\mathbf{H}_n$ then follows from Stone–von Neumann's theorem. □

Let us now state the main properties of the operators $\widehat{A}_{\mathrm{ph}}$; as we will see these are simply read from those of the usual Weyl operators using the intertwining formula $\widehat{A}_{\mathrm{ph}}U_\phi = U_\phi\widehat{A}$.

10.3.2 Properties of phase-space Weyl operators

The phase-space Weyl operators

$$\widehat{A}_{\mathrm{ph}} = \left(\tfrac{1}{2\pi\hbar}\right)^n \int a_\sigma(z_0)\widehat{T}_{\mathrm{ph}}(z_0)d^{2n}z_0$$

enjoy the same property which makes the main appeal of ordinary Weyl operators, namely that they are self-adjoint if and only if their symbols are real. This is part of the following result where we also investigate the relation between the spectra of usual and phase-space Weyl operators:

Proposition 10.12. *Let $\widehat{A}_{ph}$ and $\widehat{A}$ be the operators associated to the Weyl symbol a.*

(i) *The operator $\widehat{A}_{ph}$ is symmetric if and only if $\widehat{A}$ is, that is, if and only if $a = \overline{a}$.*

(ii) *Every eigenvalue of $\widehat{A}$ is also an eigenvalue of $\widehat{A}_{ph}$ (but the converse is not true).*

Proof. (i) By definition of $\widehat{A}_{\mathrm{ph}}$ and $\widehat{T}_{\mathrm{ph}}$ we have

$$\begin{aligned}\widehat{A}_{\mathrm{ph}}\Psi(z) &= \left(\tfrac{1}{2\pi\hbar}\right)^n \int a_\sigma(z_0)e^{-\frac{i}{2\hbar}\sigma(z,z_0)}\Psi(z-z_0)d^{2n}z_0\\ &= \left(\tfrac{1}{2\pi\hbar}\right)^n \int a_\sigma(z-z')e^{\frac{i}{2\hbar}\sigma(z,z')}\Psi(z')d^{2n}z',\end{aligned}$$

hence the kernel of the operator $\widehat{A}_{\mathrm{ph}}$ is

$$K(z,z') = \left(\tfrac{1}{2\pi\hbar}\right)^n e^{\frac{i}{2\hbar}\sigma(z,z')}a_\sigma(z-z').$$

In view of the standard theory of integral operators $\widehat{A}_{\mathrm{ph}}$ is self-adjoint if and only if $K(z,z') = \overline{K(z',z)}$; using the antisymmetry of the symplectic form we have

$$\overline{K(z',z)} = \left(\tfrac{1}{2\pi\hbar}\right)^n e^{\frac{i}{2\hbar}\sigma(z,z')}\overline{a_\sigma(z'-z)},$$

hence our claim since, by definition of the symplectic Fourier transform,

$$\begin{aligned}\overline{a_\sigma(z'-z)} &= \left(\tfrac{1}{2\pi\hbar}\right)^n \int e^{-\frac{i}{\hbar}\sigma(z-z',z'')}\overline{a(z'')}d^{2n}z'' \\ &= \mathcal{F}_\sigma\overline{a}(z-z').\end{aligned}$$

(ii) Assume that $\widehat{A}\psi = \lambda\psi$; choosing $\phi \in \mathcal{S}(\mathbb{R}^n_x)$ we have, using the intertwining formula (10.31),

$$U_\phi(\widehat{A}\psi) = \widehat{A}_{\mathrm{ph}}(U_\phi\psi) = \lambda U_\phi\psi,$$

hence λ is an eigenvalue of $\widehat{A}_{\mathrm{ph}}$. □

Notice that there is no reason for an arbitrary eigenvalue of $\widehat{A}_{\mathrm{ph}}$ to be an eigenvalue of $\widehat{A}$; this is only the case if the corresponding eigenvector belongs to the range of a Wigner wave-packet transform.

Let us next establish a composition result:

Proposition 10.13. *Let a_σ and b_σ be the twisted symbols of the Weyl operators $\widehat{A}_{ph}$ and $\widehat{B}_{ph}$. The twisted symbol c_σ of the compose $\widehat{A}_{ph}\widehat{B}_{ph}$ is the same as that of $\widehat{A}\widehat{B}$, that is*

$$c_\sigma(z) = \left(\tfrac{1}{2\pi\hbar}\right)^n \int e^{\frac{i}{2\hbar}\sigma(z,z')}a_\sigma(z-z')b_\sigma(z')d^{2n}z.$$

Proof. By repeated use of (10.31) we have

$$\begin{aligned}(\widehat{A}_{\mathrm{ph}}\widehat{B}_{\mathrm{ph}})U_\phi &= \widehat{A}_{\mathrm{ph}}(\widehat{B}_{\mathrm{ph}}U_\phi) \\ &= \widehat{A}_{\mathrm{ph}}U_\phi\widehat{B} \\ &= U_\phi(\widehat{A}\widehat{B}),\end{aligned}$$

hence $\widehat{A}_{\mathrm{ph}}\widehat{B}_{\mathrm{ph}} = (\widehat{A}\widehat{B})_{\mathrm{ph}}$; the twisted symbol of $\widehat{A}\widehat{B}$ is precisely c_σ (Theorem 6.30, Chapter 6, Subsection 6.3.2). □

In Theorem 9.21 of Chapter 9 (Subsection 9.2) we showed that the class of Weyl operators with L^2 symbols was identical to the class of Hilbert–Schmidt operators. Not very surprisingly this identification carries over to the case of phase-space Weyl operators:

Proposition 10.14. *Let $\phi \in \mathcal{S}(\mathbb{R}^{2n}_z)$ and U_ϕ the corresponding Wigner wave-packet transform. The correspondence $a \longleftrightarrow \widehat{A}_{ph}$ induces an isomorphism between $L^2(\mathbb{R}^n_x)$ and the space of Hilbert–Schmidt operators on the range $\mathcal{H}_\phi$ of U_ϕ.*

Proof. This is an immediate consequence of the aforementioned Theorem 9.21 since U_ϕ is an unitary isomorphism $L^2(\mathbb{R}^n_x) \longrightarrow \mathcal{H}_\phi$. □

We leave it to the reader to restate the continuity properties proven in Chapter 6, Subsection 6.3.1 for Weyl operators in terms of the corresponding phase-space operators.

One of the most agreeable features of standard Weyl calculus is its covariance under symplectic transformation of the symbols. In the next subsection we examine the symplectic covariance of the phase-space calculus.

10.3.3 Metaplectic covariance

Recall from Chapter 7, Section 7.4 (Proposition 7.37) that each operator $\widehat{S} \in \mathrm{Mp}(n)$ can be written as a product $\widehat{S} = \widehat{R}_\nu(S_W)\widehat{R}_{\nu'}(S_{W'})$ where $\widehat{R}_\nu(S_W)$ and $\widehat{R}_{\nu'}(S_{W'})$ are in $\mathrm{Mp}(n)$ and correspond to a factorization of $S = S_W S_{W'}$ by free symplectic matrices such that

$$\det[(S_W - I)(S_{W'} - I)] \neq 0;$$

the operators $\widehat{R}_\nu(S_W)$ and $\widehat{R}_{\nu'}(S_{W'})$ are of the type

$$\widehat{R}_\nu(S_W) = \left(\frac{1}{2\pi\hbar}\right)^n \frac{i^\nu}{\sqrt{|\det(S_W - I)|}} \int e^{\frac{i}{2\hbar}\langle M_{S_W} z, z\rangle} \widehat{T}(z) d^{2n}z$$

for $\det(S_W - I) \neq 0$ where

$$M_{S_W} = \frac{1}{2} J(S_W + I)(S_W - I)^{-1}$$

is the symplectic Cayley transform of S.

In conformity with what we have done above we associate to each operator

$$\widehat{R}_\nu(S) = \left(\frac{1}{2\pi\hbar}\right)^n \frac{i^\nu}{\sqrt{|\det(S - I)|}} \int e^{\frac{i}{2\hbar}\langle M_S z, z\rangle} \widehat{T}(z) d^{2n}z,$$

such that $\det(S - I) \neq 0$, a phase-space operator by the formula

$$\widehat{R}_\nu(S)_{\mathrm{ph}} = \left(\frac{1}{2\pi\hbar}\right)^n \frac{i^\nu}{\sqrt{|\det(S - I)|}} \int e^{\frac{i}{2\hbar}\langle M_S z, z\rangle} \widehat{T}_{\mathrm{ph}}(z) d^{2n}z.$$

The operators $\widehat{R}_\nu(S_W)_{\mathrm{ph}}$ generate a group of operators acting on $L^2(\mathbb{R}^{2n}_z)$ which is isomorphic to $\mathrm{Mp}(n)$; it is the same group as that generated by the $\widehat{R}_\nu(S)_{\mathrm{ph}}$ with $\det(S - I) \neq 0$.

Definition 10.15. The group generated by the operators $\widehat{R}_\nu(S)_{\mathrm{ph}}$ is denoted by $\mathrm{Mp}_{\mathrm{ph}}(n)$; we will call it the group of metaplectic phase-space operators, and denote by $\widehat{S}_{\mathrm{ph}}$ the element of $\mathrm{Mp}_{\mathrm{ph}}(n)$ corresponding to $\widehat{S} \in \mathrm{Mp}(n)$.

We have:

Lemma 10.16. *Let* $S \in \operatorname{Sp}(n)$ *be such that* $\det(S-I) \neq 0$. *The operator* $\widehat{R}_\nu(S)_{ph}$ *can be written as*

$$\widehat{R}_\nu(S)_{ph} = \left(\frac{1}{2\pi\hbar}\right)^n \frac{i^\nu}{\sqrt{|\det(S-I)|}} \int e^{-\frac{i}{2\hbar}\sigma(Sz,z)} \widehat{T}_{ph}((S-I)z) d^{2n}z \tag{10.34}$$

or, equivalently,

$$\widehat{R}_\nu(S)_{ph} = \left(\frac{1}{2\pi\hbar}\right)^n i^\nu \sqrt{|\det(S-I)|} \int \widehat{T}_{ph}(Sz)\widehat{T}_{ph}(-z) d^{2n}z. \tag{10.35}$$

Proof. It is, *mutatis mutandis*, the same as the proof of Lemma 7.32 in Section 7.4 of Chapter 7 for the Weyl operators $\widehat{R}_\nu(S)$. □

We are going to show in a simple way that the well-known "metaplectic covariance" relation

$$\widehat{A \circ S} = \widehat{S}^{-1}\widehat{A}\widehat{S} \tag{10.36}$$

for standard Weyl operators (Theorem 7.13, Subsection 7.1.3 of Chapter 7) extends to the phase-space Weyl operators $\widehat{A}_{\mathrm{ph}}$, provided one replaces $\operatorname{Mp}(n)$ with $\operatorname{Mp}_{\mathrm{ph}}(n)$.

Proposition 10.17. *Let* S *be a symplectic matrix and* $\widehat{S}_{ph}$ *any of the two operators in* $\operatorname{Mp}_{ph}(n)$ *associated with* S. *The following phase-space metaplectic covariance formulae hold:*

$$\widehat{S}_{ph}\widehat{T}_{ph}(z_0)\widehat{S}_{ph}^{-1} = \widehat{T}_{ph}(Sz) \quad , \quad \widehat{A \circ S}_{ph} = \widehat{S}_{ph}^{-1}\widehat{A}_{ph}\widehat{S}_{ph}. \tag{10.37}$$

Proof. To prove the first formula (10.37) it is sufficient to assume that $\widehat{S}_{\mathrm{ph}} = \widehat{R}_\nu(S)_{\mathrm{ph}}$ with $\det(S-I) \neq 0$ since these operators generate $\operatorname{Mp}_{\mathrm{ph}}(n)$. Let us thus prove that

$$\widehat{T}_{\mathrm{ph}}(Sz_0)\widehat{S}_{\mathrm{ph}} = \widehat{S}_{\mathrm{ph}}\widehat{T}_{\mathrm{ph}}(z_0) \text{ if } \det(S-I) \neq 0 \tag{10.38}$$

where, in view of (10.35) the operator $\widehat{S}_{\mathrm{ph}}$is the Bochner integral

$$\widehat{S}_{\mathrm{ph}} = C_S \int \widehat{T}_{\mathrm{ph}}(Sz)\widehat{T}_{\mathrm{ph}}(-z) d^{2n}z$$

with

$$C_S = \left(\tfrac{1}{2\pi\hbar}\right)^n i^{\nu(S)} \sqrt{|\det(S-I)|}.$$

We have

$$\widehat{T}_{\mathrm{ph}}(Sz_0)\widehat{S}_{\mathrm{ph}} = C_S \int \widehat{T}_{\mathrm{ph}}(Sz_0)\widehat{T}_{\mathrm{ph}}(Sz)\widehat{T}_{\mathrm{ph}}(-z) d^{2n}z$$

and, similarly

$$\widehat{S}_{\mathrm{ph}}\widehat{T}_{\mathrm{ph}}(z_0) = C_S \int \widehat{T}_{\mathrm{ph}}(Sz)\widehat{T}_{\mathrm{ph}}(-z)\widehat{T}_{\mathrm{ph}}(z_0) d^{2n}z.$$

Since the constant C_S does not play any special role in the argument we set

$$A(z_0) = \int \widehat{T}_{\rm ph}(Sz_0)\widehat{T}_{\rm ph}(Sz)\widehat{T}_{\rm ph}(-z)d^{2n}z,$$

$$B(z_0) = \int \widehat{T}_{\rm ph}(Sz)\widehat{T}_{\rm ph}(-z)\widehat{T}_{\rm ph}(z_0)d^{2n}z.$$

We have, by repeated use of formula (10.15),

$$A(z_0) = \int e^{\frac{i}{2\hbar}\Phi_1(z,z_0)}\widehat{T}_{\rm ph}(Sz_0 + (S-I)z)d^{2n}z,$$

$$B(z_0) = \int e^{\frac{i}{2\hbar}\Phi_2(z,z_0)}\widehat{T}_{\rm ph}(z_0 + (S-I)z)d^{2n}z$$

where the phases Φ_1 and Φ_2 are given by

$$\Phi_1(z,z_0) = \sigma(z_0,z) - \sigma(S(z+z_0),z),$$
$$\Phi_2(z,z_0) = -\sigma(Sz,z) + \sigma((S-I)z,z_0).$$

Performing the change of variables $z' = z + z_0$ in the integral defining $A(z_0)$ we get

$$A(z_0) = \int e^{\frac{i}{2\hbar}\Phi_1(z'-z_0,z_0)}\widehat{T}_{\rm ph}(z_0 + (S-I)z')d^{2n}z'$$

and we have

$$\begin{aligned}\Phi_1(z'-z_0,z_0) &= \sigma(z_0,z'-z_0) - \sigma(Sz',z'-z_0)\\ &= \sigma((S-I)z',z_0) - \sigma(Sz',z')\\ &= \Phi_2(z',z_0),\end{aligned}$$

hence $A(z_0) = B(z_0)$ proving (10.38). The second formula (10.37) easily follows from the first: noting that the symplectic Fourier transform satisfies

$$\begin{aligned}\mathcal{F}_\sigma[A\circ S](z) &= \left(\tfrac{1}{2\pi\hbar}\right)^n \int e^{-\frac{i}{\hbar}\sigma(z_0,z')}A(Sz')d^{2n}z'\\ &= \left(\tfrac{1}{2\pi\hbar}\right)^n \int e^{-\frac{i}{\hbar}\sigma(Sz_0,z')}A(z')d^{2n}z'\\ &= \mathcal{F}_\sigma A(Sz),\end{aligned}$$

we have

$$\begin{aligned}\widehat{A\circ S}_{\rm ph} &= \left(\tfrac{1}{2\pi\hbar}\right)^n \int \mathcal{F}_\sigma A(Sz)\widehat{T}_{\rm ph}(z)d^{2n}z\\ &= \left(\tfrac{1}{2\pi\hbar}\right)^n \int \mathcal{F}_\sigma A(z)\widehat{T}_{\rm ph}(S^{-1}z)d^{2n}z\\ &= \left(\tfrac{1}{2\pi\hbar}\right)^n \int \mathcal{F}_\sigma A(z)\widehat{S}_{\rm ph}^{-1}\widehat{T}_{\rm ph}(z)\widehat{S}_{\rm ph}d^{2n}z\end{aligned}$$

which concludes the proof. □

10.4 Schrödinger Equation in Phase Space

We now have all the material we need to derive the phase-space Schrödinger equation

$$i\hbar\frac{\partial \Psi}{\partial t} = \widehat{H}_{\mathrm{ph}}\Psi \tag{10.39}$$

formally written in the beginning of this chapter as

$$i\hbar\frac{\partial \Psi}{\partial t} = H(\tfrac{1}{2}x + i\hbar\partial_p, \tfrac{1}{2}x - i\hbar\partial_x)\Psi;$$

the operator on the right-hand side is actually the phase-space Weyl operator $\widehat{H}_{\mathrm{ph}}$ with symbol the Hamiltonian H.

10.4.1 Derivation of the equation (10.39)

The following consequence of Theorem 10.10 above links standard "configuration space" quantum mechanics to phase-space quantum mechanics via the Wigner wave-packet transform and the extended Heisenberg group studied in the previous sections.

Proposition 10.18. *Let U_ϕ, $\phi \in \mathcal{S}(\mathbb{R}^n_x)$, be an arbitrary Wigner wave-packet transform.*

(i) *If ψ is a solution of the standard Schrödinger's equation*

$$i\hbar\frac{\partial \psi}{\partial t} = \widehat{H}\psi,$$

then $\Psi = U_\phi\psi$ is a solution of the phase-space Schrödinger equation

$$i\hbar\frac{\partial \Psi}{\partial t} = \widehat{H}_{ph}\Psi. \tag{10.40}$$

(ii) *Assume that Ψ is a solution of this equation and that $\Psi_0 = \Psi(\cdot, 0)$ belongs to the range $\mathcal{H}_\phi$ of U_ϕ. Then $\Psi(\cdot, t) \in \mathcal{H}_\phi$ for every t for which Ψ is defined.*

Proof. Since time-derivatives obviously commute with U_ϕ we have, using (10.31),

$$i\hbar\frac{\partial \Psi}{\partial t} = U_\phi(\widehat{H}\psi) = \widehat{H}_{\mathrm{ph}}(U_\phi\psi) = \widehat{H}_{\mathrm{ph}}\Psi,$$

hence (i).

Statement (ii) follows. □

The result above raises some interesting physical questions: since the solutions of the phase-space Schrödinger equation (10.40) exist independently of the choice of any isometry U_ϕ, what is the relation between the corresponding configuration-space wave-functions $\psi = U_\phi^*\Psi$ and $\psi' = U_{\phi'}^*\Psi$?

Proposition 10.19. *Let Ψ be a solution of the phase space Schrödinger equation* (10.40) *with initial condition Ψ_0 and define functions ψ_1 and ψ_2 in $L^2(\mathbb{R}^n_x)$ by*

$$\Psi = U_{\phi_1}\psi_1 = U_{\phi_2}\psi_2.$$

We assume that $\Psi_0 \in \mathcal{H}_{\phi_1} \cap \mathcal{H}_{\phi_2}$.

(i) *We have $\Psi(\cdot, t) \in \mathcal{H}_{\phi_1} \cap \mathcal{H}_{\phi_2}$ for all t .*

(ii) *If $(\phi_1, \phi_2)_{L^2(\mathbb{R}^n_x)} = 0$ then ψ_1 and ψ_2 are orthogonal quantum states:*

$$(\psi_1, \psi_2)_{L^2(\mathbb{R}^n_x)} = 0.$$

Proof. The statement (i) follows from Theorem 10.18(iii). In view of formula (10.26) we have

$$(U_{\phi_1}\psi_1, U_{\phi_2}\psi_2)_{L^2(\mathbb{R}^{2n}_z)} = \overline{(\phi_1, \phi_2)_{L^2(\mathbb{R}^n_x)}}(\psi_1, \psi_2)_{L^2(\mathbb{R}^n_x)},$$

that is

$$||\Psi||^2_{L^2(\mathbb{R}^{2n}_z)} = \lambda(\psi_1, \psi_2)_{L^2(\mathbb{R}^n_x)} \quad , \quad \lambda = (\phi_1, \phi_2)_{L^2(\mathbb{R}^n_x)}.$$

Property (ii) follows. □

Suppose now that $(\psi_j)_j$ and $(\phi_k)_k$ are complete orthonormal systems in $L^2(\mathbb{R}^n_x)$ and define vectors Σ_{jk} in $L^2(\mathbb{R}^{2n}_z)$ by $\Sigma_{jk} = U_{\phi_k}\psi_j$. Since the U_{ϕ_k} are isometries and

$$(\Sigma_{jk}, \Sigma_{j'k'})_{L^2(\mathbb{R}^{2n}_z)} = \overline{(\phi_k, \phi_{k'})_{L^2(\mathbb{R}^n_x)}}(\psi_j, \psi_{j'})_{L^2(\mathbb{R}^n_x)} = \delta_{jj'}\delta_{kk'},$$

it follows that $(\Sigma_{jk})_{jk}$ is an orthonormal system in $L^2(\mathbb{R}^{2n}_z)$. It is legitimate to ask whether this system is complete; equivalently could it be that the subspace

$$\mathcal{H} = \mathcal{H}_1 \oplus \mathcal{H}_2 \oplus \cdots \oplus H_N \oplus \cdots$$

with $\mathcal{H}_N$ the range of U_{ϕ_N}, is identical to $L^2(\mathbb{R}^{2n}_z)$? The answer is *no*, because there are square-integrable functions which do not belong to the range of any of the Wigner wave-packet transforms: we have proven in Subsection 10.2.2 (Theorem 10.9) that the only non-degenerate Gaussians

$$\Psi_G(z) = e^{-\frac{1}{2\hbar}\langle Gz,z\rangle}.$$

which belong to the range $\mathcal{H}_\phi$ of some Wigner wave-packet transform U_ϕ are those for which we have $G \in \mathrm{Sp}(n)$.

10.4.2 The case of quadratic Hamiltonians

There is an interesting application of the theory of the metaplectic group to Schrödinger's equation in phase space. Assume that H is a quadratic Hamiltonian (for instance the harmonic oscillator Hamiltonian); the flow determined by the associated Hamilton equations is linear and consists of symplectic matrices S_t. Letting time vary, thus obtain a curve $t \longmapsto S_t$ in the symplectic group $\mathrm{Sp}(n)$ passing through the identity I at time $t = 0$; following general principles to that curve we can associate (in a unique way) a curve $t \longmapsto \widehat{S}_t$ of metaplectic operators. Let now $\psi_0 = \psi_0(x)$ be some square integrable function and set $\psi(x,t) = \widehat{S}_t\psi_0(x)$. Then ψ is just the solution of the standard Schrödinger's equation

$$i\hbar\frac{\partial\psi}{\partial t} = \widehat{H}\psi \quad , \quad \psi(\cdot, 0) = \psi_0 \tag{10.41}$$

associated to the quadratic Hamiltonian function H. (Equivalently, $\widehat{S}_t$ is just the propagator for (10.41).) This observation allows us to solve explicitly the phase-space Schrödinger equation for any such H. Here is how. Since the wave-packet transform U automatically takes the solution ψ of (10.41) to a solution of the phase-space Schrödinger equation

$$i\hbar\frac{\partial\Psi}{\partial t} = \widehat{H}_{\mathrm{ph}}\Psi,$$

we have

$$\Psi(z,t) = (\widehat{S}_t)_{\mathrm{ph}}\Psi(z,0).$$

Assume now that the symplectic matrix S_t is free and $\det(S_t - I) \neq 0$; then

$$\Psi(z,t) = \widehat{R}_\nu(S_t)_{\mathrm{ph}}\Psi(z,0) \tag{10.42}$$

where

$$\widehat{R}_\nu(S_t)_{\mathrm{ph}} = \left(\frac{1}{2\pi\hbar}\right)^{n/2} \frac{i^{m(t)-\operatorname{Inert} W_S(t)}}{\sqrt{|\det(S_t - I)|}} \int e^{\frac{i}{2\hbar}\langle M_S(t)z_0, z_0\rangle} \widehat{T}_{\mathrm{ph}}(z_0) d^{2n}z_0,$$

$m(t)$, $W_S(t)$, and $M_S(t)$ corresponding to S_t. If t is such that S_t is not free, or $\det(S_t - I) = 0$, then it suffices to write the propagator $\widehat{S}_t$ as the product of two operators of the type $\widehat{R}_\nu(S)_{\mathrm{ph}}$; note however that such values of t are exceptional, and that the solution (10.42) can be extended by taking the limit near such t provided that one takes some care in calculating the Maslov and Conley–Zehnder indices.

Here is a simple but nevertheless interesting illustration. Let H be the Hamiltonian function of the one-dimensional harmonic oscillator put in normal form

$$H = \frac{\omega}{2}(p^2 + x^2).$$

An immediate calculation, which is left to the reader as an exercise, shows that the associated phase-space Weyl operator is

$$\widehat{H}_{\mathrm{ph}} = -\frac{\hbar^2\omega}{2}\left(\frac{\partial^2}{\partial x^2} + \frac{\partial^2}{\partial p^2}\right) - i\frac{\hbar\omega}{2}\left(p\frac{\partial}{\partial x} - x\frac{\partial}{\partial p}\right) + \frac{\omega}{8}(p^2 + x^2).$$

The one-parameter group (S_t) is in this case given by

$$S_t = \begin{bmatrix} \cos\omega t & \sin\omega t \\ -\sin\omega t & \cos\omega t \end{bmatrix}$$

and the Hamilton principal function by

$$W(x, x'; t) = \frac{1}{2\sin\omega t}((x^2 + x'^2)\cos\omega t - 2xx').$$

A straightforward calculation yields

$$M_S(t) = \begin{bmatrix} \frac{\sin\omega t}{-2\cos\omega t+2} & 0 \\ 0 & \frac{\sin\omega t}{-2\cos\omega t+2} \end{bmatrix} = \frac{1}{2}\begin{bmatrix} \cot(\frac{\omega t}{2}) & 0 \\ 0 & \cot(\frac{\omega t}{2}) \end{bmatrix}$$

and

$$\det(S_t - I) = 2(1 - \cos\omega t) = 4\sin^2(\tfrac{\omega t}{2});$$

moreover

$$W_{xx}(t) = -\tan(\tfrac{\omega t}{2}).$$

Insertion in formula (10.42) yields the explicit solution

$$\Psi(z,t) = \frac{i^{\nu(t)}}{2\left|2\pi\hbar\sin(\frac{\omega t}{2})\right|^{1/2}} \times \int \exp\left[\frac{i}{4\hbar}(x_0^2 + p_0^2)\cot(\frac{\omega t}{2})\right] \widehat{T}_{\mathrm{ph}}(z_0)\Psi(z,0)d^2z_0,$$

where the Conley–Zehnder index $\nu(t)$ is given by

$$\nu(t) = \begin{cases} 0 \text{ if } 0 < t < \frac{\pi}{\omega}, \\ -2 \text{ if } -\frac{\pi}{\omega} < t < 0. \end{cases}$$

10.4.3 Probabilistic interpretation

Let us begin by discussing the probabilistic interpretation of the solutions Ψ of the phase-space Schrödinger equation

$$i\hbar\frac{\partial\Psi}{\partial t} = \widehat{H}_{\mathrm{ph}}\Psi$$

where $\widehat{H}_{\mathrm{ph}}$ is a symmetric operator.

Let ψ be in $L^2(\mathbb{R}^n_x)$; if ψ is normalized, then so is $\Psi = U_\phi\psi$ in view of the Parseval formula (10.24):

$$||\psi||_{L^2(\mathbb{R}^n_x)} = 1 \Longleftrightarrow ||\Psi||_{L^2(\mathbb{R}^{2n}_z)} = 1.$$

It follows that $|\Psi|^2$ is a probability density in phase space. That this property is conserved during the time-evolution is straightforward:

Proposition 10.20. *Let $\rho(z,t) = |\Psi(z,t)|^2$. We have*

$$\int \rho(z,t)d^{2n}z = \int \rho(z,0)d^{2n}z$$

for all $t \in \mathbb{R}$. In particular $\rho(z,t)$ is a probability density if and only if $\rho(z,0)$ is.

Proof. The argument is exactly the same as when one establishes that the L^2-norm of the solution ψ of Schrödinger's equation is conserved in time: we have

$$i\hbar\frac{\partial \rho}{\partial t} = (\widehat{H}_{\mathrm{ph}}\Psi)\overline{\Psi} - \overline{(\widehat{H}_{\mathrm{ph}}\Psi)}\Psi$$

and hence, since $\widehat{H}_{\mathrm{ph}} = \widehat{H}_{\mathrm{ph}}^*$:

$$\frac{\partial}{\partial t}\int \rho(z,t)d^{2n}z = \frac{1}{i\hbar}\left[(\widehat{H}_{\mathrm{ph}}\Psi,\Psi)_{L^2(\mathbb{R}^{2n}_z)} - (\widehat{H}_{\mathrm{ph}}\Psi,\Psi)_{L^2(\mathbb{R}^{2n}_z)}\right] = 0$$

so that $\int\rho(z,t)d^{2n}z$ is constant; taking $t=0$ this constant is precisely $\int\rho(z,0)d^{2n}z$. □

Exercise 10.21. Give another proof of the proposition above when Ψ is in the range of the wave-packet transform.

It turns out that by an appropriate choice of ϕ the marginal probabilities can be chosen arbitrarily close to $|\psi|^2$ and $|F\psi|^2$:

Theorem 10.22. *Let $\psi \in L^2(\mathbb{R}^n_x)$ and set $\Psi = U_\phi\psi$.*

(i) *We have*

$$\int |\Psi(x,p)|^2 d^n p = (|\phi|^2 * |\psi|^2)(x), \tag{10.43}$$

$$\int |\Psi(x,p)|^2 d^n x = (|F\phi|^2 * |F\psi|^2)(p). \tag{10.44}$$

(ii) *Let $\langle A\rangle_\psi = (\widehat{A}\psi,\psi)$ be the mathematical expectation of the symbol A in the normalized quantum state ψ. We have*

$$\langle A\rangle_\psi = (\widehat{A}_{ph}\Psi,\Psi)_{L^2(\mathbb{R}^{2n}_z)} \quad , \quad \Psi = U_\phi\psi. \tag{10.45}$$

Proof. We have, by definition of Ψ,

$$|\Psi(z)|^2 = \left(\tfrac{1}{2\pi\hbar}\right)^n \iint e^{-\frac{i}{\hbar}\langle p,x'-x''\rangle}\phi(x-x')\phi(x-x'')\psi(x')\overline{\psi}(x'')d^n x' d^n x''.$$

Since we have, by the Fourier inversion formula,

$$\int e^{-\frac{i}{\hbar}\langle p,x'-x''\rangle}d^n p = (2\pi\hbar)^n\delta(x'-x''),$$

it follows that

$$\begin{aligned}\int |\Psi(z)|^2 d^n p &= \iiint \delta(x'-x'')|\phi(x-x')|^2|\psi(x')|^2 d^n x' d^n x'' \\ &= \int \left[\int \delta(x'-x'')d^n x''\right] |\phi(x-x')|^2|\psi(x')|^2 d^n x' \\ &= \int |\phi(x-x')|^2|\psi(x')|^2 d^n x',\end{aligned}$$

hence formula (10.43). To prove (10.44) we note that in view of the metaplectic covariance formula (10.23) for the wave-packet transform we have

$$U_{\widehat{J}\phi}(\widehat{J}\psi)(x,p) = U_\phi\psi(-p,x)$$

where $\widehat{J} = i^{-n/2}F$ is the metaplectic Fourier transform. It follows that

$$U_{F\phi}(F\psi)(x,p) = i^{-n}U_\phi\psi(-p,x)$$

and hence changing $(-p,x)$ into (x,p):

$$U_\phi\psi(x,p) = i^n U_{F\phi}(F\psi)(p,-x).$$

And hence, using (10.43),

$$\begin{aligned}\int |\Psi(x,p)|^2 d^n x &= \int |U_{F\phi}(F\psi)(p,-x)|^2 d^n x \\ &= \int |U_{F\phi}(F\psi)(p,x)|^2 d^n x = (|F\phi|^2 * |F\psi|^2)(p)\end{aligned}$$

which concludes the proof of (10.44). To prove (10.45) it suffices to note that, in view of the intertwining formula (10.31) and the fact that $U_\phi^* = U_\phi^{-1}$, we have

$$\begin{aligned}(A_{\mathrm{ph}}\Psi,\Psi)_{L^2(\mathbb{R}^{2n}_z)} &= (\widehat{A}_{\mathrm{ph}}U_\phi\psi, U_\phi\psi)_{L^2(\mathbb{R}^{2n}_z)} \\ &= (U_\phi^*\widehat{A}_{\mathrm{ph}}U_\phi\psi,\psi)_{L^2(\mathbb{R}^n_x)} = (\widehat{A}\psi,\psi)_{L^2(\mathbb{R}^n_x)}\end{aligned}$$

proving (10.45). □

The result above shows that the marginal probabilities of $|\Psi|^2$ are just the traditional position and momentum probability densities $|\psi|^2$ and $|F\psi|^2$ "smoothed out" by convoluting them with $|\phi|^2$ and $|F\phi|^2$ respectively.

Let us investigate the limit $\hbar \to 0$. Choose for ϕ the Gaussian (10.20):

$$\phi(x) = \phi_\hbar(x) = \left(\tfrac{1}{\pi\hbar}\right)^{n/4} e^{-\frac{1}{2\hbar}|x|^2}.$$

The Fourier transform of ϕ is identical to ϕ,

$$F\phi_\hbar(p) = \left(\tfrac{1}{\pi\hbar}\right)^{n/4} e^{-\frac{1}{2\hbar}|p|^2} = \phi_\hbar(p),$$

hence, setting $\Psi_\hbar = U_{\phi_\hbar}\psi$, and observing that $|\phi_\hbar|^2 \to \delta$ when $\hbar \to 0$:

$$\lim_{\hbar\to 0}\int |\Psi_\hbar(x,p)|^2 d^n p = (|\psi|^2 * |\phi_\hbar|^2)(x) = |\psi(x)|^2,$$

$$\lim_{\hbar\to 0}\int |\Psi_\hbar(x,p)|^2 d^n x = (|F\psi|^2 * |\phi_\hbar|^2)(p) = |F\psi(p)|^2.$$

Thus, in the limit $\hbar \to 0$ the square of the modulus of the phase-space wavefunction becomes a true joint probability density for the probability densities $|\psi|^2$ and $|F\psi|^2$.

Let us now return to the notion of density operator which was discussed in detail in Chapter 9. We showed in particular that an operator $\widehat{\rho}$ on $L^2(\mathbb{R}^n_x)$ was a density operator if and only if its Weyl symbol is a convex sum of Wigner transforms of functions $\psi_j \in L^2(\mathbb{R}^n_x)$:

$$\rho = \sum_j \lambda_j W\psi_j \ , \ \lambda_j \geq 0 \ , \ \sum_j \lambda_j = 1.$$

To $\widehat{\rho}$ we can associate the phase-space operator $\widehat{\rho}_{\mathrm{ph}}$; notice that $\widehat{\rho}_{\mathrm{ph}}$ is automatically self-adjoint since $\widehat{\rho}$ is (Proposition 10.12(i)). The following result is almost obvious; we nevertheless give its proof in detail.

Proposition 10.23. *Let $\widehat{\rho}$ be a density operator on $L^2(\mathbb{R}^n_x)$. The restriction of $\widehat{\rho}_{\mathrm{ph}}$ to any of the subspaces $\mathcal{H}_\phi = \mathrm{Range}(\widehat{\rho})$, $\phi \in \mathcal{S}(\mathbb{R}^n_x)$, is a density operator on $\mathcal{H}_\phi$.*

Proof. Since $\widehat{\rho}_{\mathrm{ph}}$ is self-adjoint there remains to prove that $\widehat{\rho}_{\mathrm{ph}}$ is non-negative and that $\widehat{\rho}_{\mathrm{ph}}$ is of trace-class with $Tr(\widehat{\rho}_{\mathrm{ph}}) = 1$. Since we have $\widehat{\rho}_{\mathrm{ph}}\Psi = U_\phi\widehat{\rho}U_\phi^*\Psi$ for every $\Psi \in U_\phi$ we have for such a Ψ,

$$\begin{aligned}(\widehat{\rho}_{\mathrm{ph}}\Psi, \Psi)_{L^2(\mathbb{R}^{2n}_z)} &= (U_\phi\widehat{\rho}U_\phi^*\Psi, \Psi)_{L^2(\mathbb{R}^{2n}_z)} \\ &= (\widehat{\rho}U_\phi^*\Psi, U_\phi^*\Psi)_{L^2(\mathbb{R}^n_x)}\end{aligned}$$

and hence $(\widehat{\rho}_{\mathrm{ph}}\Psi, \Psi)_{L^2(\mathbb{R}^{2n}_z)} \geq 0$ since $(\widehat{\rho}\psi, \psi)_{L^2(\mathbb{R}^n_x)} \geq 0$ for all $\psi \in L^2(\mathbb{R}^n_x)$; $\widehat{\rho}_{\mathrm{ph}}$ is thus a non-negative operator. Let us show that $\widehat{\rho}_{\mathrm{ph}}$ is of trace class and has trace 1. Choose an orthonormal basis $(\Psi_j)_j$ of $\mathcal{H}_\phi$; then $(\psi_j)_j$ with $\psi_j = U_\phi^*\Psi_j$ is an orthonormal basis of $L^2(\mathbb{R}^n_x)$ (because U_ϕ is an isometry of $L^2(\mathbb{R}^n_x)$ onto $\mathcal{H}_\phi$). We have

$$\begin{aligned}(\widehat{\rho}_{\mathrm{ph}}\Psi_j, \Psi_j)_{L^2(\mathbb{R}^{2n}_z)} &= (\widehat{\rho}U_\phi^*\Psi_j, U_\phi^*\Psi_j)_{L^2(\mathbb{R}^n_x)} \\ &= (\widehat{\rho}\psi_j, \psi_j)_{L^2(\mathbb{R}^n_x)}\end{aligned}$$

and hence

$$\sum_j (\widehat{\rho}_{\mathrm{ph}}\Psi_j, \Psi_j)_{L^2(\mathbb{R}^{2n}_z)} = \sum_j (\widehat{\rho}\psi_j, \psi_j)_{L^2(\mathbb{R}^n_x)} = 1$$

since $\widehat{\rho}$ is of trace class and has trace 1. □

10.5 Conclusion

We have sketched in this last section a theory of Schrödinger equation in symplectic phase space which is consistent with the Stone–von Neumann theorem; by the properties of the wave-packet transform, the theory of the ordinary Schrödinger equation on "configuration space" becomes a particular case of our constructions. One must however be aware of the fact that the quantization scheme

$$H \overset{\text{Weyl}}{\longrightarrow} \widehat{H} \longrightarrow \widehat{H}_{\text{ph}}$$

we have been using is not the only possible. While our choice was dictated by considerations of maximal symplectic covariance, there are, however, many other possibilities. For instance, while it is accepted by a majority of mathematicians and physicists that Weyl quantization $H \overset{\text{Weyl}}{\longrightarrow} \widehat{H}$ is the most natural in standard quantum mechanics, there are other ways to define the quantized Hamiltonian; see for instance [128] for a discussion of some of these. Secondly, the phase-space quantization scheme

$$x_j \longmapsto \widehat{X}_{j,\text{ph}} = \tfrac{1}{2}x_j + i\hbar\tfrac{\partial}{\partial p_j} \quad , \quad p_j \longmapsto \widehat{P}_{j,\text{ph}}\tfrac{1}{2}p_j - i\hbar\tfrac{\partial}{\partial x_j}$$

we have been using, and which leads to the correspondence $\widehat{H} \longrightarrow \widehat{H}_{\text{ph}}$ can be replaced by the more general scheme

$$x_j \longmapsto \widehat{X}^{\alpha\beta}_{j,\text{ph}} = \alpha x_j + i\hbar\beta\tfrac{\partial}{\partial p_j} \quad , \quad p_j \longmapsto \widehat{P}^{\alpha\beta}_{j,\text{ph}} = \gamma p_j - i\hbar\delta\tfrac{\partial}{\partial x_j}$$

where $\alpha, \beta, \gamma, \delta$ are any real numbers such that $\beta\gamma - \alpha\delta = 1$, and this without altering the position-momentum commutation relations; in fact one verifies by an immediate calculation that

$$\left[\widehat{X}^{\alpha\beta}_{j,\text{ph}}, \widehat{P}^{\alpha\beta}_{j,\text{ph}}\right] = i\hbar$$

for all such choices. For instance the rule $\widehat{H} \longrightarrow \widehat{H}_{\text{ph}}$ we have been using corresponds to the case $\alpha = \gamma = \frac{1}{2}$, $\beta = 1$, $\delta = -1$, while the choice $\alpha = \beta = 1$, $\gamma = 0$, $\delta = -1$ leads the the quantization rules

$$x_j \longmapsto x_j - i\hbar\tfrac{\partial}{\partial p_j} \quad , \quad p_j \longmapsto -i\hbar\tfrac{\partial}{\partial x_j}$$

which have been considered by some physicists (see e.g. [162, 163]). Notice that this choice is in a sense very natural, because, as is easily verified, it corresponds to extending the Heisenberg–Weyl operators to phase space functions by the formula

$$\widehat{T}(z_0) = e^{\frac{i}{\hbar}(\langle p_0, x\rangle - \frac{1}{2}\langle p_0, x_0\rangle)}T(z_0)$$

where $T(z_0)$ is the usual translation operator $T(z_0)\Psi(z) = \Psi(z - z_0)$.

There are furthermore several topics we have not discussed in this section because of time and space. For instance, what is the physical meaning of functions $\Psi \in L^2(\mathbb{R}_z^{2n})$ that do not belong to the range of any wave-packet transform? These functions are certainly (at least when they are not orthogonal to the range of every U_ϕ) related to the density matrix, but in which way? It would be interesting to give a precise correspondence between these objects. Another topic which we have not mentioned is that of the semi-classical approximations to the solutions of the phase space Schrödinger equation

$$i\hbar\frac{\partial\Psi}{\partial t} = \widehat{H}_{\mathrm{ph}}\Psi.$$

I have in particular in mind the "nearby orbit method" (see Littlejohn [112] for a nice description in the context of semi-classical analysis). The idea is the following: to each $z_0 \in \mathbb{R}_z^{2n}$ one associates the quadratic (inhomogeneous) Hamiltonian function defined by

$$H_{z_0}(z,t) = H(z_t,t) + \langle H'(z_t,t), z - z_t\rangle + \frac{1}{2}\langle H''(z_t,t)(z-z_t), z - z_t\rangle$$

where $t \longmapsto z_t$ is the solution of Hamilton's equations $\dot{z} = J\partial_z H(z,t)$ passing through z_0 at time $t = 0$ (H_{z_0} is thus the truncated Taylor series of H at z_t obtained by discarding all terms $O(|z - z_t|^3)$). The corresponding flow $(f_t^{z_0})$ is then expressed in terms of translations and linear symplectic transformations; lifting this flow to the inhomogeneous group $\mathrm{IMp}(n)$ one obtains a one-parameter family of operators $\widehat{f_t^{z_0}}$ such that $\widehat{f_0^{z_0}}$ is the identity. Let now $\psi_{z_0} = \widehat{T}(z_0)\psi_0$ where ψ_0 is the Gaussian defined by

$$\psi_0(x) = (\pi\hbar)^{-n/4}e^{-\frac{1}{\hbar}|x|^2};$$

such functions are called coherent states . It turns out that one proves that the function $\psi = \widehat{f_t^{z_0}}\psi_{z_0}$ is a very good approximation (for small t or small $\hbar$) to the solution of the Cauchy problem

$$i\hbar\frac{\partial\psi}{\partial t} = \widehat{H}\psi \ \ , \ \ \psi(\cdot,0) = \psi_{z_0}$$

(see Littlejohn [112] for a discussion and properties; very precise estimates are given in Combescure and Robert [24]). It would certainly be interesting (and perhaps not very difficult) to extend these results to the phase-space Schrödinger equation.

Appendix A

Classical Lie Groups

A.1 General Properties

A *Lie group* is a group G for which the mappings $g \longmapsto g^{-1}$ and $(g, g') \longmapsto gg'$ are continuous. A *classical Lie group* is a closed subgroup of a general linear group $\mathrm{GL}(m, \mathbb{K})$ ($\mathbb{K} = \mathbb{R}$ or $\mathbb{C}$). Every classical Lie group is a Lie group; the converse is not true: there are non-closed subgroups of $\mathrm{GL}(m, \mathbb{K})$ which are Lie groups.

Let G be a classical Lie group and define

$$\mathfrak{g} = \left\{X \in M(m, K) : e^{tX} \in G \text{ for all } t \in \mathbb{R}\right\}.$$

If X and Y belong to vector space $\mathfrak{g}$, then the commutator $[X, Y] = XY - YX$ also belongs to $\mathfrak{g}$.

Definition A.1. $\mathfrak{g}$ is called the Lie algebra of the classical Lie group G.

The following holds for every classical Lie group G:

- There exists a neighborhood $\mathcal{U}$ of 0 in $\mathfrak{g}$ and a neighborhood $\mathcal{V}$ of I in G such that the exponential mapping $\exp : X \longmapsto e^X$ is a diffeomorphism $\mathcal{U} \longrightarrow \mathcal{V}$.
- If G is connected, then the set $\exp(\mathfrak{g})$ generates G.

Cartan's theorem, which we state below, is a refinement of the usual polar decomposition result. Let us begin by defining a notion of logarithm for invertible matrices:

Proposition A.2. *Let $\mathbb{K} = \mathbb{R}$ or $\mathbb{C}$ and let M be an invertible $m \times m$ matrix with entries in $\mathbb{K}$. There exists an $m \times m$ matrix L such that $M = e^L$.*

Proof. Let $\lambda_1, \ldots, \lambda_r$ be the eigenvalues of M (counted with their multiplicities $m(\lambda_1), \ldots, m(\lambda_r)$), and set $E_k = \mathrm{Ker}(M - \lambda_k I)^{m(\lambda_k)}$. Let M_k be the restriction $M_{|E_k}$; there exists a nilpotent matrix N_k such that $M_k = \lambda_k I + N_k$ (see *e.g.* the first chapter of Kato [100]). Since $\mathbb{C}^m = E_1 \oplus \cdots \oplus E_r$ it is sufficient to prove the proposition when $M = M_k$, that is we may as well assume that $M = \lambda I + N$ where $\lambda \neq 0$ and $N^k = 0$ if $k \geq k_0$ for some integer $k_0 \geq 0$. Define L by

$$L = (\log \lambda)I + \log(I + \lambda^{-1}N) \tag{A.1}$$

where we have set

$$\log(I + \lambda^{-1}N) = \sum_{k=0}^{k_0} (-1)^{-k+1} \frac{\lambda^{-k}N^k}{k} \tag{A.2}$$

where $\log \lambda$ is any choice of ordinary (complex logarithm). Direct substitution in the power series defining the exponential shows, after some lengthy but straightforward calculations, that $M = e^L$. □

We will write $L = \log M$. Notice that even when M is real, $\log M$ is usually complex; its definition actually depends on the choice of a determination of the logarithm of a complex number via $\log \lambda$ in formula (A.1). However:

Corollary A.3. *Assume that M has a real square root $\sqrt{M}$: $M = (\sqrt{M})^2$. Then $\log M$ is a real matrix.*

Proof. As in the proof of the proposition above it is no restriction to assume that $\sqrt{M} = \lambda I + N$ where $\lambda \neq 0$ and N is nilpotent, and we thus have

$$\sqrt{M} = e^L \quad , L = (\log \lambda)I + \log(I + \lambda^{-1}N)$$

where $\log(I + \lambda^{-1}N)$ is defined by a series of the type (A.2). Since $\sqrt{M}$ is real we have $\sqrt{M} = \overline{\sqrt{M}}$ and thus

$$M = e^L \overline{e^L} = e^L e^{\bar{L}}.$$

Now, conjugating the series (A.2) we get

$$\bar{L} = \sum_{k=0}^{k_0} (-1)^{-k+1} \frac{\bar{\lambda}^{-k}\bar{N}^k}{k}$$

hence the result. □

Let us now state Cartan's theorem:

Cartan's Theorem Let M be a real or complex invertible matrix and $X = \frac{1}{2}\operatorname{Log}(M^T M)$. Then $R = Me^{-X}$ is orthogonal: $R^T R = I$ and the mapping $\mathcal{C} : M \longmapsto (R, X)$ is a diffeomorphism

$$\mathcal{C} : \mathrm{GL}(m, \mathbb{C}) \longrightarrow O(m, \mathbb{C}) \times \mathrm{Sym}(m, \mathbb{C}).$$

A.2 The Baker–Campbell–Hausdorff Formula

The exponential mapping $\exp : \mathfrak{g} \longrightarrow G$ does not satisfy the relation $\exp X \exp Y = \exp(X+Y)$ if $XY \neq YX$. The Baker–Campbell–Hausdorff formula says that, however, under some conditions, there exists $C(X,Y)$ in $\mathfrak{g}$ such that

$$e^X e^Y = e^{C(X,Y)}.$$

More precisely:

Theorem A.4. *Let $||\cdot||$ be a submultiplicative norm on $M(2n,\mathbb{R})$ and V the subset of $M(2n,\mathbb{R})$ consisting of all M such that $||X|| < \pi/2$.*

(i) *There exists a unique analytic function $(X,Y) \longmapsto C(X,Y)$ defined on V^2 such that* $\exp X \exp Y = \exp C(X,Y)$.

(ii) *If X and Y belong to some Lie algebra $\mathfrak{g}$ then so does $C(X,Y)$.*

For a complete proof see for instance Varadarajan [170]; it is based on the following algorithm for constructing the analytic function C:

$$C(X,Y) = C_1(X,Y) + C_2(X,Y) + \cdots + C_j(x+Y) + \cdots$$

where:

$$C_1(X,Y) = X+Y \quad , \quad C_2(X,Y) = \frac{1}{2}[X,Y]$$

and $C_j(X,Y)$ is a linear combination of commutators of higher order; for instance

$$C_3(X,Y) = \frac{1}{12}[[X,Y],Y] - \frac{1}{12}[[X,Y],X],$$

$$C_4(X,Y) = -\frac{1}{48}[Y,[X,[X,Y]]] - [X,[Y,[X,Y]]].$$

The following immediate consequence of the Campbell–Hausdorff formula is useful when dealing with the Heisenberg group:

Corollary A.5. *Under the conditions in the theorem above assume that all commutators of order superior to 2 are equal to zero. Then*

$$\exp X \exp Y = \exp(X + Y + \tfrac{1}{2}[X,Y]).$$

A.3 One-parameter Subgroups of $\mathrm{GL}(m,\mathbb{R})$

Let us review the notion of continuous one-parameter subgroups of the general linear group $\mathrm{GL}(m,\mathbb{R})$ (m any integer ≥ 1). Such a subgroup is determined by a continuous homomorphism $t \longmapsto \varphi(t)$ of the additive group $(\mathbb{R},+)$ into $\mathrm{GL}(m,\mathbb{R})$: it is thus a continuous mapping $\varphi : \mathbb{R} \longrightarrow \mathrm{GL}(m,\mathbb{R})$ such that

$$\varphi(t+t') = \varphi(t)\varphi(t') \quad \text{for all} \quad t,t' \in \mathbb{R}.$$

It turns out that continuity here implies differentiability:

Proposition A.6. *The mapping φ is infinitely differentiable and there exists an $m \times m$ matrix X such that*

$$\varphi(t) = \exp(tX) \tag{A.3}$$

for every $t \in \mathbb{R}$; in fact $X = D\varphi(0)$.

Proof. Let us begin by showing that φ is C^∞. Choose a smooth function $\theta : \mathbb{R} \longrightarrow \mathbb{R}_+$ with support contained in some closed interval $[-a, a]$ $(a > 0)$, and such that

$$\int_{-a}^{a} \theta(x)dx = 1.$$

Consider now the convolution product $\theta * \varphi$; since $\varphi(t - x) = \varphi(t)\varphi(x)^{-1}$ we have

$$(\theta * \varphi)(t) = \int_{-a}^{a} \theta(x)\varphi(t - x)dx = \varphi(t)M(\theta, \varphi)$$

where $M(\theta, \varphi)$ is the matrix

$$M(\theta, \varphi) = \int_{-a}^{a} \theta(x)\varphi(x)^{-1}dx.$$

Let us show that if a is small enough, then $M(\theta, \varphi)$ is invertible; since $\theta * \varphi$ is differentiable, φ will also be differentiable since

$$\varphi(t) = (\theta * \varphi)(t)M(\theta, \varphi)^{-1}.$$

Let $||\cdot||$ be any norm on the space of all $m \times m$ matrices; we claim that if a (and hence the support of θ) is small enough, then

$$||M(\theta, \varphi) - I_m|| < 1. \tag{A.4}$$

The invertibility of $M(\theta, \varphi)$ will follow since the series with general term

$$S_k = \sum_{j=0}^{k} (-1)^k (M - I_m)^k$$

will then converge towards a limit S such that $MS = SM = I_m$. Now,

$$\begin{aligned} ||M(\theta, \varphi) - I_d|| &\leq \int_{-a}^{a} \theta(x)||\varphi(x)^{-1} - I_m||dx \\ &\leq \sup_{-a \leq x \leq a} ||\varphi(x)^{-1} - I_m||. \end{aligned}$$

Since φ is continuous we will have $||\varphi(x)^{-1} - I_m|| < 1$ if a is small enough, hence (A.4), and we have proven that φ is differentiable. Let us next show that there exists a real matrix X such that $\varphi(t) = \exp(tX)$. Differentiating both sides of the equality $\varphi(t+t') = \varphi(t) + \varphi(t')$ with respect to t' and setting thereafter $t' = 0$, we get

$$\frac{d}{dt}\varphi(t) = \varphi(t)X$$

where X is the derivative of φ at 0. This is equivalent to the equation

$$\frac{d}{dt}\left(\exp(-tX)\varphi(t)\right) = 0$$

and hence

$$\varphi(t) = \exp(tX)\varphi(0) = \exp(tX)$$

as claimed. □

Appendix B

Covering Spaces and Groups

We briefly review the elementary theory of covering spaces. For complete proofs one can consult any book on algebraic topology (a few good references are Seifert–Threlfall [148], Spanier [157], or Singer and Thorpe [154]).

Let M be a topological manifold, that is, a topological space which is locally homeomorphic to some Euclidean space $\mathbb{R}^m$, and G will be a topological manifold with an additional compatible group structure.

Put (very) concisely, a covering is a locally trivial fibre bundle with discrete fibre. Let us unfold this definition a bit. Choosing a base point m_0 in M one denotes by $\tilde{M}$ the set of all homotopy classes of continuous paths joining m_0 to the points of M (the homotopy relation considered here is the usual homotopy "with fixed endpoints"). Let us denote the class of a path joining m_0 to m by $\tilde{m}$, and define a mapping $\pi : \tilde{M} \longrightarrow M$ by $\pi(\tilde{m}) = m$. One proves that there exists a topology on $\tilde{M}$ such that $\tilde{M}$ is simply connected and π becomes a continuous function such that

- for every point $m \in M$ there exists an open neighborhood U_m of m in M and a discrete set F_m such that $\pi^{-1}(U_m)$ is the union of pairwise disjoint open subsets $U_m^{(k)}$ $(k \in F_m)$ of $\tilde{M}$;
- the restriction of π to each $U_m^{(k)}$ is a homeomorphism $U_m^{(k)} \longrightarrow U_m$ (in particular, π is a local homeomorphism).

One says, committing a slight, but convenient, abuse of terminology, that $\tilde{M}$ is the universal covering of M (where it is understood that the base point is fixed once for all); $\pi : \tilde{M} \longrightarrow M$ is called the "covering mapping" and the inverse image $\pi^{-1}(m)$ is called the fiber over m. When M is connected, all fibers have the same cardinality; one moreover shows that when M is a differential manifold, then $\tilde{M}$ is equipped with a differentiable structure for which π becomes a local diffeomorphism such that $d_m\pi$ has maximal rank m at each point m.

There is a natural action of the Poincaré group $\pi_1[M] = \pi_1[M, m_0]$ on $\tilde{M}$: let $\tilde{\gamma}$ be the homotopy class of a loop γ in M originating and ending at m_0, and let $\tilde{m}$ be the homotopy class of a path μ joining m_0 to m in M. The homotopy class of the concatenation $\gamma * \mu$ (*i.e.*, the loop γ followed by the path μ) is denoted by $\tilde{\gamma}\tilde{m}$; the action

$$\pi_1[M] \times \tilde{M} \longrightarrow \tilde{M}$$

thus defined is transitive on the fibers.

Regular coverings

Let Γ be a subgroup of $\pi_1[M]$; we denote by $\Gamma\tilde{m}$ the set $\{\tilde{\gamma}\tilde{m} : \tilde{\gamma} \in \Gamma\}$, and

$$M_\Gamma = \{\Gamma\tilde{m} : \tilde{m} \in \tilde{M}\}.$$

The mapping

$$\pi_\Gamma : M_\Gamma \longrightarrow M \quad , \quad \Gamma\tilde{m} \longmapsto m$$

is called the covering of M associated with the subgroup Γ of $\pi_1[M]$; we will use the shorthand notation

$$M_\Gamma = \tilde{M}/\Gamma \quad (\text{hence } \ M = \tilde{M}/\pi_1[M]).$$

If Γ is a normal subgroup of $\pi_1[M]$, then $\tilde{\gamma}(\Gamma\tilde{m}) = \Gamma(\tilde{\gamma}\tilde{m})$ for every $\tilde{\gamma} \in \Gamma$, hence $\pi_1[M]$, or rather $\pi_1[M]/\Gamma$, acts on $M_\Gamma = \tilde{M}/\Gamma$ and

$$\pi_1[M]/\Gamma = \pi_1(M_\Gamma).$$

The covering M_Γ is in this case called a regular covering of M. Notice that

$$M = M_\Gamma/(\pi_1[M]/\Gamma) = (\tilde{M}/\Gamma)/(\pi_1[M]/\Gamma) = \tilde{M}/\pi_1[M].$$

The order of a covering M_Γ is the (constant) number of elements of each fibre $\pi_\Gamma^{-1}(m)$. It is equal to the order of the group $\pi_1[M]/\Gamma$. We have:

Proposition B.1. *If $\pi_1[M] = (\mathbb{Z}, +)$, then the only covering of M having infinite order is its universal covering.*

Proof. Let M_Γ be a covering; Γ is thus a subgroup of $(\mathbb{Z}, +)$ and hence consists of the multiples of some integer k. If $k \neq 0$ then the quotient group $\mathbb{Z}/k\mathbb{Z}$ is finite, and so is the order of M_Γ. If $k = 0$, the covering it defines is $\tilde{M}$. □

Appendix C

Pseudo-Differential Operators

In traditional pseudo-differential calculus, as practiced by most mathematicians working in the theory of partial differential equations, one associates to a suitable "symbol" $a \in C^\infty(\mathbb{R}_z^{2n})$ an operator A defined, for $\psi \in C_o^\infty(\mathbb{R}_x^n)$ (or $\mathcal{S}(\mathbb{R}_x^n)$), by the formula

$$A\psi(x) = \left(\tfrac{1}{2\pi}\right)^{n/2} \int e^{i\langle p,x\rangle} a(x,p)\widehat{\psi}(p)d^n p \tag{C.1}$$

where $\widehat{\psi}$ is the Fourier transform of ψ defined by

$$\widehat{\psi}(p) = \left(\tfrac{1}{2\pi}\right)^{n/2} \int e^{-i\langle p,y\rangle} \psi(y)d^n y.$$

Definition (C.1) is motivated by the fact that if a is a polynomial in the variables $p_1, \ldots, p_n$ with coefficients depending on x, then A is an ordinary partial differential operator which can be immediately "read" from a by replacing the powers p^α, $\alpha \in \mathbb{N}^n$, by $i^{-|\alpha|}\partial_x^\alpha$ (we refer to the Preface for the multi-index notations that we use here).

Thus, setting $D_x^\alpha = i^{-|\alpha|}\partial_x^\alpha$, to the polynomial

$$a(x,p) = \sum_{|\alpha| \leq m} a_\alpha(x)p^\alpha \quad , \quad a_\alpha \in C^\infty(\mathbb{R}_x^n)$$

(m an integer) corresponds the operator

$$A = \sum_{|\alpha| \leq m} a_\alpha(x)D_x^\alpha.$$

C.1 The Classes $S^m_{\rho,\delta}$, $L^m_{\rho,\delta}$

One of the most used class of symbols is exhibited in the following definition:

Definition C.1. We say that a is a classical pseudo-differential symbol on $\mathbb{R}^{2n}_z$ if $a \in C^\infty(\mathbb{R}^{2n}_z)$ and if there exist real numbers m, ρ, δ with $0 \leq \rho < \delta \leq 1$ and such that for all multi-indices α and β in $\mathbb{N}^n$ and every compact subset K of $\mathbb{R}^n_x$ we can find a constant $C_{\alpha,\beta,K} > 0$ such that

$$|\partial_\alpha \partial_\beta a(x,p)| \leq C_{\alpha,\beta,K}(1+|p|)^{m-|\beta|} \tag{C.2}$$

for all $(x,p) \in \mathbb{R}^{2n}_z$. The vector space of all a satisfying (C.2) is denoted by $S^m_{\rho,\delta}(\mathbb{R}^{2n}_z)$.

We have of course the trivial inclusions

$$\mathcal{S}(\mathbb{R}^{2n}_z) \subset S^m_{\rho,\delta}(\mathbb{R}^{2n}_z).$$

The vector space of pseudo-differential operators (C.1) with symbols in $S^m_{\rho,\delta}(\mathbb{R}^{2n}_z)$ is denoted by $L^m_{\rho,\delta}(\mathbb{R}^{2n}_z)$; when $\delta = 0$ and $\rho = 1$, we use the notations $S^m(\mathbb{R}^{2n}_z)$ and $L^m(\mathbb{R}^{2n}_z)$

A classical result is then the following:

Theorem C.2. *If $A \in L^m_{\rho,\delta}(\mathbb{R}^{2n}_z)$, then A is a continuous operator $\mathcal{S}(\mathbb{R}^n_x) \longrightarrow \mathcal{S}(\mathbb{R}^n_x)$ which extends into a continuous operator $\mathcal{S}'(\mathbb{R}^n_x) \longrightarrow \mathcal{S}'(\mathbb{R}^n_x)$.*

Notice that in general $A \in L^m_{\rho,\delta}(\mathbb{R}^{2n}_z)$ does not map the space $C_0^\infty(\mathbb{R}^n_x)$ of compactly supported C^∞ functions into itself. One however proves that there exists an operator $R \in L^{-\infty}(\mathbb{R}^{2n}_z)$ (*i.e.*, $R \in L^m(\mathbb{R}^{2n}_z)$ for every $m \in \mathbb{R}$) such that $A = A_0 + R$ and $A_0 : C_0^\infty(\mathbb{R}^n_x) \longrightarrow C_0^\infty(\mathbb{R}^n_x)$). The operator R is "smoothing" in the sense that $R : \mathcal{E}'(\mathbb{R}^{2n}_z) \longrightarrow C_0^\infty(\mathbb{R}^n_x)$.

C.2 Composition and Adjoint

Let us introduce the following notation: given an $a \in S^m_{\rho,\delta}(\mathbb{R}^{2n}_z)$, $a_0 \in S^m_{\rho,\delta}(\mathbb{R}^{2n}_z)$ and a sequence $(a_j)_{j\in\mathbb{N}}$ with $a_j \in S^{m_j}_{\rho,\delta}(\mathbb{R}^{2n}_z)$ where $m > m_1 > m_2 > \cdots$ and $\lim_{j\to\infty} m_j = -\infty$, we write $a \sim \sum_{j\geq 0} a_j$ when

$$a - \sum_{j=0}^{N-1} a_j \in S^{m_N}_{\rho,\delta}(\mathbb{R}^{2n}_z) \quad \text{for} \quad N \geq 1.$$

Let $A \in L^m_{\rho,\delta}(\mathbb{R}^{2n}_z)$ have symbol a. The adjoint A^* is also a pseudo-differential operator and its symbol b is then determined by the asymptotic expansion

$$b(x,p) \sim \sum_{\alpha\in\mathbb{N}^n} \frac{1}{\alpha!} \partial_p^\alpha D_x^\alpha a(x,p).$$

Note that this formula is very complicated compared to the easy rule used when one deals with Weyl pseudo-differential operators.

Let $A \in L^{m_1}_{\rho_1,\delta_1}(\mathbb{R}^{2n}_z)$, $B \in L^{m_2}_{\rho_2,\delta_2}(\mathbb{R}^{2n}_z)$ have respective symbols a and b. Assume that the composed operator $A \circ B$ exists (this can always be assumed to be true replacing A by A_0 such that $A - A_0 = R \in L^{-\infty}(\mathbb{R}^{2n}_z)$). Then $C = A \circ B$ is a pseudo-differential operator

$$C \in L^{m_1+m_2}_{\rho,\delta}(\mathbb{R}^{2n}_z) \ , \ \rho = \min\{\rho_1, \rho_2\} \ , \ \delta = \max\{\delta_1, \delta_2\}$$

and symbol c is determined by

$$c(x,p) \sim \sum_{\alpha \in \mathbb{N}^n} \frac{1}{\alpha!} D^\alpha_x a(x,p) \partial^\alpha_p b(x,p).$$

Appendix D

Basics of Probability Theory

Let us begin by introducing some notation and definitions.

D.1 Elementary Concepts

A *probability density* on $\mathbb{R}^m$ is any integrable function $\rho : \mathbb{R}^m \longrightarrow \mathbb{R}$ such that

$$\rho \geq 0 \quad \text{and} \quad \int \rho(z) d^m z = 1.$$

We will assume in what follows that $z_j^k \rho \in L^1(\mathbb{R}^m)$ for $k = 1, 2$; this property holds for instance when $\rho \in \mathcal{S}(\mathbb{R}^m)$. Let Z be a continuous function $\mathbb{R}^m \longrightarrow \mathbb{R}$; we will view Z as a real-valued *random variable* associated with the probability density ρ. That is, if Ω is a Borel subset of $\mathbb{R}^m$, the number

$$\Pr(Z \in \Omega) = \int_\Omega \rho(z) d^m z$$

is the "probability that the value of z is in Ω".

Definition D.1. The "*mathematical expectation*" (also called "*mean value*") of the random variable Z is

$$\langle Z \rangle = \int Z(z) \rho(z) d^m z$$

and the "*variance*" of Z is

$$\operatorname{Var}(Z) = \langle (Z - \langle Z \rangle)^2 \rangle = \langle Z^2 \rangle - \langle Z \rangle^2 .$$

The square root $\Delta Z = \sqrt{\operatorname{Var}(Z)}$ is called the "*standard deviation*" of Z.

Here are a few other concepts we will use. The convolution product

$$(\rho * \rho')(z) = \int \rho(z-u)\rho'(u)d^n u \tag{D.1}$$

of two probability densities ρ and ρ' is again a probability density: obviously $\rho(x-u)\rho'(u)$ is non-negative for all z and u, and we have

$$\int (\rho * \rho')(z)d^m z = \int \left(\int \rho(z-u)d^m z \right) \rho'(u)d^m u = 1.$$

In fact:

Proposition D.2. *Let ρ and ρ' be probability densities corresponding to independent random variables X and X'. The probability density of the sum $X + X'$ is the convolution $\rho * \rho'$.*

The function $\varphi_\rho : \mathbb{R}^m \longrightarrow \mathbb{R}$ defined by

$$\varphi_\rho(\lambda) = \int e^{i\langle \lambda, z \rangle} \rho(z)d^m z$$

is called the *characteristic function* of the probability density ρ; it is essentially its Fourier transform, and we have

$$\varphi_{\rho * \rho'}(\lambda) = \varphi_\rho(\lambda)\varphi_{\rho'}(\lambda). \tag{D.2}$$

Let now $Z_1, \ldots, Z_m$ be a finite sequence of random variables of the type above. We will call $Z = (Z_1, \ldots, Z_m)$ a (real) continuous vector-valued random variable. By definition

$$\mathrm{Cov}(Z_j, Z_k) = \langle (Z_j - \langle Z_j \rangle)(Z_k - \langle Z_k \rangle) \rangle$$

is the *covariance* of the pair (Z_j, Z_k); this can be alternatively written as

$$\mathrm{Cov}(Z_j, Z_k) = \langle Z_j Z_k \rangle - \langle Z_j \rangle \langle Z_k \rangle .$$

Obviously, for every random variable Z, $\mathrm{Cov}(Z, Z) = \mathrm{Var}(Z)$. The quotient

$$\rho(Z_j, Z_k) = \frac{\mathrm{Cov}(Z_j, Z_k)}{\Delta Z_j \Delta Z_j} = \rho(Z_k, Z_j)$$

is its *correlation coefficient* of the pair Z_j, Z_k; we always have $-1 \leq \rho(Z_k, Z_j) \leq 1$. In fact:

- We have $|\rho(Z_j, Z_k)| \leq 1$ for all j, k and equality occurs if and only if $Z_j = aZ_k + b$ for some $a, b \in \mathbb{R}$ (and hence that $\rho(Z_j, Z_j) = 1$).
- Let $Z = (Z_1, \ldots, Z_m)$ and $U = (U_1, \ldots, U_m)$ be vector-valued random variables and A, B invertible $m \times m$ matrices. We have

  $$\langle AU + BZ \rangle = A \langle U \rangle + B \langle Z \rangle$$

  where $\langle U \rangle = (\langle U_1 \rangle, \ldots, \langle U_n \rangle)$ and $\langle Z \rangle = (\langle Z_1 \rangle, \ldots, \langle Z_n \rangle)$.

D.2 Gaussian Densities

Proposition D.3. *Let $\Sigma > 0$ and define, for $z \in \mathbb{R}^m$,*

$$\rho(z) = \left(\tfrac{1}{2\pi}\right)^{m/2} \det(\Sigma^{-1/2}) e^{-\frac{1}{2}\langle \Sigma^{-1}(z-\bar{z}), z-\bar{z}\rangle}. \tag{D.3}$$

(i) *The function ρ is a probability density on $\mathbb{R}^m$; (ii) Let Z be the vector-valued random variable associated to ρ; we have $\langle Z\rangle = \bar{z}$;*
(iii) *The covariance matrix of Z is Σ.*

Proof. (i) Clearly $\rho > 0$; let us show that the integral of ρ over $\mathbb{R}^m$ is equal to one. We can diagonalize Σ by an orthogonal matrix R, and the proof thus reduces to showing that

$$\frac{1}{\sqrt{2\pi\lambda}} \int_{-\infty}^{\infty} e^{-u^2/2\lambda} du = 1 \quad \text{for} \quad \lambda > 0;$$

this equality immediately follows, changing variables, from the classical Gauss integral

$$\frac{1}{\sqrt{2\pi}} \int_{-\infty}^{\infty} e^{-u^2/2} du = 1.$$

(ii) Setting $y = z - \bar{z}$ we have

$$\begin{aligned}
\langle Z_i\rangle - \bar{z}_i &= \int z_i \rho(z) d^m z - \bar{z}_i \int z_i \rho(z) d^m z \\
&= \left(\tfrac{1}{2\pi}\right)^{m/2} \det(\Sigma^{-1/2}) \int y_i e^{-\frac{1}{2}\langle \Sigma^{-1}y, y\rangle} d^m y
\end{aligned}$$

and the last integral is zero since the integrand is an odd function; hence $\langle Z_i\rangle = \bar{z}_i$ as claimed.

(iii) Set $\xi_{ij} = \operatorname{Cov}(Z_i, Z_j)$; by definition $\xi_{ij} = \langle Z_i Z_j\rangle - \langle Z_i\rangle\langle Z_j\rangle$ hence, performing again the change of variables $y = z - \bar{z}$,

$$\xi_{ij} = \left(\tfrac{1}{2\pi}\right)^{m/2} \det(\Sigma^{-1/2}) \int y_i y_j e^{-\frac{1}{2}\langle \Sigma^{-1}y, y\rangle} d^m y.$$

Let now $R = (r_{ij})_{1\leq i,j\leq m}$ be an orthogonal matrix such that $D^{-1} = R\Sigma^{-1}R^T$ is diagonal; setting $y = Ru$, we have

$$\xi_{ij} = \left(\tfrac{1}{2\pi}\right)^{m/2} \det(D^{-1/2}) \int f_{ij}(u) e^{-\frac{1}{2}\langle D^{-1}u, u\rangle} d^m u,$$

where the functions f_{ij} are given by

$$f_{ij}(u) = \sum_{k,\ell=1}^{m} r_{ik} r_{j\ell} u_k u_\ell.$$

Noting that

$$\int u_k u_\ell e^{-\frac{1}{2}\langle \Sigma^{-1}u,u\rangle} d^m u = 0 \quad \text{for } k \neq \ell$$

the formula for ξ_{ij} reduces to

$$\xi_{ij} = \left(\tfrac{1}{2\pi}\right)^{m/2} \det(D^{-1/2}) \sum_{\ell=1}^{m} r_{i\ell} r_{j\ell} \int u_\ell^2 e^{-\frac{1}{2}\langle D^{-1}u,u\rangle} d^m u. \tag{D.4}$$

Writing $D = \operatorname{diag}[\delta_1, \ldots, \delta_m]$ we have

$$\int u_k^2 e^{-\frac{1}{2}\langle D^{-1}u,u\rangle} d^m u = \int e^{-\frac{1}{2}\langle \hat{D}^{-1}\hat{u},\hat{u}\rangle} d^{m-1}\hat{u} \int_{-\infty}^{\infty} t^2 e^{-t^2/2d} dt$$

where $\hat{v} = (u_1, \ldots, \hat{u}_k, \ldots, u_m)$, $\hat{D} = \operatorname{diag}[\delta_1, \ldots, \hat{\delta}_k, \ldots, \delta_m]$ (the cap ^ suppressing the term it covers); using the elementary formula

$$\int_{-\infty}^{\infty} t^2 e^{-t^2/2\delta} dt = \delta^{3/2}\sqrt{2}\Gamma(\tfrac{3}{2}) = \delta^{3/2}\sqrt{2\pi}$$

we thus have

$$\int u_k^2 e^{-\frac{1}{2}\langle D^{-1}u,u\rangle} d^m u = (2\pi)^{m/2}\delta_k^{3/2}$$

and hence (D.4) becomes $\xi_{ij} = \sum_{\ell=1}^{m} r_{i\ell} r_{j\ell}\delta$. The sum on the right-hand side being the ith row and jth column entry of $RDR^T = \Sigma$, this formula concludes the proof. □

Proposition D.4. *Let ρ_Σ and $\rho_{\Sigma'}$ be two Gaussian probability densities centered at $\bar{z}$ and $\bar{z}'$, respectively; then $\rho_\Sigma * \rho_{\Sigma'} = \rho_{\Sigma''}$ where $\Sigma'' = \Sigma + \Sigma'$ and $\rho_{\Sigma''}$ is centered at $\bar{z}'' = \bar{z} + \bar{z}'$.*

Proof. It is sufficient to consider the case $\bar{z} = \bar{z}' = 0$. The Fourier transform of ρ_Σ is

$$F\rho_\Sigma(\zeta) = \left(\frac{1}{2\pi}\right)^{m/2} \int e^{-i\langle\zeta,z\rangle}\rho_\Sigma(z) d^m z = \left(\frac{1}{2\pi}\right)^{m} e^{-\frac{1}{2}\langle\Sigma\zeta,\zeta\rangle},$$

hence

$$F\rho_\Sigma(\zeta)F\rho_{\Sigma'}(\zeta) = \left(\frac{1}{2\pi}\right)^{2m} e^{-\frac{1}{2}\langle(\Sigma+\Sigma')\zeta,\zeta\rangle}.$$

We have

$$F(\rho_\Sigma * \rho_{\Sigma'})(\zeta) = (2\pi)^m F\rho_\Sigma(\zeta)F\rho_{\Sigma'}(\zeta)$$

so that $\rho_\Sigma * \rho_{\Sigma'} = \rho_{\Sigma+\Sigma'}$ as claimed. □

Solutions to Selected Exercises

Solution of Exercise 1.12 (ii). Suppose indeed that there exists a symplectic form σ on S^{2n}, $n > 1$. Then $\sigma^{\wedge n}$ would be a volume form. Since $H^k(S^{2n}) = 0$ for $k \neq 0$ and $k \neq 2n$ the symplectic form is exact: $\sigma = d\beta$ for some one-form β on S^{2n}; it follows that $\sigma^{\wedge n}$ must also be exact, in fact $\sigma^{\wedge n} = d(\beta \wedge \sigma^{\wedge(n-1)})$. In view of Stoke's theorem we would then have

$$\int_{S^{2n}} \sigma^{\wedge n} = \int_{\partial S^{2n}} \beta \wedge \sigma^{\wedge(n-1)} = 0$$

which is absurd. (This example generalizes to any $2n$-dimensional compact manifold such that $H^k(M) = 0$ for $k \neq 0$ and $k \neq 2n$.)

Solution of Exercise 1.6. Let us prove that if $U = A+iB$ is any matrix, symplectic, or not (A and B being real $n \times n$ matrices), then the determinant of the $2n \times 2n$ block matrix

$$U = \begin{bmatrix} A & -B \\ B & A \end{bmatrix}$$

is given by the simple formula

$$\det U = |\det(A + iB)|^2. \tag{D.5}$$

This is easily seen by block-diagonalizing U as follows:

$$\begin{bmatrix} I_n & -iI_n \\ -iI_n & I_n \end{bmatrix} \begin{bmatrix} A & -B \\ B & A \end{bmatrix} \begin{bmatrix} I_n & iI_n \\ iI_n & I_n \end{bmatrix} = 4^n \begin{bmatrix} A+iB & 0 \\ 0 & A-iB \end{bmatrix}$$

and computing the determinants. In fact,

$$\det \begin{bmatrix} I_n & -iI_n \\ -iI_n & I_n \end{bmatrix} \begin{bmatrix} I_n & iI_n \\ iI_n & I_n \end{bmatrix} = \det \begin{bmatrix} 2I_n & 0 \\ 0 & 2I_n \end{bmatrix} = 4^n$$

hence

$$\begin{aligned} \det \begin{bmatrix} A & -B \\ B & A \end{bmatrix} &= \det \begin{bmatrix} A+iB & 0 \\ 0 & A-iB \end{bmatrix} \\ &= \det(A+iB)\det(A-iB) = \det U \det \overline{U} \end{aligned}$$

which is just (D.5).

Solution of Exercise 5.15. (i) Writing S in block-matrix form

$$S = \begin{bmatrix} A & B \\ C & D \end{bmatrix},$$

the condition that S is symplectic implies that A^TC and B^TD are symmetric, and that $A^TD - C^TB = I$. Setting $x_S = Ax + Bp$, $p_S = Cx + Dp$, and expanding the products, we get

$$\begin{aligned} p_S dx_S - x_S dp_S &= (A^TCx + A^TDp - C^TAx - C^TBp)dx \\ &+ (B^TCx + B^TDp - D^TAx - D^TBp)dp \\ &= pdx - xdp \end{aligned}$$

proving (5.8). (Notice that in general we do *not have* $p_S dx_S = pdx$.).
(ii) Differentiating the right-hand side of (5.9) we get, since $d\varphi(\check{z}) = pdx$,

$$\begin{aligned} d\varphi_S(\check{z}) &= \tfrac{1}{2}(pdx - xdp) + \tfrac{1}{2}d\langle p_S, x_S\rangle \\ &= \tfrac{1}{2}(p_S dx_S - x_S dp_S) + \tfrac{1}{2}d\langle p_S, x_S\rangle \\ &= p_S dx_S. \end{aligned}$$

Solution of Exercise 2.58. Let us construct explicitly a homotopy of the first path on the second, that is, a continuous mapping

$$h : [0,1] \times [0,1] \longrightarrow \operatorname{Ham}(n)$$

such that $h(t,0) = f_t^H f_t^K$ and $h(t,1) = f_t$. Define h by $h(t,s) = a(t,s)b(t,s)$ where a and b are functions,

$$a(t,s) = \begin{cases} I & \text{for } 0 \leq t \leq \frac{s}{2}, \\ f^H_{(2t-s)/(2-s)} & \text{for } \frac{s}{2} \leq t \leq 1, \end{cases}$$

$$b(t,s) = \begin{cases} f^K_{2t/(2-s)} & \text{for } 0 \leq t \leq 1 - \frac{s}{2}, \\ f_1^K & \text{for } \frac{s}{2} \leq t \leq 1. \end{cases}$$

We have $a(t,0) = f_t^H$, $b(t,0) = f_t^K$ hence $h(t,0) = f_t^H f_t^K$; similarly

$$h(t,1) = \begin{cases} f^K_{2t} & \text{for } 0 \leq t \leq \frac{1}{2}, \\ f^H_{2t-1} f_1^K & \text{for } \frac{1}{2} \leq t \leq 1, \end{cases}$$

that is $h(t,1) = f_t$.

Solution of Exercise 5.48. The phase of $T(z_a)\mathbb{V}^n$ is

$$\varphi_a(\check{z}) = \varphi(\check{z}) + \tfrac{1}{2}\langle p_a, x_a\rangle + \langle p_a, x\rangle,$$

hence that of $S_t^H(T(z_a)\mathbb{V}^n)$ is (using (5.36) and the linearity of S_t^H):

$$A(t) = \varphi(\check{z}_0) + \tfrac{1}{2}\langle p_a, x_a\rangle + \langle p_a, x\rangle \\ + \tfrac{1}{2}\langle p_{0,t} + p_{a,t}, x_t + x_{a,t}\rangle - \tfrac{1}{2}\langle p + p_a, x + x_a\rangle$$

where $z_{0,t} = S_t^H z_0$, $z_{a,t} = S_t^H z_a$. Similarly, the Hamiltonian phase of $S_t^H\mathbb{V}^n$ is

$$\varphi(\check{z}, t) = \varphi(\check{z}_0) + \tfrac{1}{2}(\langle p_t, x_t\rangle - \langle p, x\rangle),$$

hence that of $T(S_t^H(z_a))\mathbb{V}^n$ is

$$B(t) = \varphi(\check{z}) + \tfrac{1}{2}(\langle p_t, x_t\rangle - \langle p, \rangle) + \tfrac{1}{2}\langle p_{a,t}, x_{a,t}\rangle + \langle p_{a,t}, x_t\rangle$$

and thus

$$\begin{aligned} A(t) - B(t) &= \tfrac{1}{2}(\langle p_a, x\rangle - \langle p, x_a\rangle) - \tfrac{1}{2}(\langle p_{a,t}, x_t\rangle - \langle p_t, x_{a,t}\rangle) \\ &= \tfrac{1}{2}(\sigma(z_a, z) - \sigma(z_{a,t}, z_t)) \\ &= \tfrac{1}{2}(\sigma(z_a, z) - \sigma(S_t^H z_a, S_t^H z)). \end{aligned}$$

Since $S_t^H \in \operatorname{Sp}(n)$ we have $\sigma(S_t^H z_a, S_t^H z) = \sigma(z_a, z)$ and hence $A(t) = B(t)$.

Solution of Exercise 2.47. The condition $X_t \in \mathfrak{sp}(n)$ is equivalent to JX_t being symmetric. Hence

$$\frac{d}{dt}(S_t^T J S_t) = S_t^T X_t^T J S_t + S_t^T J X_t S_t = 0$$

so that $S_t^T J S_t = S_0^T J S_0 = J$ and $S_t \in \operatorname{Sp}(n)$ as claimed.

Solution of Exercise 8.29. Write the covariance matrix in the form

$$\Sigma = \begin{bmatrix} \Sigma_{XX} & \Sigma_{XP} \\ \Sigma_{PX} & \Sigma_{PP} \end{bmatrix} \quad , \quad \Sigma_{PX} = \Sigma_{XP}^T,$$

that is

$$\begin{aligned} \Sigma_{XX} &= \operatorname{Cov}(X_i, X_j)_{1\leq i,j\leq n} \quad , \quad (\Delta X_j)^2 = \operatorname{Cov}(X_j, X_j), \\ \Sigma_{PP} &= \operatorname{Cov}(P_i, P_j)_{1\leq i,j\leq n} \quad , \quad (\Delta P_j)^2 = \operatorname{Cov}(P_j, P_j) \end{aligned}$$

and $\Sigma_{XP} = \operatorname{Cov}(X_i, P_j)_{1\leq i,j\leq n}$. We have

$$\begin{vmatrix} (\Delta X_i)^2 + \varepsilon & \operatorname{Cov}(X_i, P_i) + \frac{i}{2}\hbar \\ \operatorname{Cov}(X_i, P_i) - \frac{i}{2}\hbar & (\Delta P_i)^2 + \varepsilon \end{vmatrix} > 0$$

for every $\varepsilon > 0$, hence

$$(\Delta X_i)^2(\Delta P_i)^2 - (\operatorname{Cov}(X_i, P_i)^2 + \tfrac{1}{4}\hbar^2) \geq 0.$$

Solution of Exercise 6.7 Set

$$g(z) = \mathcal{F}_\sigma f(z) = \left(\tfrac{1}{2\pi\hbar}\right)^n \int e^{\frac{i}{\hbar}\sigma(z,z')} f(z')d^{2n}z'.$$

We have

$$g(z) = \left(\tfrac{1}{2\pi\hbar}\right)^n \int e^{-\frac{i}{\hbar}\langle z,z'\rangle} f(-Jz')d^{2n}z' = F(f \circ (-J))$$

and hence

$$f(-Jz) = F^{-1}g(z) = \left(\tfrac{1}{2\pi\hbar}\right)^n \int e^{\frac{i}{\hbar}\langle z,z'\rangle} g(z')d^{2n}z'$$

so that

$$f(z) = \left(\tfrac{1}{2\pi\hbar}\right)^n \int e^{\frac{i}{\hbar}\sigma(z,z')} g(z')d^{2n}z' = \mathcal{F}_\sigma g(z).$$

Solution of Exercise 9.25. We have

$$\widehat{A}\Psi = \sum_{j=1}^{\infty} c_j(\Psi)\lambda_j\psi_j \quad , \quad c_j(\Psi) = (\Psi, \psi_j)_{\mathcal{H}}$$

and hence

$$\rho_\Psi \widehat{A}(\psi_j) = (\widehat{A}\psi_j, \Psi)_{\mathcal{H}}\Psi = \lambda_j(\psi_j, \Psi)_{\mathcal{H}}\Psi,$$

that is

$$\rho_\Psi \widehat{A}(\psi_j) = \sum_{k=1}^{\infty} \lambda_j \overline{c_j(\Psi)} c_k(\Psi).$$

The trace of $\widehat{\rho}_\Psi \widehat{A}$ is the convergent series

$$\operatorname{Tr}(\widehat{\rho}_\Psi \widehat{A}) = \sum_{k=1}^{\infty} \lambda_j |c_j(\Psi)|^2;$$

but this is just the expectation $\langle \widehat{A} \rangle_\Psi$.

Solution of Exercise 8.46. It is clear that $\rho \geq 0$. To prove that

$$\int_{-\infty}^{\infty} \rho(z)dpdx = 1$$

it suffices to use the change of variables defined by

$$du = \rho_X(x)dx \quad , \quad dv = \rho_X(p)dp.$$

Solution of Exercise 5.20. The function $\Phi_0(x) = x\sqrt{2mE}$ is a solution of the time-independent Hamilton–Jacobi equation and thus $\Phi(x,t) = x\sqrt{2mE} - Et$. Setting $\alpha = \sqrt{2mE}$ this yields the complete solution

$$\Phi(x,t) = \alpha x - \frac{\alpha^2}{2m}t$$

which depends on the parameter α.

Solution of Exercise 6.34. We have, by definition of $\mathcal{F}_\sigma$ and $*_{-2\hbar}$,

$$(\mathcal{F}_\sigma a) *_\sigma b(z) = \left(\tfrac{1}{2\pi\hbar}\right)^n \int e^{-\frac{i}{\hbar}\sigma(z,z')} \left(\int e^{-\frac{i}{\hbar}\sigma(z-z',z'')} a(z'')d^{2n}z''\right) b(z')d^{2n}z'$$
$$= \left(\tfrac{1}{2\pi\hbar}\right)^n \iint e^{-\frac{i}{\hbar}(\sigma(z,z')+\sigma(z-z',z''))} a(z'')b(z')d^{2n}z'd^{2n}z'';$$

setting $z'' = u - v$ and $z' = v$ we have $d^{2n}z'd^{2n}z'' = d^{2n}ud^{2n}v$ and hence

$$(\mathcal{F}_\sigma a) *_\sigma b(z) = \left(\tfrac{1}{2\pi\hbar}\right)^n \iint e^{-\frac{i}{\hbar}(\sigma(z,u)+\sigma(u,v))} a(u-v)b(v)d^{2n}ud^{2n}v$$
$$= \left(\tfrac{1}{2\pi\hbar}\right)^n \int e^{-\frac{i}{\hbar}\sigma(z,u)} \left[\int e^{-\frac{i}{\hbar}\sigma(u,v)} a(u-v)b(v)d^{2n}v\right] d^{2n}u$$
$$= \mathcal{F}_\sigma(a *_\sigma b)(z)$$

which proves the first equality (6.50). The second equality is proven likewise; alternatively it follows from the first using (6.48). Formula (6.51) follows since $\mathcal{F}_\sigma$ is involutive.

Solution of Exercise 6.42. We have, by definition of $\widehat{T}(z_0)$:

$$(\widehat{T}(z_0)\Psi)(x + \tfrac{1}{2}y) = e^{\frac{i}{\hbar}(\langle p_0, x+\frac{1}{2}y\rangle - \frac{1}{2}\langle p_0, x_0\rangle)} \Psi(x - x_0 + \tfrac{1}{2}y),$$
$$\overline{(\widehat{T}(z_0)\Psi)(x - \tfrac{1}{2}y)} = e^{-\frac{i}{\hbar}(\langle p_0, x-\frac{1}{2}y\rangle - \frac{1}{2}\langle p_0, x_0\rangle)} \overline{\Psi(x - x_0 + \tfrac{1}{2}y)},$$

and hence

$$W(\widehat{T}(z_0)\Psi)(z) = \left(\tfrac{1}{2\pi\hbar}\right)^n \int e^{-\frac{i}{\hbar}\langle p-p_0, y\rangle} \Psi(x - x_0 + \tfrac{1}{2}y)\overline{\Psi(x - x_0 + \tfrac{1}{2}y)}d^n y,$$

that is

$$T(z_0)W\Psi(z) = W(\widehat{T}(z_0)\Psi).$$

Solution of Exercise 8.38. We have

$$S = \begin{bmatrix} \lambda_1 & 0 & 0 & 0 \\ 0 & \lambda_2 & 0 & 0 \\ 0 & 0 & 1/\lambda_1 & 0 \\ 0 & 0 & 0 & 1/\lambda_2 \end{bmatrix}$$

so $S(B^{2n}(R))$ is the ellipsoid

$$\frac{1}{\lambda_1}x_1^2 + \frac{1}{\lambda_2}x_2^2 + \lambda_1 p_1^2 + \lambda_2 p_2^2 \leq R^2.$$

The intersection of that ellipsoid with the x_2, p_1 plane (which is not conjugate) is the ellipse

$$\frac{1}{\lambda_1}x_1^2 + \lambda_2 p_2^2 \leq R^2$$

which has area $\pi R^2\sqrt{\lambda_1/\lambda_2} \neq \pi R^2$.

Solution of Exercise 8.39. We have

$$S' = \begin{bmatrix} \lambda_1 & 0 & 0 & 0 \\ 0 & \lambda_2 & 0 & 0 \\ 0 & 0 & 1/\lambda_2 & 0 \\ 0 & 0 & 0 & 1/\lambda_1 \end{bmatrix}$$

and the blocks

$$A = \begin{bmatrix} \lambda_1 & 0 \\ 0 & \lambda_2 \end{bmatrix} \ , \ D = \begin{bmatrix} 1/\lambda_2 & 0 \\ 0 & 1/\lambda_1 \end{bmatrix}$$

do not satisfy the condition $A^T D = D^T A$ which is necessary for S' to be symplectic. The section $S'(B^{2n}(R))$ by the symplectic x_2, p_2 plane is the ellipse

$$\frac{1}{\lambda_1} x_1^2 + \lambda_2 p_1^2 \leq R^2$$

which has area $\pi R^2 \sqrt{\lambda_1/\lambda_2} \neq \pi R^2$.

Solution of Exercise 6.3 (ii). The isomorphism (6.9) is C^∞ and induces a Lie algebra isomorphism $d\phi(0,0) : \mathfrak{h}_n^{\mathrm{pol}} \longrightarrow \mathfrak{h}_n$; the Jacobian of ϕ at $(z,t) = (0,0)$ being the identity it follows that $\mathfrak{h}_n = \mathfrak{h}_n^{\mathrm{pol}}$. Let us determine $\mathfrak{h}_n^{\mathrm{pol}}$. We have $M(z,t) = I + m(z,t)$ where

$$m(z,t) = \begin{bmatrix} 0 & p^T & t \\ 0 & 0 & x \\ 0 & 0 & 0 \end{bmatrix};$$

$m(z,t)$ is nilpotent: $m(z,t)^k = 0$ for $k > 2$ and $m(z,t)^2 = m(0, \langle p, x \rangle)$ and hence

$$e^{m(z,t)} = I + m(z,t) + \frac{1}{2} m(0, \langle p, x \rangle = M(z, t + \tfrac{1}{2} \langle p, x \rangle)$$

so that $\mathfrak{h}_n^{\mathrm{pol}}$ consists of all matrices

$$X^{\mathrm{pol}}(z,t) = \begin{bmatrix} 1 & p^T & t - \frac{1}{2} \langle p, x \rangle \\ 0 & 1 & x \\ 0 & 0 & 1 \end{bmatrix}.$$

Solution of Exercise 6.3. This immediately follows from Proposition 10.18, (i), using the fact that U_ϕ is an isometry:

$$\int \rho(z,t) d^{2n} z = ||U_\phi \psi(\cdot, t)||_{L^2(\mathbb{R}_z^{2n})} = ||\psi(\cdot, t)||_{L^2(\mathbb{R}_x^n)}$$

and hence, since $||\psi(\cdot, t)||_{L^2(\mathbb{R}_x^n)}$ is conserved in time,

$$\int \rho(z,t) d^{2n} z = ||\psi(\cdot, 0)||_{L^2(\mathbb{R}_x^n)} = ||U_\phi \psi(\cdot, 0)||_{L^2(\mathbb{R}_z^{2n})} = \int \rho(z, 0) d^{2n} z.$$

Bibliography

[1] R. Abraham and J.E. Marsden. *Foundations of Mechanics.* The Benjamin/Cummings Publishing Company, Second Edition, 1978.

[2] R. Abraham, J.E. Marsden, and T. Ratiu. *Manifolds, Tensor Analysis, and Applications.* Applied Mathematical Sciences **75**, Springer, 1988.

[3] V.I. Arnold. *Mathematical Methods of Classical Mechanics.* Graduate Texts in Mathematics, second edition, Springer-Verlag, 1989.

[4] V.I. Arnold. A characteristic class entering in quantization conditions. *Funkt. Anal. i. Priloz.* **1**(1):1–14 (in Russian), 1967; *Funct. Anal. Appl.* **1**:1–14 (English translation), 1967.

[5] Arvind, B. Dutta, N. Mukunda and R. Simon. 1995 *The Real Symplectic Groups in Quantum Mechanics and Optics*; e-print arXiv:quant-ph/9509002 v3.

[6] A. Banyaga. Sur la structure du groupe des difféomorphismes qui préservent une forme symplectique. *Comm. Math. Helv.* **53**:174–227, 1978.

[7] F. Bayen, M. Flato, C. Fronsdal, A. Lichnerowicz, and D. Sternheimer. Deformation Theory and Quantization. I. Deformation of Symplectic Structures. *Annals of Physics* **111**, 6–110, 1978; II Physical Applications **110**, 111–151, 1978.

[8] S. Bergia and L. Navarro. On the early history of Einstein's quantization rule of 1917 (Preprint).

[9] E. Binz, S. Pods, and W. Schempp. Heisenberg groups – a unifying structure of signal theory, holography and quantum information theory. *J. Appl. Math. Comput.* **11**, 1–57, 2003.

[10] E. Binz, S. Pods, and W. Schempp. Heisenberg groups – the fundamental ingredient to describe information, its transmission and quantization. *J. Phys. A: Math. Gen.* **36**, 6401–6421, 2003.

[11] D. Bleecker and B. Booss-Bavnbek. *Index Theory with Applications to Mathematics and Physics.* Preprint 2004.

[12] D. Bohm. *Quantum Theory.* Prentice Hall, New York, 1951.

[13] M. Born and P. Jordan. Zur Quantenmechanik. *Zeitschrift für Physik* **34**, 858–888, 1925.

[14] B. Booss-Bavnbek and K. Furutani. The Maslov Index: a Functional Analytical Definition and the Spectral Flow Formula. *Tokyo J. Math.* **21**(1), 1998.

[15] B. Booss-Bavnbek and K.P. Wojciechowski. *Elliptic Boundary problems for Dirac Operators*, Birkhäuser, 1993.

[16] M. Brown. *The Symplectic and Metaplectic Groups in Quantum Mechanics and the Bohm Interpretation.* Ph.D. Thesis, University of London, 2004.

[17] N.G. de Bruijn. A theory of generalized functions, with applications to Wigner distribution and Weyl correspondence. *Nieuw Archiev voor Wiskunde,* **21**, 205–280, 1973.

[18] F.A. Berezin and M.A. Shubin. *The Schrödinger Equation.* Kluwer Publishers, 1991.

[19] V.C. Buslaev. Quantization and the W.K.B method. *Trudy Mat. Inst. Steklov* **110**:5–28, 1978 [in Russian].

[20] J. Butterfield. On Hamilton–Jacobi Theory as a Classical Root of Quantum Theory. Preprint submitted to the proceedings of *Quo Vadis Quantum Mechanics?* (Conference at Temple University, Sept. 2002, Ed. A. Elitzur and N. Kolenda, 2003.

[21] A. Cannas da Silva. *Lectures on Symplectic Geometry.* Springer Verlag, Berlin Heidelberg New York, 2001.

[22] S.E. Cappell, R. Lee, and E.Y. Miller. On the Maslov index. *Comm. Pure and Appl. Math.* **17** 121–186, 1994.

[23] Y. Choquet-Bruhat, C. DeWitt-Morette. *Analysis, Manifolds and Physics. Part II – Revised and Enlarged.* Elsevier Science Publishing Company, 2001.

[24] M. Combescure and D. Robert. Semiclassical spreading of quantum wave packets and applications to near unstable fixed points of the classical flow. Asymptotic Analysis **14** (1997) 377–404

[25] C. Conley and E. Zehnder. Morse-type index theory for flows and periodic solutions of Hamiltonian equations. *Comm. Pure and Appl. Math.* **37**:207–253, 1984.

[26] A. Crumeyrolle. *Orthogonal and Symplectic Clifford Algebras.* Kluwer Academic Publishers, 1990.

[27] R.H. Cushman and L.M. Bates. *Global Aspects of Classical Integrable Systems,* Birkhäuser, 1997.

[28] P. Dazord. Invariants homotopiques attachés aux fibrés symplectiques. *Ann. Inst. Fourier* **29**(2):25–78, 1979.

[29] M. Demazure. *Classe de Maslov II,* Exposé numéro 10, Séminaire sur le fibré cotangent, Orsay (1975–76).

[30] J. Dieudonné. *Foundations of Modern Analysis.* Translated from the French Academic Press.

[31] P.A.M. Dirac. *The Principles of Quantum Mechanics* (Oxford Science Publications, fourth revised edition, 1999).

[32] D. Dragoman. Phase space formulation of quantum mechanics. Insight into the measurement problem. Phys. Scripta **72**, 290–296, 2005.

[33] D.A. Dubin, M.A. Hennings, and T.B. Smith. *Mathematical Aspects of Weyl Quantization and Phase.* World Scientific, 2000.

[34] J.J. Duistermaat. On the Morse index in variational calculus. *Adv. in Math.* **21** 173–195, 1976.

[35] J.J. Duistermaat. On operators of trace class in $L^2(X, \mu)$. In "*Geometry and Analysis. Papers Dedicated to the Memory of V.K. Patodi*", Ed. Indian Academy of Science, 29–32, Springer-Verlag, 1981.

[36] Ju.V. Egorov and B.-W. Schulze. *Pseudo-differential operators, singularities, applications.* Vol. 93 of Operator Theory, Advances, and Applications. Birkhäuser, Basel, 1997.

[37] A. Einstein. Zum Quantensatz von Sommerfeld und Epstein. *Verhandlungen der Deutschen Phys. Ges.* 9/10, 1917.

[38] Y. Elskens and D. Escande. *Microscopic Dynamics of Plasmas and Chaos.* IOP Publ. Ltd., 2003.

[39] I. Ekeland. *Convexity Methods in Hamiltonian Mechanics.* Springer-Verlag, Berlin, 1990.

[40] U. Fano. *Description of States of Quantum Mechanics and Operator Techniques.* Rev. Mod. Phys. **29**(74), 1967.

[41] C. Fefferman and D.H. Phong. *The uncertainty principle and sharp Gårding inequalities.* Comm. Pure Appl. Math. **75**, 285–331, 1981.

[42] G.B. Folland. *Harmonic Analysis in Phase space.* Annals of Mathematics studies, Princeton University Press, Princeton, N.J., 1989.

[43] T. Frankel. *The Geometry of Physics, An Introduction.* Cambridge University Press, 1997.

[44] A. Friedman. *Foundations of Modern Analysis.* Dover, 1982.

[45] B. Gaveau, P. Greiner, and Vauthier. Intégrales de Fourier quadratiques et calcul symbolique exact sur le groupe de Heisenberg. *J. Funct. Anal. 68,* 1986, 248–272.

[46] *Geometry and Quantum Field Theory.* D.S. Freed and K.K. Uhlenbeck editors. IAS/Park City Mathematics series, Volume 1, Amer. Math. Soc. and Institute for Advanced Studies, 1995.

[47] G. Giacchetta, L. Mangiarotti, and G. Sardanashvily. Geometric quantization of completely integrable Hamiltonian systems in the action-angle variables. *Phys. Lett. A* **301**, 53–57, 2002.

[48] G. Giachetta, L. Mangiarotti, G. Sardanashvily. *Geometric and Algebraic Topological Methods in Quantum Mechanics,* World Scientific, Singapore, 2005.

[49] G. Giedke, J. Eisert, J.I. Cirac, and M.B. Plenio. Entanglement transformations of pure Gaussian states. E-print arXiv:quant-ph/0301038 v1, 2003.

[50] C. Godbillon. *Géométrie Différentielle et Mécanique Analytique,* Hermann, Paris, 1969.

[51] C. Godbillon. *Éléments de Topologie Algébrique.* Hermann, Paris, 1971.

[52] I. Gohberg and S. Goldberg. *Basic Operator Theory.* Birkhäuser, 1981.

[53] H. Goldstein. *Classical Mechanics.* Addison–Wesley, 1950; second edition, 1980; third edition, 2002.

[54] M. de Gosson. La définition de l'indice de Maslov sans hypothèse de transversalité. C. R. *Acad. Sci., Paris, Série I.* **309**:279–281, 1990.

[55] M. de Gosson. La relation entre Sp_∞, revêtement universel du groupe symplectique Sp et Sp $\times\mathbb{Z}$. *C. R. Acad. Sci., Paris,* Série I, **310**:245–248, 1990.

[56] M. de Gosson. Maslov Indices on Mp(n). *Ann. Inst. Fourier,* Grenoble, **40**(3):537–555, 1990.

[57] M. de Gosson. The structure of q-symplectic geometry. *J. Math. Pures et Appl.* **71**, 429–453, 1992.

[58] M. de Gosson. Cocycles de Demazure–Kashiwara et Géométrie Métaplectique. *J. Geom. Phys.* **9**:255–280, 1992.

[59] M. de Gosson. On the Leray–Maslov quantization of Lagrangian submanifolds. *J. Geom. Phys.* **13**(2), 158–168, 1994

[60] M. de Gosson. On half-form quantization of Lagrangian manifolds and quantum mechanics in phase space. *Bull. Sci. Math.* **121**:301–322, 1997.

[61] M. de Gosson. *Maslov Classes, Metaplectic Representation and Lagrangian Quantization.* Research Notes in Mathematics **95**, Wiley–VCH, Berlin, 1997.

[62] M. de Gosson. On the classical and quantum evolution of Lagrangian half-forms in phase space. *Ann. Inst. H. Poincaré*, **70**(6):547–73, 1999.

[63] M. de Gosson. Lagrangian path intersections and the Leray index: Aarhus Geometry and Topology Conference. *Contemp. Math., Amer. Math. Soc., Providence, RI,* **258**:177–184, 2000.

[64] M. de Gosson. *The Principles of Newtonian and Quantum Mechanics.* Imperial College Press, London, 2001.

[65] M. de Gosson. The symplectic camel and phase space quantization. *J. Phys. A: Math. Gen.* **34**:10085–10096, 2001.

[66] M. de Gosson. The 'symplectic camel principle' and semiclassical mechanics, *J. Phys. A: Math. Gen.* **35**:6825–6851, 2002.

[67] M. de Gosson and S. de Gosson. The cohomological interpretation of the indices of Robbin and Salamon. *Jean Leray' 99 Conference Proceedings, Math. Phys. Studies* **4**, Kluwer Academic Publishers, 2003.

[68] M. de Gosson and S. de Gosson. The Maslov indices of periodic Hamiltonian orbits. *J. Phys. A:Math. Gen.* **36**:615–622, 2003.

[69] M. de Gosson. Phase Space Quantization and the Uncertainty Principle. *Phys. Lett. A*, **317**:365–369, 2003.

[70] M. de Gosson. On the notion of phase in mechanics. *J. Phys. A: Math. Gen.* **37**(29), 7297–7314, 2004.

[71] M. de Gosson. The optimal pure Gaussian state canonically associated to a Gaussian quantum state. *Phys. Lett. A*, **330**:3–4, 161–167, 2004

[72] M. de Gosson. Cellules quantiques symplectiques et fonctions de Husimi–Wigner. *Bull. sci. math.*, Paris, **129**:211–226, 2005

[73] M. de Gosson. Extended Weyl calculus and application to the Schrödinger equation. *J. Phys. A: Math. Gen.* **38**, L325–L329, 2005.

[74] S. de Gosson. Ph.D. thesis, Växjö, 2005.

[75] M.J. Gotay. Functorial geometric quantization and van Hove's theorem. *Internat. J. Phys.* **19**:139–161, 1980.

[76] M.J. Gotay and G.A. Isenberg. The Symplectization of Science. *Gazette des Mathématiciens* **54**:59–79, 1992.

[77] J.M. Gracia-Bondia. Generalized Moyal Quantization on Homogeneous Symplectic spaces. *Contemporary Mathematics*, **134**, *Math., Amer. Math. Soc., Providence,* RI, 93–114, 1992.

[78] K. Gröchenig. *Foundations of Time-Frequency Analysis.* Birkhäuser, Boston, 2000.

[79] K. Gröchenig. An uncertainty principle related to the Poisson summation formula. *Studia Math.* **121**, 1996, 87–104.

[80] H.J. Groenewold. On the principles of elementary quantum mechanics. *Physics* **12**, 405–460, 1946.

[81] M. Gromov. Pseudoholomorphic curves in symplectic manifolds. *Invent. Math.* **82**:307–347, 1985.

[82] A. Grossmann. Parity operators and quantization of δ-functions. *Commun. Math. Phys.* **48** 191–193,1976.

[83] A. Grossmann, G. Loupias, and E.M. Stein. *An algebra of pseudo-differential operators and quantum mechanics in phase space*, Ann. Inst. Fourier, Grenoble **18**(2) 343–368, 1968.

[84] V. Guillemin and S. Sternberg. *Geometric Asymptotics*. Math. Surveys Monographs **14**, Amer. Math. Soc., Providence R.I., 1978.

[85] V. Guillemin and S. Sternberg. *Symplectic Techniques in Physics*. Cambridge University Press, Cambridge, Mass., 1984.

[86] M.C. Gutzwiller. *Chaos in Classical and Quantum Mechanics*. Interdisciplinary Applied Mathematics, Springer-Verlag, 1990.

[87] K. Habermann and L. Habermann. Introduction to Symplectic Spinors (preprint), 2005.

[88] *Quantum Implications: Essays in Honour of David Bohm*. Edited by B.J. Hiley and F. David Peat. Routledge, London and New York, 1987.

[89] W. Heisenberg. Über den anschaulichen Inhalt der quantentheoretischen Kinematik und Mechanik. *Zeitschrift für Physik* **43**, 172–198, 1927.

[90] H. Hofer, K. Wysocki, and E. Zehnder. Properties of pseudoholomorphic curves in symplectizations II: Embedding controls and algebraic invariants. *Geometric and Functional Analysis*, **2**(5), 270–328, 1995.

[91] H. Hofer and E. Zehnder. *Symplectic Invariants and Hamiltonian Dynamics*. Birkhäuser Advanced texts (Basler Lehrbücher, Birkhäuser Verlag, 1994.

[92] L. Hörmander. *The Analysis of Linear Partial differential Operators*. Springer-Verlag, III, 1985.

[93] L. Hörmander. The Weyl calculus of pseudo-differential operators. *Comm. Pure Appl. Math.* **32**, 359–443, 1979

[94] R.L. Hudson. When is the Wigner quasi-probability density non-negative? *Rep. Math. Phys.* **6**:249–252, 1974.

[95] J. Igusa. *Theta Functions*. Springer-Verlag, New-York, 1972.

[96] C.J. Isham. *Lectures on Quantum Theory: Mathematical and Structural Foundations*. Imperial College Press, 1995.

[97] M. Jammer. *The Conceptual Development of Quantum Mechanics,* Int. Series in Pure and Appl. Physics (McGraw-Hill Book Company, *1966)*.

[98] A.J.E.M. Jansen. A note on Hudson's theorem about functions with nonnegative Wigner distributions. *Siam. J. Math. Anal.* **15**(1), 170–176, 1984.

[99] J.M. Jauch. *Foundations of Quantum Mechanics*. Addison–Wesley Publishing Company, 1968.

[100] T. Kato. *Perturbation Theory for Linear Operators*. Springer-Verlag, New York, 1966.

[101] M. Kauderer. *Symplectic Matrices: First-Order Systems and Special Relativity*. World Scientific, 1994.

[102] J.B. Keller. Corrected Bohr-Sommerfeld Quantum Conditions for Nonseparable Systems. *Ann. of Physics* **4**:180–188, 1958.

[103] A.Y. Khinchin. *Mathematical Foundations of Quantum Statistics.* Dover Publ. Inc. 1998; originally published in 1980 by Graylock Press.

[104] J.J. Kohn and L. Nirenberg. On the algebra of pseudo-differential operators. *Comm. Pure. Appl. Math.* **19**, 269–305, 1965.

[105] W. Lisiecki. Coherent state representations. A survey. *Rep. Math. Phys.* **35**(2/3), 327–358, 1995.

[106] L.D. Landau and E.M. Lifschitz. *Statistical Physics.* Pergamon Press (translated from the Russian), 1980.

[107] J. Leray. *Lagrangian Analysis and Quantum Mechanics, a mathematical structure related to asymptotic expansions and the Maslov index* (the MIT Press, Cambridge, Mass., 1981); translated from *Analyse Lagrangienne* RCP 25, Strasbourg Collège de France, 1976–1977.

[108] J. Leray. The meaning of Maslov's asymptotic method the need of Planck's constant in mathematics. In *Bull. of the Amer. Math. Soc.*, Symposium on the Mathematical Heritage of Henri Poincaré, 1980.

[109] J. Leray. Complément à la théorie d' Arnold de l'indice de Maslov. *Convegno di geometrica simplettica et fisica matematica,* Instituto di Alta Matematica, Roma, 1973.

[110] P. Libermann and C.-M. Marle. *Symplectic Geometry and Analytical Mechanics.* D. Reidel Publishing Company, 1987.

[111] G. Lion, and M. Vergne. *The Weil representation, Maslov index and Theta series.* Progress in mathematics **6**, Birkhäuser, 1980.

[112] R.G. Littlejohn. The semiclassical evolution of wave packets. *Physics Reports* **138**(4–5):193–291, 1986.

[113] Y. Long. *Index Theory for Symplectic Paths with Applications.* Progress in Mathematics **207**, Birkhäuser, Basel, 2002.

[114] D. McDuff, and D. Salamon. *Symplectic Topology.* Oxford Science Publications, 1998.

[115] G.W. Mackey. *The Mathematical Foundations of Quantum Mechanics.* Benjamin, Inc., New York, Amsterdam, 1963.

[116] G.W. Mackey. *Unitary Group Representations.* The Benjamin/Cummings Publ. Co., Inc., Reading, Mass., 1978.

[117] G.W. Mackey. The Relationship Between Classical and Quantum Mechanics. In *Contemporary Mathematics* **214**, Amer. Math. Soc., Providence, RI, 1998.

[118] M.A. Marchiolli. *Mecânica Quântica no Espaço de Fase: I. Formulação de Weyl-Wigner.* Revista Brasileira de Ensino de Fisica, **24**(4) , 421–436, 2002

[119] V.P. Maslov. *Théorie des Perturbations et Méthodes Asymptotiques.* Dunod, Paris, 1972; translated from Russian [original Russian edition 1965].

[120] V.P. Maslov and M.V. Fedoriuk. *Semi-Classical Approximations in Quantum Mechanics.* Reidel, Boston, 1981.

[121] B. Mehlig and M. Wilkinson. Semiclassical trace formulae using coherent states. *Ann. Phys.* **18**(10), 6–7, 541–555, 2001.

[122] E. Merzbacher. *Quantum Mechanics*, 2nd ed., Wiley, New York, 1970.

[123] A. Messiah. *Quantum Mechanics* (two volumes). North–Holland Publ. Co., 1991; translated from the French; original title: *Mécanique Quantique*. Dunod, Paris, 1961.

[124] A.S. Mischenko, V.E. Shatalov, and B.Yu. Sternin. *Lagrangian Manifolds and the Maslov Operator*, Springer-Verlag, Heidelberg, 1990.

[125] W. Moore. *Schrödinger: life and thought* Cambridge University Press, 1989.

[126] M. Morse. *The Calculus of Variations in the Large*. AMS, Providence, R. I., 1935.

[127] J.E. Moyal. Quantum mechanics as a statistical theory. *Proc. Camb. Phil. Soc.* **45**:99–124, 1947.

[128] V. Nazaikiinskii, B.-W. Schulze, and B. Sternin. *Quantization Methods in Differential Equations*. Differential and Integral Equations and Their Applications, Taylor & Francis, 2002.

[129] D. Park. *Introduction to the Quantum Theory*, 3d edition. McGraw–Hill, 1992.

[130] M.A. Javaloyes and P. Piccione. Conjugate points and Maslov index in locally symmetric semi-Riemannian manifolds. Preprint 2005.

[131] R.C. Nostre Marques, P. Piccione, and D. Tausk. On the Morse and the Maslov index for periodic geodesics of arbitrary causal character. *Differential Geometry and its Applications*. Proc. Conf. Opava, 2001.

[132] L. Polterovich. *The Geometry of the Group of Symplectic Diffeomorphisms*. Lectures in Mathematics, Birkhäuser, 2001.

[133] J.C.T. Pool. Mathematical aspects of the Weyl correspondence. *J. Math. Phys.* **7**, 66–76, 1966.

[134] M. Reed and B. Simon. *Methods of Modern Mathematical Physics*. Academic Press, New York, 1972.

[135] J. Robbin and D. Salamon. The Maslov index for paths. *Topology* **32**(4), 827–844, 1993.

[136] H.P. Robertson. The uncertainty principle. *Phys. Rev.* 34, 163–164, 1929.

[137] A. Royer. Wigner functions as the expectation value of a parity operator. *Phys. Rev. A* **15**, 449–450, 1977.

[138] D. Ruelle. *Statistical mechanics: rigorous results*. Benjamin, 1974.

[139] M. Schechter. *Operator Methods in Quantum Mechanics*. North-Holland, 1981.

[140] W. Schempp. *Harmonic Analysis on the Heisenberg Nilpotent Lie Group*. Pitman Research Notes in Mathematics **147**, Longman Scientifical and Technical, 1986.

[141] W. Schempp. *Magnetic Resonance Imaging: Mathematical Foundations and Applications*. Wiley, 1997.

[142] W.P. Schleich. *Quantum Optics in Phase Space*. Wiley-VCH, Berlin, 2001.

[143] E. Schrödinger. Zum Heisenbergschen Unschärfeprinzip. *Berliner Berichte*, 296–303, 1930.

[144] B.-W. Schulze. *Pseudo.differential operators on manifolds with singularities*. North-Holland, Amsterdam, 1991.

[145] B.-W. Schulze. Operators with symbol hierarchies and iterated asymptotics. *Publications of RIMS*, Kyoto University, **38**(4), 735–802, 2002.

[146] B.-W. Schulze. *Boundary value problems, conical singularities, and asymptotics*. Akademie Verlag Berlin, 1994.

[147] I.E. Segal. Foundations of the theory of dynamical systems of infinitely many degrees of freedom (I). *Mat. Fys. Medd., Danske Vid. Selsk.*, **31**(12):1–39, 1959.

[148] H. Seifert and W. Threlfall. *A Textbook on Topology.* Academic Press, 1980 [Original version: *Lehrbuch der Topologie.* Teubner, 1934]

[149] D. Shale. Linear Symmetries of free Boson fields. *Trans. Amer. Math. Soc.* **103**:149–167, 1962.

[150] B. Simon. *Trace Ideals and Their Applications.* Cambridge University Press, 1979.

[151] B. Simon. *The Weyl transform and L^p functions on phase space.* Amer. Math. Soc. 116, 1045–1047, 1992.

[152] R. Simon, N. Mukunda and B. Dutta. Quantum Noise Matrix for Multimode Systems: U(n)-invariance, squeezing and normal forms. *Phys. Rev. A* **49**, 1567–1583, 1994.

[153] R. Simon, E.C.G. Sudarshan, and N. Mukunda. Gaussian–Wigner distributions in quantum mechanics and optics. *Phys. Rev. A* **36**(8), 3868–3880, 1987

[154] I.M. Singer and J.A. Thorpe. *Lecture Notes on Elementary Geometry and Topology.* Springer-Verlag, 1987.

[155] V.A. Sobolev. Absolute continuity of the periodic magnetic Schrödinger operator. *Invent. Math.* **137**(1) 85–112, 1999.

[156] J.-M. Souriau. Indice de Maslov des variétés lagrangiennes orientables. *C. R. Acad. Sci.*, Paris, Série A, **276**:1025–1026, 1973.

[157] E. Spanier. *Algebraic Topology.* McGraw–Hill Book company, 1966.

[158] E.M. Stein. *Harmonic Analysis: Real Variable Methods, Orthogonality, and Oscillatory Integrals.* Princeton University Press, 1993.

[159] E.M. Stein and G. Weiss. *Fourier Analysis on Euclidean Spaces.* Princeton University Press, 1971.

[160] J. Toft. Regularizations, decompositions and lower bound problems in the Weyl calculus. *Comm. Partial Diff. Eq.*, **25**, 1201–1234, 2000.

[161] J. Toft. Continuity properties in non-commutative convolution algebras, with applications in pseudo-differential calculus. *Bull. Sci. math.* **126**, 115–142, 2002.

[162] G. Torres-Vega and J.H. Frederick. Quantum mechanics in phase space: New approaches to the correspondence principle. *J. Chem. Phys.* **93**(12), 8862–8874, 1990.

[163] G. Torres-Vega and J.H. Frederick. A quantum mechanical representation in phase space. *J. Chem. Phys.* **98**(4), 3103–3120, 1993.

[164] F. Trèves. *Introduction to Pseudo-differential and Fourier Integral Operators* (two Volumes). University Series in Mathematics, Plenum Press, 1980.

[165] F. Trèves. *Topological Vector Spaces, Distributions, and Kernels.* Academic Press, New York, 1967.

[166] G. Tuynman. What is prequantization, and what is geometric quantization? *Proceedings, Seminar 1989–1990, Mathematical Structures in field theory, 1–28. CWI Syllabus 39. CWI*, Amsterdam, 1996.

[167] A. Unterbeger. *Automorphic pseudo-differential Analysis and Higher Level Weyl Calculi.* Progress in Mathematics, Birkhäuser, 2003.

[168] I. Vaisman. *Symplectic Geometry and Secondary Characteristic Classes.* Birkhäuser, Progress in Mathematics **72**, 1987.

[169] L. van Hove. Sur certaines représentations unitaires d'un groupe fini de transformations. *Mém. Acad. Roy. Belg. Classe des Sci.* **26**(6), 1952.

[170] V.S. Varadarajan. *Lie groups, Lie algebras, and their Representations.* Prentice Hall, 1974; Springer Verlag 1984.

[171] J. von Neumann. *Mathematical Foundations of Quantum Mechanics.* Princeton University Press, reedition, 1996

[172] A. Voros. Asymptotic $\hbar$-expansions of stationary quantum states. *Ann. Inst. H. Poincaré, Sect. A,* **26**:343–403, 1977.

[173] A. Voros. An algebra of pseudo-differential operators and the asymptotics of quantum mechanics. *J. Funct. Anal.* **29**:104–132, 1978.

[174] C.T.C. Wall. Nonadditivity of the signature. *Invent. Math.* 7, 1969, 269–274.

[175] N. Wallach, *Lie Groups: History, Frontiers and Applications*, **5**. Symplectic Geometry and Fourier Analysis, Math. Sci. Press, Brookline, MA, 1977.

[176] A. Weil. Sur certains groupes d'opérateurs unitaires. *Acta Math.* **111**:143–211, 1964; also in *Collected Papers*, Vol. III:1–69, Springer-Verlag, Heidelberg, 1980.

[177] A. Weinstein. Symplectic Geometry. *Bull. Amer. Math. Soc.* **5**(1):1–11, 1981.

[178] A. Weinstein. *Lectures on symplectic manifolds.* CBMS Conf. Series **29**, Amer. Math. Soc., Providence R.I., 1977.

[179] H. Weyl. *The Theory of Groups and Quantum Mechanics.* Methuen, London, 1931 [reprinted by Dover Publ., New York, 1950].

[180] E. Wigner. On the quantum correction for thermodynamic equilibrium, *Phys. Rev.* **40**, 799–755, 1932.

[181] M.W. Wong. *Weyl Transforms.* Springer, 1998.

[182] M.W. Wong. *Trace-Class Weyl Transforms.* Integr. equ. oper. theory **99**, Birkhäuser Verlag, Basel, 2003.

Index

Subseries Advances in Partial Differential Equations

Subseries editors:

Bert-Wolfgang Schulze **Institut für Mathematik** **Universität Potsdam** **14415 Potsdam** **Germany**	**Jerome A. Goldstein** **The University of Memphis** **Mathematical Sciences** **273 Dunn Hall** **Memphis, TN 38152-3240** **USA**
Sergio Albeverio **Institut für Angewandte Mathematik** **Universität Bonn** **53115 Bonn** **Germany**	**Noboyuki Tose** **COE of Mathematics** **Keio University** **4-1-1 Hiyoshi** **Kohoku** **Yokohama 223-8521** **Japan**
Michael Demuth **Institut für Mathematik** **Technische Universität Clausthal** **38678 Clausthal-Zellerfeld** **Germany**	

The subseries *Advances in Partial Differential Equations* is intended to report on recent developments in all areas of partial differential equations, in particular, on those belonging to the tradition defined by the series *Operator Theory: Advances and Applications* and new areas. The *Advances* volumes will consist primarily of expository research articles, presenting both an overview of the state of a field and new results, in such areas as microlocal analysis and its applications to asymptotic phenomena, mathematical physics, spectral theory, as well as symplectic geometry, the analysis of singularities of solutions, and geometric analysis. Selected articles of this kind may also appear as a separate monographic volume. A certain number of short communications will also be published.

Proposals for manuscripts may be sent to B.-W. Schulze or to one of the other subseries editors.

www.ingramcontent.com/pod-product-compliance
Lightning Source LLC
Chambersburg PA
CBHW082131181125
35618CB00005B/212

9783764375744